Approximation, Probability, and Related Fields

Approximation, Probability, and Related Fields

Edited by

George Anastassiou
Memphis State University
Memphis, Tennessee

and

Svetlozar T. Rachev
University of California
Santa Barbara, California

Springer Science+Business Media, LLC

Library of Congress Cataloging-in-Publication Data

Approximation, probability, and related fields / edited by George
Anastassiou and Svetlozar T. Rachev.
p. cm.
"Proceedings of a Conference on Approximation, Probability, and
Related Fields, held May 20-22, 1993, in Santa Barbara, California"-
-T.p. verso.
Includes bibliographical references and index.
ISBN 978-1-4613-6063-6 ISBN 978-1-4615-2494-6 (eBook)
DOI 10.1007/978-1-4615-2494-6
1. Approximation theory--Congresses. 2. Probabilities-
-Congresses. 3. Stochastic processes--Congresses. I. Anastassiou,
George A., 1952- . II. Rachev, S. T. (Svetlozar Todorov)
III. Conference on Approximation, Probability, and Related Fields
(1993 : Santa Barbara, Calif.)
QA221.A643 1994
511.4--dc20 94-27693
CIP

Proceedings of a conference on Approximation, Probability, and Related Fields, held May 20–22, 1993, in Santa Barbara, California

ISBN 978-1-4613-6063-6

DEDICATED TO

LEOPOLDO NACHBIN

1922-1993

DEDICATION

Leopoldo Nachbin was born on January 7, 1922 in the city of Recife, Brazil, and died on April 3, 1993. It is not easy to summarize in a few paragraphs Nachbin's manifold contributions to mathematics. Fortunately, the interested reader can consult the analysis that Professor John Horváth made for the volume honoring Nachbin on the occasion of his sixtieth birthday, published by North-Holland under the title *Aspects of Mathematics and its Applications* (Editor: J.A. Barroso). This volume appeared in 1986 and Professor Horváth's paper is entitled "The Life and Works of Leopoldo Nachbin," and occupies pages 1 to 75.

Nachbin moved from Recife to Rio de Janeiro in 1938 and received an engineering degree in 1943. But already in 1941, while still an undergraduate student, he had published his first paper in mathematics, in Volume 13 of the Proceedings of the Brazilian Academy of Sciences, and in 1942 he published two more papers, one in Italy and the other in Argentina. All three papers were reviewed in *Mathematical Reviews.* In 1946, his eighth paper appeared in the *Comptes Rendus de l'Academie des Sciences Paris.* In that same year, two other papers were published, one in the *Transactions of the American Mathematical Society* and the other in *Summa Brasiliensis Mathematicae.* In 1948 he published his first book. The subject was topological vector spaces. Horváth remarks that this was the first text ever written on the theory of general topological vector spaces. But the mathematical activity of Nachbin covered a wide range of subjects and in each of them he left his imprint: Boolean algebras, general topology, ordered topological spaces, topological vector spaces, an extension of the Hahn-Banach theorem for linear transformations, ordered topological vector spaces, approximation theory, harmonic analysis, infinite-dimensional holomorphy.

Nachbin's work on approximation theory started with a 1949 paper establishing the conditions for density of subalgebras of r-times continuously differentiable functions on open subsets of n-dimensional Euclidean space, or more generally on n-dimensional differentiable manifolds, for the topology of uniform convergence of order r on compact subsets, extending to this case the so-called Stone-Weierstrass theorem. In 1948 H. Whitney had described the closure in this topology of an ideal, and Nachbin solved the problem of density of subalgebras, but the problem of describing the closure of a subalgebra remains open.

The Stone-Weierstrass theorem for modules and the theorem of Whitney on the closure of an ideal led Nachbin to his work on algebras of finite differentiable order and the operational calculus, which appeared in the *Annals of Mathematics* of 1959. His next interest was the study of the Bernstein weighted approximation problem. His first paper in this area appeared in 1961. At this time, more exactly during the two academic years

1961/62 and 1962/63, Nachbin was a visiting professor at the University of Paris. He was an invited lecturer at the 1962 International Congress of Mathematicians held in Stockholm, and his lecture dealt with algebraic problems in approximation theory. What he did with the Bernstein weighted approximation problem is similar to what Stone had done with the Wierstrass approximation problem. His work in this area can be found in his book *Elements of Approximation Theory*, published by Van Nostrand in 1967, and reprinted by Krieger in 1976. Since the academic year 1963/64, Nachbin worked in Rochester, NY, and since 1967 he had the title of George Eastman Professor at the University of Rochester. Previously, he had spent the years 1948/49 and 1949/50 at the University of Chicago, and 1956/58 again in Chicago and at the Institute for Advanced Study in Princeton.

Already in 1967 his first paper in infinite-dimensional holomorphy introduced a new topology for the space of all holomorphic functions on a given open subset. From then on, most of his efforts were in infinite-dimensional holomorphy, with occasional looks at approximation theory. See, for example, his contribution to the volume *Approximation Theory and Functional Analysis*, Proceedings of an International Symposium on Approximation Theory, held in Campinas in 1977, North-Holland Math. Studies 35.

Nachbin developed during his entire life and intensive work as teacher and advisor of many students from several countries. His editorial work is impressive. The well-known *Notas de Matematica* are the best testimonial for his dedication to mathetimatics, and will for a long time recall his name to those perusing those volumes. For a long period Nachbin dedicated most of his time to make IMPA at Rio de Janeiro one of the most active and prestigious mathematical centers of Latin America.

Leopoldo married Maria de Graça Mousinho in 1956. They had three children: André, Léa, and Luís. Graça died in 1982 after a long chronic illness.

João B. Prolla

PREFACE

These proceedings contain selected papers presented at the Conference on Approximation, Probability and Related Fields held in Santa Barbara, California, on May 20-22, 1993.

The main topics of the conference were:

1) approximation of functions by polynomials, splines, and operators, and applications to stochastics
2) numerical methods for approximation of deterministic and stochastic integrals
3) orthogonal polynomials and stochastic processes
4) positive linear operators and related deterministic and stochastic inequalities
5) multivariate approximation and interpolation
6) rate of convergence in probability theory
7) approximations and martingales
8) deterministic and stochastic inequalities
9) stability of deterministic and stochastic models
10) signal analysis
11) prediction theory
12) wavelets and approximations based on wavelets

The Conference was very successful and received many compliments. We quote some of the letters sent by the participants:

- *"Many thanks for the wonderful conference and the exemplary organization!"*
- *"Many thanks for your good and effective work"*

We would like to thank the international organizing committee consisting of Paul Butzer, Stamatis Cambanis and Zuhair Nashed as well as the local organizing committee consisting of Bessy Athanasopoulos, Raisa Feldman, and Gleb Haynatzki for their superb work and for their contribution to the success of the Conference.

The chairmen:

George Anastassiou

Svetlozar Rachev

CONTENTS

PRESERVATION OF MODULI OF CONTINUITY FOR BERNSTEIN-TYPE OPERATORS *

José A. Adell,[1] and Jesús de la Cal[2]

[1]Departamento de Métodos Estadísticos
Universidad de Zaragoza
50009 Zaragoza, Spain
[2]Departamento de Matemática Aplicada
y Estadística e Investigación Operativa
Universidad del País Vasco
Apdo. 644, 48080 Bilbao, Spain

1. INTRODUCTION

It is well known that many Bernstein-type operators preserve some properties of the functions on which they act, such as monotonicity, convexity, Lipschitz constants, etc. (cf. for instance [**2**]). In this paper, attention is focused on preservation of global smoothness, as measured by the usual moduli of continuity of first and second order.

To the best of our knowledge, this problem has been studied by Kratz and Standtmüller in [**11**] for the first time. In this work, the authors consider sequences $(L_n)_{n\geq 1}$ of one-dimensional discrete operators satisfying certain moment assumptions and obtain estimates of the form

$$\omega(L_n f; h) \leq c\, \omega(f; h), \tag{1}$$

where $\omega(f;.)$ stands for the usual first modulus of continuity of the function f, and c is a positive constant which depends on the particular family of operators considered, but not upon f, nor n and h. They provide the estimate $c \leq 4$ in some important examples, such as Bernstein, Szász and Baskakov operators.

Using a different approach based on K-functional techniques, Anastassiou, Cottin and Gonska give in [**4,5**] general results of type (1) for bounded linear operators acting on $C(X)$, the space of all real-valued continuous functions defined on a compact metric space X. In particular, they show that $c = 2$ is the best possible absolute constant, in the case of Bernstein polynomials on the standard m-simplex.

Finally, a probabilistic approach has been developed by the authors in [**2**]. In this paper, we consider families $(L_t)_{t\in T}$ of one-dimensional operators which are representable in terms of stochastic processes satisfying simple structural properties. The

* Research supported by the University of the Basque Country and by the grant PB92-0437 of the Spanish DGICYT.

Approximation, Probability, and Related Fields, Edited by
G. Anastassiou and S.T. Rachev, Plenum Press, New York 1994

setting is general enough to include many classical examples and allows us to derive estimates of type (1) in a fairly easy way. In particular, it is shown that (1) holds with $c = 2$ for the Bernstein, Szász and Baskakov operators.

The aim of the present paper is to get deep into the probabilistic methods proposed in [**2**]. The next section contains a description of the mathematical framework together with some necessary notations. In section 3, we deal with preservation of the first modulus of continuity and obtain general results which extend and improve those given in [**2**], as well as asymptotic results. The analogous results for the second modulus of continuity are established in section 5. Finally, sections 4 and 6 contain various applications of the general results to particular families of approximation operators.

2. MATHEMATICAL FRAMEWORK

Let I be an interval of the real line and let $F(I)$ be the set of all real-valued continuous functions defined on I such that

$$\omega(f;h) := \sup_{\substack{x,y\in I\\ |x-y|\leq h}} |f(x)-f(y)| < \infty, \qquad 0 \leq h < l(I), \tag{2}$$

where $l(I)$ stands for the lenght of I. Throughout the paper, we shall consider families of operators $(L_t)_{t\in T}$ having the form

$$L_t(f,x) = Ef(Z_t^x), \qquad x\in I,\ t\in T,\ f\in F(I), \tag{3}$$

where E denotes mathematical expectation, T is either the set of positive integers or the set of positive real numbers, and $Z := (Z_t^x)_{(x,t)\in I\times T}$ is a double-indexed stochastic process defined on a probability space $(\Omega, \mathcal{F}, P)$ such that each Z_t^x is I-valued and integrable. Observe that (2) together with the integrability of Z_t^x implies

$$E|f(Z_t^x)| < \infty$$

and, therefore, the right-hand side of (3) is well-defined.

We shall estimate the moduli of continuity of $L_t f$ in terms of the moduli of continuity of f and certain quantities depending on the structure of the process Z, specially the following:

$$a_p(h,t) := \sup_{\substack{x,y\in I\\ |x-y|\leq h}} E^{1/p}|Z_t^x - Z_t^y|^p, \qquad p\geq 1,\ t\in T,\ 0\leq h < l(I), \tag{4}$$

and

$$b(h,t) := \sup_{\substack{x,y\in I\\ |x-y|\leq h}} P(|Z_t^x - Z_t^y| > 0), \qquad t\in T,\ 0\leq h < l(I). \tag{5}$$

In order to obtain sharp estimates, the following condition concerning the paths of the process Z turns out to be crucial:

$$(H_1): \qquad Z_t^x \leq Z_t^y \quad \text{a.s.}, \qquad t\in T,\ x,\ y\in I,\ x\leq y.$$

Condition (H_1) is closely related with the 'splitting' method introduced by Khan and Peters [**9**]. As we shall see in sections 4 and 6, this condition is fulfilled in the usual cases. On the other hand, the condition

$$EZ_t^x = x, \qquad t \in T,\ x \in I, \tag{6}$$

is satisfied in many important examples. Observe that (H_1) and (6) imply that

$$a_1(h,t) = h, \qquad 0 \le h < l(I),\ t \in T, \tag{7}$$

where $a_1(h,t)$ is defined in (4).

3. FIRST MODULUS

Our first result is implicitely contained in [**2**, proof of Theorem 1]. Despite its simplicity, it has a great deal of interesting consequences.

THEOREM 1. *For $t \in T$, $f \in F(I)$ and $0 < h < l(I)$, we have*

$$\omega(L_t f; h) \le \sup_{\substack{x,y \in I \\ |x-y| \le h}} E\omega(f; |Z_t^x - Z_t^y|). \tag{8}$$

Proof. It follows from (3) and the obvious inequality

$$|f(Z_t^x) - f(Z_t^y)| \le \omega(f; |Z_t^x - Z_t^y|), \qquad x, y \in I. \tag{9}$$

REMARK 1. Assume that, for $t \in T$, the process $(Z_t^x)_{x \in I}$ takes values in a subinterval J of I (which may depend on t). Then, from (9) it is clear that $\omega(f;.)$ can be replaced by $\omega^J(f;.)$ on the right-hand side in (8), where $\omega^J(f;.)$ is the modulus of continuity given by

$$\omega^J(f;h) := \sup_{\substack{x,y \in J \\ |x-y| \le h}} |f(x) - f(y)|. \tag{10}$$

Such a circumstance occurs, for instance, in the case of Bleimann-Butzer-Hahn operators (see example (C) in section 4).

REMARK 2. Under (H_1), the right-hand side in (8) is equal to

$$\sup_{x,\, x+h \in I} E\omega(f; Z_t^{x+h} - Z_t^x),$$

and this quantity equals to $E\omega(f; Z_t^h - Z_t^0)$ if $I = [0,a]$ or $[0,\infty)$ and the process $(Z_t^x)_{x \in I}$ has stationary increments.

COROLLARY 1. *Let t, f and h be as in Theorem 1. We have:*
(a) If $\omega(f;.)$ is concave, then

$$\omega(L_t f; h) \le \omega(f; a_1(h,t)),$$

where $a_1(h,t)$ is defined in (4).
(b) If $f \in$ Lip(A,μ), that is, if $\omega(f;h) \le Ah^\mu$, where $A > 0$ and $\mu \in (0,1]$, then

$$\omega(L_t f; h) \le A\,(a_1(h,t))^\mu .$$

In particular, $L_t f \in \mathrm{Lip}(A, \mu)$ *whenever* (H_1) *and* (6) *are fulfilled.*

Proof. Use (8) and Jensen's inequality.

COROLLARY 2. *Let* t, f *and* h *be as in Theorem 1. Then*

$$\omega(L_t f; h) \leq \left(\frac{a_1(h,t)}{h} + b(h,t)\right) \omega(f; h), \tag{11}$$

where $a_1(h,t)$ *and* $b(h,t)$ *are defined in* (4) *and* (5) , *respectively.*

Proof. Use (8) and the following argument: Let U be a non-negative random variable. Then by the well-known property

$$\omega(f; u) \leq (1 + [u/v])\omega(f; v), \qquad u \geq 0,\ v > 0, \tag{12}$$

where brackets indicate integral part, we have

$$\begin{aligned} E\omega(f; U) &= E\omega(f; U)\, I(U > 0) \leq \omega(f; h)\, E(1 + U/h)\, I(U > 0) \\ &= \left(P(U > 0) + \frac{EU}{h}\right) \omega(f; h), \end{aligned}$$

where $I(C)$ denotes the indicator function of the event C.

REMARK 3. Using the same notations as in the former proof, we also have for any $v > 0$

$$\begin{aligned} E\omega(f; U) &\leq \left(1 + v^{-1} EU\, I(U > v)\right) \omega(f; v) \\ &\leq \left(1 + v^{-1} E^{1/2} U^2\, P^{1/2}(U > v)\right) \omega(f; v), \end{aligned}$$

the last step by Hölder's inequality. Applying the preceding inequality to the right-hand side of (8), we get

$$\omega(L_t f; h) \leq \left(1 + \frac{a_2(h,t)}{v} \sup_{\substack{x,y \in I \\ |x-y| \leq h}} P^{1/2}(|Z_t^x - Z_t^y| > v)\right) \omega(f; v), \tag{13}$$

where $a_2(h,t)$ is defined in (4). The interesting point is that, in some cases, suitable choices of v provide sharp estimates (see example (F) in section 4).

COROLLARY 3. *Let* t, f *and* h *be as in Theorem 1. Under* (H_1), *we have*

$$\omega(L_t f; h) \leq \omega(f; h) + 2\omega(f; c(h,t)),$$

where

$$c(h,t) := \sup_{x,\, x+h \in I} E|Z_t^{x+h} - Z_t^x - h|. \tag{14}$$

Proof. Let x, $x + h \in I$. Since

$$\omega(\omega(f; .); .) = \omega(f; .),$$

we have, by (12)

$$\begin{aligned} E\omega(f; Z_t^{x+h} - Z_t^x) &\leq \omega(f; h) + E|\omega(f; Z_t^{x+h} - Z_t^x) - \omega(f; h)| \\ &\leq \omega(f; h) + \omega(f; c(h,t)) \left(1 + \frac{E|Z_t^{x+h} - Z_t^x - h|}{c(h,t)}\right) \\ &\leq \omega(f; h) + 2\omega(f; c(h,t)), \end{aligned}$$

where $c(h,t)$ is defined in (14). The conclusion follows from Theorem 1 and Remark 2.

We shall complete this section with two asymptotic results.

THEOREM 2. *Let* $0 < h < l(I)$ *and let* $f \in F(I)$ *be uniformly continuous. Assume that:*

(a) $\lim_{t\to\infty} L_t f = f$.

(b) Condition (H_1) *is satisfied.*

(c) $\lim_{t\to\infty} c(h,t) = 0$, *where* $c(h,t)$ *is defined in* (14).

Then

$$\lim_{t\to\infty} \omega(L_t f; h) = \omega(f; h).$$

Proof. Since f is uniformly continuous, we obtain from Corollary 3 and assumptions (b) and (c)

$$\limsup_{t\to\infty} \omega(L_t f; h) \le \omega(f; h). \tag{15}$$

On the other hand, let x, $y \in I$ such that $|x - y| \le h$. By the triangle inequality, we have

$$|f(x) - f(y)| \le \omega(L_t f; h) + |L_t(f, x) - f(x)| + |L_t(f, y) - f(y)|, \qquad t \in T$$

and, therefore, by assumption (a)

$$\omega(f; h) \le \liminf_{t\to\infty} \omega(L_t f; h). \tag{16}$$

The conclusion follows from (15) and (16).

REMARK 4. Using standard methods, we have for $x \in I$ and $t \in T$

$$|L_t(f, x) - f(x)| \le 2\,\omega(f; E|Z_t^x - x|). \tag{17}$$

Thus, if f is uniformly continuous, then the condition

$$\lim_{t\to\infty} E|Z_t^x - x| = 0, \qquad x \in I, \tag{18}$$

implies assumption (a) in Theorem 2. Moreover, if (H_1) is fulfilled and the process $(Z_t^x)_{x\in I}$ has stationary increments, then (18) also implies assumption (c).

THEOREM 3. *Let* $0 < h < l(I)$, $f \in F(I)$ *and let* J *be a compact subinterval of* I. *Then*

$$|\omega^J(L_t f; h) - \omega^J(f; h)| \le 4\,\omega(f; d(J, t)), \qquad t \in T, \tag{19}$$

where $\omega^J(f; .)$ *is defined in* (10) *and*

$$d(J, t) := \sup_{x\in J} E|Z_t^x - x|. \tag{20}$$

Therefore, $\omega^J(L_t f; .)$ *converges uniformly to* $\omega^J(f; .)$, *as* $t \to \infty$, *whenever* f *is uniformly continuous and*

$$\lim_{t\to\infty} d(J, t) = 0. \tag{21}$$

Proof. Inequality (19) follows from (17) and the well-known inequality

$$|\omega^J(L_t f; h) - \omega^J(f; h)| \le 2 \sup_{x\in J} |L_t(f, x) - f(x)|.$$

REMARK 5. In many classical examples, we have

$$\lim_{t\to\infty} \sup_{x\in J} E(Z_t^x - x)^2 = 0,$$

which implies (21).

4. EXAMPLES

(A) *Bernstein operators.* Let $(X_n)_{n\geq 1}$ be a sequence of independent and on the interval $[0,1]$ uniformly distributed random variables. The random variable

$$S_n(x) := \sum_{i=1}^{n} I(X_i \leq x), \qquad x \in [0,1],\ n = 1, 2, \ldots, \tag{22}$$

has the binomial distribution with parameters n, x. Therefore, we can write

$$B_n(f,x) = Ef\left(\frac{S_n(x)}{n}\right),$$

where

$$B_n(f,x) := \sum_{k=0}^{n} f(k/n)\binom{n}{k} x^k (1-x)^{n-k}.$$

It is clear that conditions (H_1) and (6) are satisfied. Hence, (7) also holds. Moreover, for $0 \leq x < y \leq 1$, the difference $S_n(y) - S_n(x)$ has the binomial distribution with parameters n, $y - x$. Consequently, the application of Corollary 2 to this case gives

$$\omega(B_n f; h) \leq (2 - (1-h)^n)\,\omega(f,h), \qquad h \in [0,1],\ n = 1, 2, \ldots$$

On the other hand, Corollary 3 applies with

$$c(h,n) = E\left|\frac{S_n(h)}{n} - h\right| \leq E^{1/2}\left(\frac{S_n(h)}{n} - h\right)^2 = \sqrt{\frac{h(1-h)}{n}}.$$

Finally, from Theorem 3 we have

$$|\omega(B_n f; h) - \omega(f;h)| \leq 4\,\omega(f; 1/2\sqrt{n}).$$

(B) *Stancu operators.* Let P_n^α be the linear polynomial operator defined by

$$P_n^\alpha(f,x) := \sum_{k=0}^{n} f(k/n)\binom{n}{k}\frac{\prod_{i=0}^{k-1}(x + i\alpha)\prod_{j=0}^{n-k-1}(1 - x + j\alpha)}{(1+\alpha)(1+2\alpha)\ldots(1+(n-1)\alpha)}, \qquad x \in [0,1],\ n \geq 1,$$

where $f \in C[0,1]$ and α is a non-negative parameter. This operator was introduced by Stancu in **[13]**. Observe that P_n^0 is the Bernstein operator. A suitable representation for P_n^α with $\alpha > 0$ is constructed in the following way: Let $(X_n)_{n\geq 1}$ and $(U_t)_{t\geq 0}$ be two independent stochastic processes defined on the same probability space, such that $(X_n)_{n\geq 1}$ is the same process as in the preceding example and $(U_t)_{t\geq 0}$ is a gamma process, that is, a stochastic process starting at the origin, having stationary independent increments, and such that each U_t with $t > 0$ has the gamma density given by

$$d_t(\theta) := \frac{\theta^{t-1}e^{-\theta}}{\Gamma(t)}, \qquad \theta > 0.$$

For $\alpha > 0$ and $x \in (0,1)$, the random variable

$$W_\alpha^x := \frac{U_{x\alpha^{-1}}}{U_{\alpha^{-1}}}$$

has the beta distribution with parameters x/α, $(1-x)/\alpha$, and the random variable $S_n(W_\alpha^x)$, where $S_n(.)$ is defined in (22), has the Pólya distribution with paramenters n, x, $1-x$, α (cf. [**8**]). Therefore, we have the representation

$$P_n^\alpha(f,x) = Ef\left(\frac{S_n(W_\alpha^x)}{n}\right).$$

It is clear that (H_1) and (6) are fulfilled. Moreover, it is not hard to check that, for $n \geq 1$ and $\alpha > 0$, the process $(S_n(W_\alpha^x))_{0\leq x\leq 1}$ has stationary increments. We conclude that Corollary 2 applies with

$$a_1(h,n) = h, \qquad b(h,n) = 1 - \prod_{k=0}^{n-1} \frac{1-h+k\alpha}{1+k\alpha},$$

and Corollary 3 applies with

$$c(h,n) = E\left|\frac{S_n(W_\alpha^h)}{n} - h\right| \leq E^{1/2}\left(\frac{S_n(W_\alpha^h)}{n} - h\right)^2 = \sqrt{\frac{h(1-h)(1+n\alpha)}{n(1+\alpha)}}.$$

Finally, from Theorem 3 we have

$$|\omega(P_n^\alpha f; h) - \omega(f;h)| \leq 4\,\omega\left(f; \sqrt{\frac{1+n\alpha}{4n(1+\alpha)}}\right).$$

(C) *Bleimann-Butzer-Hahn operators.* Here, we shall consider the linear rational operator L_n, introduced by Bleimann, Butzer and Hahn in [**6**], which is defined by

$$L_n(f,x) := \sum_{k=0}^{n} f\left(\frac{k}{n-k+1}\right)\binom{n}{k}\frac{x^k}{(1+x)^n}, \qquad x \geq 0,\ n = 1,2,\ldots,$$

where $f \in C[0,\infty)$. In this case, we have the representation

$$L_n(f,x) = Ef(h_n(S_n(p(x)))),$$

where S_n is the same as in example (A) and h_n and p are, respectively, the functions given by

$$h_n(v) := \frac{v}{n-v+1}, \qquad v \in [0,n]$$

and

$$p(x) := \frac{x}{1+x}, \qquad x \geq 0.$$

Observe that, for each $n \geq 1$, the process $(Z_n^x)_{x\geq 0}$, where $Z_n^x := h_n(S_n(p(x)))$, takes values in the interval $[0,n]$. Since both h_n and p are non-decreasing, we see that condition (H_1) is satisfied. Moreover,

$$EZ_n^x = x - x\left(\frac{x}{1+x}\right)^n, \qquad x \geq 0,\ n \geq 1,$$

and it is readily checked that

$$a_1(h,n) = h - h\left(\frac{h}{1+h}\right)^n, \qquad h \geq 0,\ n \geq 1.$$

On the other hand, since $(S_n(x))_{0\leq x\leq 1}$ has stationary increments, we have for $0 \leq x < y$

$$\begin{aligned} P\left(Z_n^y - Z_n^x > 0\right) &= P\left(S_n(p(y)) - S_n(p(x)) > 0\right) \\ &= P\left(S_n(p(y) - p(x)) > 0\right), \end{aligned}$$

and, by the concavity of p, we obtain

$$b(h,n) = P\left(S_n(p(h)) > 0\right) = 1 - \left(\frac{1}{1+h}\right)^n.$$

Thus, from Corollary 2 and Remark 1, we get

$$\omega(L_nf;h) \leq \left(2 - \left(\frac{1}{1+h}\right)^n - \left(\frac{h}{1+h}\right)^n\right)\omega^{[0,n]}(f;h), \qquad h \geq 0,\ n \geq 1.$$

Finally, using the estimate (cf. **[10]**)

$$E(Z_n^x - x)^2 \leq \frac{4x(1+x)^2}{n}, \qquad x \geq 0,\ n \geq 1,$$

Theorem 3 yields

$$|\omega^{[0,a]}(L_nf;h) - \omega^{[0,a]}(f;h)| \leq 4\omega\left(f;\sqrt{\frac{4a(1+a)^2}{n}}\right), \qquad h \geq 0, n \geq 1, a > 0.$$

(D) *Szász-Mirakyan operators.* The Szász-Mirakyan operator S_t is defined by

$$S_t(f,x) := e^{-tx}\sum_{k=0}^{\infty} f(k/t)\frac{(tx)^k}{k!}, \qquad t > 0, x \geq 0, f \in F([0,\infty)).$$

A suitable probabilistic representation is given by

$$S_t(f,x) = Ef\left(\frac{N_{tx}}{t}\right),$$

where $(N_t)_{t\geq 0}$ is the standard Poisson process. Conditions (H_1) and (6) are obviously satisfied. Furthermore, since the standard Poisson process has stationary increments, we obtain from Corollary 2

$$\omega(S_t f; h) \leq (2 - e^{-th})\, \omega(f; h), \qquad h \geq 0,\, t > 0,$$

and, from Theorem 2

$$\omega(S_t f; h) \leq \omega(f; h) + 2\omega(f; \sqrt{h/t}), \qquad h \geq 0,\, t > 0.$$

Finally, Theorem 3 gives in this case

$$|\omega^{[0,a]}(S_t f; h) - \omega^{[0,a]}(f; h)| \leq 4\omega\left(f; \sqrt{a/t}\right), \qquad h \geq 0,\, t > 0,\, a > 0.$$

(E) *Baskakov operators.* The Baskakov operator B_t^* is defined by

$$B_t^*(f,x) := \sum_{k=0}^{\infty} f(k/t) \binom{t+k-1}{k} \frac{x^k}{(1+x)^{t+k}}, \qquad t > 0,\, x \geq 0,\, f \in F([0,\infty)).$$

Let $(N_t)_{t\geq 0}$ and $(U_t)_{t\geq 0}$ be two independent stochastic processes defined on the same probability space, the former (resp. the later) being a standard Poisson process (resp. a gamma process). Then, for $x \geq 0$ and $t > 0$, the random variable N_{xU_t} has the negative binomial distribution with parameters t, $x/(1+x)$, and therefore we can write

$$B_t^*(f,x) = Ef\left(\frac{N_{xU_t}}{t}\right).$$

Again, conditions (H_1) and (6) are fulfilled. Moreover, for fixed $t > 0$, the process $(N_{xU_t})_{x\geq 0}$ has stationary increments. As in the former example, we obtain

$$\omega(B_t^* f; h) \leq (2 - (1+h)^{-t})\, \omega(f; h),$$

$$\omega(B_t^* f; h) \leq \omega(f; h) + 2\omega\left(f; \sqrt{\frac{h(1+h)}{t}}\right),$$

and

$$|\omega^{[0,a]}(B_t^* f; h) - \omega^{[0,a]}(f; h)| \leq 4\omega\left(f; \sqrt{\frac{a(1+a)}{t}}\right).$$

(F) *Gamma operators.* The Gamma operator G_t is defined by

$$G_t(f,x) := \frac{1}{\Gamma(t)} \int_0^{\infty} f\left(\frac{x\theta}{t}\right) \theta^{t-1} e^{-\theta} d\theta, \qquad t > 0,\, x \geq 0,\, f \in F([0,\infty)).$$

It is clear that

$$G_t(f,x) = Ef\left(\frac{xU_t}{t}\right),$$

where $(U_t)_{t\geq 0}$ is a gamma process. From the properties of this process, it can be easily shown that

$$\omega(G_t f; h) \leq 2\omega(f; h),$$

$$\omega(G_t f; h) \le \omega(f; h) + 2\omega(f; h/\sqrt{t})$$

and

$$|\omega^{[0,a]}(G_t f; h) - \omega^{[0,a]}(f; h)| \le 4\omega\left(f; a/\sqrt{t}\right).$$

On the other hand, as an application of Remark 3 we shall show that, for $h > 0$, $t \ge 1$ and $0 < \alpha < 1$, we have

$$\omega(G_t f; h) \le \left(1 + \sqrt{2}\, e^{-t^\alpha/16}\right) \omega\left(f; h\left(1 + t^{(\alpha-1)/2}\right)\right). \tag{23}$$

Firstly, we note that in this case

$$a_2(h,t) = \frac{h}{t} E^{1/2} U_t^2 = \frac{h}{t}\sqrt{t + t^2} \le \sqrt{2}h. \tag{24}$$

Secondly, using the estimate (cf. [**3**])

$$P\left(\frac{U_t}{t} \ge x\right) \le \left(x e^{1-x}\right)^t, \qquad x \ge 1, t \ge 1,$$

we obtain, for $0 \le \epsilon \le 1$

$$P\left(\frac{U_t}{t} \ge 1 + \epsilon\right) \le \exp\left(-t\epsilon + t\log(1+\epsilon)\right) \le \exp\left(-t\epsilon^2/8\right). \tag{25}$$

Therefore, inequality (23) follows from (13), by choosing

$$v = h\left(1 + t^{(\alpha-1)/2}\right)$$

and applying (24) and (25) with $\epsilon = t^{(\alpha-1)/2}$.

Analogously, using the representations given in [**2**], it is possible to apply the results in the preceding section to the following operators: Weierstrass, Beta, Lupas Beta, Inverse Beta, Double-indexed Beta, Müller Gamma and Meyer-König and Zeller operators. Details are omitted.

5. SECOND MODULUS

In this section, we shall investigate the preservation of the second modulus of continuity as defined by

$$\omega_2(f; h) := \sup_{\substack{x,y \in I \\ |x-y| \le h}} \left| f(x) + f(y) - 2f\left(\frac{x+y}{2}\right)\right|, \qquad 0 \le h < l(I),$$

where $f \in F(I)$. In order to do this, we shall need some additional assumptions on the process $Z = (Z_t^x)_{(x,t)\in I\times T}$. Specifically, throughout this section we shall assume that Z satisfies (H_1) and

(H_2): For all $t \in T$ and x, $y \in I$ with $x < y$, there is a σ-algebra $\mathcal{G}^y_{(t,x)} \subset \mathcal{F}$ such that both Z_t^x and Z_t^y are $\mathcal{G}^y_{(t,x)}$-measurable and

$$E\left(Z_t^u - Z_t^x | \mathcal{G}^y_{(t,x)}\right) = \frac{u-x}{y-x}\left(Z_t^y - Z_t^x\right) \qquad \text{a.s.,} \qquad x < u < y,$$

where $E(.|.)$ denotes conditional expectation.

Observe that these hypotheses guarantee preservation of monotonicity and convexity (cf. [2]). On the other hand, a general inequality of the form

$$\omega_2(L_t f; h) \leq c\,\omega_2(f; h),$$

where $c > 0$ may depend on h and t, is only possible if the function

$$l_t(x) := E Z_t^x, \qquad x \in I,$$

is an affine function; this is not true, for instance, for the Bleimann-Butzer-Hahn operator (see example (C) in the preceding section). Observe, finally, that the martingale-type assumption (H_2) guarantees that $l_t(.)$ is an affine function.

The main result in this section is the following:

THEOREM 4. *Let $t \in T$, $0 < h < l(I)$ and $f \in F(I)$. Then*
(a)

$$\omega_2(L_t f; h) \leq 2 \sup_{x,\, x+h \in I} E\omega_2(f; Z_t^{x+h} - Z_t^x).$$

(b) If, in addition, f is convex, then

$$\omega_2(L_t f; h) \leq \sup_{x,\, x+h \in I} E\omega_2(f; Z_t^{x+h} - Z_t^x).$$

Furthermore, if the process $(Z_t^x)_{x \in I}$ takes values in the set $\{n/t : n = 0, 1, \ldots\}$, then $\omega_2(f; Z_t^{x+h} - Z_t^x)$ can be replaced by $\omega_2(f; Z_t^{x+h} - Z_t^x) I(Z_t^{x+h} - Z_t^x \geq 2/t)$ in assertions (a) and (b).

Before proving Theorem 4, we shall introduce some necessary technical tools. For a function $f \in F(I)$, we define the auxiliary modulus of smoothness $\tilde{\omega}_2(f; .)$ by

$$\tilde{\omega}_2(f; h) := \sup |s f(x) + (1 - s) f(y) - f(sx + (1 - s)y)|, \qquad 0 \leq h < l(I),$$

where the supremum is taken over all $s \in [0, 1]$ and $x, y \in I$ with $|x - y| \leq h$. The relationship between ω_2 and $\tilde{\omega}_2$ is given in the following

LEMMA 1. *For $f \in F(I)$ and $0 \leq h < l(I)$, we have*

$$\frac{1}{2}\omega_2(f; h) \leq \tilde{\omega}_2(f; h) \leq \omega_2(f; h). \tag{26}$$

Proof of Lemma 1. The first inequality in (26) is trivial. To show the second, fix $x, y \in I$ with $0 < y - x \leq h$. Since for any affine function g we have $\omega_2(f + g; .) = \omega_2(f; .)$ and $\tilde{\omega}_2(f + g; .) = \tilde{\omega}_2(f; .)$, it can be assumed that $f(x) = f(y)$. Denote by x_0 and y_0 the points where f attains its minimum and maximum values on $[x, y]$ respectively. Then

$$\sup_{s \in [0,1]} |s f(x) + (1-s) f(y) - f(sx + (1-s)y)| = \max(f(x) - f(x_0),\, f(y_0) - f(x)). \tag{27}$$

For the sake of concreteness, suppose that $f(x) - f(x_0) \geq f(y_0) - f(x)$ and $x_0 \in [x, (x + y)/2]$. Then, (27) is not greater than

$$f(x) - f(x_0) \leq f(x) + f(2x_0 - x) - 2f(x_0) \leq \omega_2(f; h).$$

The remaining alternative cases can be treated similarly and, therefore, the conclusion follows.

REMARK 6. The constants $1/2$ and 1 appearing in Lemma 1 are the best possible. To see this, let $I = [0,1]$, $0 < a \le 1/2$ and take the function f_a defined by

$$f_a(x) := \begin{cases} (a-x)/a & \text{if } 0 \le x \le a \\ (x-a)/(1-a) & \text{if } a \le x \le 1. \end{cases}$$

It is readily seen that

$$\tilde{\omega}_2(f_a;1) = 1, \qquad \omega_2(f_a;1) = (1-a)^{-1}.$$

Therefore, $\omega_2(f_a;1)$ approaches to 1 or 2 according to $a \to 0$ or $a \to 1/2$.

Now, we are in a position to prove Theorem 4.

Proof of Theorem 4. Let $x, y \in I$ with $0 < y - x \le h$ and set $u := (x+y)/2$. In view of (H_1), we can write Z_t^u as the following convex linear combination

$$Z_t^u = UZ_t^x + (1-U)Z_t^y \qquad \text{a.s.},$$

where

$$U := \frac{Z_t^y - Z_t^u}{Z_t^y - Z_t^x} I(Z_t^x < Z_t^y). \tag{28}$$

Therefore, we have from (H_2)

$$|L_t(f,x) + L_t(f,y) - 2L_t(f,u)| = 2\left| E\left(\frac{1}{2}f(Z_t^x) + \frac{1}{2}f(Z_t^y) - f(Z_t^u)\right)\right|$$

$$= 2\left|E\left(E\left(Uf(Z_t^x) + (1-U)f(Z_t^y) - f(UZ_t^x + (1-U)Z_t^y) \mid \mathcal{G}_{(t,x)}^y\right)\right)\right|. \tag{29}$$

By definition of $\tilde{\omega}_2(f;.)$, Lemma 1 and (H_1), the right-hand side in (29) does not exceed

$$2\,E\tilde{\omega}_2(f;Z_t^y - Z_t^x) \le 2\,E\omega_2(f;Z_t^y - Z_t^x)$$

and the conclusion in part (a) follows. In order to show part (b), observe that if f is convex, then we have

$$f(UZ_t^x + (1-U)Z_t^y) \le Uf(Z_t^x) + (1-U)f(Z_t^y)$$

and, by the conditional version of Jensen's inequality

$$E\left(f(UZ_t^x + (1-U)Z_t^y) \mid \mathcal{G}_{(t,x)}^y\right) \ge f\left(E\left(UZ_t^x + (1-U)Z_t^y \mid \mathcal{G}_{(t,x)}^y\right)\right)$$
$$= f\left(\frac{1}{2}Z_t^x + \frac{1}{2}Z_t^y\right) \qquad \text{a.s.},$$

the last equality by (H_2). Therefore, the right hand side in (29) is bounded above by

$$E\left(f(Z_t^x) + f(Z_t^y) - 2f\left(\frac{1}{2}Z_t^x + \frac{1}{2}Z_t^y\right)\right) \le E\omega_2(f;Z_t^y - Z_t^x),$$

and the conclusion follows as in part (a). Finally, in view of the former proofs, the validity of the last assertion in Theorem 4 is an easy consequence of the following fact: If the process $(Z_t^x)_{x\in I}$ takes values in the set $\{n/t : n = 0,1,\ldots\}$, then, on the event $\{Z_t^y - Z_t^x \le 1/t\}$, the random variable U defined in (29) takes only the values 0 or 1; therefore, by the $\mathcal{G}^y_{(t,x)}$-measurability of Z_t^x and Z_t^y, the conditional expectation on the right-hand side in (29) can be replaced by

$$E\left(Uf(Z_t^x) + (1-U)f(Z_t^y) - f(UZ_t^x + (1-U)Z_t^y) \mid \mathcal{G}^y_{(t,x)}\right) I(Z_t^y - Z_t^x \ge 2/t).$$

The proof of Theorem 4 is complete.

As we did for the first modulus of continuity, we shall derive from Theorem 4 some interesting consequences by using standard properties of the second modulus of continuity. In the following estimates, the coefficient 2 can be dropped if the function f is convex.

COROLLARY 4. *If $\omega_2(f;.)$ is concave, then*

$$\omega_2(L_t f; h) \le 2\omega_2(f; a_1(h,t)),$$

where $a_1(h,t)$ is defined in (4).

Proof. Use Jensen's inequality.

COROLLARY 5. *Assume that $f \in \mathrm{Lip}_2(A,\mu)$, that is, $\omega_2(f;h) \le Ah^\mu$, where $A > 0$ and $\mu \in (0,2]$. Then we have*

$$\omega_2(L_t f; h) \le 2A\,(a_1(h,t))^\mu, \tag{30}$$

or

$$\omega_2(L_t f; h) \le 2A\,(a_2(h,t))^\mu, \tag{31}$$

according to $\mu \in (0,1]$ or $\mu \in (1,2]$, where $a_i(h,t)$, $(i = 1,2)$, is defined in (4).

Proof. Inequality (30) (resp. (31)) follows from Jensen's inequality (resp. Hölder's inequality).

COROLLARY 6. *For $t \in T$, $0 < h < l(I)$ and $f \in F(I)$, we have*

$$\omega_2(L_t f; h) \le e(h,t)\,\omega_2(f;h),$$

where

$$e(h,t) := 2 \sup_{x,\, x+h\in I} E\left(1 + \left[\frac{Z_t^{x+h} - Z_t^x}{h}\right]\right)^2, \tag{32}$$

brackets indicating integral part. Moreover, if the process $(Z_t^x)_{x\in I}$ takes values in the set $\{n/t : n = 0,1,\ldots\}$, then the right-hand side in (32) *can be replaced by*

$$2 \sup_{x,\, x+h\in I} E\left(1 + \left[\frac{Z_t^{x+h} - Z_t^x}{h}\right]\right)^2 I(Z_t^{x+h} - Z_t^x \ge 2/t).$$

Proof. Use the well-known property

$$\omega_2(f;u) \le (1 + [u/v])^2\,\omega_2(f;v), \qquad u \ge 0,\ v > 0.$$

We shall need the following

LEMMA 2. *Let $0 \le h < l(I)$ and $f \in F(I)$. Then*

$$\omega(\omega_2(f;.);h) \le 2\,\omega(f;h/2).$$

Proof. Let $x,\ y \in I$ with $0 < u < y - x \le u + h$ and set $x' := (x+y-u)/2$, $y' := (x+y+u)/2$. We have

$$\left|f(x)+f(y)-2f\left(\frac{x+y}{2}\right)\right| \le \left|f(x')+f(y')-2f\left(\frac{x'+y'}{2}\right)\right| \\ + |f(x)-f(x')| + |f(y)-f(y')|,$$

which implies

$$\omega_2(f;u+h) \le \omega_2(f;u) + 2\,\omega(f;h/2).$$

The conclusion follows.

THEOREM 5. *Let $t \in T$, $0 < h < l(I)$ and $f \in F(I)$. Then*

$$\omega_2(L_tf;h) \le \omega_2(f;h) + 3\,\omega(f;c(h,t)) + 3\,\omega(f;k(h,t)),$$

where $c(h,t)$ is defined in (14) and

$$k(h,t) := \sup_{\substack{x,y\in I\\ |x-y|\le h}} E|Z_t^x + Z_t^y - 2Z_t^{(x+y)/2}|. \tag{33}$$

Proof. Let $x,\ y \in I$ with $0 < y - x \le h$ and set $u := (x+y)/2$. We can write

$$|L_t(f,x)+L_t(f,y)-2L_t(f,u)| \le E\left|f(Z_t^x)+f(Z_t^y)-2f\left(\frac{1}{2}Z_t^x+\frac{1}{2}Z_t^y\right)\right| \\ + 2E\left|f(Z_t^u)-f\left(\frac{1}{2}Z_t^x+\frac{1}{2}Z_t^y\right)\right|$$

$$\le \sup_{x,\,x+h\in I} E\omega_2(f;Z_t^{x+h}-Z_t^x) + 2\sup_{\substack{x,y\in I\\ |x-y|\le h}} E\omega\left(f;\frac{1}{2}|Z_t^x+Z_t^y-2Z_t^u|\right). \tag{34}$$

Now, we add and substract $\omega_2(f;h)$ on the right-hand side in (34) and then proceed as in the proof of Corollary 3. We have

$$E\omega\left(f;\frac{1}{2}|Z_t^x+Z_t^y-2Z_t^u|\right) \le \frac{3}{2}\omega(f;k(h,t)),$$

where $k(h,t)$ is defined in (33). Likewise, using Lemma 2, we obtain

$$E|\omega_2(f;Z_t^{x+h}-Z_t^x)-\omega_2(f;h)| \le 2\,E\omega\left(f;\frac{1}{2}|Z_t^{x+h}-Z_t^x-h|\right), \\ \le 3\,\omega(f;c(h,t))$$

where $c(h,t)$ is defined in (14). The conclusion follows.

We shall close this section by giving two asymptotic results which are analogous to Theorems 2 and 3, respectively. Recall that, throughout this section, it is assumed that the process Z satisfies hypotheses (H_1) and (H_2). Such requirements are not necessary for the validity of Theorem 7 below.

THEOREM 6. *Let $0 \leq h < l(I)$ and let $f \in F(I)$ be uniformly continuous. Assume that*

(a) $\lim_{t\to\infty} L_t f = f$.

(b) $\lim_{t\to\infty} c(h,t) = \lim_{t\to\infty} k(h,t) = 0$,

where $c(h,t)$ and $k(h,t)$ are the same as in Theorem 5. Then

$$\lim_{t\to\infty} \omega_2(L_t f; h) = \omega_2(f; h).$$

Proof. Since f is uniformly continuous, the inequality

$$\limsup_{t\to\infty} \omega_2(L_t f; h) \leq \omega_2(f; h)$$

follows from Theorem 5 and assumption (b). On the other hand, the inequality

$$\omega_2(f; h) \leq \liminf_{t\to\infty} \omega_2(L_t f; h)$$

follows from assumption (a) (see the proof of Theorem 2).

THEOREM 7. *Let $0 \leq h < l(I)$, $f \in F(I)$ and let J be a compact subinterval of I. Then*

$$|\omega_2^J(L_t f; h) - \omega_2^J(f; h)| \leq 8\omega(f; d(J,t)), \qquad t \in T,$$

where $\omega_2^J(f;.)$ is defined analogously to $\omega^J(f;.)$ and $d(J,t)$ is defined in (20). *Therefore, $\omega_2^J(L_t f;.)$ converges uniformly to $\omega_2^J(f;.)$, as $t \to \infty$, whenever f is uniformly continuous and condition* (21) *is satisfied.*

Proof. It is easy to see that

$$|\omega_2^J(L_t f; h) - \omega_2^J(f; h)| \leq 4 \sup_{x \in J} |L_t(f,x) - f(x)|,$$

and, then, the same argument used in the proof of Theorem 3 works.

6. EXAMPLES

We shall apply the results in the preceding section to some of the operators considered in section 4. We shall keep the same representations and notations. Also, recall that such representations satisfy hypothesis (H_1). In what follows, $\sigma(U, V, \ldots)$ will denote the σ-algebra generated by the random variables U, $V, \ldots$

(A) *Bernstein operators.* It is not hard to see that, for $0 \leq x < u < y \leq 1$ and $n \geq 1$,

$$E(S_n(u) - S_n(x) \,|\, S_n(x),\, S_n(y)) = \frac{u-x}{y-x}(S_n(y) - S_n(x)) \qquad \text{a.s.},$$

so that hypothesis (H_2) is fulfilled if we choose

$$\mathcal{G}^y_{(n,x)} = \sigma(S_n(x),\, S_n(y)).$$

Since the process $(S_n(x))_{x\in[0,1]}$ has stationary increments and takes integer values, the quantity $e(h,n)$ in Corollary 6 is bounded above by*

$$2\,E\left(1+\frac{S_n(h)}{nh}\right)^2 I(S_n(h)\geq 2)=2\left(4-\frac{1}{n}-(3+(n-1)h)(1-h)^{n-1}\right)\leq 8.$$

On the other hand, for $0\leq x<y\leq 1$, we have

$$E\,(S_n(x)+S_n(y)-2S_n((x+y)/2))^2=n(y-x),$$

so that Theorem 5 applies with $k(h,n)\leq\sqrt{h/n}$ (and $c(h,n)\leq\sqrt{h(1-h)/n}$, as it is shown in section 4). Finally, from Theorem 7 we have, for any $f\in C[0,1]$

$$|\omega_2(B_nf;h)-\omega_2(f;h)|\leq 8\,\omega(f;1/2\sqrt{n}).$$

(B) *Szász-Mirakyan operators.* Without loss of generality, it can be assumed that the standard Poisson process $(N_t)_{t\geq 0}$ is defined on a *complete* probability space $(\Omega,\mathcal{F},P)$. For $t>0$ and $0\leq x<y$, let $\mathcal{G}^y_{(t,x)}$ be the augmentation with the P-null sets of the σ-algebra $\sigma(N_{tu}:\,0\leq u\leq x,\,u\geq y)$. Then, the adapted process

$$\left(\frac{N_{ty}-N_{tx}}{y-x},\mathcal{G}^y_{(t,x)}\right)_{y>x}$$

is a reverse martingale, as it follows from [**1**, proof of Lemma 1]. Thus, the representation given in section 4 for the Szász-Mirakyan operators satisfies hypothesis (H_2). Moreover, the quantity $e(h,t)$ defined in (32) satisfies

$$\begin{aligned}e(h,t)&\leq 2\,E\left(1+\frac{N_{th}}{th}\right)^2 I(N_{th}\geq 2)\\&=2\left(4-(3+th)e^{-th}+\frac{1-e^{-th}}{th}\right)\leq 10.\end{aligned}$$

On the other hand, it is easy to see that, for $0\leq x<y$,

$$E(N_{tx}+N_{ty}-2N_{t(x+y)/2})^2=t(y-x),$$

and, therefore, we have in this case

$$k(h,t)\leq\sqrt{\frac{h}{t}},$$

where $k(h,t)$ is defined in (33). Finally, from Theorems 6 and 7 we conclude

$$\lim_{t\to\infty}\omega_2(S_tf;h)=\omega_2(f;h)$$

and

$$|\omega_2^{[0,a]}(S_tf;h)-\omega_2^{[0,a]}(f;h)|\leq 8\,\omega\left(f;\sqrt{a/t}\right),$$

respectively.

* Using a different approach, H. Gonska has shown that $\omega_2(B_nf;h)\leq 5.5\,\omega_2(f;h)$. We learnt this inequality at the congress held in Acquafredda di Maratea, September 1992. As far as we know, this result has not been published yet.

(C) *Baskakov operators.* Recall that the Poisson process and the gamma process involved in the probabilistic representation of Baskakov operators are independent. Using this independence and the facts mentioned in the former example, it is not hard to see that such representation fulfills (H_2) if we take

$$\mathcal{G}^y_{(t,x)} := \sigma\left(N_{xU_t}, N_{yU_t}, U_t\right).$$

In this case

$$e(h,t) \le 2E\left(1+\frac{N_{hU_t}}{th}\right)^2 I(N_{hU_t} \ge 2)$$

$$= 2\left(4+\frac{1}{t}-\frac{3+(t+1)h}{(1+h)^{t+1}}+\frac{1}{th}\left(1-\left(\frac{1}{1+h}\right)^{t+1}\right)\right) \le 2\left(5+\frac{2}{t}\right).$$

Moreover, for $0 \le x < y$ and $z := (x+y)/2$

$$E\left(N_{xU_t}+N_{yU_t}-2N_{zU_t}\right)^2 = t(y-x),$$

which implies

$$k(h,t) \le \sqrt{\frac{h}{t}}.$$

(D) *Gamma operators.* The representation given in section 4 trivially fulfills hypothesis (H_2). In this case, we have

$$e(h,t) \le 2E\left(1+\frac{U_t}{t}\right)^2 = 2\left(4+\frac{1}{t}\right). \tag{35}$$

However, it can be shown that the coefficient 2 on the right-hand side in (35) can be dropped without any additional assumption (see the next section). On the other hand, we have

$$k(h,t) = 0.$$

7. CONCLUDING REMARKS

For some concrete operators, the general results given above are not the best possible. Actually, sharper estimates can be obtained by the only consideration of its particular probabilistic representation. Take, for instance, a convolution operator L which can be represented by

$$L(f,x) = Ef(x+U), \qquad -\infty < x < \infty,$$

where U is an integrable random variable whose distribution does not depend upon x. As typical examples, we can mention Weierstrass operators (cf. **[2]**), where U has a gaussian distribution. In this case, it is immediate that

$$\omega_k(Lf;h) \le E\omega_k(f;h) = \omega_k(f;h), \qquad k = 1,2,\ldots,$$

where $\omega_k(f;.)$ is the usual modulus of continuity of order k (cf. [**7**]). Similarly, let L be an operator having the form

$$L(f,x) = Ef(xU), \qquad x \geq 0,$$

where U is a non-negative integrable random variable whose distribution does not depend upon x, (examples: Gamma operators and Müller Gamma operators, see [**2**,**12**]). Then

$$\omega_k(Lf;h) \leq E\omega_k(f;hU) \leq \omega_k(f;h)E(1+U)^k, \qquad k = 1,2,\ldots,$$

where the second inequality follows by the well-known property

$$\omega_k(f;hu) \leq (1+[u])^k \omega_k(f;h).$$

Note that second inequality provides significant bounds only if U^k is integrable.

Finally, we think that the methods used in the present paper can be adapted to deal with other moduli of continuity (e.g., the modulus Ω_σ considered by Kratz and Stadtmüller in [**11**]) or similar problems concerning multivariate operators.

REFERENCES

1. J.A. Adell, F.G. Badía and J. de la Cal, Beta-type operators preserve shape properties, *Stochastic Process. Appl.* (to appear).
2. J.A. Adell and J. de la Cal, Using stochastic processes for studying Bernstein-type operators, *Rend. Circ. Mat. Palermo*, (to appear).
3. J.A. Adell and J. de la Cal, Approximating gamma distributions by normalized negative binomial distributions, *J. Appl. Probab.*, (to appear).
4. G.A. Anastassiou, C. Cottin and H.H. Gonska, Global smoothness of approximating functions, *Analysis* **11** (1991), 43-57.
5. G.A. Anastassiou, C. Cottin and H.H. Gonska, Global smoothness preservation by multivariate approximation operators, *in* : "Israel Math. Conf. Proceedings" (1991).
6. G. Bleimann, P.L. Butzer and L. Hahn, A Bernstein-type operator approximating continuous functions on the semi-axis, *Indag. Math.* **42** (1980), 255-262.
7. Z. Ditzian and V. Totik. "Moduli of Smoothness," Springer, New York, 1987.
8. N.L. Johnson and S. Kotz. "Discrete Distributions," Houghton-Mifflin, Boston, 1969.
9. M.K. Khan and M.A. Peters, Lipschitz constants for some approximation operators of a Lipschitz continuous function, *J. Approx. Theory* **59** (1989), 307-315.
10. R.A. Khan, A note on a Bernstein-type operator of Bleimann, Butzer and Hahn, *J. Approx. Theory* **53** (1988), 295-303.
11. W. Kratz and U. Stadtmüller, On the uniform modulus of continuity of certain discrete approximation operators, *J. Approx. Theory* **54** (1988), 326-337.
12. M. Müller, "Die Folge der Gammaoperatoren," Dissertation, Universität Stuttgart, 1967.
13. D.D. Stancu, Approximation of functions by a new class of linear polynomial operators, *Rev. Roumaine Math. Pures Appl.* **13** (1968), 1173-1194.

L_p-KOROVKIN TYPE INEQUALITIES FOR POSITIVE LINEAR OPERATORS

G.A. Anastassiou

Department of Mathematical Sciences
Memphis State University
Memphis, Tennessee 38152

Abstract: The author, in recent years, has produced several quantitative type of results for estimating the rate of convergence of a sequence of positive linear operators to the unit. These involve the modulus of continuity of the associated function or its derivative of certain order, and they are *pointwise Korovkin type inequalities*, most of them sharp. Using these inequalities, we are able to produce a great variety of general L_p $(1 \leq p \leq +\infty)$ analogs, covering most of the expected cases of the convergence of positive linear operators with rates to the unit. In the same inequality we achieve to combine different L_p-norms.

1. INTRODUCTION

The study of the convergence of a sequence of positive linear operators $\{L_j\}_{j\in\mathbb{N}}$ to the identity operator I acting on $C([a,b])$, $[a,b] \subset \mathbb{R}$, became serious with Korovkin (1960). Later on in Shisha and Mond (1968), this convergence was put in a quantitative form. In there, the associated inequality involves the modulus of continuity of the related function, it is with respect to the supremum norm and contains the basic Korovkin Theorem, but it is not a sharp inequality. In the following years came a lot of researchers with a big set of Shisha–Mond type inequalities, reflecting related Korovkin type theorems in all sorts of different directions of research, but still none of these proved a sharp inequality. Finally the author in Anastassiou (1985), (1986), (1991), among many other related papers, with the use of geometric moment methods from probability theory, was able to produce several sharp or nearly sharp *pointwise* inequalities of Shisha–Mond type related to Korovkin Theory. In this paper, we use some of these pointwise results, see Corollaries A, B; Theorems A, B and Remark A, in order to produce L_p $(1 \leq p \leq +\infty)$ Shisha–Mond type inequalities containing basic Korovkin arguments. We work over one-dimensional, multi-dimensional, and abstract compact convex domains. In our study we include also positive stochastic operators.

The functions we are dealing with are either differentiable or just continuous. Let M be a compact convex subset of $(V, \|\cdot\|)$ a real normed vector space. Let L be a positive linear operator from $C(M)$ into itself and $f \in C(M)$. In this paper we estimate $\|Lf - f\|_p$ $(1 \leq p \leq +\infty)$ in many different directions. The $\|\cdot\|_p$-norm is taken over a general measure space of finite positive measure. One of the techniques used in the proofs here is transferring through the Riesz representation theorem, Anastassiou-weak convergence related results on finite positive measures to positive linear operators.

Our inequalities involve the standard modulus of continuity of f or its certain order derivative (when $M = [a,b] \subset \mathbb{R}$), and combine different L_p-norms at a time. Our work has been greatly motivated by the very important related article of Popov (1982). Other important and interesting articles on this topic are of Quak (1988), (1989). In their papers they use the averaged modulus of smoothness, the so-called τ-modulus of orders one and two, and they work on real domains with bounded Lebesque measurable functions.

2. AUXILLIARY RESULTS

We would need the following lemmas and remarks.

Lemma 1. *Let M be a compact subset of $(V, \|\cdot\|)$ a real normed vector space. And let L be a positive linear operator from $C(M)$ into itself. Here $x, x_0 \in M$. Consider $r > 0$. Then $(L(\|x - x_0\|^r))(x_0)$ is a continuous function in x_0.*

Proof. Let $x_n, x_0 \in M$ be such that $x_n \to x_0$, as $n \to +\infty$. Observe that

$$|(L(\|x - x_n\|^r))(x_n) - (L(\|x - x_0\|^r))(x_0)| \leq |(L(\|x - x_n\|^r))(x_n) - (L(\|x - x_0\|^r))(x_n)| + |(L(\|x - x_0\|^r))(x_n) - (L(\|x - x_0\|^r))(x_0)| := (*).$$

Note that $\|x - x_0\|^r$ is a continuous function of x, so is $(L(\|x - x_0\|^r))(x)$. That is $(L(\|x - x_0\|^r))(x_n) \to (L(\|x - x_0\|^r))(x_0)$, as $x_n \to x_0$, i.e.,

$$|(L(\|x - x_0\|^r))(x_n) - (L(\|x - x_0\|^r))(x_0)| < \varepsilon_1,$$

where $\varepsilon_1 > 0$ is small. Therefore

$$(*) < \|L(\|x - x_n\|^r) - L(\|x - x_0\|^r)\|_\infty + \varepsilon_1 =: (**).$$

Let $x_n, x_0, x \in M$ such that $x_n \to x_0$, as $n \to +\infty$. Note that

$$\sup_{x \in M} |\|x - x_n\| - \|x - x_0\|| \leq \|x_n - x_0\| \to 0,$$

hence

$$\|\|x - x_n\| - \|x - x_0\|\|_\infty \to 0,$$

i.e.,

$$\|x - x_n\| \xrightarrow{u} \|x - x_0\|, \text{ uniformly.}$$

Here $\|x - y\| \leq \Delta$ – the diameter of M, $\forall x, y \in M$, i.e., $\|x - x_n\|$, $\|x - x_0\| \leq \Delta$. Call

$$f_n(x) := \|x - x_n\|, \quad f(x) := \|x - x_0\|,$$

then

$$0 \leq f_n(x), f(x) \leq \Delta, \qquad \forall x \in M.$$

Consider $g(t) := t^r$, $t \in [0, \Delta]$, $r > 0$. Note that g is a uniformly continuous function on $[0, \Delta]$, i.e., given $\varepsilon_2 > 0$ $\exists \delta > 0$: whenever $|u_1 - u_2| < \delta$, $u_1, u_2 \in [0, \Delta]$ we get that $|g(u_1) - g(u_2)| < \varepsilon_2$. For the above $\delta > 0$ $\exists \mathcal{N}_0$: $\forall n \geq \mathcal{N}_0$ we get $|f_n(x) - f(x)| < \delta$, $\forall x \in M$. Thus for $\varepsilon_2 > 0$ $\exists \mathcal{N}_0$: $\forall n \geq \mathcal{N}_0$ we get

$$|g(f_n(x)) - g(f(x))| < \varepsilon_2.$$

That is, $\|x - x_n\|^r \xrightarrow{u} \|x - x_0\|^r$, uniformly. Here L is a positive linear operator from $C(M)$ into itself, therefore L is a continuous operator and $L(\|x - x_n\|^r) \xrightarrow{u} L(\|x - x_0\|^r)$, uniformly. I.e., for sufficiently large n we have

$$\|L(\|x - x_n\|^r) - L(\|x - x_0\|^r)\|_\infty < \varepsilon_1,$$

$\varepsilon_1 > 0$ as above small.

Finally we obtain

$$(**) < 2 \cdot \varepsilon_1.$$

I.e.,

$$|(L(\|x - x_n\|^r))(x_n) - (L(\|x - x_0\|^r))(x_0)| < 2\varepsilon_1$$

$\varepsilon_1 > 0$ arbitrarily small, as $x_n \to x_0$. Therefore

$$(L(\|x - x_n\|^r))(x_n) \to (L(\|x - x_0\|^r))(x_0),$$

as $x_n \to x_0$. Q.E.D.

Remark 1. (i) Let $r \geq 1$, obviously from Lemma 1 we have that $((L(\|x - x_0\|^r))(x_0))^{1/r}$ is a continuous function in $x_0 \in M$. Therefore it is Borel measurable on $M \subseteq (V, \|\cdot\|)$, and thus Lebesgue measurable on $M \subseteq \mathbb{R}^k$, $k \geq 1$.

(ii) Let L be a positive linear operator from $C([a,b])$ into itself. Let $x, t \in [a,b]$, $n \in \mathbb{N}$. Then from Lemma 1, we have that $(L(|t - x|^n))(x)$ is a continuous function in x and thus Lebesgue measurable on $[a,b] \subseteq \mathbb{R}$.

Lemma 2. *Let L be a positive linear operator from $C([a,b])$ into itself and $x, t \in [a,b]$, $n \in \mathbb{N}$. Then $(L((t - x)^n))(x)$ is a continuous function in x.*

Proof. From

$$(t - x)^n = \sum_{k=0}^{n} \binom{n}{k} \cdot (-1)^k \cdot t^{n-k} \cdot x^k,$$

we get that

$$(L((t - x)^n))(x) = \sum_{k=0}^{n} (-1)^k \cdot \binom{n}{k} \cdot x^k \cdot (L(t^{n-k}))(x)$$

is continuous in $x \in [a,b]$. Q.E.D.

Remark 2. Let $f \in C([0,1])$ and

$$B_n(f;x) := \sum_{k=0}^{n} f\left(\frac{k}{n}\right) \cdot \binom{n}{k} \cdot x^k \cdot (1 - x)^{n-k}$$

the nth Bernstein polynomial for f, $n \in \mathbb{N}$, which is a positive linear operator from $C([0,1])$ into itself. It is known that

$$(B_n(1))(x_0) = 1 > 0, \quad \forall x_0 \in [0,1],$$
$$(B_n(x))(x_0) = x_0,$$

and

$$(B_n(x^2))(x_0) = \frac{x_0}{n} + \left(1 - \frac{1}{n}\right) \cdot x_0^2, \quad n \in \mathbb{N}.$$

Note that

$$(B_n((x - x_0)^2))(x_0) = \frac{x_0 \cdot (1 - x_0)}{n} \begin{cases} = 0, & \text{when } x_0 = 0, 1 \\ > 0, & x_0 \in (0,1). \end{cases}$$

Therefore, for L a positive linear operator from $C(M)$ into itself, where M is a compact subset of $(V, \|\cdot\|)$, it is reasonable to assume that

(i) $(L(1))(x_0) > 0, \forall x_0 \in M$,

(ii) $(L(\|x - x_0\|^r))(x_0) > 0, \forall x_0 \in M, r > 0$, or

(ii)′ $(L(\|x - x_0\|^r))(x_0) > 0$, μ-a.e., i.e.,

$$\mu(x_0 \in M : (L(\|x - x_0\|^r))(x_0) = 0) = 0,$$

where μ is a finite positive measure on M with $\mu(M) > 0$.

From (ii)′, we get that $(1 \leq p \leq +\infty)$

$$0 < \|(L(\|x - x_0\|^r))(x_0)\|_{L_p(\mu)} < +\infty.$$

From the Riesz Representation Theorem, see Aliprantis and Burkinshaw (1990), pp. 246–249, we have

Lemma 3. *Let M be a compact subset of a real normed vector space $(V, \|\cdot\|)$. Let $C(M)$ be the space of continuous real valued functions on M and F be a positive linear functional on $C(M)$. Then there exists a unique regular Borel positive finite measure μ on M such that*

$$F(f) = \int_M f\, d\mu, \qquad \forall f \in C(M).$$

Remark 3. Let $L \not\equiv 0$: $C(M) \hookrightarrow C(M)$ be a positive linear operator. Then $(L(\cdot))(x_0)$ is a positive linear functional on $C(M)$, where $x_0 \in M$ is fixed. From Lemma 3 we have that

$$(L(f))(x_0) = \int_M f\, d\mu_{x_0}, \qquad \forall f \in C(M),$$

where μ_{x_0} is a unique regular Borel positive finite measure on M.

Lemma 4. *Let $(M, \mathcal{B}, \mu)$ be a measure space, where M is a compact subset of $(V, \|\cdot\|)$, $\mathcal{B}$ is the Borel σ-algebra on M and μ is a positive finite measure on M. Consider $C(M)$ $(\subset L_p(M, \mathcal{B}, \mu)$, any $p > 1)$. Let $f, g \in C(M)$. Then*

$$\|f \cdot g\|_p \leq \|f\|_{p\cdot q} \cdot \|g\|_{p^2}, \tag{1}$$

where $p, q > 1$ such that $\frac{1}{p} + \frac{1}{q} = 1$.

Proof. Easy by Hölder's inequality. Q.E.D.

Remark 4. Inequality (1) is true in general, when $f \in L_{p\cdot q}(M, \mathcal{B}, \mu)$ and $g \in L_{p^2}(M, \mathcal{B}, \mu)$ then $f \cdot g \in L_p(M, \mathcal{B}, \mu)$.

Remark 5. From Anastassiou (1985), pp. 251–252, we get the following useful function ϕ_n and its properties: Let $h > 0$ be fixed. Define

$$\phi_n(x) := \int_0^{|x|} \left\lceil \frac{t}{h} \right\rceil \cdot \frac{(|x| - t)^{n-1}}{(n-1)!} \cdot dt, \quad (x \in \mathbb{R}), n \in \mathbb{N}, \tag{2}$$

where $\lceil\cdot\rceil$ is the ceiling of the number.

Define $\phi_0(t) := \lceil \frac{t}{h} \rceil$, $t \geq 0$ then

$$\phi_n(x) = \int_0^x \phi_{n-1}(t)dt, \qquad (x \in \mathbb{R}^+, n \in \mathbb{N}). \tag{3}$$

The function ϕ_n is even, continuous and convex on $\mathbb{R}$, and it is strictly increasing on $\mathbb{R}^+$, $n \in \mathbb{N}$; $\phi_n(0) = 0$. One can easily prove that $\phi_n(u^{1/n})$ is a convex function in $u \geq 0$, $n \in \mathbb{N}$.

We have the following inequality true

$$\phi_n(x) \leq \left(\frac{|x|^{n+1}}{(n+1)! \cdot h} + \frac{|x|^n}{2 \cdot n!} + \frac{h \cdot |x|^{n-1}}{8 \cdot (n-1)!} \right), \qquad \forall x \in \mathbb{R},\ n \in \mathbb{N}. \tag{4}$$

Hence for $x \neq 0$ we get

$$\frac{\phi_n(x)}{|x|^n} \leq \left(\frac{|x|}{(n+1)! \cdot h} + \frac{1}{2 \cdot n!} + \frac{h}{8 \cdot |x| \cdot (n-1)!} \right). \tag{5}$$

We also need to mention the following result from Anastassiou (1985), p. 259.

Corollary A. *Consider the positive linear operator*

$$L : C([a,b]) \longrightarrow C([a,b]).$$

For a fixed $x \in [a,b]$ define

$$c(x) := \max(x-a, b-x) \tag{6}$$

(i.e., $\left(\frac{b-a}{2}\right) \leq c(x) \leq b-a$). *Let $f \in C^n([a,b])$, $n \in \mathbb{N}$, such that the first modulus of continuity $\omega_1(f^{(n)}, h) \leq w$, where w, h are fixed positive numbers, $0 < h < b-a$. Then we have the upper bound*

$$|L(f,x) - f(x)| \leq |f(x)| \cdot |L(1,x) - 1| + \sum_{k=1}^{n} \frac{|f^{(k)}(x)|}{k!}$$

$$\cdot |L((t-x)^k, x)| + w \cdot \frac{\phi_n(c(x))}{(c(x))^n} \cdot L(|t-x|^n, x), \quad \text{all } x \in [a,b]. \tag{7}$$

Inequality (7) *is sharp.*

3. ONE-DIMENSIONAL RESULTS

These are inequalities with respect to basic norms for differentiable functions.
Next comes our first main result.

Theorem 1. *Consider the positive linear operator*

$$L : C([a,b]) \longrightarrow C([a,b]).$$

Define ($n \in \mathbb{N}$)

$$D_n := \|(L(|t-\cdot|^n))(\cdot)\|_\infty^{1/n}, \tag{8}$$

where $\|\cdot\|_\infty$ is the supremum norm. Let $f \in C^n([a,b])$. Then we obtain

$$\|Lf - f\|_\infty \leq \|f\|_\infty \cdot \|L1 - 1\|_\infty + \sum_{k=1}^{n} \frac{\|f^{(k)}\|_\infty}{k!} \cdot \|(L(t-\cdot)^k)(\cdot)\|_\infty$$

$$+ \omega_1(f^{(n)}, D_n) \cdot D_n^{n-1} \left(\frac{(b-a)}{(n+1)!} + \frac{D_n}{2 \cdot n!} + \frac{D_n^2}{8 \cdot (b-a) \cdot (n-1)!} \right). \tag{9}$$

Note that by Remark 1(ii), $D_n < +\infty$.

Proof. From $\phi_n(u^{1/n})$ being convex in $u \geq 0$ and $\phi_n(0) = 0$, we have that $\frac{\phi_n(u^{1/n})}{u}$ is increasing in $u > 0$. Therefore by $c^n(x) \leq (b-a)^n$ we get

$$\frac{\phi_n(c(x))}{c^n(x)} \leq \frac{\phi_n(b-a)}{(b-a)^n}.$$

Consequently

$$\text{R.H.S.}(7) \le \|f\|_\infty \cdot \|L1 - 1\|_\infty + \sum_{k=1}^{n} \frac{\|f^{(k)}\|_\infty}{k!} \cdot \|(L(t-\cdot)^k)(\cdot)\|_\infty$$

$$+ w \cdot \frac{\phi_n(b-a)}{(b-a)^n} \cdot \|L(|t-\cdot|^n)(\cdot)\|_\infty.$$

Hence

$$\|Lf - f\|_\infty \le \|f\|_\infty \cdot \|L1 - 1\|_\infty + \sum_{k=1}^{n} \frac{\|f^{(k)}\|_\infty}{k!} \cdot \|(L(t-\cdot)^k)(\cdot)\|_\infty$$

$$+ w \cdot \frac{\phi_n(b-a)}{(b-a)^n} \cdot \|L(|t-\cdot|^n)(\cdot)\|_\infty. \tag{10}$$

In particular from (5) we have

$$\frac{\phi_n(b-a)}{(b-a)^n} \le \left(\frac{(b-a)}{(n+1)! \cdot h} + \frac{1}{2 \cdot n!} + \frac{h}{8 \cdot (b-a) \cdot (n-1)!} \right).$$

Pick $h = D_n := \|(L(|t-\cdot|^n))(\cdot)\|_\infty^{1/n}$, i.e., $h^n = D_n^n = \|(L(|t-\cdot|^n))(\cdot)\|_\infty$. We get that

$$\begin{aligned}
\text{Remainder}(10) &= w \cdot \frac{\phi_n(b-a)}{(b-a)^n} \cdot D_n^n \le w \cdot \left(\frac{(b-a) \cdot D_n^n}{(n+1)! \cdot h} + \frac{D_n^n}{2 \cdot n!} + \frac{h \cdot D_n^n}{8 \cdot (b-a) \cdot (n-1)!} \right) \\
&= w \cdot \left(\frac{(b-a) \cdot D_n^n}{(n+1)! \cdot D_n} + \frac{D_n^n}{2 \cdot n!} + \frac{D_n \cdot D_n^n}{8 \cdot (b-a) \cdot (n-1)!} \right) \\
&= w \cdot D_n^{n-1} \cdot \left(\frac{(b-a)}{(n+1)!} + \frac{D_n}{2 \cdot n!} + \frac{D_n^2}{8 \cdot (b-a) \cdot (n-1)!} \right).
\end{aligned}$$

Also pick $w = \omega_1(f^{(n)}, D_n)$. Q.E.D.

Corollary 1. *Let L be a positive linear operator from $C([a,b])$ into itself. Here*

$$D_1 := \|(L(|t-\cdot|))(\cdot)\|_\infty < +\infty. \tag{11}$$

Let $f \in C^1([a,b])$. Then

$$\|Lf - f\|_\infty \le \|f\|_\infty \cdot \|L1 - 1\|_\infty + \|f'\|_\infty \cdot \|(L(t-\cdot))(\cdot)\|_\infty$$

$$+ \frac{1}{2} \cdot \omega_1(f', D_1) \cdot \left((b-a) + D_1 + \frac{D_1^2}{4 \cdot (b-a)} \right). \tag{12}$$

Corollary 2. *Let L be a positive linear operator from $C([a,b])$ into itself. Here*

$$D_2 := \|(L((t-\cdot)^2))(\cdot)\|_\infty^{1/2} < +\infty. \tag{13}$$

Let $f \in C^2([a,b])$. Then

$$\|Lf - f\|_\infty \le \|f\|_\infty \cdot \|L1 - 1\|_\infty + \|f'\|_\infty \cdot \|(L(t-\cdot))(\cdot)\|_\infty + \frac{\|f''\|_\infty}{2} \cdot \|(L(t-\cdot)^2)(\cdot)\|_\infty$$

$$+ \frac{1}{2} \cdot \omega_1(f'', D_2) \cdot D_2 \cdot \left(\frac{(b-a)}{3} + \frac{D_2}{2} + \frac{D_2^2}{4 \cdot (b-a)} \right). \tag{14}$$

Next comes an application of Corollary 2.

Example 1. Let B_n be the Bernstein polynomials (see Remark 2) acting on $C([0,1])$. Let $f \in C^2([0,1])$. Then, we easily obtain

$$\|B_n f - f\|_\infty \le \frac{\|f''\|_\infty}{8 \cdot n} + \frac{1}{4\sqrt{n}} \cdot \omega_1\left(f'', \frac{1}{2\sqrt{n}}\right) \cdot \left(\frac{1}{3} + \frac{1}{4\sqrt{n}} + \frac{1}{16n}\right). \tag{15}$$

An improved result for $f \in C^1([a,b])$ with respect to $\|\cdot\|_\infty$ follows.

Theorem 2. *Let $L \not\equiv 0$ be a positive linear operator from $C([a,b])$ into itself. Call*

$$\rho := \|(L(t-x_0)^2)(x_0)\|_\infty^{1/2}, \tag{16}$$

and consider $r > 0$. Let $f \in C^1([a,b])$. Then

$$\|Lf - f\|_\infty - \|f\|_\infty \cdot \|L1 - 1\|_\infty - \|f'\|_\infty \cdot \|(L((t-x_0)))(x_0)\|_\infty$$
$$\le \begin{cases} \frac{1}{8 \cdot r} \cdot (2 + \sqrt{\|L(1)\|_\infty} \cdot r)^2 \cdot \omega_1(f', r \cdot \rho) \cdot \rho, & \text{if } r \le \frac{2}{\sqrt{\|L(1)\|_\infty}}; \\ \sqrt{\|L(1)\|_\infty} \cdot \omega_1(f', r \cdot \rho) \cdot \rho, & \text{if } r > \frac{2}{\sqrt{\|L(1)\|_\infty}}. \end{cases} \tag{17}$$

Note that by Lemma 2 we have $\rho < +\infty$.

Proof. Let $L \not\equiv 0$ be a positive linear operator from $C([a,b])$ into itself. Let $x_0 \in [a,b]$. By Remark 3 we have

$$(L(f))(x_0) = \int_{[a,b]} f(t) \cdot \mu_{x_0}(dt), \quad \text{all } f \in C([a,b]),$$

where μ_{x_0} is finite non-negative measure on $[a,b]$. Let $r > 0$ and call

$$m_{x_0} := (L(1))(x_0) = \mu_{x_0}([a,b]),$$
$$d_2(x_0) := \left(\int_{[a,b]} (t-x_0)^2 \cdot \mu_{x_0}(dt)\right)^{1/2}$$

Then, working similarly as in the proof of Theorem 2.19, pp. 263–265, Anastassiou (1985), we obtain

$$\left|\int_{[a,b]} f\, d\mu_{x_0} - f(x_0)\right| \le |m_{x_0} - 1| \cdot |f(x_0)| + |f'(x_0)| \cdot \left|\int_{[a,b]} (t-x_0) \cdot \mu_{x_0}(dt)\right|$$
$$+ \frac{1}{8 \cdot r} \cdot (2 + \sqrt{m_{x_0}} \cdot r)^2 \cdot \omega_1(f', r \cdot d_2(x_0)) \cdot d_2(x_0) =: (*).$$

Note

$$d_2^2(x_0) = \int (t-x_0)^2 \cdot \mu_{x_0}(dt) = (L((x-x_0)^2)))(x_0) \le \|(L(x-x_0)^2)(x_0)\|_\infty.$$

I.e.,

$$d_2(x_0) \le \|(L(x-x_0)^2)(x_0)\|_\infty^{1/2} := \rho.$$

Thus

$$(*) \le \|L1 - 1\|_\infty \cdot \|f\|_\infty + \|f'\|_\infty \cdot \|(L(t-x_0))(x_0)\|_\infty$$
$$+ \frac{1}{8 \cdot r} \cdot \left(2 + \sqrt{\|L(1)\|_\infty \cdot r}\right)^2 \cdot \omega_1(f', r \cdot \rho) \cdot \rho.$$

We have established that

$$\|Lf - f\|_\infty \le \|f\|_\infty \cdot \|L1 - 1\|_\infty + \|f'\|_\infty \cdot \|(L(t-x_0))(x_0)\|_\infty$$

$$+\frac{1}{8\cdot r}\cdot\left(2+\sqrt{\|L(1)\|_\infty}\cdot r\right)^2\cdot\omega_1(f',r\cdot\rho)\cdot\rho,\quad r>0. \tag{18}$$

Call $M:=\|L(1)\|_\infty>0$, by $L\not\equiv 0$, and

$$g(r):=\frac{1}{8\cdot r}\cdot(2+\sqrt{M}\cdot r)^2>0,\quad \text{all } r>0.$$

Note that $g(r)$ is a convex function in r. And

$$\min_{r>0} g(r)=g\left(\frac{2}{\sqrt{M}}\right)=\sqrt{M}.$$

Hence the remainder of (18), when $r=2/\sqrt{M}$, equals

$$\left(\sqrt{\|L(1)\|_\infty}\cdot\omega_1\left(f',\frac{2}{\sqrt{\|L(1)\|_\infty}}\cdot\rho\right)\cdot\rho\right)\le$$

$$\left(\sqrt{\|L(1)\|_\infty}\cdot\omega_1(f',r\cdot\rho)\cdot\rho\right),\quad \text{all } r\ge\frac{2}{\sqrt{\|L(1)\|_\infty}}.$$

Q.E.D.

Using Corollary A we obtain some $\|\cdot\|_{L_1}-\|\cdot\|_{L_2}$ results with respect to Lebesgue measure that follow.

Theorem 3. *Consider the positive linear operator*

$$L: C([a,b])\longrightarrow C([a,b]),\qquad a\neq b.$$

Call $(n\in\mathbb{N})$

$$h_n:=\|L(|t-x|^n)(x)\|_2^{1/n}. \tag{19}$$

Let $f\in C^n([a,b])$. *Then*

$$\|Lf-f\|_1\le\|f\|_2\cdot\|L1-1\|_2+\sum_{k=1}^{n}\frac{1}{k!}\cdot\|f^{(k)}\|_2\cdot\|L((t-\cdot)^k)(\cdot)\|_2$$

$$+\frac{1}{2}\cdot\omega_1(f^{(n)},h_n)\cdot h_n^{n-1}\cdot\left\{\left(\frac{b-a}{(n+1)!}\right)\cdot\sqrt{\frac{7}{3}\cdot(b-a)}+\frac{\sqrt{b-a}}{n!}\cdot h_n\right.$$

$$\left.+\frac{h_n^2}{4\cdot(n-1)!}\cdot\sqrt{\frac{2}{b-a}}\right\}. \tag{20}$$

From Lemmas 1,2 $h_n<+\infty$ *and* R.H.S.(20) $<+\infty$.

Proof. Note that from (5) we get

$$\text{Remainder}(7)=w\cdot\frac{\phi_n(c(x))}{(c(x))^n}\cdot L(|t-x|^n,x)\le w\cdot\left(\frac{c(x)}{(n+1)!\cdot h}+\frac{1}{2\cdot n!}\right.$$

$$\left.+\frac{h}{8\cdot c(x)\cdot(n-1)!}\right)\cdot L(|t-x|^n,x),\quad \forall x\in[a,b],$$

where $c(x):=\max(x-a,b-x)$. Therefore from (7) we have

$$|L(f,x)-f(x)|\le|f(x)|\cdot|L(1,x)-1|+\sum_{k=1}^{n}\frac{|f^{(k)}(x)|}{k!}\cdot|L((t-x)^k,x)|$$

$$+w\cdot\left(\frac{c(x)}{(n+1)!\cdot h}+\frac{1}{2\cdot n!}+\frac{h}{8\cdot c(x)\cdot(n-1)!}\right)\cdot L(|t-x|^n,x).$$

Since all functions involved in the last inequality are L_1-, L_2-Lebesgue integrable functions (see Lemmas 1, 2 and assumptions of the theorem) we obtain

$$\|Lf-f\|_1 \leq \| \, |f(x)| \cdot |L(1,x)-1| \, \|_1 + \sum_{k=1}^{n} \left\| \frac{|f^{(k)}(x)|}{k!} \cdot |L((t-x)^k, x)| \right\|_1$$

$$+ w \cdot \left\| \left(\frac{c(x)}{(n+1)! \cdot h} + \frac{1}{2 \cdot n!} + \frac{h}{8 \cdot c(x) \cdot (n-1)!} \right) \cdot L(|t-x|^n, x) \right\|_1 .$$

Then, by using Cauchy–Schwarz's inequality repeatedly, we get

$$\|Lf-f\|_1 \leq \|f\|_2 \cdot \|L1-1\|_2 + \sum_{k=1}^{n} \frac{1}{k!} \|f^{(k)}\|_2 \|L((t-\cdot)^k)(\cdot)\|_2$$

$$+ w \cdot \left\{ \frac{\|c(x)\|_2}{(n+1)! \cdot h} + \frac{1}{2 \cdot n!} \cdot \sqrt{b-a} + \frac{h}{8 \cdot (n-1)!} \cdot \left\| \frac{1}{c(x)} \right\|_2 \right\} \cdot \|L(|t-\cdot|^n)(\cdot)\|_2. \tag{21}$$

Observe that

$$\|c(x)\|_2 = \sqrt{\int_a^b c^2(x) dx} = \sqrt{\int_a^{\frac{a+b}{2}} (b-x)^2 \cdot dx + \int_{\frac{a+b}{2}}^b (x-a)^2 \cdot dx}$$

$$= \left(\frac{b-a}{2} \right) \cdot \sqrt{\frac{7}{3} \cdot (b-a)}.$$

I.e.,

$$\|c(x)\|_2 = \left(\frac{b-a}{2} \right) \cdot \sqrt{\frac{7}{3} \cdot (b-a)}. \tag{22}$$

Also

$$\left\| \frac{1}{c(x)} \right\|_2 = \sqrt{\int_a^b \frac{1}{c^2(x)} dx} = \sqrt{\int_a^b (c(x))^{-2} dx}$$

$$= \sqrt{\int_a^{\frac{a+b}{2}} (b-x)^{-2} dx + \int_{\frac{a+b}{2}}^b (x-a)^{-2} dx} = \sqrt{\frac{2}{b-a}}.$$

I.e.,

$$\left\| \frac{1}{c(x)} \right\|_2 = \sqrt{\frac{2}{b-a}}. \tag{23}$$

Putting (22), (23) into (21) we have

$$\|Lf-f\|_1 \leq \|f\|_2 \cdot \|L1-1\|_2 + \sum_{k=1}^{n} \frac{1}{k!} \cdot \|f^{(k)}\|_2 \cdot \|L((t-\cdot)^k)(\cdot)\|_2$$

$$+ w \cdot \left\{ \frac{\left(\frac{b-a}{2}\right) \cdot \sqrt{\frac{7}{3} \cdot (b-a)}}{(n+1)! \cdot h} + \frac{1}{2 \cdot n!} \cdot \sqrt{b-a} \right.$$

$$\left. + \frac{h}{8 \cdot (n-1)!} \cdot \sqrt{\frac{2}{b-a}} \right\} \cdot \|L(|t-\cdot|^n)(\cdot)\|_2.$$

Choosing

$$h = h_n := \|L(|t-x|^n)(x)\|_2^{1/n}, \qquad n \in \mathbb{N},$$

i.e., $h^n = \|L(|t-x|^n)(x)\|_2$, we obtain (20). Q.E.D.

Corollary 3. *Consider the positive linear operator*

$$L : C([a,b]) \longrightarrow C([a,b]).$$

Call

$$h_1 := \|L(|t-x|)(x)\|_2. \tag{24}$$

Let $f \in C^1([a,b])$. Then

$$\|Lf - f\|_1 \le \|f\|_2 \cdot \|L1 - 1\|_2 + \|f'\|_2 \cdot \|L((t-\cdot))(\cdot)\|_2 + \frac{\omega_1(f', h_1)}{2}$$

$$\cdot \left\{ \left(\frac{b-a}{2}\right) \cdot \sqrt{\frac{7}{3}(b-a)} + (\sqrt{b-a}) \cdot h_1 + \frac{h_1^2}{4} \cdot \sqrt{\frac{2}{b-a}} \right\}. \tag{25}$$

Corollary 4. *Consider the positive linear operator*

$$L : C([a,b]) \longrightarrow C([a,b]).$$

Call

$$h_2 := \|L((t-x)^2)(x)\|_2^{1/2}. \tag{26}$$

Let $f \in C^2([a,b])$. Then

$$\|Lf - f\|_1 \le \|f\|_2 \cdot \|L1 - 1\|_2 + \|f'\|_2 \cdot \|L((t-\cdot))(\cdot)\|_2 + \frac{\|f''\|_2}{2} \cdot h_2^2$$

$$+ \frac{1}{4} \cdot \omega_1(f'', h_2) \cdot h_2 \cdot \left\{ \left(\frac{b-a}{3}\right) \cdot \sqrt{\frac{7}{3} \cdot (b-a)} + (\sqrt{b-a}) \cdot h_2 + \frac{h_2^2}{2} \cdot \sqrt{\frac{2}{b-a}} \right\}. \tag{27}$$

Remark 6 (on Corollary 4). Using Lemmas 3 and Cauchy–Schwarz's inequality we obtain

$$((L(t-x))(x))^2 \le (L((t-x)^2)(x)) \cdot ((L(1))(x)).$$

Then

$$\int_a^b ((L(t-x))(x))^2 \cdot dx \le \int_a^b ((L(t-x)^2)(x) \cdot (L(1))(x)) \cdot dx$$
$$\le \|(L(t-\cdot)^2)(\cdot)\|_2 \cdot \|L(1)\|_2,$$

the last inequality comes by Cauchy–Schwarz's inequality. We have established that

$$\|(L(t-\cdot))(\cdot)\|_2 \le h_2 \cdot \sqrt{\|L(1)\|_2}. \tag{28}$$

Next we obtain some general L_p results.

Theorem 4. *Let $([a,b], \mathcal{B}, \mu)$, $a \ne b$, be a measure space, where $\mathcal{B}$ is the Borel σ-algebra on $[a,b]$ and μ is a positive finite measure on $[a,b]$. (Note that $C([a,b]) \subset L_p([a,b], \mathcal{B}, \mu)$, any $p > 1$.) Here $\|\cdot\|_p$ stands for the related L_p norm with respect to μ. Let $p, q > 1$ such that $\frac{1}{p} + \frac{1}{q} = 1$. Consider the positive linear operator*

$$L : C([a,b]) \longrightarrow C([a,b]).$$

Call ($n \in \mathbb{N}$)

$$h_n := \|(L(|t-\cdot|^n))(\cdot)\|_p^{1/n}. \tag{29}$$

Let $f \in C^n([a,b])$. Then

$$\|Lf - f\|_p \le \|f\|_{p\cdot q} \cdot \|L1 - 1\|_{p^2} + \sum_{k=1}^{n} \frac{1}{k!} \cdot \|f^{(k)}\|_{p\cdot q} \cdot \|L(t - \cdot)^k(\cdot)\|_{p^2} + \omega_1(f^{(n)}, h_n)$$

$$\cdot h_n^{n-1} \cdot \left[\frac{(b-a)}{(n+1)!} + \frac{h_n}{2 \cdot n!} + \frac{h_n^2}{4 \cdot (n-1)! \cdot (b-a)} \right]. \tag{30}$$

Note that from Lemmas 1,2 $h_n < +\infty$ and R.H.S.(30) $< +\infty$.

Proof. In the proof of Theorem 3 we found that

$$|L(f,x) - f(x)| \le |f(x)| \cdot |L(1,x) - 1| + \sum_{k=1}^{n} \frac{|f^{(k)}(x)|}{k!} \cdot |L((t-x)^k, x)|$$

$$+ w \cdot \left(\frac{c(x)}{(n+1)! \cdot h} + \frac{1}{2 \cdot n!} + \frac{h}{8 \cdot c(x) \cdot (n-1)!} \right) \cdot L(|t-x|^n, x). \tag{31}$$

Note that $0 < c(x) \le b - a$ and

$$\frac{1}{c(x)} \le \frac{2}{b-a}, \qquad \forall x \in [a,b].$$

Therefore from (31) we find

$$|L(f,x) - f(x)| \le |f(x)| \cdot |L(1,x) - 1| + \sum_{k=1}^{n} \frac{|f^{(k)}(x)|}{k!} \cdot |L((t-x)^k, x)|$$

$$+ w \cdot \left(\frac{(b-a)}{(n+1)! \cdot h} + \frac{1}{2 \cdot n!} + \frac{h}{4 \cdot (b-a) \cdot (n-1)!} \right) \cdot L(|t-x|^n, x). \tag{32}$$

Hence

$$\|Lf - f\|_p \le \| \, |f(x)| \cdot |L(1,x) - 1| \, \|_p + \sum_{k=1}^{n} \frac{1}{k!} \cdot \| \, |f^{(k)}(x)| \cdot |L((t-x)^k, x)| \, \|_p$$

$$+ w \cdot \left[\frac{(b-a)}{(n+1)! \cdot h} + \frac{1}{2 \cdot n!} + \frac{h}{4 \cdot (n-1)! \cdot (b-a)} \right] \cdot \|(L(|t - \cdot|^n))(\cdot)\|_p. \tag{33}$$

In (33) we apply Lemma 4, namely inequality (1), and we get

$$\|Lf - f\|_p \le \|f\|_{p\cdot q} \cdot \|L1 - 1\|_{p^2} + \sum_{k=1}^{n} \frac{1}{k!} \cdot \|f^{(k)}\|_{p\cdot q} \cdot \|L(t - \cdot)^k(\cdot)\|_{p^2}$$

$$+ w \cdot \left[\frac{(b-a)}{(n+1)! \cdot h} + \frac{1}{2 \cdot n!} + \frac{h}{4 \cdot (n-1)! \cdot (b-a)} \right] \cdot \|(L(|t - \cdot|^n))(\cdot)\|_p. \tag{34}$$

Choose

$$h = h_n := \|(L(|t - \cdot|^n))(\cdot)\|_p^{1/n}, \qquad n \in \mathbb{N},$$

i.e., $h^n = h_n^n = \|(L(|t - \cdot|^n))(\cdot)\|_p$. From (34) we now have that

$$\|Lf - f\|_p \le \|f\|_{p\cdot q} \cdot \|L1 - 1\|_{p^2} + \sum_{k=1}^{n} \frac{1}{k!} \cdot \|f^{(k)}\|_{p\cdot q} \cdot \|L(t - \cdot)^k(\cdot)\|_{p^2} + \omega_1(f^{(n)}, h_n)$$

$$\cdot \left[\frac{(b-a)}{(n+1)!} \cdot h_n^{n-1} + \frac{1}{2 \cdot n!} \cdot h_n^n + \frac{h_n^{n+1}}{4 \cdot (n-1)! \cdot (b-a)} \right].$$

Q.E.D.

Remark 7 (on Theorem 4). (i) Case of $n = 1$. Then

$$\|Lf - f\|_p \leq \|f\|_{p\cdot q} \cdot \|L1 - 1\|_{p^2} + \|f'\|_{p\cdot q} \cdot \|L(t - \cdot)(\cdot)\|_{p^2}$$

$$+ \frac{\omega_1(f', h_1)}{2} \cdot \left[(b - a) + h_1 + \frac{h_1^2}{2 \cdot (b - a)}\right]. \tag{35}$$

(ii) Case of $n = 2$. Then

$$\|Lf - f\|_p \leq \|f\|_{p\cdot q} \cdot \|L1 - 1\|_{p^2} + \|f'\|_{p\cdot q} \cdot \|L(t - \cdot)(\cdot)\|_{p^2} + \frac{\|f''\|_{p\cdot q}}{2} \cdot \|L(t - \cdot)^2(\cdot)\|_{p^2}$$

$$+ \frac{\omega_1(f'', h_2)}{2} \cdot h_2 \cdot \left[\frac{(b - a)}{3} + \frac{h_2}{2} + \frac{h_2^2}{2 \cdot (b - a)}\right]. \tag{36}$$

Theorem 5. *Let $([a, b], \mathcal{B}, \mu)$, $a \neq b$, be a measure space, where $\mathcal{B}$ is the Borel σ-algebra on $[a, b]$ and μ is a positive finite measure on $[a, b]$. (Note that $C([a, b]) \subset L_p([a, b], \mathcal{B}, \mu)$, any $1 \leq p < +\infty$.) Here $\|\cdot\|_p$ stands for the related L_p norm with respect to μ and $\|\cdot\|_\infty$ stands for the supremum norm. Consider the positive linear operator*

$$L : C([a, b]) \longrightarrow C([a, b]).$$

Call ($n \in \mathbb{N}$)

$$h_n := \|L(|t - x|^n, x)\|_p^{1/n}. \tag{37}$$

Let $f \in C^n([a, b])$. Then

$$\|Lf - f\|_p \leq \|f\|_\infty \cdot \|L1 - 1\|_p + \sum_{k=1}^{n} \frac{\|f^{(k)}\|_\infty}{k!} \cdot \|L((t - x)^k, x)\|_p + \omega_1(f^{(n)}, h_n)$$

$$\cdot h_n^{n-1} \cdot \left[\frac{(b - a)}{(n + 1)!} + \frac{h_n}{2 \cdot n!} + \frac{h_n^2}{4 \cdot (n - 1)! \cdot (b - a)}\right]. \tag{38}$$

Note that from Lemmas 1, 2 $h_n < +\infty$ and R.H.S.(38) $< +\infty$.

Proof. From (32), in the proof of Theorem 4, we obtain

$$|L(f, x) - f(x)| \leq \|f\|_\infty \cdot |L(1, x) - 1| + \sum_{k=1}^{n} \frac{\|f^{(k)}\|_\infty}{k!} \cdot |L((t - x)^k, x)| +$$

$$+ w \cdot \left[\frac{(b - a)}{(n + 1)! \cdot h} + \frac{1}{2 \cdot n!} + \frac{h}{4 \cdot (n - 1)! \cdot (b - a)}\right] \cdot (L(|t - x|^n, x)), \quad \forall x \in [a, b],\ n \in \mathbb{N}. \tag{39}$$

Therefore $(1 \leq p < +\infty)$

$$\|Lf - f\|_p \leq \|f\|_\infty \cdot \|L1 - 1\|_p + \sum_{k=1}^{n} \frac{\|f^{(k)}\|_\infty}{k!} \cdot \|L((t - x)^k, x)\|_p$$

$$+ w \cdot \left[\frac{(b - a)}{(n + 1)! \cdot h} + \frac{1}{2 \cdot n!} + \frac{h}{4 \cdot (n - 1)! \cdot (b - a)}\right] \cdot \|L(|t - x|^n, x)\|_p. \tag{40}$$

Choose

$$h = h_n := \|L(|t - x|^n, x)\|_p^{1/n},$$

i.e.,

$$h^n = h_n^n = \|L(|t - x|^n, x)\|_p.$$

Then

$$\|Lf - f\|_p \le \|f\|_\infty \cdot \|L1 - 1\|_p + \sum_{k=1}^{n} \frac{\|f^{(k)}\|_\infty}{k!} \cdot \|L((t-x)^k, x)\|_p$$

$$+\omega_1(f^{(n)}, h_n) \cdot \left[\frac{(b-a) \cdot h_n^{n-1}}{(n+1)!} + \frac{h_n^n}{2 \cdot n!} + \frac{h_n^{n+1}}{4 \cdot (n-1)! \cdot (b-a)}\right].$$

Q.E.D.

Definition 1. Let $f, g \in C([a,b])$ such that $|f(x)| \le |g(x)|$, $\forall x \in [a,b]$. A norm $\|\cdot\|$ on $C([a,b])$ is called *monotone* iff $\|f\| \le \|g\|$. We would denote a monotone norm by $\|\cdot\|_m$, e.g. L_p norms in general $(1 \le p \le +\infty)$, Orlicz norms, etc.

In a similar way as in the proof of Theorem 5 we get

Corollary 5. *Consider the positive linear operator*

$$L : C([a,b]) \longrightarrow C([a,b]), \qquad a \ne b.$$

Let $\|\cdot\|_m$ *be a monotone norm on* $C([a,b])$. *Call* $(n \in \mathbb{N})$

$$h_n := \|L(|t-x|^n, x)\|_m^{1/n}. \tag{41}$$

Let $f \in C^n([a,b])$. *Then*

$$\|Lf - f\|_m \le \|f\|_\infty \cdot \|L1 - 1\|_m + \sum_{k=1}^{n} \frac{\|f^{(k)}\|_\infty}{k!} \cdot \|L((t-x)^k, x)\|_m + \omega_1(f^{(n)}, h_n) \cdot h_n^{n-1}$$

$$\cdot \left[\frac{(b-a)}{(n+1)!} + \frac{h_n}{2 \cdot n!} + \frac{h_n^2}{4 \cdot (n-1)! \cdot (b-a)}\right]. \tag{42}$$

Remark 8 (on Corollary 5). *Special case of* $n = 1$. We have $(f \in C^1([a,b]))$

$$\|Lf - f\|_m \le \|f\|_\infty \cdot \|L1 - 1\|_m + \|f'\|_\infty \cdot \|L((t-x), x)\|_m$$

$$+\frac{\omega_1(f', h_1)}{2} \cdot \left[(b-a) + h_1 + \frac{h_1^2}{2 \cdot (b-a)}\right], \tag{43}$$

where

$$h_1 := \|L(|t-x|, x)\|_m. \tag{44}$$

4. ABSTRACT RESULTS

Next we present results for just continuous functions.

We need

Definition 2. Let M be a non-empty compact subset of the real normed vector space $(V, \|\cdot\|)$. Let $f : M \to \mathbb{R}$ be a Borel measurable function and $h > 0$. We call

$$\omega_1(f, h) := \sup_{\substack{x, y \in M \\ \|x-y\| \le h}} |f(x) - f(y)|$$

the (first) modulus of continuity of f.

We would also need to use the following result from Anastassiou (1) (1986), p. 44.

Corollary B. Let μ be a positive finite measure of mass $m \neq 0$ on the Borel σ-algebra of the nonempty convex and compact subset M of the real normed vector space $(V, \|\cdot\|)$. Consider $x_0 \in M$. Call

$$\left(\int_M \|x - x_0\|^r \mu(dx)\right)^{1/r} := D_r(x_0), \tag{45}$$

where $r \geq 1$. Assume $D_r(x_0) > 0$. Let $f : M \to \mathbb{R}$ be a Borel measurable function with $\omega_1(f, h)$ its first modulus of continuity, where $h > 0$ is given. Then

$$\left|\int_M f\, d\mu - f(x_0)\right| \leq |m - 1| \cdot |f(x_0)| + \omega_1(f, h) \cdot \left(m + \frac{D_r(x_0)}{h} \cdot m^{(1-\frac{1}{r})}\right). \tag{46}$$

Note. If $D_r(x_0) = 0$, then $\mu(M - \{x_0\}) = 0$ and $\mu = m \cdot \delta_{x_0}$. Then (46) holds trivially. It is also trivial when $m = 0$. When $r = 1$, then one can prove (independently) in an easier way inequality (46).

Remark 9. In Corollary B, we need to consider only regular Borel measures μ on M, these among others are positive measures such that $\mu(M) < +\infty$. Let L be a positive linear operator from $C(M)$ into itself, where M is a nonempty convex and compact subset of $(V, \|\cdot\|)$. Consider $x_0 \in M$. By Remark 3 we have

$$(Lf)(x_0) = \int_M f\, d\mu_{x_0}, \qquad \forall f \in C(M), \tag{47}$$

where μ_{x_0} is a unique regular Borel positive finite measure on M. Note that

$$\infty + > m_{x_0} := \mu_{x_0}(M) = (L(1))(x_0) \geq 0.$$

We would assume

$$(L(1))(x_0) > 0, \qquad \forall x_0 \in M. \tag{48}$$

Also see that

$$(L(\|x - x_0\|^r))(x_0) = \int_M \|x - x_0\|^r \cdot \mu_{x_0}(dx) := D_r^r(x_0), \qquad r \geq 1.$$

We would also assume

$$(L(\|x - x_0\|^r))(x_0) > 0, \qquad \forall x_0 \in M,\ r \geq 1. \tag{49}$$

For a justification of the assumptions (48), (49) see Remark 2. By (47), (48), (49) and Corollary B, we get for $f \in C(M)$ that

$$\begin{aligned} |(L(f))(x_0) - f(x_0)| &\leq |(L(1))(x_0) - 1| \cdot |f(x_0)| + \omega_1(f, h) \\ &\cdot \left[(L(1))(x_0) + \frac{((L(\|x - x_0\|^r))(x_0))^{1/r}}{h} \cdot ((L(1))(x_0))^{(1-\frac{1}{r})}\right], \qquad h > 0. \end{aligned} \tag{50}$$

When $r = 1$ we get

$$\begin{aligned} |(Lf)(x_0) - f(x_0)| &\leq |(L(1))(x_0) - 1| \cdot |f(x_0)| + \omega_1(f, h) \\ &\cdot \left[(L(1))(x_0) + \frac{L(\|x - x_0\|)(x_0)}{h}\right]. \end{aligned} \tag{51}$$

Therefore

$$\begin{aligned} |(Lf)(x_0) - f(x_0)| &\leq |(L(1))(x_0) - 1| \cdot \|f\|_\infty + \omega_1(f, h) \\ &\cdot \left[(L(1))(x_0) + \frac{L(\|x - x_0\|)(x_0)}{h}\right]. \end{aligned} \tag{52}$$

For our convenience we call

$$\varphi_{x_0}(x) := \left[\frac{(L(\|x-x_0\|^r))(x)}{(L(1))(x)}\right]^{1/r}, \qquad \forall x \in M,\ x_0 \in M \text{ fixed}, r \geq 1. \tag{53}$$

Note that $\varphi_{x_0} \in C(M)$, $\forall x_0 \in M$.

We also call

$$F(x_0) := \varphi_{x_0}(x_0), \qquad \forall x_0 \in M. \tag{54}$$

I.e.,

$$F(x_0) = \left(\frac{(L(\|x-x_0\|^r))(x_0)}{(L(1))(x_0)}\right)^{1/r}, \qquad \forall x_0 \in M. \tag{55}$$

From Lemma 1 and assumption (48) we have that $F \in C(M)$. By assumptions (48), (49) we get that $F > 0$, thus $\|F\|_\infty > 0$.

Next comes our $\|\cdot\|_\infty$-result for continuous functions.

Theorem 6. *Let M be a nonempty convex and compact subset of the real normed vector space $(V, \|\cdot\|)$. Let L be a positive linear operator from $C(M)$ into itself. Assume that $(L(1))(x_0)$, $(L(\|x-x_0\|^r))(x_0) > 0$, $\forall x_0 \in M$, $r \geq 1$. Call*

$$F(x_0) := \left(\frac{(L(\|x-x_0\|^r))(x_0)}{(L(1))(x_0)}\right)^{1/r}, \qquad \forall x_0 \in M.$$

Consider $f \in C(M)$. Then

$$\|Lf - f\|_\infty \leq \|f\|_\infty \cdot \|L1 - 1\|_\infty + \omega_1(f, r \cdot \|F\|_\infty) \cdot \|L(1)\|_\infty \cdot \left(1 + \frac{1}{r}\right). \tag{56}$$

Note that $0 < \|F\|_\infty < +\infty$.

Proof. Directly from (50) by taking the supremum norms everywhere and by choosing $h := r \cdot \|F\|_\infty$. Q.E.D.

It follows the case $r = 1$.

Corollary 6. *Using the same terms and assumptions of Theorem 6, we get:*

$$\|Lf - f\|_\infty \leq \|f\|_\infty \cdot \|L1 - 1\|_\infty + \omega_1(f; \|L(\|x-x_0\|)(x_0)\|_\infty)$$

$$\cdot (1 + \|L1\|_\infty), \qquad \forall f \in C(M). \tag{57}$$

(*Note that $0 < \|L(\|x-x_0\|)(x_0)\|_\infty < +\infty$.*)

Proof. Directly from (51) by taking the supremum norms everywhere and by choosing $h := \|L(\|x-x_0\|)(x_0)\|_\infty$. Q.E.D.

Next we obtain some general L_p results for continuous functions ($1 \leq p < +\infty$).

Proposition 1. *Let M be a nonempty compact convex subset of the real normed vector space $(V, \|\cdot\|)$. Let $(M, \mathcal{B}, \mu)$ be a measure space, where $\mathcal{B}$ is the Borel σ-algebra on M and μ is a positive finite measure on M. (Note that $C(M) \subset L_p(M, \mathcal{B}, \mu)$, any $1 \leq p < +\infty$.) Here $\|\cdot\|_p$ stands for the related L_p norm with respect to μ and $\|\cdot\|_\infty$ stands for the supremum norm. Consider the positive linear operator*

$$L : C(M) \longrightarrow C(M).$$

Assume that $(L(\|x - x_0\|))(x_0) > 0$, $\forall x_0 \in M$. Call

$$h^* := \|L(\|x - \cdot\|)(\cdot)\|_p. \tag{58}$$

Let $f \in C(M)$. Then

$$\|Lf - f\|_p \leq \|f\|_\infty \cdot \|L1 - 1\|_p + \omega_1(f, h^*) \cdot (1 + \|L1\|_p). \tag{59}$$

Note that by Lemma 1 we get $0 < h^* < +\infty$.

Proof. By monotonicity, subadditivity and linearity properties of $\|\cdot\|_p$ applied on (52), we obtain

$$\|Lf - f\|_p \leq \|L1 - 1\|_p \cdot \|f\|_\infty + \omega_1(f, h) \cdot \left[\|L1\|_p + \frac{1}{h} \cdot \|L(\|x - \cdot\|)(\cdot)\|_p\right]. \tag{60}$$

Then pick $h = h^*$ as in (58). Q.E.D.

Theorem 7. *Let M be a nonempty compact convex subset of the real normed vector space $(V, \|\cdot\|)$. Let $(M, \mathcal{B}, \mu)$ be a measure space, where $\mathcal{B}$ is the Borel σ-algebra on M and μ is a positive finite measure on M. (Note that $C(M) \subset L_p(M, \mathcal{B}, \mu)$, any $1 \leq p < +\infty$.) Consider the positive linear operator*

$$L : C(M) \longrightarrow C(M).$$

Assume

$$\begin{aligned} &(L(1))(x_0) > 0, \qquad \forall x_0 \in M, \\ &(L(\|x - x_0\|^r))(x_0) > 0, \qquad \forall x_0 \in M,\ r > 1. \end{aligned} \tag{61}$$

Call

$$F(x_0) := \left(\frac{(L(\|x - x_0\|^r))(x_0)}{(L(1))(x_0)}\right)^{1/r}, \qquad \forall x_0 \in M. \tag{62}$$

Let $f \in C(M)$. Then

$$\begin{aligned} \|Lf - f\|_p \leq \|f\|_\infty \cdot \|L1 - 1\|_p + \omega_1(f; r \cdot \|F\|_p) \cdot \|L(1)\|_\infty \\ \cdot \left[(\mu(M))^{1/p} + \frac{1}{r}\right]. \end{aligned} \tag{63}$$

Note that $0 < \|F\|_p < +\infty$.

Proof. Let $f \in C(M)$. Then from inequality (50) we have

$$\begin{aligned} |(Lf)(x_0) - f(x_0)| \leq |(L1)(x_0) - 1| \cdot |f(x_0)| + \omega_1(f, h) \cdot (L1)(x_0) \\ \cdot \left[1 + \frac{1}{h} \cdot F(x_0)\right], \qquad \forall x_0 \in M,\ h > 0,\ r > 1. \end{aligned} \tag{64}$$

Hence

$$\begin{aligned} |(Lf)(x_0) - f(x_0)| \leq |(L1)(x_0) - 1| \cdot \|f\|_\infty + \omega_1(f, h) \cdot \|L(1)\|_\infty \\ \cdot \left[1 + \frac{1}{h} \cdot F(x_0)\right], \qquad \forall x_0 \in M,\ h > 0,\ r > 1. \end{aligned} \tag{65}$$

Thus

$$\begin{aligned} \|Lf - f\|_p \leq \|f\|_\infty \cdot \|L1 - 1\|_p + \omega_1(f, h) \cdot \|L(1)\|_\infty \\ \cdot \left[(\mu(M))^{1/p} + \frac{1}{h} \cdot \|F\|_p\right], \qquad h > 0,\ r > 1. \end{aligned} \tag{66}$$

Now choose in (66)

$$h := r \cdot \|F\|_p.$$

Q.E.D.

5. MULTIDIMENSIONAL RESULTS

Here we would need Theorem 5 from Anastassiou (2) (1986), p. 97.

Theorem A. *Take $Q \subset \mathbb{R}^k$ of the form $Q := \{\mathbf{x} \in \mathbb{R}^k: \|\mathbf{x}\| \le 1\}$, where $\|\cdot\|$ the ℓ_1-norm in $\mathbb{R}^k$, $k \ge 1$, and let $\mathbf{x}_0 := (x_{01}, \dots, x_{0k}) \in Q$ be fixed. Let the positive measure μ satisfy $\mu(Q) = 1$ and*

$$\int_Q \|\mathbf{x} - \mathbf{x}_0\| \cdot \mu(d\mathbf{x}) := d^*.$$

Also, let $f \in C^n(Q)$ and suppose that each of its nth partial derivatives f_α has a modulus of continuity $\omega_1(f_\alpha, h) \le w$, where h and w are fixed positive numbers. Then we have the estimate

$$\left|\int_Q f\,d\mu - f(\mathbf{x}_0)\right| \le \left|\sum_{j=1}^n \frac{1}{j!} \cdot \int_Q g_{\mathbf{x}}^{(j)}(0) \cdot \mu(d\mathbf{x})\right| + w \cdot d^* \cdot \frac{\phi_n(1 + \|\mathbf{x}_0\|)}{(1 + \|\mathbf{x}_0\|)}, \tag{67}$$

where $g_{\mathbf{x}}(t) := f(\mathbf{x}_0 + t \cdot (\mathbf{x} - \mathbf{x}_0))$, $t \ge 0$.

Remark A. From $n = 1$, (67) yields the inequality ($\mathbf{x}_0 \in Q$):

$$\left|\int_Q f\,d\mu - f(\mathbf{x}_0)\right| \le \left|\sum_{i=1}^k \frac{\partial f}{\partial x_i}(\mathbf{x}_0) \cdot \int_Q (x_i - x_{0i}) \cdot \mu(d\mathbf{x})\right|$$

$$+ w \cdot d^* \cdot \frac{\phi_1(1 + \|\mathbf{x}_0\|)}{(1 + \|\mathbf{x}_0\|)}. \tag{68}$$

Next come our related results in $\mathbb{R}^k$, $k \ge 1$.

Theorem 8. *Take $Q \subset \mathbb{R}^k$ of the form $Q := \{\mathbf{x} \in \mathbb{R}^k: \|\mathbf{x}\| \le 1\}$, where $\|\cdot\|$ is the ℓ_1-norm in $\mathbb{R}^k$, $k \ge 1$. Let $(Q, \mathcal{B}, \mu)$ be a measure space, where $\mathcal{B}$ is the Borel σ-algebra on Q and μ is a positive finite measure on Q. (Note that $C(Q) \subset L_p(Q, \mathcal{B}, \mu)$, $1 \le p < +\infty$.) Consider the positive linear operator*

$$L : C(Q) \longrightarrow C(Q),$$

such that

$$(L(1))(\mathbf{x}_0) = 1, \qquad \forall \mathbf{x}_0 \in Q. \tag{69}$$

Assume that

$$h_1 := \|(L(\|\mathbf{x} - \mathbf{x}_0\|))(\mathbf{x}_0)\|_\infty > 0, \tag{70}$$

$$h_2 := \|(L(\|\mathbf{x} - \mathbf{x}_0\|))(\mathbf{x}_0)\|_p > 0, \qquad 1 \le p < +\infty. \tag{71}$$

(Note that $h_1, h_2 < +\infty$.)

Let $f \in C^n(Q)$, $n \ge 2$, and suppose that each of its nth partial derivatives f_α has a modulus of continuity $\omega_1(f_\alpha, h_i) \le w_i$, where w_i, $i = 1, 2$ are given positive numbers.

Call

$$g_{(\mathbf{x}, \mathbf{x}_0)}(t) := f(\mathbf{x}_0 + t \cdot (\mathbf{x} - \mathbf{x}_0)), \qquad t \ge 0,\ \mathbf{x}_0, \mathbf{x} \in Q. \tag{72}$$

Then we find the following estimates

$$\text{(i)} \quad \|Lf - f\|_\infty \le \sum_{j=1}^n \frac{1}{j!} \cdot \|(L(g_{(\mathbf{x}, \mathbf{x}_0)}^{(j)}(0)))(\mathbf{x}_0)\|_\infty$$

$$+ w_1 \cdot \left(\frac{2^n}{(n+1)!} + \frac{h_1 \cdot 2^{n-2}}{n!} + \frac{h_1^2 \cdot 2^{n-5}}{(n-1)!}\right), \tag{73}$$

$$\text{(ii)} \quad \|Lf - f\|_p \le \sum_{j=1}^n \frac{1}{j!} \cdot \|(L(g_{(\mathbf{x}, \mathbf{x}_0)}^{(j)}(0)))(\mathbf{x}_0)\|_p$$

$$+ w_2 \cdot \left(\frac{2^n}{(n+1)!} + \frac{h_2 \cdot 2^{n-2}}{n!} + \frac{h_2^2 \cdot 2^{n-5}}{(n-1)!} \right). \tag{74}$$

Note that R.H.S.'s (73), (74) $< +\infty$.

Proof. Here $n \geq 2$, $n \in \mathbb{N}$ and the fixed point $\mathbf{x}_0 \in Q := \{\mathbf{x} \in \mathbb{R}^k\colon \|\mathbf{x}\| \leq 1\}$, where $\|\cdot\|$ is the ℓ_1 norm in $\mathbb{R}^k$. Note that $1 + \|\mathbf{x}_0\| \leq 2$. Thus, from (4) we get

$$\frac{\phi_n(1+\|\mathbf{x}_0\|)}{(1+\|\mathbf{x}_0\|)} \leq \left(\frac{(1+\|\mathbf{x}_0\|)^n}{(n+1)! \cdot h} + \frac{(1+\|\mathbf{x}_0\|)^{n-1}}{2 \cdot n!} + \frac{h \cdot (1+\|\mathbf{x}_0\|)^{n-2}}{8 \cdot (n-1)!} \right)$$

$$\leq \left(\frac{2^n}{(n+1)! \cdot h} + \frac{2^{n-1}}{2 \cdot n!} + \frac{h \cdot 2^{n-2}}{8 \cdot (n-1)!} \right).$$

I.e.,

$$\frac{\phi_n(1+\|\mathbf{x}_0\|)}{(1+\|\mathbf{x}_0\|)} \leq \left(\frac{2^n}{(n+1)! \cdot h} + \frac{2^{n-2}}{n!} + \frac{h \cdot 2^{n-5}}{(n-1)!} \right), \qquad h > 0. \tag{75}$$

Let L be a positive linear operator from $C(Q)$ into itself, such that

$$(L(1))(\mathbf{x}_0) = 1, \qquad \forall \mathbf{x}_0 \in Q.$$

Let $f \in C^n(Q)$ such that $\omega_1(f_\alpha, h_i) \leq w_i$. By Remark 3 we have that

$$(L(f))(\mathbf{x}_0) = \int_Q f(t) \cdot \mu_{\mathbf{x}_0}(dt), \qquad \forall f \in C(Q),$$

where $\mu_{\mathbf{x}_0}(Q) = (L(1))(\mathbf{x}_0) = 1$. We call

$$(L(\|\mathbf{x} - \mathbf{x}_0\|))(\mathbf{x}_0) := d^*(\mathbf{x}_0)$$

and

$$g_{(\mathbf{x},\mathbf{x}_0)}(t) := f(\mathbf{x}_0 + t \cdot (\mathbf{x} - \mathbf{x}_0)), \qquad t \geq 0,\ \mathbf{x} \in Q.$$

Note that $g^{(j)}_{(\mathbf{x},\mathbf{x}_0)}(0)$ is a continuous function of $\mathbf{x}$ over Q. Furthermore $(L(g^{(j)}_{(\mathbf{x},\mathbf{x}_0)}(0)))(\mathbf{x}_0)$ is a continuous function in $\mathbf{x}_0$ over Q. From Theorem A and (75) we obtain

$$|(Lf)(\mathbf{x}_0) - f(\mathbf{x}_0)| \leq \sum_{j=1}^{n} \frac{1}{j!} \cdot \left| \int_Q g^{(j)}_{(\mathbf{x},\mathbf{x}_0)}(0) \cdot \mu_{\mathbf{x}_0}(dx) \right|$$

$$+ w_i \cdot d^*(\mathbf{x}_0) \cdot \left(\frac{2^n}{(n+1)! \cdot h_i} + \frac{2^{n-2}}{n!} + \frac{h_i \cdot 2^{n-5}}{(n-1)!} \right).$$

I.e.,

$$|(Lf)(\mathbf{x}_0) - f(\mathbf{x}_0)| \leq \sum_{j=1}^{n} \frac{1}{j!} \cdot \left| (L(g^{(j)}_{(\mathbf{x},\mathbf{x}_0)}(0)))(\mathbf{x}_0) \right| + w_i \cdot (L(\|\mathbf{x} - \mathbf{x}_0\|))(\mathbf{x}_0)$$

$$\cdot \left(\frac{2^n}{(n+1)! \cdot h_i} + \frac{2^{n-2}}{n!} + \frac{h_i \cdot 2^{n-5}}{(n-1)!} \right), \qquad h_i > 0,\ i = 1,2. \tag{76}$$

Q.E.D.

Proposition 2. *Take* $Q := \{\mathbf{x} \in \mathbb{R}^k\colon \|\mathbf{x}\| \leq 1\}$, *where* $\|\cdot\|$ *the* ℓ_1*-norm in* $\mathbb{R}^k$. *Let* $(Q, \mathcal{B}, \mu)$ *be a measure space, where* $\mathcal{B}$ *is the Borel* σ*-algebra on* Q *and* μ *is a positive finite measure on* Q. *Consider the positive linear operator* L *from* $C(Q)$ *into itself, such that*

$$(L(1))(\mathbf{x}_0) = 1, \qquad \forall \mathbf{x}_0 \in Q.$$

Assume that

$$h_1 := \|(L(\|\mathbf{x}-\mathbf{x}_0\|))(\mathbf{x}_0)\|_\infty > 0, \tag{77}$$

$$h_2 := \|(L(\|\mathbf{x}-\mathbf{x}_0\|))(\mathbf{x}_0)\|_p > 0, \qquad 1 \le p < +\infty. \tag{78}$$

Let $f \in C^1(Q)$, *and suppose that each of its first partial derivatives* f_j, $j = 1, \ldots, k$ *has a modulus of continuity* $\omega_1(f_j, h_i) \le w_i$, *where* w_i, $i = 1, 2$ *are given positive numbers. Then*

$$\text{(i)} \quad \|Lf - f\|_\infty \le \sum_{j=1}^{k} \left\|\frac{\partial f}{\partial x_j}\right\|_\infty \cdot \|(L(x_j - x_{0j}))(\mathbf{x}_0)\|_\infty + w_1 \cdot \left(1 + \frac{h_1}{2} + \frac{h_1^2}{8}\right), \tag{79}$$

and

$$\text{(ii)} \quad \|Lf - f\|_p \le \sum_{j=1}^{k} \left\|\frac{\partial f}{\partial x_j}\right\|_\infty \cdot \|(L(x_j - x_{0j}))(\mathbf{x}_0)\|_p + w_2 \cdot \left(1 + \frac{h_2}{2} + \frac{h_2^2}{8}\right). \tag{80}$$

Note that R.H.S.'s (79), (80) $< +\infty$.

Proof. Let $x > 0$, then by (4) we get

$$\frac{\phi_1(x)}{x} \le \left(\frac{x}{2\cdot h} + \frac{1}{2} + \frac{h}{8\cdot x}\right).$$

Since $1 + \|\mathbf{x}_0\| \ge 1$, then $\frac{1}{1+\|\mathbf{x}_0\|} \le 1$. Hence

$$\frac{\phi_1(1+\|\mathbf{x}_0\|)}{(1+\|\mathbf{x}_0\|)} \le \left(\frac{(1+\|\mathbf{x}_0\|)}{2\cdot h} + \frac{1}{2} + \frac{h}{8\cdot(1+\|\mathbf{x}_0\|)}\right) \le \left(\frac{1}{h} + \frac{1}{2} + \frac{h}{8}\right).$$

I.e.,

$$\frac{\phi_1(1+\|\mathbf{x}_0\|)}{(1+\|\mathbf{x}_0\|)} \le \left(\frac{1}{h} + \frac{1}{2} + \frac{h}{8}\right), \qquad h > 0. \tag{81}$$

Let L be a positive linear operator from $C(Q)$ into itself, such that $(L(1))(\mathbf{x}_0) = 1$, $\forall \mathbf{x}_0 \in Q$. Let $f \in C^1(Q)$ such that $\omega_1(f_j, h_i) \le w_i$, $j = 1, \ldots, k$; (f_j the first partial derivative of f), $i = 1, 2$.

By Remark 3 we have that

$$(Lf)(\mathbf{x}_0) = \int_Q f(t) \cdot \mu_{\mathbf{x}_0}(dt), \qquad \forall f \in C(Q),$$

where $\mu_{\mathbf{x}_0}(Q) = (L(1))(\mathbf{x}_0) = 1$. We call

$$(L(\|\mathbf{x}-\mathbf{x}_0\|))(\mathbf{x}_0) := d^*(\mathbf{x}_0)$$

and

$$pr_j(x) := x_j, \qquad j = 1, \ldots, k$$

the projection map, which is continuous.

From (68) and (81) we obtain

$$|(Lf)(\mathbf{x}_0) - f(\mathbf{x}_0)| \le \sum_{j=1}^{k} \left\|\frac{\partial f}{\partial x_j}\right\|_\infty \cdot |L(pr_j(\mathbf{x}-\mathbf{x}_0))(\mathbf{x}_0)| + w \cdot ((L(\|\mathbf{x}-\mathbf{x}_0\|))(\mathbf{x}_0))$$

$$\cdot \frac{\phi_1(1+\|\mathbf{x}_0\|)}{(1+\|\mathbf{x}_0\|)} \le \sum_{j=1}^{k} \left\|\frac{\partial f}{\partial x_j}\right\|_\infty \cdot |L(pr_j(\mathbf{x}-\mathbf{x}_0))(\mathbf{x}_0)|$$

$$+ w \cdot ((L(\|\mathbf{x}-\mathbf{x}_0\|))(\mathbf{x}_0)) \cdot \left(\frac{1}{h} + \frac{1}{2} + \frac{h}{8}\right).$$

I.e.,

$$|(Lf)(\mathbf{x}_0) - f(\mathbf{x}_0)| \le \sum_{j=1}^{k} \left\|\frac{\partial f}{\partial x_j}\right\|_\infty \cdot |L(pr_j(\mathbf{x}-\mathbf{x}_0))(\mathbf{x}_0)|$$

$$+ w_i \cdot \left(\frac{1}{h_i} + \frac{1}{2} + \frac{h_i}{8}\right) \cdot ((L(\|\mathbf{x}-\mathbf{x}_0\|))(\mathbf{x}_0)), \qquad i = 1,2. \tag{82}$$

Note that $(L(pr_j(\mathbf{x}-\mathbf{x}_0)))(\mathbf{x}_0)$, $(L(\|\mathbf{x}-\mathbf{x}_0\|))(\mathbf{x}_0)$ are continuous functions in $\mathbf{x}_0$. Then

$$\|Lf - f\|_\infty \le \sum_{j=1}^{k} \left\|\frac{\partial f}{\partial x_j}\right\|_\infty \cdot \|(L(x_j - x_{0j}))(\mathbf{x}_0)\|_\infty$$

$$+ w_1 \cdot \left(\frac{1}{h_1} + \frac{1}{2} + \frac{h_1}{8}\right) \cdot \|L(\|\mathbf{x}-\mathbf{x}_0\|)(\mathbf{x}_0)\|_\infty, \tag{83}$$

and $(1 \le p < +\infty)$

$$\|Lf - f\|_p \le \sum_{j=1}^{k} \left\|\frac{\partial f}{\partial x_j}\right\|_\infty \cdot \|(L(x_j - x_{0j}))(\mathbf{x}_0)\|_p$$

$$+ w_2 \cdot \left(\frac{1}{h_2} + \frac{1}{2} + \frac{h_2}{8}\right) \cdot \|(L(\|\mathbf{x}-\mathbf{x}_0\|))(\mathbf{x}_0)\|_p. \tag{84}$$

Q.E.D.

6. STOCHASTIC RESULTS

Let $(\underline{\Omega}, \mathcal{A}, P)$ denote a probability space and $L^1(\underline{\Omega}, \mathcal{A}, P)$ be the space of all real-valued random variables $Y = Y(\omega)$ with

$$\int_{\underline{\Omega}} |Y(\omega)| \cdot P(d\omega) < +\infty.$$

Let $X = X(t,\omega)$ denote a stochastic proces with index set K, a compact convex subset of $(V, \|\cdot\|)$ a real normed vector space, and real state space $(\mathbb{R}, \mathcal{B})$, $\mathcal{B}$ is the Borel σ-algebra on $\mathbb{R}$.

Denote by $C_{\underline{\Omega}}(K)$ the space of L^1-continuous stochastic processes in t and

$$B_{\underline{\Omega}}(K) := \left\{X : \sup_{t \in K} \int_{\underline{\Omega}} |X(t,\omega)| \cdot P(d\omega) < +\infty\right\};$$

obviously $C_{\underline{\Omega}}(K) \subset B_{\underline{\Omega}}(K)$, $C(K) \subset C_{\underline{\Omega}}(K)$.

With E we denote the expectation operator

$$(EX)(t) := \int_{\underline{\Omega}} X(t,\omega) \cdot P(d\omega).$$

Consider the linear operator

$$T : C_{\underline{\Omega}}(K) \longrightarrow B_{\underline{\Omega}}(K).$$

If $X \in C_{\underline{\Omega}}(K)$ is nonnegative and TX, too, then T is called a positive operator. If $ET = TE$, then T is called an E-commutative operator.

If T is E-commutative then $T(C(K)) \subset B(K)$ - the bounded real valued functions on K. Also, if $X(t,\omega) \in C_{\underline{\Omega}}(K)$ then $(EX)(t) \in C(K)$. When $K = [a,b] \subset \mathbb{R}$, we define $C^n_{\underline{\Omega}}([a,b]) := \{X$: there exists $X^{(k)}(t,\omega) \in C_{\underline{\Omega}}([a,b])$ and it is continuous in t for each $\omega \in \underline{\Omega}$, $k = 0,1,\ldots,n\}$.

From Anastassiou (1991), p. 368, Theorem 1, we get

Theorem B. *Consider the positive E-commutative linear operator*

$$T: C_{\underline{0}}([a,b]) \longrightarrow C_{\underline{0}}([a,b]).$$

Let

$$c(t_0) := \max(t_0 - a, b - t_0), \qquad t_0 \in [a,b],$$

and

$$d_n(t_0) := ((T(|t-t_0|^n))(t_0))^{1/n}, \qquad n \in \mathbb{N}.$$

Let $X \in C^n_{\underline{0}}([a,b])$ such that $\omega_1(EX^{(n)}, h) \le w$, where w, h are given positive numbers, $0 < h \le b-a$. Then

$$|(E(TX))(t_0) - (EX)(t_0)| \le |(EX)(t_0)| \cdot |(T(1))(t_0) - 1| + \sum_{k=1}^{n} \frac{|(EX^{(k)})(t_0)|}{k!}$$

$$\cdot |(T(t-t_0)^k)(t_0)| + w \cdot \frac{\phi_n(c(t_0))}{(c(t_0))^n} \cdot (d_n(t_0))^n, \qquad \forall t_0 \in [a,b]. \tag{85}$$

Remark 10 (on Theorem B). From (85) we obtain

$$|(E(TX))(t_0) - (EX)(t_0)| \le \|EX\|_\infty \cdot |(T(1))(t_0) - 1| + \sum_{k=1}^{n} \frac{\|EX^{(k)}\|_\infty}{k!}$$

$$\cdot |(T(t-t_0)^k)(t_0)| + w \cdot \frac{\phi_n(c(t_0))}{(c(t_0))^n} \cdot d_n^n(t_0), \qquad \forall t_0 \in [a,b]. \tag{86}$$

Call

$$R_n(t_0) := w \cdot \frac{\phi_n(c(t_0))}{(c(t_0))^n} \cdot d_n^n(t_0). \tag{87}$$

Let $f \in C([a,b]) \subset C_{\underline{0}}([a,b])$, and assume T as E-commutative. Then $Ef = f$ and $T(Ef) = T(f)$, i.e., $E(Tf) = Tf$, here $Tf \in C_{\underline{0}}([a,b])$. Therefore $Tf \in C([a,b])$. We have proved that, if

$$T: C_{\underline{0}}([a,b]) \longrightarrow C_{\underline{0}}([a,b]),$$

then

$$T: C([a,b]) \longrightarrow C([a,b]).$$

Let $1 \le p \le +\infty$.

Let $([a,b], \mathcal{B}, \mu)$ be a measure space, where $\mathcal{B}$ is the Borel σ-algebra on $[a,b]$, μ is a positive finite measure on $[a,b]$. Here $\|\cdot\|_p$ $(1 \le p < +\infty)$ stands for the $L_p([a,b], \mathcal{B}, \mu)$-norm, while $\|\cdot\|_\infty$ stands for the supremum norm.

From (86), (87) we find

$$\|E(TX) - EX\|_p \le \|EX\|_\infty \cdot \|T1 - 1\|_p + \sum_{k=1}^{n} \frac{\|EX^{(k)}\|_\infty}{k!}$$

$$\cdot \|(T(t-t_0)^k)(t_0)\|_p + \|R_n(t_0)\|_p. \tag{88}$$

The remainder $\|R_n(t_0)\|_p$ $(1 \le p \le +\infty)$ now can be treated exactly the same way as in §3 and get similar $\|\cdot\|_p$-results.

In general, let T be an E-commutative operator from $C_{\underline{0}}(K)$ into itself, then T maps $C(K)$ into itself. Also note that

$$|E(TX)(t_0) - (EX)(t_0)| = |(T(EX))(t_0) - (EX)(t_0)|,$$

and if $X \in C^n_{\underline{0}}([a,b])$, then $X^{(n)}(t,\omega) \in C_{\underline{0}}([a,b])$, which is continuous in t for every $\omega \in \underline{0}$, thus for each $\omega \in \underline{0}$ we have $X(t,\omega) \in C^n([a,b])$.

In that case, one can find various estimates for $\|(E(TX))(t_0) - (EX)(t_0)\|_p$, $1 \leq p \leq +\infty$, using results from Anastassiou (1991): Theorem 3, p. 374, Theorem 6, p. 378, Corollary 9′, p. 381, in a similar manner as in §§3–5, etc.

REFERENCES

1. C.D. Aliprantis and O. Burkinshaw, "Principles of Real Analysis," Second Edition, Academic Press, Boston, New York, San Diego (1990).
2. G.A. Anastassiou, A study of positive linear operators by the method of moments, one-dimensional case, *Journal of Approximation Theory*, **45**, 247–270 (1985).
3. G.A. Anastassiou, Korovkin type inequalities in real normed vector spaces, *Approximation Theory and Its Applications*, **2**, No. 2, 39–53 (1986).
4. G.A. Anastassiou, Multi-dimensional quantitative results for probability measures approximating the unit measure, *Approximation Theory and Its Applications*, **2**, No. 4, 93–103 (1986).
5. G.A. Anastassiou, Korovkin inequalities for stochastic processes, *Journal of Mathematical Analysis and Applications*, **157**, No. 2, 366–384 (1991).
6. P.P. Korovkin, *Linear Operators and Approximation Theory*, Hindustan Publ. Corp., Delhi, India (1960).
7. V.A. Popov, On the quantitative Korovkin theorems in L_p, *Comptes rendus de l'Académie Bulgare des Sciences, Mathématiques Théorie des approximations*, Tome 35, No. 7, 897–900 (1982).
8. E. Quak, L_p-error estimates for positive linear operators using the second-order τ-modulus, *Analysis Mathematica*, **14**, 259–272 (1988).
9. E. Quak, Multivariate L_p-error estimates for positive linear operators via the first-order τ modulus, *Journal of Approximation Theory*, **56**, 277–286 (1989).
10. O. Shisha and B. Mond, The degree of convergence of sequences of linear positive operators, *Nat. Acad. of Sci., U.S.*, **60**, 1196–1200 (1968).

ON SOME SHIFT-INVARIANT INTEGRAL OPERATORS, MULTIVARIATE CASE

George A. Anastassiou[1] and Heinz H. Gonska[2]

[1]Department of Mathematical Sciences
Memphis State University
Memphis, Tennessee 38152, U.S.A.
[2]Department of Mathematics
University of Duisburg
D-47048 Duisburg, Germany

Abstract: In recent papers, the authors studied global smoothness preservation by certain univariate and multivariate linear operators over compact domains. Also very recently, they studied the same property over $\mathbb{R}$, along with other characteristics, for some particular family of general shift-invariant integral operators. Here a generalization to $\mathbb{R}^d$, $d \geq 1$, is given and everything can be transferred there from the univariate case. Namely, a general positive linear multivariate integral type operator is introduced through a convolution-like iteration of another general positive linear multivariate operator with a multivariate scaling type function. For this sufficient conditions are given for shift invariance, global smoothness preservation and its sharpness, convergence to the unit with rates, shape preserving and preservation of continuous probabilistic functions. Additionally, four examples of general specialized multivariate operators are given fulfilling all the above properties; especially, the inequalities for global smoothness preservation are sharp. In this article global smoothness preservation and convergence to the unit with rates involve a naturally arising suitable multivariate modulus of continuity.

1. INTRODUCTION

In approximating a function $f \in C(\mathbb{R}^d)$, $d \geq 1$, by means of multivariate approximation operators $\mathcal{L}_k$, it is of interest to check which properties of f are preserved by the approximants $\mathcal{L}_k f$. For instance some of these could be: preservation of global smoothness by involving a multivariate modulus of continuity, shape preservation, and preservation of properties of a

[1,2]Supported in part by NATO Grant CRG 891013.

multivariate probabilistic distribution function. Shift invariance is another property of operators $\mathcal{L}_k$ to be checked. Also of importance is their convergence to the unit operator with rates.

In this paper we introduce a general family of multivariate operators $\mathcal{L}_k$, see Section 2. Furthermore, we study the above characteristics by giving sufficient conditions, so they become true and we give several examples of multivariate operators with all these properties. Our research has been mainly motivated by the works of Anastassiou, Cottin and Gonska, see [1–2] and Anastassiou and Gonska, see [3]. We would like to present some results from there.

Theorem A ([1]). *Let (X,d) be a compact metric space, and $L: C(X) \to C(X)$, $L \neq 0$, be a bounded linear operator mapping* $\mathrm{Lip}(X)$ *to* $\mathrm{Lip}(X) := \bigcup_{M>0} \mathrm{Lip}_M(1;X)$ *such that for all* $g \in \mathrm{Lip}(X)$,

$$|Lg|_{\mathrm{Lip}} \leq c \cdot |g|_{\mathrm{Lip}},$$
$$\left(|g|_{\mathrm{Lip}} := \sup_{d(x,y)>0} \frac{|g(x)-g(y)|}{d(x,y)}\right)$$

with constant c possibly depending on L, but independent of g. Then for all $f \in C(X)$ and $t \geq 0$,

$$\omega_1(Lf;t) \leq \|L\| \cdot \tilde{\omega}_1\left(f; \frac{ct}{\|L\|}\right), \tag{$*$}$$

where $\tilde{\omega}_1$ is the least concave majorant of the modulus of continuity ω_1.

In the univariate case we get

Theorem B ([1]). *Let I be a compact interval, and $L: C(I) \to C(I)$, $L \neq 0$, be a bounded linear operator mapping $C^1(I)$ to $C^1(I)$. Then the estimate $(*)$ of Theorem A is true for all $f \in C(I)$ if the condition*

$$\|(Lg)'\|_I \leq c \cdot \|g'\|_I$$

is satisfied for all $g \in C^1(I)$.

In the multivariate case we obtain

Theorem C ([2]). *Let X be a compact convex subset of $\mathbb{R}^d$, $d \geq 1$, equipped with the metric d_p, $1 \leq p \leq \infty$, and let $L: C(X) \to C(X)$ be an operator satisfying the assumption of Theorem A. Then for all $f \in C(X)$ and $t \geq 0$ we have*

$$\omega_{1,d_p}(Lf;t) \leq (\|L\| + c) \cdot \omega_{1,d_p}(f,t),$$

where ω_{1,d_p} is the modulus of continuity with respect to d_p, and

$$d_p((x_1,\ldots,x_d),(y_1,\ldots,y_d)) = \left(\sum_{i=1}^{d} |x_i - y_i|^p\right)^{1/p}, \quad 1 \leq p < \infty,$$

$$d_\infty((x_1,\ldots,x_d),(y_1,\ldots,y_d)) = \max_{1\leq i\leq d} |x_i - y_i|.$$

Definition A ([3]). Let $X := C_U(\mathbb{R})$ be the space of uniformly continuous real valued functions on $\mathbb{R}$ and $C(\mathbb{R})$ be the space of continuous functions from $\mathbb{R}$ into itself. Let $\{\ell_k\}_{k\in\mathbb{Z}}$ be a sequence of positive linear operators that map X into $C(\mathbb{R})$ with the property:

$$\ell_k(f,x) = \ell_0(f(2^{-k}\cdot);x), \quad \text{all } x \in \mathbb{R}, f \in X. \tag{$**$}$$

For fixed $a > 0$ we assume that

$$\sup_{\substack{u,y\in\mathbb{R}\\ |u-y|\leq a}} |\ell_0(f,u) - f(y)| \leq \omega_1\left(f; \frac{ma+n}{2^r}\right) \tag{$***$}$$

is true for any $f \in X$, where $m \in \mathbb{N}$, $n \in \mathbb{Z}_+$, $r \in \mathbb{Z}$, ω_1 is the modulus of continuity on $\mathbb{R}$.

Let φ be a real valued function of compact support $\subseteq [-a, a]$, $\varphi \geq 0$, φ Lebesgue measurable and

$$\int_{-\infty}^{\infty} \varphi(x-u)du = 1, \qquad \text{for all } x \in \mathbb{R}. \tag{4*}$$

Let $\{\mathcal{L}_k\}_{k\in\mathbb{Z}}$ be the sequence of positive linear operators acting on X and defined by

$$\mathcal{L}_k(f;x) := \int_{-\infty}^{\infty} \ell_k(f;u)\varphi(2^k x - u)du. \tag{5*}$$

Under a mild assumption we have proved that $\mathcal{L}_k$ is a shift invariant operator, all $k \in \mathbb{Z}$, see [3].

Theorem D ([3]). *For any $f \in X$ we assume that for all $u \in \mathbb{R}$:*

$$|\ell_0(f;x-u) - \ell_0(f;y-u)| \leq \omega_1(f;|x-y|), \quad \text{any } x, y \in \mathbb{R}.$$

Then

$$\omega_1(\mathcal{L}_k f, \delta) \leq \omega_1(f;\delta), \tag{6*}$$

any $\delta > 0$. In specific cases we proved in [3] that inequality (6*) holds as equality.

Theorem E ([3]). *For $f \in X$, under the assumptions $(***)$, it holds*

$$|\mathcal{L}_k(f;x) - f(x)| \leq \omega_1\left(f; \frac{ma+n}{2^{k+r}}\right),$$

where $m \in \mathbb{N}$, $n \in \mathbb{Z}_+$, $k, r \in \mathbb{Z}$.

From [3] we also see that the operators $\mathcal{L}_k$ maintain higher order convexity and that they preserve continuous probabilistic distribution functions.

2. MAIN RESULTS

Let $X := C_U(\mathbb{R}^d)$, $d \geq 1$ be the space of uniformly continuous functions from $\mathbb{R}^d$ into $\mathbb{R}$. A function f is in X if, for an arbitrary $\varepsilon \geq 0$, there is a $\delta > 0$ such that $\|\mathbf{x} - \mathbf{y}\| \leq \delta$ implies $|f(\mathbf{x}) - f(\mathbf{y})| < \varepsilon$. For $f \in X$, define the first order modulus of continuity of f by

$$\omega_1(f;\delta) := \sup_{\substack{\mathbf{x},\mathbf{y}\in\mathbb{R}^d \\ \|\mathbf{x}-\mathbf{y}\|\leq\delta}} |f(\mathbf{x}) - f(\mathbf{y})|, \quad \delta > 0,$$

Here $\|\cdot\|$ is an arbitrary norm in $\mathbb{R}^d$. Let $\varepsilon_0 > 0$ be arbitrary but fixed. Then there is $\delta_0 > 0$ so that $\|\mathbf{x} - \mathbf{y}\| \leq \delta_0$ implies $|f(\mathbf{x}) - f(\mathbf{y})| < \varepsilon_0$; i.e., $\omega_1(f;\delta_0) \leq \varepsilon_0 < +\infty$. Let $\delta > 0$ be arbitrary and $\mathbf{x}, \mathbf{y} \in \mathbb{R}^d$ be such that $\|\mathbf{x} - \mathbf{y}\| \leq \delta$. Choose $n \in \mathbb{N}$ so that $n\delta_0 > \delta$, and put $\mathbf{x}_i := \mathbf{x} + \frac{i}{n}(\mathbf{y} - \mathbf{x})$, $0 \leq i \leq n$.

Then

$$\begin{aligned} |f(\mathbf{x}) - f(\mathbf{y})| &= \left| \sum_{i=0}^{n-1} f(\mathbf{x}_i) - f(\mathbf{x}_{i+1}) \right| \\ &\leq \sum_{i=0}^{n-1} |f(\mathbf{x}_i) - f(\mathbf{x}_{i+1})| \\ &\leq n \cdot \omega_1(f;\delta_0) \leq n \cdot \varepsilon_0 < +\infty, \end{aligned}$$

since $\|x_i - x_{i+1}\| = \frac{1}{n} \cdot \|x - y\| \leq \frac{1}{n} \cdot \delta < \delta_0$. Thus

$$\omega_1(f;\delta) \leq n \cdot \varepsilon_0 < +\infty,$$

showing that $\omega_1(f;\delta)$ is finite for all $\delta > 0$. Furthermore, by $C(\mathbb{R}^d)$ we will denote the space of continuous functions from $\mathbb{R}^d$ into $\mathbb{R}$.

Let $\{\ell_k\}_{k\in\mathbb{Z}}$ be a sequence of positive linear operators that map X into $C(\mathbb{R}^d)$ with the property:

$$\ell_k(f;\mathbf{x}) = \ell_0(f(2^{-k}\cdot);\mathbf{x}), \quad \text{all } \mathbf{x} \in \mathbb{R}^d, f \in X. \tag{1}$$

For fixed $a > 0$ we assume that

$$\sup_{\substack{\mathbf{u},\mathbf{y}\in\mathbb{R}^d \\ \|\mathbf{u}-\mathbf{y}\|_\infty \le a}} |\ell_0(f;\mathbf{u}) - f(\mathbf{y})| \le \omega_{1,\infty}\left(f; \frac{ma+n}{2^r}\right) \tag{2}$$

is true for any $f \in X$, $m \in \mathbb{N}$, $n \in \mathbb{Z}_+$, $r \in \mathbb{Z}$, where $\omega_{1,\infty}$ is the modulus of continuity ω_1 defined with respect to $\|\cdot\|_\infty$.

Let φ be a real valued function of compact support $\subseteq \times_{i=1}^d[-a_i, a_i]$, $a_i > 0$. We assume that $\varphi \ge 0$, φ is Lebesgue measurable and

$$\underbrace{\int_{-\infty}^{+\infty}\int_{-\infty}^{+\infty}\int_{-\infty}^{+\infty}\cdots\int_{-\infty}^{+\infty}}_{d\text{-fold}} \varphi(x_1 - u_1, x_2 - u_2, \ldots, x_d - u_d)\cdot du_1 \ldots du_d = 1,$$

$$\text{for all } (x_1, \ldots, x_d) \in \mathbb{R}^d. \tag{3}$$

One can easily find that

$$\int_{-\infty}^{+\infty}\int_{-\infty}^{+\infty}\int_{-\infty}^{+\infty}\cdots\int_{-\infty}^{+\infty} \varphi(u_1, \ldots, u_d)\cdot du_1 \ldots du_d = 1. \tag{4}$$

Examples

(i) For $i = 1, \ldots, d$ take the characteristic function

$$\varphi_i(x) := \chi_{[-\frac{1}{2},\frac{1}{2})}(x) = \begin{cases} 1, & x \in [-\frac{1}{2}, \frac{1}{2}), \\ 0, & \text{else.} \end{cases}$$

Define

$$\varphi^*(\mathbf{x}) := \prod_{i=1}^d \varphi_i(x_i), \quad \text{all } \mathbf{x} := (x_1, \ldots, x_d) \in \mathbb{R}^d.$$

Note that supp $\varphi^* \subseteq \left[-\frac{1}{2}, \frac{1}{2}\right]^d$, $\varphi^* \ge 0$ and φ^* is Lebesgue measurable. See also that

$$\int_{-\infty}^{+\infty}\int_{-\infty}^{+\infty}\cdots\int_{-\infty}^{+\infty}\prod_{i=1}^d \varphi_i(x_i - u_i)\cdot du_1 \ldots du_d = \prod_{i=1}^d \left(\int_{-\infty}^{\infty} \varphi_i(x_i - u_i) du_i\right) = 1,$$

i.e.,

$$\int_{\mathbb{R}^d} \varphi^*(\mathbf{x} - \mathbf{u}) \cdot d\mathbf{u} = 1, \quad \text{for all } \mathbf{x} \in \mathbb{R}^d.$$

(ii) For $i = 1, \ldots, d$ consider the hat functions

$$\varphi_i(x_i) := \begin{cases} 1 + x_i, & -1 \le x_i \le 0, \\ 1 - x_i, & 0 \le x_i \le 1. \end{cases}$$

Define

$$\tilde{\varphi}(x_1, x_2, \ldots, x_d) := \prod_{i=1}^d \varphi_i(x_i) \ge 0, \quad \text{all } (x_1, \ldots, x_d) \in \mathbb{R}^d,$$

which is a continuous scale function; it has a support $\subseteq [-1, 1]^d$.

Observe that

$$\int_{-\infty}^{+\infty}\int_{-\infty}^{+\infty}\cdots\int_{-\infty}^{+\infty}\tilde{\varphi}(x_1-u_1,\ldots,x_d-u_d)\cdot du_1\ldots du_d$$

$$=\int_{-\infty}^{+\infty}\int_{-\infty}^{+\infty}\cdots\int_{-\infty}^{+\infty}\prod_{i=1}^{d}\varphi_i(x_i-u_i)du_1\ldots du_d=\left(\int_{-\infty}^{+\infty}\varphi_i(x_i-u_i)du_i\right)^d=1.$$

Let $\{\mathcal{L}_k\}_{k\in\mathbb{Z}}$ be the sequence of positive linear operators acting on X and defined by

$$\mathcal{L}_k(f;x_1,\ldots,x_d):=\int_{-\infty}^{+\infty}\int_{-\infty}^{+\infty}\cdots\int_{-\infty}^{+\infty}(\ell_k f)(u_1,\ldots,u_d)$$
$$\cdot\varphi(2^k x_1-u_1,\ldots,2^k x_d-u_d)\cdot du_1\ldots du_d, \tag{5}$$

all $(x_1,\ldots,x_d)\in\mathbb{R}^d$. In particular,

$$\mathcal{L}_0(f;x_1,\ldots,x_d)=\int_{-\infty}^{+\infty}\int_{-\infty}^{+\infty}\cdots\int_{-\infty}^{+\infty}\ell_0(f;u_1,\ldots,u_d)\cdot\varphi(x_1-u_1,\ldots,x_d-u_d)\cdot du_1\ldots du_d, \tag{6}$$

all $(x_1,\ldots,x_d)\in\mathbb{R}^d$. By (1) we have that

$$\mathcal{L}_k(f;\mathbf{x})=\mathcal{L}_0(f(2^{-k}\cdot);2^k\mathbf{x}),\quad \text{for all } \mathbf{x}\in\mathbb{R}^d. \tag{7}$$

Definition 1. Let $f_{\boldsymbol{\alpha}}(\cdot):=f(\cdot+\boldsymbol{\alpha})$, $\boldsymbol{\alpha}\in\mathbb{R}^d$, and ϕ be an operator. If $\phi(f_{\boldsymbol{\alpha}})=(\phi f)_{\boldsymbol{\alpha}}$, then ϕ is called a *shift invariant operator.*

Proposition 1. *Assume that*

$$\ell_0(f(2^{-k}\cdot+\boldsymbol{\alpha});2^k\cdot\mathbf{u})=\ell_0(f(2^{-k}\cdot);2^k\cdot(\mathbf{u}+\boldsymbol{\alpha})), \tag{8}$$

for all $k\in\mathbb{Z}$, $\boldsymbol{\alpha}\in\mathbb{R}^d$ *fixed, all* $\mathbf{u}\in\mathbb{R}^d$; *any* $f\in X$. *Then* $\mathcal{L}_k$ *is a shift invariant operator, all* $k\in\mathbb{Z}$.

Proof. Observe that

$$(\mathcal{L}_0 f)(x_1,\ldots,x_d)=\int_{-\infty}^{+\infty}\int_{-\infty}^{+\infty}\cdots\int_{-\infty}^{+\infty}(\ell_0 f)(u_1,\ldots,u_d)$$
$$\cdot\varphi(x_1-u_1,\ldots,x_d-u_d)\cdot du_1\ldots du_d$$
$$=\int_{-\infty}^{+\infty}\int_{-\infty}^{+\infty}\cdots\int_{-\infty}^{+\infty}(\ell_0 f)(x_1-u_1,\ldots,x_d-u_d)\cdot\varphi(u_1,\ldots,u_d)\cdot du_1\ldots du_d.$$

From (7) we have

$$\begin{aligned}\mathcal{L}_k(f(\cdot+\boldsymbol{\alpha});\mathbf{x})&=\mathcal{L}_k(f_{\boldsymbol{\alpha}};\mathbf{x})=\mathcal{L}_0(f_{\boldsymbol{\alpha}}(2^{-k}\cdot);2^k\mathbf{x})\\&=\int_{\mathbb{R}^d}(\ell_0 f(2^{-k}\cdot+\boldsymbol{\alpha}))(2^k\mathbf{x}-\mathbf{u})\cdot\varphi(\mathbf{u})\cdot d\mathbf{u}\\&=\int_{\mathbb{R}^d}(\ell_0 f(2^{-k}\cdot+\boldsymbol{\alpha}))(2^k\cdot(\mathbf{x}-2^{-k}\cdot\mathbf{u}))\cdot\varphi(\mathbf{u})\cdot d\mathbf{u}\\&=\int_{\mathbb{R}^d}(\ell_0(f(2^{-k}\cdot)))(2^k\cdot(\mathbf{x}-2^{-k}\mathbf{u}+\boldsymbol{\alpha}))\cdot\varphi(\mathbf{u})\cdot d\mathbf{u}\\&=\int_{\mathbb{R}^d}(\ell_0(f(2^{-k}\cdot)))(2^k(\mathbf{x}+\boldsymbol{\alpha})-\mathbf{u})\cdot\varphi(\mathbf{u})\cdot d\mathbf{u}\\&=\mathcal{L}_0(f(2^{-k}\cdot);2^k(\mathbf{x}+\boldsymbol{\alpha}))=\mathcal{L}_k(f;\mathbf{x}+\boldsymbol{\alpha}).\end{aligned}$$

I.e.,

$$\mathcal{L}_k(f_{\boldsymbol{\alpha}})=(\mathcal{L}_k f)_{\boldsymbol{\alpha}}.$$

Q.E.D.

In the following we study the property of *global smoothness preservation* by the operators $\mathcal{L}_k$.

Theorem 1. *For any $f \in X$ we assume that for all $\mathbf{u} \in \mathbb{R}^d$:*

$$|\ell_0(f;\mathbf{x}-\mathbf{u}) - \ell_0(f;\mathbf{y}-\mathbf{u})| \le \omega_1(f;\|\mathbf{x}-\mathbf{y}\|), \tag{9}$$

any $\mathbf{x},\mathbf{y} \in \mathbb{R}^d$. Then

$$\omega_1(\mathcal{L}_k f;\delta) \le \omega_1(f;\delta), \tag{10}$$

any $\delta > 0$.

Proof. Observe that

$$\begin{aligned}
|\mathcal{L}_0(f;\mathbf{x}) - \mathcal{L}_0(f;\mathbf{y})| &= \left|\int_{\mathbb{R}^d} \ell_0(f;\mathbf{u})\cdot\varphi(\mathbf{x}-\mathbf{u})d\mathbf{u} - \int_{\mathbb{R}^d} \ell_0(f;\mathbf{u})\varphi(\mathbf{y}-\mathbf{u})d\mathbf{u}\right| \\
&= \left|\int_{\mathbb{R}^d} \ell_0(f;\mathbf{x}-\mathbf{u})\varphi(\mathbf{u})d\mathbf{u} - \int_{\mathbb{R}^d} \ell_0(f;\mathbf{y}-\mathbf{u})\varphi(\mathbf{u})d\mathbf{u}\right| \\
&= \left|\int_{\mathbb{R}^d} (\ell_0(f;\mathbf{x}-\mathbf{u}) - \ell_0(f;\mathbf{y}-\mathbf{u}))\varphi(\mathbf{u})d\mathbf{u}\right| \\
&\le \int_{\mathbb{R}^d} |\ell_0(f;\mathbf{x}-\mathbf{u}) - \ell_0(f;\mathbf{y}-\mathbf{u})|\varphi(\mathbf{u})d\mathbf{u} \\
&\le \left(\int_{\mathbb{R}^d} \varphi(\mathbf{u})d\mathbf{u}\right)\cdot \sup_{\mathbf{u}\in\mathbb{R}^d} |\ell_0(f;\mathbf{x}-\mathbf{u}) - \ell_0(f;\mathbf{y}-\mathbf{u})| \le 1\cdot\omega_1(f;\|\mathbf{x}-\mathbf{y}\|),
\end{aligned}$$

any $\mathbf{x},\mathbf{y} \in \mathbb{R}^d$, by (4) and (9). I.e., we have proved that

$$|\mathcal{L}_0(f;\mathbf{x}) - \mathcal{L}_0(f;\mathbf{y})| \le \omega_1(f;\|\mathbf{x}-\mathbf{y}\|),$$

any $\mathbf{x},\mathbf{y} \in \mathbb{R}^d$. From (7) we get

$$\begin{aligned}
|\mathcal{L}_k(f;\mathbf{x}) - \mathcal{L}_k(f;\mathbf{y})| &= |\mathcal{L}_0(f(2^{-k}\cdot);2^k\mathbf{x}) - \mathcal{L}_0(f(2^{-k}\cdot);2^k\mathbf{y})| \\
&\le \omega_1(f(2^{-k}\cdot);2^k\|\mathbf{x}-\mathbf{y}\|) \\
&= \omega_1(f;\|\mathbf{x}-\mathbf{y}\|).
\end{aligned}$$

We have shown that

$$|\mathcal{L}_k(f;\mathbf{x}) - \mathcal{L}_k(f;\mathbf{y})| \le \omega_1(f;\|\mathbf{x}-\mathbf{y}\|).$$

I.e., global smoothness preservation by the $\mathcal{L}_k$ has been established. Q.E.D.

When $d = 1$ we have

$$\mathcal{L}_k(f;x) = \int_{-\infty}^{\infty} \ell_k(f;u)\varphi(2^k x - u)du,$$

where

$$\ell_k(f;u) = \ell_0(f(2^{-k}\cdot);u), \qquad \text{for all } u \in \mathbb{R}.$$

Corollary 1. ($d = 1$) *Assume that for all $u, x, y \in \mathbb{R}$ one has*

$$\ell_0(id(2^{-k}\cdot);2^k x - u) - \ell_0(id(2^{-k}\cdot);2^k y - u) = x - y.$$

Then for all $x, y \in \mathbb{R}$ one has

$$\mathcal{L}_k(id;x) - \mathcal{L}_k(id;y) = x - y,$$

i.e.,

$$\omega_1(\mathcal{L}_k(id);\delta) = \omega_1(id;\delta),$$

any $\delta > 0$, where $id(x) = x$, for all $x \in \mathbb{R}$. That is inequality (10) *for $d = 1$ is sharp.*

Proof. We have

$$\begin{aligned}\mathcal{L}_k(f;x) &:= \int_{-\infty}^{\infty} \ell_k(f;u)\varphi(2^k x - u)du \\ &= \int_{-\infty}^{\infty} \ell_0(f(2^{-k}\cdot);u)\varphi(2^k x - u)du \\ &= \int_{-\infty}^{\infty} \ell_0(f(2^{-k}\cdot);2^k x - u)\varphi(u)du.\end{aligned}$$

Hence,

$$\mathcal{L}_k(f;x) - \mathcal{L}_k(f;y) = \int_{-\infty}^{\infty} [\ell_0(f(2^{-k}\cdot);2^k x - u) - \ell_0(f(2^{-k}\cdot);2^k y - u)]\varphi(u)du.$$

When $f(x) = x$ and using the assumption of the statement, we obtain

$$\mathcal{L}_k(id;x) - \mathcal{L}_k(id;y) = \int_{-\infty}^{\infty} (x-y)\varphi(u)du = (x-y)\int_{-\infty}^{\infty} \varphi(u)du = x - y. \qquad \text{Q.E.D.}$$

A generalization of Corollary 1 follows.

Corollary 2. ($d \geq 2$) *For $i \in \{1,\dots,d\}$, let $pr_i : \mathbb{R}^d \ni (x_1,\dots,x_d) \to x_i \in \mathbb{R}$, denote the projection onto the ith coordinate. Assume that for at least one i one has*

$$\ell_0(pr_i(2^{-k}\cdot);2^k\mathbf{x} - \mathbf{u}) - \ell_0(pr_i(2^{-k}\cdot);2^k\mathbf{y} - \mathbf{u}) = x_i - y_i, \quad \textit{for all } \mathbf{x},\mathbf{y},\mathbf{u} \in \mathbb{R}^d.$$

Then

$$\omega_1(\mathcal{L}_k pr_i;\delta) = \omega_1(pr_i;\delta),$$

any $\delta > 0$, proving inequality (10) *to be sharp for $d \geq 2$.*

Proof. Note that pr_i is a uniformly continuous function on $\mathbb{R}^d$. Furthermore,

$$\begin{aligned}\mathcal{L}_k(pr_i;\mathbf{x}) - \mathcal{L}_k(pr_i;\mathbf{y}) &= \int_{\mathbb{R}^d} [\ell_0(pr_i(2^{-k}\cdot);2^k\mathbf{x} - \mathbf{u}) - \ell_0(pr_i(2^{-k}\cdot);2^k\mathbf{y} - \mathbf{u})]\varphi(\mathbf{u})d\mathbf{u} \\ &= \int_{\mathbb{R}^d} (x_i - y_i)\varphi(\mathbf{u})d\mathbf{u} = (x_i - y_i)\cdot 1 = x_i - y_i, \quad \text{by (4).}\end{aligned}$$

Thus

$$|\mathcal{L}_k(pr_i;\mathbf{x}) - \mathcal{L}_k(pr_i;\mathbf{y})| = |x_i - y_i| = |pr_i(\mathbf{x}) - pr_i(\mathbf{y})|,$$

proving

$$\omega_1(\mathcal{L}_k pr_i;\delta) = \omega_1(pr_i;\delta),$$

any $\delta > 0$. Q.E.D.

The convergence of $\mathcal{L}_k$ to I as $k \to +\infty$, I being the unit operator, with rates is studied next.

Theorem 2. *For $f \in X$, under the assumption* (2), *it holds*

$$|\mathcal{L}_k(f;\mathbf{x}) - f(\mathbf{x})| \leq \omega_{1,\infty}\left(f;\frac{ma+n}{2^{k+r}}\right), \tag{11}$$

where $m \in \mathbb{N}$, $n \in \mathbb{Z}_+$, $k, r \in \mathbb{Z}$, and $a := \max(a_i)$, $i = 1,\dots,n$.

Proof. From (7), (3) and supp $\varphi \subseteq \prod_{i=1}^{d}[-a_i, a_i]$, we have that

$$
\begin{aligned}
|\mathcal{L}_k(f; \mathbf{x}) - f(\mathbf{x})| &= |\mathcal{L}_0(f(2^{-k}\cdot); 2^k\mathbf{x}) - f(2^{-k}(2^k\mathbf{x}))| \\
&= \left| \int_{\mathbb{R}^d} \ell_0(f(2^{-k}\cdot); \mathbf{u}) \cdot \varphi(2^k\mathbf{x} - \mathbf{u}) \cdot d\mathbf{u} - \int_{\mathbb{R}^d} f(2^{-k}(2^k\mathbf{x})) \cdot \varphi(2^k\mathbf{x} - \mathbf{u}) \cdot d\mathbf{u} \right| \\
&= \left| \int_{\mathbb{R}^d} [\ell_0(f(2^{-k}\cdot); \mathbf{u}) - f(2^{-k}(2^k\mathbf{x}))] \cdot \varphi(2^k\mathbf{x} - \mathbf{u}) \cdot d\mathbf{u} \right| \\
&\leq \int \cdots \int_{\substack{2^k x_i - a_i, \\ \text{all } i=1,\ldots,d}}^{2^k x_i + a_i} \int |\ell_0(f(2^{-k}\cdot); \mathbf{u}) - f(2^{-k}(2^k\mathbf{x}))| \cdot \varphi(2^k\mathbf{x} - \mathbf{u}) \cdot d\mathbf{u} \\
&\leq \sup_{\mathbf{u} \in \prod_{i=1}^{d}[2^k x_i - a_i, 2^k x_i + a_i]} |\ell_0(f(2^{-k}\cdot); \mathbf{u}) - f(2^{-k}(2^k\mathbf{x}))| \cdot \\
&\qquad\qquad \cdot \int \cdots \int_{\substack{2^k x_i - a_i \\ \text{all } i=1,\ldots,d}}^{2^k x_i + a_i} \int \varphi(2^k\mathbf{x} - \mathbf{u}) d\mathbf{u}.
\end{aligned}
$$

I.e., by (3)

$$
|\mathcal{L}_k(f; \mathbf{x}) - f(\mathbf{x})| \leq \sup_{\mathbf{u} \in \prod_{i=1}^{d}[2^k x_i - a_i, 2^k x_i + a_i]} |\ell_0(f(2^{-k}\cdot); \mathbf{u}) - f(2^{-k}(2^k\mathbf{x}))|. \tag{12}
$$

Consider $g := f(2^{-k}\cdot) \in X$ and $y_i := 2^k x_i$. Thus, the right-hand side of (12) equals

$$
\begin{aligned}
&\sup_{\substack{\text{all } \mathbf{u} \in \mathbb{R}^d: \\ \mathbf{u} - 2^k\mathbf{x} \in \prod_{i=1}^{d}[-a_i, a_i]}} |\ell_0(f(2^{-k}\cdot); \mathbf{u}) - f(2^{-k}(2^k\mathbf{x}))| \\
&= \sup_{\substack{\text{all } \mathbf{u} \in \mathbb{R}^d: \\ |u_i - 2^k x_i| \leq a_i}} |\ell_0(f(2^{-k}\cdot); \mathbf{u}) - f(2^{-k}(2^k\mathbf{x}))| \\
&\leq \sup_{\substack{\text{all } \mathbf{u} \in \mathbb{R}^d: \\ |u_i - 2^k x_i| \leq \max_{i=1,\ldots,d}(a_i) =: a,}} |\ell_0(f(2^{-k}\cdot); \mathbf{u}) - f(2^{-k}(2^k\mathbf{x}))| \\
&= \sup_{|u_i - y_i| \leq a} |\ell_0(g; \mathbf{u}) - g(\mathbf{y})| \\
&= \sup_{\|\mathbf{u} - \mathbf{y}\|_\infty \leq a} |\ell_0(g; \mathbf{u}) - g(\mathbf{y})| \\
&\leq \omega_{1,\infty}\left(g; \frac{ma + n}{2^r}\right) \quad \text{(by assumption (2))} \\
&= \omega_{1,\infty}\left(f; \frac{ma + n}{2^{k+r}}\right).
\end{aligned}
$$

Therefore (11) is established. Q.E.D.

In the next we need the following.

Corollary 3 (H. Bauer [4], pp. 103–104) *Let $(\underline{0}, \mathcal{A}, \mu)$ be a measure space. Let U be an open subset of $\mathbb{R}^d$, $d \geq 1$, and let $f: U \times \underline{0} \to \mathbb{R}$ be a function with the properties:*

(a) *$\omega \to f(x, \omega)$ is μ-integrable for all $x \in U$,*

(b) *$x \to f(x, \omega)$ is at each point of U partially differentiable with respect to x_i,*

(c) *there exists a μ-integrable function $h \geq 0$ on $\underline{0}$ such that*

$$
\left| \frac{\partial f}{\partial x_i}(x, \omega) \right| \leq h(\omega), \qquad \text{for all } (x, \omega) \in U \times \underline{0}.
$$

Then the function φ *defined on* U *as*

$$\varphi(x) = \int_{\Omega} f(x,\omega)\mu(d\omega)$$

is partially differentiable with respect to x_i *on all of* U. *The mapping*

$$\omega \to \frac{\partial f}{\partial x_i}(x,\omega) \text{ is } \mu\text{-integrable,}$$

and we have

$$\frac{\partial \varphi}{\partial x_i}(x) = \int_{\Omega} \frac{\partial f}{\partial x_i}(x,\omega)\mu(d\omega), \quad \text{all } x \in U.$$

Remark 1. (§1) Here assume furthermore that φ is bounded. Consider the measure space $(\Omega, \mathcal{A}, \mu) := (\times_{i=1}^{d}[-a_i, a_i]$, σ-field of Lebesgue measurable sets, multidimensional Lebesgue measure). Let W be a box in $\mathbb{R}^d$ and $(x_1, \ldots, x_d)$ be a point of the interior $(W) := W^0 := U$. Set

$$\tilde{f}(x_1, \ldots, x_d; u_1, \ldots, u_d) := (\ell_0 f)(x_1 - u_1, x_2 - u_2, \ldots, x_d - u_d) \cdot \varphi(u_1, \ldots, u_d).$$

(a) We have that the function

$$\gamma(u_1, \ldots, u_d) := (\ell_0 f)(x_1 - u_1, x_2 - u_2, \ldots, x_d - u_d) \cdot \varphi(u_1, \ldots, u_d)$$

is Lebesgue integrable, all $(x_1, \ldots, x_d) \in U$.

(b) The function

$$\lambda(x_1, \ldots, x_d) := \lambda_{\mathbf{u}}(\mathbf{x}) := (\ell_0 f)(x_1 - u_1, x_2 - u_2, \ldots, x_d - u_d) \cdot \varphi(u_1, \ldots, u_d)$$

is partially differentiable on $\mathbb{R}^d$ with respect to x_i for $i \in \{1, \ldots, d\}$, given that we assume that $\frac{\partial}{\partial x_i}\ell_0(f, x_1, \ldots, x_d)$ exists.

(c) Assume that $\frac{\partial}{\partial x_i}\ell_0(f, x_1, \ldots, x_d)$ is continuous. Then

$$\left|\frac{\partial \tilde{f}}{\partial x_i}(\mathbf{x}, \mathbf{u})\right| = \left|\frac{\partial}{\partial x_i}\ell_0(f, \mathbf{x} - \mathbf{u}) \cdot \varphi(\mathbf{u})\right| \le M \cdot \varphi(\mathbf{u}), \quad \text{for all } (\mathbf{x}, \mathbf{u}) \in U \times \times_{i=1}^{d}[-a_i, a_i],$$

where

$$M := \left\|\frac{\partial}{\partial x_i}\ell_0(f, \mathbf{x} - \mathbf{u})\right\|_{\infty, W \times \times_{i=1}^{d}[-a_i, a_i]} < +\infty.$$

But $M \cdot \varphi(\mathbf{u})$ is a Lebesgue integrable function, since φ is bounded and of compact support. Thus by Corollary 3 we get that

$$\frac{\partial}{\partial x_i}\mathcal{L}_0(f, \mathbf{x}) = \int_{-a_1}^{a_1} \cdots \int_{-a_d}^{a_d} \frac{\partial}{\partial x_i}\ell_0(f; \mathbf{x} - \mathbf{u}) \cdot \varphi(\mathbf{u}) \cdot d\mathbf{u} \tag{13}$$

exists. Therefore $\frac{\partial}{\partial x_i}\mathcal{L}_k(f, \mathbf{x})$ exists, under the additional assumptions that φ is bounded and $\frac{\partial}{\partial x_i}\ell_0(f, \mathbf{x})$ exists and is continuous, $i \in \{1, \ldots, d\}$.

(§2) Working as in (§1) and assuming that φ is bounded, $\frac{\partial}{\partial x_i}\ell_0(f, \mathbf{x})$ exists and is continuous, $i \in \{1, \ldots, d\}$, we examine the existence of mixed partial derivatives of $\mathcal{L}_k(f, \mathbf{x})$. Here $(\Omega, \mathcal{A}, \mu)$ is chosen again as in (§1). Set

$$\tilde{f}(x, u) := \frac{\partial}{\partial x_i}\ell_0(f; \mathbf{x} - \mathbf{u}) \cdot \varphi(\mathbf{u}).$$

(a) We have that $\gamma(\mathbf{u}) := \frac{\partial}{\partial x_i}\ell_0(f; \mathbf{x} - \mathbf{u}) \cdot \varphi(\mathbf{u})$ is integrable, since $\frac{\partial}{\partial x_i}\ell_0(f; \mathbf{x} - \mathbf{u})$ is continuous in $\mathbf{u}$, all $\mathbf{x} \in U$.

(b) The function

$$\lambda(\mathbf{x}) := \lambda_{\mathbf{u}}(\mathbf{x}) := \frac{\partial}{\partial x_i}\ell_0(f;\mathbf{x}-\mathbf{u})\cdot\varphi(\mathbf{u})$$

is at each point of $\mathbb{R}^d$ partially differentiable with respect to x_j, given that we assume that $\frac{\partial^2}{\partial x_i \partial x_j}\ell_0(f,\mathbf{x})$ exists for $j \in \{1,\dots,d\}$, all $\mathbf{x} \in \mathbb{R}^d$.

(c) Assume that $\frac{\partial^2}{\partial x_i \partial x_j}\ell_0(f,\mathbf{x})$ is continuous everywhere. Then

$$\left|\frac{\partial \tilde{f}(\mathbf{x},\mathbf{u})}{\partial x_j}\right| = \left|\frac{\partial^2}{\partial x_i \partial x_j}\ell_0(f;\mathbf{x}-\mathbf{u})\cdot\varphi(\mathbf{u})\right| \le M^*\cdot\varphi(\mathbf{u}),$$

where

$$M^* := \left\|\frac{\partial^2}{\partial x_i \partial x_j}\ell_0(f,\mathbf{x}-\mathbf{u})\right\|_{\infty, W\times \times_{i=1}^d[-a_i,a_i]} < +\infty.$$

Here $M^*\cdot\varphi(\mathbf{u})$ is an integrable function. Then, by Corollary 3, we get

$$\frac{\partial^2}{\partial x_i \partial x_j}\mathcal{L}_0(f,\mathbf{x}) = \int_{-a_1}^{a_1}\cdots\int_{-a_d}^{a_d}\frac{\partial^2}{\partial x_i \partial x_j}\ell_0(f;\mathbf{x}-\mathbf{u})\cdot\varphi(\mathbf{u})\cdot d\mathbf{u}. \tag{14}$$

Therefore $\frac{\partial^2}{\partial x_i \partial x_j}\mathcal{L}_k(f,\mathbf{x})$ exists, under the additional assumptions that φ is bounded and $\frac{\partial^2 \ell_0(f,\mathbf{x})}{\partial x_i \partial x_j}$ exists and is continuous, $i,j \in \{1,\dots,d\}$. Iterating this argument we can conclude

Theorem 3. *Additionally assume that φ is bounded. Let $i_1,\dots,i_k \in \{1,\dots,d\}$ be such that $i_1 < \cdots < i_k$, and $j_1,\dots,j_k \in \mathbb{Z}_+$. Suppose that $\frac{\partial^{\sum_{k=1}^k j_k}}{\partial x_{i_1}^{j_1}\dots\partial x_{i_k}^{j_k}}\ell_0(f;\mathbf{x})$ exists and is continuous for all $\mathbf{x} \in \mathbb{R}^d$. Then*

$$\frac{\partial^{\sum_{k=1}^k j_k}}{\partial x_{i_1}^{j_1}\dots\partial x_{i_k}^{j_k}}\mathcal{L}_0(f;\mathbf{x}) = \int_{-a_1}^{a_1}\cdots\int_{-a_d}^{a_d}\frac{\partial^{\sum_{k=1}^k j_k}}{\partial x_{i_1}^{j_1}\dots\partial x_{i_k}^{j_k}}\ell_0(f;\mathbf{x}-\mathbf{u})\cdot\varphi(\mathbf{u})\cdot d\mathbf{u}. \tag{15}$$

Thus $\frac{\partial^{\sum_{k=1}^k j_k}}{\partial x_{i_1}^{j_1}\dots\partial x_{i_k}^{j_k}}\mathcal{L}_k(f;\mathbf{x})$ exists.

Remark 2. Notice that: If

$$\frac{\partial^{\sum_{k=1}^k j_k}}{\partial x_{i_1}^{j_1}\dots\partial x_{i_k}^{j_k}}\ell_0(f;\mathbf{x}) \ge 0 \qquad \text{for all } \mathbf{x} \in \mathbb{R}^d,$$

then

$$\frac{\partial^{\sum_{k=1}^k j_k}}{\partial x_{i_1}^{j_1}\dots\partial x_{i_k}^{j_k}}\mathcal{L}_0(f;\mathbf{x}) \ge 0,$$

and thus

$$\frac{\partial^{\sum_{k=1}^k j_k}}{\partial x_{i_1}^{j_1}\dots\partial x_{i_k}^{j_k}}\mathcal{L}_k(f;\mathbf{x}) \ge 0$$

Again, for the above we assumed that φ is bounded.

Note 1. For the next we need the following notions and the definition of a multidimensional probability distribution function from Shiryayev ([6], pp. 157–158): Let P be a probability measure on the measurable space $(\mathbb{R}^d, \mathcal{B}(\mathbb{R}^d))$, where $\mathcal{B}$ stands for the Borel σ-algebra.

Write

$$F_d(x_1,\dots,x_d) := P((-\infty,x_1]\times\cdots\times(-\infty,x_d]),$$

or, in a compact form,

$$F_d(\mathbf{x}) := P(-\infty, \mathbf{x}],$$

where

$$\mathbf{x} := (x_1, \ldots, x_d), \; (-\infty, \mathbf{x}] := (-\infty, x_1] \times \cdots \times [-\infty, x_d].$$

Here $\Delta_{a_i,b_i}: \mathbb{R}^d \to \mathbb{R}$ denotes the difference operator defined by the formula

$$\Delta_{a_i,b_i} F_d(x_1, \ldots, x_d) := F_d(x_1, \ldots, x_{i-1}, b_i, x_{i+1}, \ldots, x_d) - F_d(x_1, \ldots, x_{i-1}, a_i, x_{i+1}, \ldots, x_d),$$

where $a_i \le b_i$ and $i \in \{1, \ldots, d\}$. Easily one can see that

$$\Delta_{a_1,b_1} \cdots \Delta_{a_d,b_d} F_d(x_1, \ldots, x_d) = P(\mathbf{a}, \mathbf{b}],$$

where

$$(\mathbf{a}, \mathbf{b}] := (a_1, b_1] \times \cdots \times (a_d, b_d]; \; \mathbf{a} := (a_1, \ldots, a_d), \; \mathbf{b} := (b_1, \ldots, b_d).$$

By $P(\mathbf{a}, \mathbf{b}] \ge 0$ we get

(i) $\Delta_{a_1,b_1} \cdot \Delta_{a_d,b_d} F_d(x_1, \ldots, x_d) \ge 0$ for arbitrary $\mathbf{a}$, $\mathbf{b}$ with $a_i \le b_i$, $i = 1, \ldots, d$.

From the continuity of P it follows that $F_d(x_1, \ldots, x_d)$ is continuous on the right with respect to all variables collectively, i.e.,

(ii) if $\mathbf{x}^{(k)} \downarrow \mathbf{x}$, $\mathbf{x}^{(k)} := (x_1^{(k)}, \ldots, x_d^{(k)})$, then $F_d(\mathbf{x}^{(k)}) \downarrow F_d(\mathbf{x})$, $\quad k \to +\infty$.

It is also clear that

(iii) $F_d(+\infty, \ldots, +\infty) = 1$

and

(iv) $\lim_{\mathbf{x} \downarrow \mathbf{y}} F_d(x_1, \ldots, x_d) = 0$,

if at least one coordinate of $\mathbf{y}$ is equal to $-\infty$.

Definition 2. A d-dimensional probability distribution function on $\mathbb{R}^d$, $d \ge 1$, is a function $F = F(x_1, \ldots, x_d)$ fulfilling the properties (i)–(iv) of Note 1.

We also need the

Dominated Convergence Theorem (Lebesgue). (See Kingman & Taylor [5], p. 121).

(i) *If $g: \Omega \to \mathbb{R}^+$ is integrable, where $(\Omega, \mathcal{A}, \mu)$ is a measure space, $\{f_n\}$ is a sequence of measurable functions $\Omega \to \mathbb{R} \cup \{-\infty, +\infty\}$ such that $|f_n| \le g$, $n = 1, 2, \ldots$ and $f_n \to f$ as $n \to +\infty$, then f is integrable and*

$$\int f_n \, d\mu \to \int f \, d\mu \quad \text{as } n \to +\infty.$$

(ii) *Suppose $g: \Omega \to \mathbb{R}^+$ is integrable, $-\infty \le a < b \le +\infty$, and for each $t \in (a, b)$, f_t is a measurable function from Ω into $\mathbb{R} \cup \{-\infty, +\infty\}$. Then if $|f_t| \le g$ for all $t \in (a, b)$ and $f_t \to f$ as $t \to a+$ or $t \to b-$, then f is integrable and*

$$\int f_t \, d\mu \to \int f \, d\mu.$$

Problem 1. If f is a continuous probabilistic distribution function on $\mathbb{R}^d$, $d \ge 1$, and $(\ell_0 f)$ is the same, $\varphi \ge 0$, φ continuous on $\times_{i=1}^{d} [-a_i, a_i]$, then can we conclude that $(\mathcal{L}_0 f)$ is a continuous distribution function?

The answer to the above question is to the affirmative and is given by

Theorem 4. *Let ℓ_k be a positive linear operator from $C(\mathbb{R}^d)$ into itself as in (1), i.e. $\ell_k(f, \mathbf{x}) = \ell_0(f(2^{-k} \cdot), \mathbf{x})$, for all $\mathbf{x} \in \mathbb{R}^d$, all $f \in X$, $k \in \mathbb{Z}$. Assume that f is a probabilistic distribution function from $\mathbb{R}^d$ into $\mathbb{R}$ that is continuous. Assume that $\ell_0(f)$ is also a continuous distribution* *function, where f is any continuous distribution function. Assume furthermore, that $\varphi \geq 0$ is continuous on $\times_{i=1}^{d}[-a_i, a_i]$, $a_i > 0$, supp $\varphi \subseteq \times_{i=1}^{d}[-a_i, a_i]$,*

$$\int_{\mathbb{R}^d} \varphi(\mathbf{x} - \mathbf{u}) d\mathbf{u} = 1, \qquad \textit{for all } \mathbf{x} \in \mathbb{R}^d.$$

Then the operator

$$\mathcal{L}_k(f; \mathbf{x}) = \int_{-\mathbf{a}}^{\mathbf{a}} (\ell_k f)(2^k \mathbf{x} - \mathbf{u}) \cdot \varphi(\mathbf{u}) \cdot d\mathbf{u}, \tag{16}$$

$k \in \mathbb{Z}$, where $\mathbf{a} := (a_1, \ldots, a_d)$, when applied on f as above produces a continuous probabilistic distribution function from $\mathbb{R}^d$ to $\mathbb{R}$.

Note 2. Observe that $\omega_1(F, \delta) < +\infty$ for any probability distribution function $F: \mathbb{R}^d \to [0,1]$. And of course all estimates involving ω_1 in this article apply also for just continuous probabilistic distribution functions on $\mathbb{R}^d$.

Proof of Theorem 4. (i) If f is a continuous probability distribution function from $\mathbb{R}^d$ into $\mathbb{R}$, then so is $f(2^{-k}\cdot)$, a trivial proof. Thus, by the assumption of the theorem $\ell_0(f(2^{-k}\cdot))$ is a continuous distribution function, too. That is, by the defining property (1) of ℓ_k we get that $(\ell_k f)$ is a continuous distribution function.

Let now $\mathbf{x}^{(\tilde{k})} \to \mathbf{x}$, where $\mathbf{x}^{(\tilde{k})} := (x_1^{(\tilde{k})}, \ldots, x_d^{(\tilde{k})})$, $\mathbf{x} := (x_1, \ldots, x_d)$, $\tilde{k} \in \mathbb{N}$. Then $2^k \cdot \mathbf{x}^{(\tilde{k})} - \mathbf{u} \to 2^k \cdot \mathbf{x} - \mathbf{u}$, that is $(\ell_k f)(2^k \cdot \mathbf{x}^{(\tilde{k})} - \mathbf{u}) \to (\ell_k f)(2^k \cdot \mathbf{x} - \mathbf{u})$, as $\tilde{k} \to +\infty$, by continuity of $(\ell_k f)$, $k \in \mathbb{Z}$. Hence

$$\mathcal{F}_{\tilde{k}}(\mathbf{u}) := (\ell_k f)(2^k \cdot \mathbf{x}^{(\tilde{k})} - \mathbf{u}) \cdot \varphi(\mathbf{u}) \to \mathcal{F}(\mathbf{u}) := (\ell_k f)(2^k \cdot \mathbf{x} - \mathbf{u}) \cdot \varphi(\mathbf{u}),$$

as $\tilde{k} \to +\infty$, for each $\mathbf{u} \in \times_{i=1}^{d}[-a_i, a_i]$. Note that $\mathcal{F}_{\tilde{k}}$, $\mathcal{F}$ are continuous functions of $\mathbf{u}$.

Consider now that $\mathbf{x}^{(\tilde{k})}(\tilde{k} \in \mathbb{N})$, $\mathbf{x}$ belong to U^0 (interior of U), where U is a box in $\mathbb{R}^d$, $d \geq 1$. Then

$$|\mathcal{F}_{\tilde{k}}(\mathbf{u})|, |\mathcal{F}(\mathbf{u})| \leq \left\| (\ell_k f)\left(2^k \cdot \begin{Bmatrix} \mathbf{x}^{(\tilde{k})} \\ \mathbf{x} \end{Bmatrix} - \mathbf{u}\right) \right\|_{\infty, U \times \times_{i=1}^{d}[-a_i, a_i]} \cdot \varphi(\mathbf{u}) \leq \varphi(\mathbf{u}) \leq M,$$

all $\tilde{k} \in \mathbb{N}$.

We apply the Dominated Convergence Theorem; the associated measure here is the Lebesgue one on $\times_{i=1}^{d}[-a_i, a_i]$. Therefore

$$\mathcal{L}_k(f; \mathbf{x}^{(\tilde{k})}) \to \mathcal{L}_k(f; \mathbf{x}),$$

as $\tilde{k} \to +\infty$. That is, $\mathcal{L}_k(f; \mathbf{x})$ is continuous in $\mathbf{x}$.

(ii) As before (in (i)) $(\ell_k f)$ is a continuous distribution function. Let now $\mathbf{x}^{(\tilde{k})} \to +\boldsymbol{\infty}$, where $\mathbf{x}^{(\tilde{k})} := (x_1^{(\tilde{k})}, \ldots, x_d^{(\tilde{k})})$, $+\boldsymbol{\infty} := (+\infty, \ldots, +\infty)$, $\tilde{k} \in \mathbb{N}$. Then $2^k \cdot \mathbf{x}^{(\tilde{k})} - \mathbf{u} \to +\boldsymbol{\infty}$. That is,

$$(\ell_k f)(2^k \cdot \mathbf{x}^{(\tilde{k})} - \mathbf{u}) \to (\ell_k f)(+\boldsymbol{\infty}) = 1.$$

Hence

$$\mathcal{F}_{\tilde{k}}(\mathbf{u}) := (\ell_k f)(2^k \cdot \mathbf{x}^{(\tilde{k})} - \mathbf{u}) \cdot \varphi(\mathbf{u}) \to \varphi(\mathbf{u}),$$

for each $\mathbf{u} \in \times_{i=1}^{d}[-a_i, a_i]$. Obviously $\mathcal{F}_{\tilde{k}}(\mathbf{u})$ is continuous in $\mathbf{u}$ (φ is continuous by assumption). Furthermore,

$$|(\ell_k f)(2^k \cdot \mathbf{x}^{(\tilde{k})} - \mathbf{u}) \cdot \varphi(\mathbf{u})| \leq \varphi(\mathbf{u}) \leq M,$$

all $\tilde{k} \in \mathbb{N}$.

I.e.,

$$|\mathcal{F}_{\tilde{k}}| \leq M, \quad \text{all } \tilde{k} \in \mathbb{N}.$$

Again by the Dominated Convergence Theorem, where the associated measure μ is the Lebesgue one on $\times_{i=1}^d[-a_i, a_i]$, we obtain

$$\int_{\times_{i=1}^d[-a_i,a_i]} \mathcal{F}_{\tilde{k}}(\mathbf{u}) \cdot \mu(d\mathbf{u}) \to \int_{\times_{i=1}^d[-a_i,a_i]} \varphi(\mathbf{u}) \cdot \mu(d\mathbf{u}) = 1.$$

The last equality comes from (4). That is, $\mathcal{L}_k(f; \mathbf{x}^{(\tilde{k})}) \to 1$, as $\tilde{k} \to +\infty$. I.e.,

$$\mathcal{L}_k(f; +\infty, +\infty, \ldots, +\infty) = 1. \tag{17}$$

(iii) Using again that $(\ell_k f)$ is a continuous distribution function we have the following: let $x_j^{\tilde{k}} \to -\infty$, i.e.

$$\mathbf{x}^{(\tilde{k})} := (x_1, \ldots, x_{j-1}, x_j^{(\tilde{k})}, x_{j+1}, \ldots, x_d) \to \mathbf{x}_0 := (x_1, \ldots, x_{j-1}, -\infty, x_{j+1}, \ldots, x_d).$$

One can see easily that

$$(\ell_k f)(2^k \cdot \mathbf{x}^{(\tilde{k})} - \mathbf{u}) \to (\ell_k f)(2^k x_1 - u_1, \ldots, 2^k x_{j-1} - u_{j-1}, -\infty, 2^k x_{j+1} - u_{j+1}, \ldots, 2^k x_d - u_d) = 0.$$

Hence

$$\mathcal{F}_{\tilde{k}}(\mathbf{u}) := (\ell_k f)(2^k \cdot \mathbf{x}^{(\tilde{k})} - \mathbf{u}) \cdot \varphi(\mathbf{u}) \to 0,$$

for each $\mathbf{u} \in \times_{i=1}^d[-a_i, a_i]$. Obviously $\mathcal{F}_{\tilde{k}}(\mathbf{u})$ is continuous in $\mathbf{u}$. Again

$$|\mathcal{F}_{\tilde{k}}| \leq M, \quad \text{all } \tilde{k} \in \mathbb{N}.$$

One more time by the Dominated Convergence Theorem, where the associated measure μ is the Lebesgue one on $\times_{i=1}^d[-a_i, a_i]$, we get

$$\int_{\times_{i=1}^d[-a_i,a_i]} \mathcal{F}_{\tilde{k}}(\mathbf{u})\mu(d\mathbf{u}) \to 0.$$

That is, $\mathcal{L}_k(f; \mathbf{x}^{(\tilde{k})}) \to 0$, as $\tilde{k} \to +\infty$. I.e.,

$$\mathcal{L}_k(f; x_1, \ldots, -\infty, \ldots, x_d) = 0. \tag{18}$$

(iv) Here denote by $\Delta_h g(x) := g(x+h) - g(x)$, the difference operator, g a function, $h > 0$. It remains to show that

$$(\Delta_{h_d} \circ \cdots \circ \Delta_{h_1}) \mathcal{L}_k(f; x_1, \ldots, x_d) \geq 0,$$

for any $h_\delta \geq 0$, $\delta = 1, \ldots, d$; all $\mathbf{x} \in \mathbb{R}^d$. Here $\Delta_{h_i}^{x_i}$ means that Δ_{h_i} is applied in the ith entry.

We have

$$\begin{aligned}
&(\Delta_{h_d} \circ \cdots \circ \Delta_{h_1}) \mathcal{L}_k(f; x_1, \ldots, x_d) \\
&= (\Delta_{h_d} \circ \cdots \circ \Delta_{h_1}) \int_{-a_1}^{a_1} \cdots \int_{-a_d}^{a_d} (\ell_k f)((2^k x_1 - u_1, \ldots, 2^k x_d - u_d) \cdot \varphi(u_1, \ldots, u_d) \cdot du_d \cdots du_1 \\
&= (\Delta_{h_d} \circ \cdots \circ \Delta_{h_2}) \circ \Delta_{h_1}^{x_1} \int_{-a_1}^{a_1} \cdots \int_{-a_d}^{a_d} \ell_0(f(2^{-k}\cdot); 2^k x_1 - u_1, \ldots, 2^k x_d - u_d) \cdot \varphi(\mathbf{u}) \cdot d\mathbf{u} \\
&= (\Delta_{h_d} \circ \cdots \circ \Delta_{h_2}) \Big[\int_{-a_1}^{a_1} \cdots \int_{-a_d}^{a_d} \ell_0(f(2^{-k}\cdot); 2^k x_1 \\
&\quad + h_1 - u_1, 2^k x_2 - u_2, \ldots, 2^k x_d - u_d) \cdot \varphi(\mathbf{u}) \cdot d\mathbf{u} \\
&\quad - \int_{-a_1}^{a_1} \cdots \int_{-a_d}^{a_d} \ell_0(f(2^{-k}\cdot); 2^k x_1 - u_1, 2^k x_2 - u_2, \ldots, 2^k x_d - u_d) \cdot \varphi(\mathbf{u}) \cdot d\mathbf{u} \Big] \\
&= (\Delta_{h_d} \circ \cdots \circ \Delta_{h_2}) \int_{-a_1}^{a_1} \cdots \int_{-a_d}^{a_d} \Delta_{h_1}^{x_1} \ell_0(f(2^{-k}\cdot); 2^k x_1 - u_1, \ldots, 2^k x_d - u_d) \cdot \varphi(\mathbf{u}) \cdot d\mathbf{u} \\
&= \cdots = \int_{-a_1}^{a_1} \cdots \int_{-a_d}^{a_d} (\Delta_{h_d} \circ \cdots \circ \Delta_{h_1}) \ell_0(f(2^{-k}\cdot); 2^k x_1 - u_1, \ldots, 2^k x_d - u_d) \cdot \varphi(\mathbf{u}) \cdot d\mathbf{u} \geq 0,
\end{aligned}$$

because $\varphi(\mathbf{u}) \geq 0$, and

$$(\Delta_{h_d} \circ \cdots \circ \Delta_{h_1}) \ell_0(f(2^{-k}\cdot); 2^k x_1 - u_1, \ldots, 2^k x_d - u_d) \geq 0,$$

by $\ell_0(f(2^{-k}\cdot))$ being a probability distribution function.

We have proved that

$$(\Delta_{h_d} \circ \cdots \circ \Delta_{h_1}) \mathcal{L}_k(f; x_1, \ldots, x_d) \geq 0, \tag{19}$$

all $h_\delta \geq 0$, $\delta = 1, \ldots, d$, $d \geq 1$ and all $\mathbf{x} \in \mathbb{R}^d$, $\mathbf{x} := (x_1, \ldots, x_d)$. So we have established that $\mathcal{L}_k(f)$ is a continuous distribution function from $\mathbb{R}^d$ into $\mathbb{R}$. Q.E.D.

3. APPLICATIONS

In the following we present four examples of shift invariant multivariate integral operators where ℓ_k is specified. It will be shown that they verify exactly the theory presented in Section 2. The basic function φ will be as in Section 2. In particular for the case of operators $(A_k)_{k \in \mathbb{Z}}$, to be defined next, φ will be assumed additionally to be an even continuous function, $\varphi(-\mathbf{x}) = \varphi(\mathbf{x})$, $\forall \mathbf{x} \in \mathbb{R}^d$. So, the properties of our specific multivariate operators will be presented according to the order of properties of general multivariate operators $\mathcal{L}_k$, $k \in \mathbb{Z}$, in Section 2. However, first we need to introduce these multivariate operators, all defined for each $k \in \mathbb{Z}$.

(i)

$$(A_k f)(\mathbf{x}) := \int_{\mathbb{R}^d} r_k^f(\mathbf{u}) \cdot \varphi(2^k \mathbf{x} - \mathbf{u}) \cdot d\mathbf{u}, \tag{20}$$

where

$$r_k^f(\mathbf{u}) := 2^{kd} \cdot \int_{\mathbb{R}^d} f(\mathbf{t}) \cdot \varphi(2^k \mathbf{t} - \mathbf{u}) dt \tag{21}$$

is continuous in $\mathbf{u}$ (see Lemma 1 later). I.e., here

$$\ell_k(f, \mathbf{u}) = r_k^f(\mathbf{u}), \qquad \forall \mathbf{u} \in \mathbb{R}^d. \tag{22}$$

(ii)

$$(B_k f)(\mathbf{x}) := \int_{\mathbb{R}^d} f\left(\frac{\mathbf{u}}{2^k}\right) \cdot \varphi(2^k \mathbf{x} - \mathbf{u}) \cdot d\mathbf{u} \tag{23}$$

i.e., here

$$\ell_k(f, \mathbf{u}) = f\left(\frac{\mathbf{u}}{2^k}\right), \qquad \forall \mathbf{u} \in \mathbb{R}^d, \tag{24}$$

is continuous in $\mathbf{u}$.

(iii)

$$(L_k f)(\mathbf{x}) := \int_{\mathbb{R}^d} c_k^f(\mathbf{u}) \cdot \varphi(2^k \mathbf{x} - \mathbf{u}) \cdot d\mathbf{u}, \tag{25}$$

where

$$c_k^f(\mathbf{u}) := 2^{kd} \cdot \int \cdots \int_{2^{-k} u_i}^{2^{-k}(u_i+1)} \int f(\mathbf{t}) dt, \tag{26}$$

is (by Lemma 1) continuous in $\mathbf{u}$, i.e., here

$$\ell_k(f, \mathbf{u}) = c_k^f(\mathbf{u}), \qquad \forall \mathbf{u} \in \mathbb{R}^d. \tag{27}$$

(iv)

$$(\Gamma_k f)(\mathbf{x}) := \int_{\mathbb{R}^d} \gamma_k^f(\mathbf{u}) \cdot \varphi(2^k \mathbf{x} - \mathbf{u}) \cdot d\mathbf{u}, \tag{28}$$

where

$$\gamma_k^f(\mathbf{u}) := \sum_{j_1=0}^{n_1} \cdots \sum_{j_d=0}^{n_d} w_{j_1,\ldots,j_d} \cdot f\left(\frac{u_1}{2^k} + \frac{j_1}{2^k \cdot n_1}, \ldots, \frac{u_d}{2^k} + \frac{j_d}{2^k \cdot n_d}\right), \tag{29}$$

$$(n_1, \ldots, n_d) \in \mathbb{N}^d,\ w_{j_1,\ldots,j_d} \geq 0,\ \sum_{j_1=0}^{n_1} \cdots \sum_{j_d=0}^{n_d} w_{j_1,\ldots,j_d} = 1.$$

I.e., here

$$\ell_k(f, \mathbf{u}) = \gamma_k^f(\mathbf{u}) \tag{30}$$

is continuous in $\mathbf{u} \in \mathbb{R}^d$.

First observe that the specific ℓ_k, $k \in \mathbb{Z}$ (see (22), (24), (27) and (30)) fulfill (1). Thus (7) is true for all A_k, B_k, L_k, Γ_k, so that

$$\begin{aligned} A_k(f;\mathbf{x}) &= A_0(f(2^{-k}\cdot);2^k\mathbf{x}), \\ B_k(f;\mathbf{x}) &= B_0(f(2^{-k}\cdot);2^k\mathbf{x}), \\ L_k(f;\mathbf{x}) &= L_0(f(2^{-k}\cdot);2^k\mathbf{x}), \\ \Gamma_k(f;\mathbf{x}) &= \Gamma_0(f(2^{-k}\cdot);2^k\mathbf{x}), \quad \text{all } k \in \mathbb{Z}, \text{ for all } \mathbf{x} \in \mathbb{R}^d. \end{aligned} \tag{31}$$

Note that

$$A_k(1) = B_k(1) = L_k(1) = \Gamma_k(1) = 1. \tag{32}$$

Proposition 2. *The operators A_k, B_k, L_k, Γ_k are shift invariant operators.*

Proof. For each of the above operators we need to prove (8), i.e., that

$$\ell_0(f(2^{-k}\cdot+\boldsymbol{\alpha});2^k\mathbf{u}) = \ell_0(f(2^{-k}\cdot);2^k(\mathbf{u}+\boldsymbol{\alpha})),$$

all $k \in \mathbb{Z}$, $\boldsymbol{\alpha} \in \mathbb{R}^d$ fixed, all $\mathbf{u} \in \mathbb{R}^d$; any $f \in X$.

i) For the operators A_k (φ even): Note that

$$\begin{aligned} \ell_0(f,\mathbf{x}) = r_0^f(\mathbf{x}) &= \int_{\mathbb{R}^d} f(\mathbf{t})\varphi(\mathbf{t}-\mathbf{x})d\mathbf{t} \\ &= \int_{\mathbb{R}^d} f(\mathbf{t})\varphi(\mathbf{x}-\mathbf{t})d\mathbf{t} = \int_{\mathbb{R}^d} f(\mathbf{x}-\mathbf{t})\varphi(\mathbf{t})d\mathbf{t}. \end{aligned}$$

I.e.,

$$\ell_0(f,\mathbf{x}) = \int_{\mathbb{R}^d} f(\mathbf{x}-\mathbf{t})\varphi(\mathbf{t})d\mathbf{t}. \tag{33}$$

Thus for $\boldsymbol{\alpha} \in \mathbb{R}^d$ we have

$$\begin{aligned} \ell_0(f(2^{-k}\cdot+\boldsymbol{\alpha});2^k\mathbf{u}) &= \int_{\mathbb{R}^d} f(2^{-k}\cdot(2^k\mathbf{u}-\mathbf{t})+\boldsymbol{\alpha})\varphi(\mathbf{t})d\mathbf{t} \\ &= \int_{\mathbb{R}^d} f(\mathbf{u}+\boldsymbol{\alpha}-2^{-k}\mathbf{t})\varphi(\mathbf{t})d\mathbf{t} \\ &= \int_{\mathbb{R}^d} f(2^{-k}\cdot 2^k(\mathbf{u}+\boldsymbol{\alpha})-2^{-k}\mathbf{t})\varphi(\mathbf{t})d\mathbf{t} \\ &= \int_{\mathbb{R}^d} f(2^{-k}\cdot(2^k(\mathbf{u}+\boldsymbol{\alpha})-\mathbf{t}))\varphi(\mathbf{t})d\mathbf{t} \\ &= \ell_0(f(2^{-k}\cdot);2^k\cdot(\mathbf{u}+\boldsymbol{\alpha})). \end{aligned}$$

So r_0^f fulfills (8), i.e., A_k is a shift invariant operator.

ii) For the operators B_k: Here $\ell_0 f = f$. Thus

$$\ell_0(f(2^{-k}\cdot+\boldsymbol{\alpha});2^k\cdot\mathbf{u}) = f(2^{-k}2^k\mathbf{u}+\boldsymbol{\alpha}) = f(\mathbf{u}+\boldsymbol{\alpha}) = \ell_0(f(2^{-k}\cdot);2^k\cdot(\mathbf{u}+\boldsymbol{\alpha})),$$

i.e., B_k is a shift invariant operator.

iii) For the operators L_k: Here

$$\ell_0(f,\mathbf{x}) = \int_{x_1}^{x_1+1} \cdots \int_{x_d}^{x_d+1} f(\mathbf{t})d\mathbf{t}, \quad \mathbf{x} = (x_1, \ldots, x_d), \ \mathbf{t} = (t_1, \ldots, t_d).$$

Then

$$\begin{aligned}\ell_0(f(2^{-k}\cdot+\boldsymbol{\alpha}); 2^k\mathbf{u}) &= \int_{2^k u_1}^{2^k u_1+1} \cdots \int_{2^k u_d}^{2^k u_d+1} f(2^{-k}\mathbf{t}+\boldsymbol{\alpha})d\mathbf{t} \\ &= \int_{2^k(u_1+\alpha_1)}^{2^k(u_1+\alpha_1)+1} \cdots \int_{2^k(u_d+\alpha_d)}^{2^k(u_d+\alpha_d)+1} f(2^{-k}\cdot\mathbf{t})d\mathbf{t} \\ &= \ell_0(f(2^{-k}\cdot); 2^k\cdot(\mathbf{u}+\boldsymbol{\alpha})), \quad \boldsymbol{\alpha} = (\alpha_1, \ldots, \alpha_d).\end{aligned}$$

Therefore L_k is a shift invariant operator.

iv) For the operators Γ_k:

$$\ell_0(f,\mathbf{u}) = \sum_{j_1=0}^{n_1} \cdots \sum_{j_d=0}^{n_d} w_{j_1,\ldots,j_d} \cdot f\left(u_1 + \frac{j_1}{n_1}, \ldots, u_d + \frac{j_d}{n_d}\right).$$

Thus

$$\begin{aligned}&\ell_0(f(2^{-k}\cdot+\boldsymbol{\alpha}); 2^k\cdot\mathbf{u}) \\ &= \sum_{j_1=0}^{n_1} \cdots \sum_{j_d=0}^{n_d} w_{j_1,\ldots,j_d} \cdot f\left(2^{-k}\left(2^k u_1 + \frac{j_1}{n_1}\right) + \alpha_1, \ldots, 2^{-k}\left(2^k u_d + \frac{j_d}{n_d}\right) + \alpha_d\right) \\ &= \sum_{j_1=0}^{n_1} \cdots \sum_{j_d=0}^{n_d} w_{j_1,\ldots,j_d} \cdot f\left(2^{-k}\left(2^k(u_1+\alpha_1) + \frac{j_1}{n_1}\right), \ldots, 2^{-k}\left(2^k(u_d+\alpha_d) + \frac{j_d}{n_d}\right)\right) \\ &= \ell_0(f(2^{-k}\cdot); 2^k\cdot(\mathbf{u}+\boldsymbol{\alpha})).\end{aligned}$$

That is, Γ_k is also a shift invariant operator. Q.E.D.

In the following we establish that the operators A_k, B_k, L_k, Γ_k fulfill the property of global smoothness preservation.

Theorem 5. *For all $f \in C_U(\mathbb{R}^d)$, $d \geq 1$, and all $\delta > 0$ we have*

$$\begin{aligned}&\text{i)} \quad \omega_1(A_k f;\delta) \leq \omega_1(f;\delta), \\ &\text{ii)} \quad \omega_1(B_k f;\delta) \leq \omega_1(f;\delta), \\ &\text{iii)} \quad \omega_1(L_k f;\delta) \leq \omega_1(f;\delta), \\ &\text{iv)} \quad \omega_1(\Gamma_k f;\delta) \leq \omega_1(f;\delta).\end{aligned} \tag{34}$$

Proof. Here we apply Theorem 1. We need only verify condition (9):

$$|\ell_0(f;\mathbf{x}-\mathbf{u}) - \ell_0(f;\mathbf{y}-\mathbf{u})| \leq \omega_1(f;\|\mathbf{x}-\mathbf{y}\|), \quad \text{any } \mathbf{x},\mathbf{y},\mathbf{u} \in \mathbb{R}^d.$$

i) For the operators A_k (φ even): From (33) we get

$$\begin{aligned}|\ell_0(f;\mathbf{x}-\mathbf{u}) - \ell_0(f;\mathbf{y}-\mathbf{u})| &= \left|\int_{\mathbb{R}^d} f(\mathbf{x}-\mathbf{u}-\mathbf{t})\varphi(\mathbf{t})dt - \int_{\mathbb{R}^d} f(\mathbf{y}-\mathbf{u}-\mathbf{t})\varphi(\mathbf{t})dt\right| \\ &\leq \int_{\mathbb{R}^d} |f(\mathbf{x}-\mathbf{u}-\mathbf{t}) - f(\mathbf{y}-\mathbf{u}-\mathbf{t})|\varphi(\mathbf{t})dt \\ &\leq \int_{\mathbb{R}^d} \omega_1(f;\|\mathbf{x}-\mathbf{y}\|)\varphi(\mathbf{t})dt \\ &= \omega_1(f;\|\mathbf{x}-\mathbf{y}\|)\cdot 1,\end{aligned}$$

the last equality being true by (4). I.e., (9) holds for the operators A_k.

ii) For B_k operators: Here $\ell_0 f = f$, thus

$$|\ell_0(f;\mathbf{x}-\mathbf{u}) - \ell_0(f;\mathbf{y}-\mathbf{u})| = |f(\mathbf{x}-\mathbf{u}) - f(\mathbf{y}-\mathbf{u})| \leq \omega_1(f;\|\mathbf{x}-\mathbf{y}\|),$$

proving (9) for the operators B_k.

iii) For operators L_k: we have

$$\begin{aligned}
|\ell_0(f;\mathbf{x}-\mathbf{u}) - \ell_0(f;\mathbf{y}-\mathbf{u})| &= \left|\int_{\mathbf{x}-\mathbf{u}}^{\mathbf{x}-\mathbf{u}+\mathbf{1}} f(\mathbf{t})d\mathbf{t} - \int_{\mathbf{y}-\mathbf{u}}^{\mathbf{y}-\mathbf{u}+\mathbf{1}} f(\mathbf{t})d\mathbf{t}\right| \\
&= \left|\int_0^1 \cdots \int_0^1 (f(\mathbf{x}-\mathbf{u}+\mathbf{t}) - f(\mathbf{y}-\mathbf{u}+\mathbf{t}))d\mathbf{t}\right| \\
&\leq \int_0^1 \cdots \int_0^1 |f(\mathbf{x}-\mathbf{u}+\mathbf{t}) - f(\mathbf{y}-\mathbf{u}+\mathbf{t})|d\mathbf{t} \\
&\leq \int_0^1 \cdots \int_0^1 \omega_1(f;\|\mathbf{x}-\mathbf{y}\|)d\mathbf{t} \\
&= \omega_1(f;\|\mathbf{x}-\mathbf{y}\|),
\end{aligned}$$

establishing (9) for the operators L_k.

iv) For the operators Γ_k we see that

$$\begin{aligned}
&|\ell_0(f;\mathbf{x}-\mathbf{u}) - \ell_0(f;\mathbf{y}-\mathbf{u})| \\
&= \left|\sum_{j_1=0}^{n_1} \cdots \sum_{j_d=0}^{n_d} w_{j_1,\dots,j_d} \cdot f\left(x_1 - u_1 + \frac{j_1}{n_1}, \dots, x_d - u_d + \frac{j_d}{n_d}\right)\right. \\
&\quad \left. - \sum_{j_1=0}^{n_1} \cdots \sum_{j_d=0}^{n_d} w_{j_1,\dots,j_d} \cdot f\left(y_1 - u_1 + \frac{j_1}{n_1}, \dots, y_d - u_d + \frac{j_d}{n_d}\right)\right| \\
&\leq \sum_{j_1=0}^{n_1} \cdots \sum_{j_d=0}^{n_d} w_{j_1,\dots,j_d} \cdot \omega_1(f;\|\mathbf{x}-\mathbf{y}\|) \\
&\leq \omega_1(f;\|\mathbf{x}-\mathbf{y}\|) \cdot 1,
\end{aligned}$$

proving (9) for the operators Γ_k. Q.E.D.

Theorem 6. *Inequalities* (34) *are sharp, in the sense that they are attained by the projection functions.*

Proof. i) Operators A_k: Observe that

$$\ell_0(f;\mathbf{u}) = \int_{\mathbb{R}^d} f(\mathbf{t})\varphi(\mathbf{t}-\mathbf{u})d\mathbf{t} = \int_{\mathbb{R}^d} f(\mathbf{s}+\mathbf{u})\varphi(\mathbf{s})d\mathbf{s},$$

by setting $\mathbf{s} = \mathbf{t} - \mathbf{u}$. Hence

$$\begin{aligned}
&\ell_0(pr_i(2^{-k}\cdot); 2^k\mathbf{x} - \mathbf{u}) - \ell_0(pr_i(2^{-k}\cdot); 2^k\mathbf{y} - \mathbf{u}) \\
&= \int_{-\infty}^{+\infty} \cdots \int_{-\infty}^{+\infty} [(x_i - 2^{-k}u_i) - (y_i - 2^{-k}u_i)] \cdot \varphi(\mathbf{s}) \cdot d\mathbf{s} \\
&= (x_i - y_i) \cdot \int_{\mathbb{R}^d} \varphi(\mathbf{s})d\mathbf{s} = (x_i - y_i) \cdot 1 = x_i - y_i,
\end{aligned}$$

by (4), i.e., we have established

$$\ell_0(pr_i(2^{-k}\cdot); 2^k\mathbf{x} - \mathbf{u}) - \ell_0(pr_i(2^{-k}\cdot); 2^k\mathbf{y} - \mathbf{u}) = x_i - y_i, \quad i \in \{1, \dots, d\}.$$

So, from Corollary 2 we get

$$\omega_1(A_k pr_i;\delta) = \omega_1(pr_i;\delta),$$

$i \in \{1,\dots,d\}$, any $\delta > 0$, proving that inequality (34)(i) is sharp.

ii) Operators B_k: Here $\ell_0(f;\mathbf{x}) = f(\mathbf{x})$. Thus

$$\begin{aligned}
&\ell_0(pr_i(2^{-k}\cdot);2^k\mathbf{x}-\mathbf{u}) - \ell_0(pr_i(2^{-k}\cdot);2^k\mathbf{y}-\mathbf{u}) \\
&\quad = pr_i(2^{-k}\cdot)(2^k\mathbf{x}-\mathbf{u}) - pr_i(2^{-k}\cdot)(2^k\mathbf{y}-\mathbf{u}) \\
&\quad = x_i - 2^{-k}u_i - (y_i - 2^{-k}u_i) = x_i - y_i.
\end{aligned}$$

And from Corollary 2 we obtain

$$\omega_1(B_k pr_i;\delta) = \omega_1(pr_i;\delta),$$

$i \in \{1,\dots,d\}$, any $\delta > 0$, proving that inequality (34)(ii) is sharp.

iii) Operators L_k: Here

$$\ell_0(f;\mathbf{u}) = \int_{\mathbf{u}}^{\mathbf{u}+1} f(\mathbf{t})d\mathbf{t}.$$

We have

$$\begin{aligned}
&\ell_0(pr_i(2^{-k}\cdot);2^k\mathbf{x}-\mathbf{u}) - \ell_0(pr_i(2^{-k}\cdot);2^k\mathbf{y}-\mathbf{u}) \\
&\quad = \int_{2^k\mathbf{x}-\mathbf{u}}^{2^k\mathbf{x}-\mathbf{u}+1} pr_i(2^{-k}\cdot)(\mathbf{t})d\mathbf{t} - \int_{2^k\mathbf{y}-\mathbf{u}}^{2^k\mathbf{y}-\mathbf{u}+1} pr_i(2^{-k}\cdot)(\mathbf{t})d\mathbf{t} \\
&\quad = \int_{2^k\mathbf{x}-\mathbf{u}}^{2^k\mathbf{x}-\mathbf{u}+1} 2^{-k}\cdot t_i \cdot d\mathbf{t} - \int_{2^k\mathbf{y}-\mathbf{u}}^{2^k\mathbf{y}-\mathbf{u}+1} 2^{-k}\cdot t_i \cdot d\mathbf{t} \\
&\quad = \int_0^1 \cdots \int_0^1 2^{-k}\cdot(2^k x_i - u_i + t_i)\cdot d\mathbf{t} \\
&\qquad - \int_0^1 \cdots \int_0^1 2^{-k}\cdot(2^k y_i - u_i + t_i)\cdot d\mathbf{t} \\
&\quad = \int_0^1 \cdots \int_0^1 (x_i - y_i)d\mathbf{t} = x_i - y_i.
\end{aligned}$$

Thus Corollary 2 implies that

$$\omega_1(L_k pr_i;\delta) = \omega_1(pr_i;\delta),$$

$i \in \{1,\dots,d\}$, any $\delta > 0$, establishing that inequality (34)(iii) is sharp.

iv) Operators Γ_k: Here

$$\ell_0(f;\mathbf{u}) = \sum_{j_1=0}^{n_1} \cdots \sum_{j_d=0}^{n_d} w_{j_1,\dots,j_d}\cdot f\left(u_1 + \frac{j_1}{n_1},\dots,u_d + \frac{j_d}{n_d}\right).$$

We see that

$$\begin{aligned}
&\ell_0(pr_i(2^{-k}\cdot);2^k\mathbf{x}-\mathbf{u}) - \ell_0(pr_i(2^{-k}\cdot);2^k\mathbf{y}-\mathbf{u}) \\
&\quad = \sum_{j_1=0}^{n_1} \cdots \sum_{j_d=0}^{n_d} w_{j_1,\dots,j_d}\cdot\left[\left(x_i - 2^{-k}u_i + 2^{-k}\frac{j_i}{n_i}\right) - \left(y_i - 2^{-k}u_i + 2^{-k}\frac{j_i}{n_i}\right)\right] \\
&\quad = \sum_{j_1=0}^{n_1} \cdots \sum_{j_d=0}^{n_d} w_{j_1,\dots,j_d}\cdot(x_i - y_i) = (x_i - y_i)\cdot 1 \\
&\quad = x_i - y_i,
\end{aligned}$$

by

$$\sum_{j_1=0}^{n_1}\cdots\sum_{j_d=0}^{n_d} w_{j_1,\ldots,j_d} = 1.$$

Therefore Corollary 2 implies that

$$\omega_1(\Gamma_k pr_i; \delta) = \omega_1(pr_i; \delta),$$

$i \in \{1, \ldots, d\}$, any $\delta > 0$, establishing that inequality (34)(iv) is sharp. Q.E.D.

The operators A_k, B_k, L_k, Γ_k; $k \in \mathbb{Z}$, converge to the unit operator I with rates as given next, as $k \to +\infty$.

Theorem 7. *We obtain* ($k \in \mathbb{Z}$, $a := \max_{1\le i\le d}(a_i)$):

$$|A_k(f;\mathbf{x}) - f(\mathbf{x})| \le \omega_{1,\infty}\left(f, \frac{a}{2^{k-1}}\right), \tag{35}$$

$$|B_k(f;\mathbf{x}) - f(\mathbf{x})| \le \omega_{1,\infty}\left(f, \frac{a}{2^{k}}\right), \tag{36}$$

$$|L_k(f;\mathbf{x}) - f(\mathbf{x})| \le \omega_{1,\infty}\left(f, \frac{1+a}{2^{k}}\right), \tag{37}$$

$$|\Gamma_k(f;\mathbf{x}) - f(\mathbf{x})| \le \omega_{1,\infty}\left(f, \frac{1+a}{2^{k}}\right). \tag{38}$$

Proof. According to Theorem 2 we only need to prove inequality (2), i.e.,

$$\sup_{\substack{\mathbf{u},\mathbf{y}\in\mathbb{R}^d\\ \|\mathbf{u}-\mathbf{y}\|_\infty\le a}} |\ell_0(f,\mathbf{u}) - f(\mathbf{y})| \le \omega_{1,\infty}\left(f; \frac{ma+n}{2^r}\right),$$

for appropriate values of m, n and r.

i) For operators A_k (φ even): We see that

$$\ell_0(f,\mathbf{u}) = r_0^f(\mathbf{u}) = \int_{\mathbb{R}^d} f(\mathbf{t})\varphi(\mathbf{t}-\mathbf{u})d\mathbf{t} = \int_{\mathbb{R}^d} f(\mathbf{t})\varphi(\mathbf{u}-\mathbf{t})d\mathbf{t}.$$

Hence,

$$\begin{aligned}
&\sup_{\substack{\mathbf{u},\mathbf{y}\in\mathbb{R}^d\\ \|\mathbf{u}-\mathbf{y}\|_\infty\le a}} |\ell_0(f,\mathbf{u}) - f(\mathbf{y})| \\
&= \sup_{\substack{\mathbf{u},\mathbf{y}\in\mathbb{R}^d\\ \|\mathbf{u}-\mathbf{y}\|_\infty\le a}} \left|\int_{\mathbb{R}^d} f(\mathbf{t})\varphi(\mathbf{u}-\mathbf{t})d\mathbf{t} - \int_{\mathbb{R}^d} f(\mathbf{y})\varphi(\mathbf{u}-\mathbf{t})d\mathbf{t}\right| \\
&= \sup_{\substack{\mathbf{u},\mathbf{y}\in\mathbb{R}^d\\ \|\mathbf{u}-\mathbf{y}\|_\infty\le a}} \left|\int_{\mathbb{R}^d} (f(\mathbf{t}) - f(\mathbf{y}))\varphi(\mathbf{u}-\mathbf{t})d\mathbf{t}\right| \\
&\le \sup_{\substack{\mathbf{u},\mathbf{y}\in\mathbb{R}^d\\ \|\mathbf{u}-\mathbf{y}\|_\infty\le a}} \int_{\mathbb{R}^d} |f(\mathbf{t}) - f(\mathbf{y})|\varphi(\mathbf{u}-\mathbf{t})d\mathbf{t} \\
&\le \sup_{\substack{\mathbf{u},\mathbf{y}\in\mathbb{R}^d\\ \|\mathbf{u}-\mathbf{y}\|_\infty\le a}} \int_{\mathbb{R}^d} \omega_{1,\infty}(f, \|\mathbf{t}-\mathbf{y}\|_\infty)\varphi(\mathbf{u}-\mathbf{t})d\mathbf{t} \\
&\le \int_{\mathbb{R}^d} \omega_{1,\infty}(f, 2a)\varphi(\mathbf{u}-\mathbf{t})d\mathbf{t} \\
&= \omega_{1,\infty}(f, 2a) \cdot \int_{\mathbb{R}^d} \varphi(\mathbf{u}-\mathbf{t})d\mathbf{t} \\
&= \omega_{1,\infty}(f, 2a), \quad \text{by (3)}.
\end{aligned}$$

That is

$$\sup_{\substack{\mathbf{u},\mathbf{y}\in\mathbb{R}^d\\ \|\mathbf{u}-\mathbf{y}\|_\infty\le a}} |\ell_0(f,\mathbf{u}) - f(\mathbf{y})| \le \omega_{1,\infty}(f,2a).$$

Now by Theorem 2, inequality (35) is established.

ii) For operators B_k: Here $\ell_0(f,\mathbf{u}) = f(\mathbf{u})$ and

$$\sup_{\substack{\mathbf{u},\mathbf{y}\in\mathbb{R}^d\\ \|\mathbf{u}-\mathbf{y}\|_\infty\le a}} |\ell_0(f,\mathbf{u}) - f(\mathbf{y})| = \sup_{\substack{\mathbf{u},\mathbf{y}\in\mathbb{R}^d\\ \|\mathbf{u}-\mathbf{y}\|_\infty\le a}} |f(\mathbf{u}) - f(\mathbf{y})| = \omega_{1,\infty}(f;a).$$

And from Theorem 2, inequality (36) is true.

iii) For operators L_k: Here

$$\ell_0(f,\mathbf{u}) = c_0^f(\mathbf{u}) = \int_{\mathbf{u}}^{\mathbf{u}+\mathbf{1}} f(\mathbf{t})d\mathbf{t}.$$

And

$$\begin{aligned}
&\sup_{\substack{\mathbf{u},\mathbf{y}\in\mathbb{R}^d\\ \|\mathbf{u}-\mathbf{y}\|_\infty\le a}} |\ell_0(f,\mathbf{u}) - f(\mathbf{y})| \\
&= \sup_{\substack{\mathbf{u},\mathbf{y}\in\mathbb{R}^d\\ \|\mathbf{u}-\mathbf{y}\|_\infty\le a}} \left|\int_{\mathbf{u}}^{\mathbf{u}+\mathbf{1}} f(\mathbf{t})d\mathbf{t} - f(\mathbf{y})\right| \\
&= \sup_{\substack{\mathbf{u},\mathbf{y}\in\mathbb{R}^d\\ \|\mathbf{u}-\mathbf{y}\|_\infty\le a}} \left|\int_{\mathbf{u}}^{\mathbf{u}+\mathbf{1}} (f(\mathbf{t}) - f(\mathbf{y}))d\mathbf{t}\right| \\
&\le \sup_{\substack{\mathbf{u},\mathbf{y}\in\mathbb{R}^d\\ \|\mathbf{u}-\mathbf{y}\|_\infty\le a}} \int_{\mathbf{u}}^{\mathbf{u}+\mathbf{1}} |f(\mathbf{t}) - f(\mathbf{y})|d\mathbf{t} \\
&= \sup_{\substack{\mathbf{u},\mathbf{y}\in\mathbb{R}^d\\ \|\mathbf{u}-\mathbf{y}\|_\infty\le a}} \int_0^1 \cdots \int_0^1 |f(\mathbf{u}+\mathbf{t}) - f(\mathbf{y})|d\mathbf{t} \\
&\le \sup_{\substack{\mathbf{u},\mathbf{y}\in\mathbb{R}^d\\ \|\mathbf{u}-\mathbf{y}\|_\infty\le a}} \int_0^1 \cdots \int_0^1 \omega_{1,\infty}(f;\|\mathbf{u}+\mathbf{t}-\mathbf{y}\|_\infty)d\mathbf{t} \\
&\le \sup_{\substack{\mathbf{u},\mathbf{y}\in\mathbb{R}^d\\ \|\mathbf{u}-\mathbf{y}\|_\infty\le a}} \int_0^1 \cdots \int_0^1 \omega_{1,\infty}(f;\|\mathbf{t}\|_\infty + \|\mathbf{u}-\mathbf{y}\|_\infty)d\mathbf{t} \\
&\le \sup_{\substack{\mathbf{u},\mathbf{y}\in\mathbb{R}^d\\ \|\mathbf{u}-\mathbf{y}\|_\infty\le a}} \int_0^1 \cdots \int_0^1 \omega_{1,\infty}(f;1 + \|\mathbf{u}-\mathbf{y}\|_\infty)d\mathbf{t} \\
&= \omega_{1,\infty}(f;1+a).
\end{aligned}$$

I.e., we have proved that

$$\sup_{\substack{\mathbf{u},\mathbf{y}\in\mathbb{R}^d\\ \|\mathbf{u}-\mathbf{y}\|_\infty\le a}} |\ell_0(f,\mathbf{u}) - f(\mathbf{y})| \le \omega_{1,\infty}(f;1+a).$$

Now from Theorem 2 we get the validity of (37).

iv) For operators Γ_k: Here

$$\ell_0(f,\mathbf{u}) = \gamma_0^f(\mathbf{u}) = \sum_{j_1=0}^{n_1} \cdots \sum_{j_d=0}^{n_d} w_{j_1,\dots,j_d} \cdot f\left(u_1 + \frac{j_1}{n_1}, \dots, u_d + \frac{j_d}{n_d}\right),$$

$$w_{j_1,\dots,j_d} \geq 0 \quad \text{and} \quad \sum_{j_1=0}^{n_1} \cdots \sum_{j_d=0}^{n_d} w_{j_1,\dots,j_d} = 1.$$

Thus

$$\begin{aligned}
&\sup_{\substack{\mathbf{u},\mathbf{y}\in\mathbb{R}^d \\ \|\mathbf{u}-\mathbf{y}\|_\infty \leq a}} |\ell_0(f,\mathbf{u}) - f(\mathbf{y})| \\
&= \sup_{\substack{\mathbf{u},\mathbf{y}\in\mathbb{R}^d \\ \|\mathbf{u}-\mathbf{y}\|_\infty \leq a}} \left| \sum_{j_1=0}^{n_1} \cdots \sum_{j_d=0}^{n_d} w_{j_1,\dots,j_d} \cdot f\left(u_1 + \frac{j_1}{n_1}, \dots, u_d + \frac{j_d}{n_d}\right) - f(\mathbf{y}) \right| \\
&= \sup_{\substack{\mathbf{u},\mathbf{y}\in\mathbb{R}^d \\ \|\mathbf{u}-\mathbf{y}\|_\infty \leq a}} \left| \sum_{j_1=0}^{n_1} \cdots \sum_{j_d=0}^{n_d} w_{j_1,\dots,j_d} \cdot \left(f\left(u_1 + \frac{j_1}{n_1}, \dots, u_d + \frac{j_d}{n_d}\right) - f(\mathbf{y})\right) \right| \\
&\leq \sup_{\substack{\mathbf{u},\mathbf{y}\in\mathbb{R}^d \\ \|\mathbf{u}-\mathbf{y}\|_\infty \leq a}} \sum_{j_1=0}^{n_1} \cdots \sum_{j_d=0}^{n_d} w_{j_1,\dots,j_d} \cdot \left| f\left(u_1 + \frac{j_1}{n_1}, \dots, u_d + \frac{j_d}{n_d}\right) - f(\mathbf{y}) \right| \\
&\leq \sup_{\substack{\mathbf{u},\mathbf{y}\in\mathbb{R}^d \\ \|\mathbf{u}-\mathbf{y}\|_\infty \leq a}} \sum_{j_1=0}^{n_1} \cdots \sum_{j_d=0}^{n_d} w_{j_1,\dots,j_d} \cdot \omega_1\left(f; \left\|\mathbf{u} + \frac{\mathbf{j}}{\mathbf{n}} - \mathbf{y}\right\|_\infty\right) \quad \left(\text{where } \frac{j}{n} := \left(\frac{j_1}{n_1}, \dots, \frac{j_d}{n_d}\right)\right) \\
&\leq \sup_{\substack{\mathbf{u},\mathbf{y}\in\mathbb{R}^d \\ \|\mathbf{u}-\mathbf{y}\|_\infty \leq a}} \sum_{j_1=0}^{n_1} \cdots \sum_{j_d=0}^{n_d} w_{j_1,\dots,j_d} \cdot \omega_1\left(f; \left\|\frac{\mathbf{j}}{\mathbf{n}}\right\|_\infty + \|\mathbf{u}-\mathbf{y}\|_\infty\right) \\
&\leq \left(\sum_{j_1=0}^{n_1} \cdots \sum_{j_d=0}^{n_d} w_{j_1,\dots,j_d} \right) \cdot \omega_{1,\infty}(f; 1 + a) \\
&= \omega_{1,\infty}(f; a+1).
\end{aligned}$$

I.e.,

$$\sup_{\substack{\mathbf{u},\mathbf{y}\in\mathbb{R}^d \\ \|\mathbf{u}-\mathbf{y}\|_\infty \leq a}} |\ell_0(f,\mathbf{u}) - f(\mathbf{y})| \leq \omega_{1,\infty}(f; a+1).$$

Finally, from Theorem 2 we find inequality (38). Q.E.D.

Remark 3. We have that

$$(r_0^f)(\mathbf{u}) = \int_{-\mathbf{a}}^{\mathbf{a}} f(\mathbf{u} - \mathbf{t})\varphi(\mathbf{t})d\mathbf{t}, \qquad \mathbf{a} := (a_1, \dots, a_d),$$

where φ is even and continuous, $(\ell_0 f)(\mathbf{u}) = f(\mathbf{u})$ for the operator B_0,

$$(c_0^f)(\mathbf{u}) = \int_0^1 \cdots \int_0^1 f(\mathbf{u} + \mathbf{t})d\mathbf{t},$$

and

$$\gamma_0^f(\mathbf{u}) = \sum_{j_1=0}^{n_1} \cdots \sum_{j_d=0}^{n_d} w_{j_1,\dots,j_d} \cdot f\left(u_1 + \frac{j_1}{n_1}, \dots, u_d + \frac{j_d}{n_d}\right).$$

In the last three cases we will assume that φ is bounded. Let $i_1, \dots, i_k \in \{1, \dots, d\}$ be such that $i_1 < \cdots < i_k$; $j_1, \dots, j_k \in \mathbb{Z}_+$. Assume that

$$\frac{\partial^{\sum_{k=1}^{k} j_k} f(\mathbf{x})}{\partial x_{i_1}^{j_1} \cdots \partial x_{i_k}^{j_k}}$$

exists, is continuous and ≥ 0, for all $\mathbf{x} \in \mathbb{R}^d$. Then

$$\frac{\partial^{\sum_{k=1}^{k} j_k}}{\partial x_{i_1}^{j_1} \cdots \partial x_{i_k}^{j_k}} \begin{Bmatrix} r_0^f(\mathbf{x}), \\ c_0^f(\mathbf{x}), \\ \gamma_0^f(\mathbf{x}) \end{Bmatrix}$$

exist, are continuous (by continuity of f, and Lemma 1 below and Corollary 3) and ≥ 0, for all $\mathbf{x} \in \mathbb{R}^d$. And, thus, by Theorem 3 there exist

$$\frac{\partial^{\sum_{k=1}^{k} j_k}}{\partial x_{i_1}^{j_1} \cdots \partial x_{i_k}^{j_k}} \begin{Bmatrix} (A_0 f)(\mathbf{x}), \\ (B_0 f)(\mathbf{x}), \\ (L_0 f)(\mathbf{x}), \\ (\Gamma_0 f)(\mathbf{x}) \end{Bmatrix} \geq 0, \qquad \text{for all } \mathbf{x} \in \mathbb{R}^d.$$

Conclusion. Let $i_1, \ldots, i_k \in \{1, \ldots, d\}$ be such that $i_1 < \cdots < i_k$; $j_1, \ldots, j_k \in \mathbb{Z}_+$. Assume that

$$\frac{\partial^{\sum_{k=1}^{k} j_k}}{\partial x_{i_1}^{j_1} \cdots \partial x_{i_k}^{j_k}} f(\mathbf{x})$$

exists, is continuous and ≥ 0, for all $\mathbf{x} \in \mathbb{R}^d$; $f \in C_U(\mathbb{R}^d)$. Then

$$\frac{\partial^{\sum_{k=1}^{k} j_k}}{\partial x_{i_1}^{j_1} \cdots \partial x_{i_k}^{j_k}} \begin{Bmatrix} A_k f, \\ B_k f, \\ L_k f, \\ \Gamma_k f \end{Bmatrix} \geq 0, \qquad \text{for all } k \in \mathbb{Z}.$$

Here again we assume for A_k that the function φ is even and continuous; for the other operators it only has to be Lebesgue measurable and bounded. We always assume $\varphi \geq 0$.

Lemma 1. *If $F(\mathbf{x}, \mathbf{u})$ is continuous on $\mathbb{R}^d \times \times_{i=1}^d [-a_i, a_i]$, then*

$$\int_{-a_1}^{a_1} \cdots \int_{-a_d}^{a_d} F(\mathbf{x}, \mathbf{u}) d\mathbf{u}$$

is continuous in $\mathbf{x} \in \mathbb{R}^d$.

Proof. Let $\mathbf{x} \in \mathbb{R}^d$, $\mathbf{u} \in \times_{i=1}^d [-a_i, a_i]$, then F is uniformly continuous on

$$\left(\times_{i=1}^d [x_i - 1, x_i + 1]\right) \times \left(\times_{i=1}^d [-a_i, a_i]\right), \quad \text{where } \mathbf{x} := (x_1, \ldots, x_d).$$

Choose $\|h\|_\infty \leq 1$, where $\|\cdot\|_\infty$ is the supremum norm on $\mathbb{R}^d$. Uniform continuity implies

$$|F(\mathbf{x} + \mathbf{h}, \mathbf{u}) - F(\mathbf{x}, \mathbf{u})| < \varepsilon,$$

given that

$$\|(\mathbf{x} + \mathbf{h}, \mathbf{u}) - (\mathbf{x}, \mathbf{u})\| < \delta,$$

where $\|\cdot\|$ is a norm in $\mathbb{R}^{2d}$. Note that

$$\|(\mathbf{x} + \mathbf{h}, \mathbf{u}) - (\mathbf{x}, \mathbf{u})\| = \|(\mathbf{h}, \mathbf{0})\| \leq \|\mathbf{h}\|_\infty \cdot c,$$

where $c := \sum_{i=1}^d \|(0, \ldots, 1_i, \ldots, 0, \mathbf{0})\|$; here 1_i means 1 at the i-entry. Hence

$$\left| \int \cdots \int_{-a_i}^{a_i} \cdots \int (F(\mathbf{x} + \mathbf{h}, \mathbf{u}) - F(\mathbf{x}, \mathbf{u})) d\mathbf{u} \right|$$

$$\leq \int \cdots \int_{-a_i}^{a_i} \cdots \int |F(\mathbf{x} + \mathbf{h}, \mathbf{u}) - F(\mathbf{x}, \mathbf{u})| \cdot d\mathbf{u} < \varepsilon \cdot 2^d \cdot \prod_{i=1}^d a_i. \qquad \text{Q.E.D.}$$

Remark 4. Let $\varphi \geq 0$, φ continuous on $\times_{i=1}^{d}[-a_i, a_i]$, $a_i > 0$, supp $\varphi \subseteq \times_{i=1}^{d}[-a_i, a_i]$ and $\int_{\mathbb{R}^d} \varphi(\mathbf{x} - \mathbf{u})d\mathbf{u} = 1$, for all $\mathbf{x} \in \mathbb{R}^d$. Let $f \in C(\mathbb{R}^d)$ be a probability distribution function, then r_0^f, c_0^f, γ_0^f are continuous distribution functions. Furthermore, by Theorem 4 we get that $A_k f$, $B_k f$, $L_k f$, $\Gamma_k f$ $(k \in \mathbb{Z})$ are continuous distribution functions.

Acknowledgement: The authors gratefully acknowledge the hospitality of the Mathematics Research Institute at Oberwolfach where much of the research of this article was done during a most pleasant visit. They also express their sincere thanks to the Mayor of the town of Nea Makri (Greece) for providing them with the facilities needed during the Summer of 1993 where this paper was finished.

REFERENCES

1. G. Anastassiou, C. Cottin and H. Gonska, Global smoothness of approximating functions, *Analysis*, Vol. 11, 43–57, (1991).
2. G. Anastassiou, C. Cottin and H. Gonska, Global smoothness preservation by multivariate approximation operators, Israel Mathematical Conference Proc., Weizmann Science Press, Vol. 4, 31–44, (1991).
3. G. Anastassiou and H. Gonska, On some shift-invariant integral operators, univariate case, submitted for publication.
4. H. Bauer, *Maß-und Integrationstheorie*, de Gruyter, Berlin (1990).
5. J.F.C. Kingman and S.J. Taylor, *Introduction to Measure and Probability*, Cambridge University Press, Cambridge, U.K. (1966).
6. A.N. Shiryayev, *Probability*, Springer-Verlag, New York–Berlin (1984).

MULTIVARIATE PROBABILISTIC WAVELET APPROXIMATION

G.A. Anastassiou,[1] S.T. Rachev,[2] and X.M. Yu[3]

[1]Department of Mathematical Sciences
Memphis State University
Memphis, Tennessee 38152
[2]Department of Mathematics
University of California at Santa Barbara
Santa Barbara, California 93106
[3]Department of Mathematics
Southwest Missouri State University
Springfield, Missouri 65803

Abstract. Multivariate probability laws are approximated by some naturally arising wavelet operations. They transform multivariate distribution functions to multivariable distribution functions. The degree of this approximation is estimated by establishing some sharp local Jackson type inequalities. We extend the results for distributions of stochastic processes and unbounded measures.

1. INTRODUCTION

We are interested in approximation of probability distributions by distributions with wavelet structure. We refer to the survey papers by Daubechies (1990), S. Mallat (1989), Anastassiou and Yu (1991) for the general idea of wavelets and probabilistic wavelets approximation.

We shall construct wavelets that generate a dense (in the sense of weak *-topology) family in the set of probability laws in $\mathbb{R}^r$, $1 \leq r \leq \infty$. We extend this result by studying the approximations of Borel measures finite on the ring of all bounded Borel subsets of $\mathbb{R}^r$ with respect to the vague convergence (see Rachev (1990), Ch. 10, and the references there). We apply our result to approximation of stochastic process by a process with distribution that possesses a wavelet structure.

2. APPROXIMATION OF PROBABILITY LAWS BY WAVELETS

Given a probability P on $\mathbb{R}^r$, $1 \le r \le \infty$, a Borel probability measure on $\mathbb{R}^r$, let $F = F_P$ be the corresponding distribution function,

$$F(x) = P((-\infty, x]), \qquad x \in \mathbb{R}^r. \tag{1}$$

Let φ be a bounded compactly supported function on $\mathbb{R}^r$, supp $\varphi \subset [-a, a]$, $a \in \mathbb{R}_+^r$, and right continuous with respect to all variables. We want to approximate F on $\mathbb{R}^r$ by the linear combinations of the translated dilates of φ. Define

$$B_k(F)(x) = \sum_{j \in \mathbb{Z}^r} F(2^{-k}j)\varphi(2^k x - j) \tag{2}$$

on $\mathbb{R}^r$. Since φ is compactly supported for any $x \in \mathbb{R}^r$ the summation in (2) only involves finite terms, so $B_k(F)$ is well-defined on $\mathbb{R}^r$. We shall show that under some regularity conditions on φ, $B_k(F)$ is a distribution function itself. Besides, the Kolmogorov (uniform) distance between F and $B_k(F)$

$$\rho(F, B_k(F)) = \sup_{x \in \mathbb{R}^r} |F(x) - B_k(F)(x)|$$

is bounded by the modulus of continuity

$$w(f, 2^{-k}d) = \sup_{x \in \mathbb{R}^r} w(f, 2^{-k}d, x),$$

where $d = \|a\|_\infty := \max\{a_i; i = 1, \ldots, r\}$, $a_i > 0$, and $w(f, h, x)$ stands for the local modulus of continuity

$$w(f, h, x) = \sup\{|f(x) - f(x')| : \|x - x'\|_\infty \le h\}.$$

For the local approximation we have $|F(x) - B_k(F)(x)| \le w(F, h, x)$ and the inequality is sharp for distribution functions on $\mathbb{R}^r$. Therefore, if $P_{k,d}$ is the law of $B_k(F)$ then $P_{k,d}$ converges to P in the weak *-topology, i.e.,

$$\int f\, dP_{k,d} \to \int f\, dP$$

for any $f \in C_b(\mathbb{R}^r)$.

Let us first introduce the regularity conditions on φ which will lead to the main result of this section. Define the difference operators: for $\delta_i > 0$ $(i = 1, \ldots, r)$, with $(e_1, \ldots, e_r)$ being the usual basis in $\mathbb{R}^r$,

$$\begin{aligned}
\Delta_{\delta_1} f(x) &= f(x + \delta_1 e_1) - f(x), \\
\Delta_{\delta_1,\delta_2} f(x) &= \Delta_{\delta_1} f(x + \delta_2 e_2) - \Delta_{\delta_1} f(x), \ldots, \\
\Delta_{\delta_1,\ldots,\delta_r} f(x) &= \Delta_{\delta_1,\ldots,\delta_{r-1}} f(x + \delta_r e_r) - \Delta_{\delta_1,\ldots,\delta_{r-1}} f(x).
\end{aligned}$$

We impose the following conditions on the function φ in (2.2):

(i) For each $i = 1, \ldots, r$ and any $x = (x_1, \ldots, x_r) \in \mathbb{R}^r$, $\sum_{j_i \in \mathbb{Z}} \varphi(x - j_i e_i)$ does not depend on x_i;

(ii) $\sum_{j \in \mathbb{Z}^r} \varphi(x - j) = 1$ for all $x \in \mathbb{R}^r$;

(iii) With respect to each variable the $\varphi(x_1, \ldots, x_r)$ is a two-pieces monotone function, which is first nondecreasing and then nonincreasing;

(iv) There is a point $b = (b_1, \ldots b_r) \in \mathbb{R}^r$ such that for all $\delta = (\delta_1, \ldots, \delta_r) \in \mathbb{R}^r_+$, and all $\varepsilon_i = \pm 1$, $i = 1, \ldots, r$

$$\varepsilon \cdot \Delta_{\delta_1,\ldots,\delta_r}\varphi(x) \geq 0, \qquad \varepsilon := \prod_{i=1}^{r}(-\varepsilon_i)$$

whenever x and $x + \delta$ belong to the closure $\overline{T}(\varepsilon_1, \ldots, \varepsilon_r)$ $(\varepsilon_i = \pm 1)$ of the set

$$T(\varepsilon_1, \ldots, \varepsilon_r) := \{x \in \mathbb{R}^r;\ \mathrm{sign}(x_i - b_i) = \varepsilon_i\}.$$

Two Examples. Take φ_0 to be the density of the uniform distribution on $[-\frac{1}{2}, \frac{1}{2}]^r$, i.e., $\varphi_0(x) = 1$ for $-\frac{1}{2} \leq x_i < \frac{1}{2}$, $i = 1, \ldots, r$ and zero otherwise. Next take φ_1 to be the convolution

$$\varphi_1(x) = \varphi_0 * \varphi_0(x) = \int_{\mathbb{R}^r} \varphi_0(x - u)\varphi_0(u)du,$$

i.e., $\varphi_1(x)$ is the product density of triangular laws,

$$\varphi_1(x) = \prod_{j=1}^{r} \varphi_{1,j}(x_j), \qquad x \in \mathbb{R}^r$$

where the marginal density is triangular

$$\varphi_{1,j}(u) = (u+1)\mathbf{1}_{[-1,0]}(u) + (1-u)\mathbf{1}_{(0,1]}(u), \qquad u \in \mathbb{R}.$$

Finally, let $\varphi_n(x)$ be the n-fold convolution

$$\varphi_n(x) = \int_{\mathbb{R}^r} \varphi_{n-1}(x - u)\varphi_0(u)du,$$

i.e., $\varphi_n(x)$ is the product density with ith marginal

$$\varphi_{n,i}(u) = \frac{1}{n!}\sum_{r=0}^{n}(-1)^{\nu}\binom{n}{\nu}\left(x + \frac{(n+1)}{2} - \nu\right)_+^n$$

where $(\cdot)_+ = \max(\cdot, 0)$.

Let us show that φ_0 and φ_1 satisfy (i) – (iv). Indeed for $s = 0$ and 1,

$$\sum_{j \in \mathbb{Z}} \varphi_{s,i}(u - j) = 1$$

and since φ is the product density of uniforms and triangulars $\varphi_{s,j}$ that we get (i) and (ii), (iii) is clear. To see (iv), observe that φ_0 and φ_1 may be viewed as products of signed bounded measures and so, for $\delta = (\delta_1, \ldots, \delta_r)$

$$\Delta_\delta \varphi_s(x) = \prod_{i=1}^{r}(\varphi_{s,i}(x_i + \delta_i) - \varphi_{s,i}(x_i))$$

Therefore, for $s = 0, 1$, and $x, x + \delta \in T(\varepsilon_1, \ldots, \varepsilon_r)$ we have $\mathrm{sign}[\varphi_{s,i}(x_i + \delta_i) - \varphi_{s,i}(x_i)] = -\varepsilon_i$ and

$$\mathrm{sign}\prod_{i=1}^{r}[\varphi_{s,i}(x_i + \delta_i) - \varphi_{s,i}(x_i)] = \prod_{i=1}^{r}(-\varepsilon_i),$$

which showed (iv).

Other Example. In general if $\varphi(x) = \prod_{i=1}^{r} \varphi^{(i)}(x)$ where the $\varphi^{(i)}$'s are bounded compactly supported functions on $\mathbb{R}$, right continuous and satisfy

$$\sum_{j \in \mathbb{Z}} \varphi^{(i)}(x - j) = 1 \text{ on } \mathbb{R},$$

and for each i, $i = 1,\dots,r$ there is b_i, such that $\varphi^{(i)}(x)$ is nondecreasing if $x \le b_i$ and nonincreasing of $x \ge b_i$, then φ satisfies (i) – (iv).

Theorem 1. *Suppose φ is a bounded right-continuous function on $\mathbb{R}^r$ with support in the compact interval $[-a, a]$, $a \in \mathbb{R}^r_+$, $\|a\|_\infty = d$. Suppose φ can be represented for the convolution form*

$$\varphi(x) = \tilde{\varphi} \circ \psi(x) = \int \tilde{\varphi}(x-u)\psi(du)$$

where $\psi(du)$ is a probability measure on $\mathbb{R}^r$ and $\tilde{\varphi}$ satisfies (i) – (iv). Then if F is a distribution function on $\mathbb{R}^r$, $B_k(F)(x)$ defined by (2) is also a distribution function and we have the sharp bound

$$\rho(B_k(F), F) \le w(F, 2^{-k}d).$$

Proof. Since $B_k(F)(x)$ is a finite linear combination of right-continuous functions $(\varphi(x - j))_{j\in\mathbb{Z}^r}$ with bounded support, then for any $x_0 \in \mathbb{R}^r$,

$$\begin{aligned}\lim_{x\to x_0+} B_k(F)(x) &= \lim_{x\to x_0+} \sum_{\substack{2^k x_0 - a < j < 2^k x_0 + a \\ j\in\mathbb{Z}^r}} F(2^{-k}j)\varphi(2^k x - j) \\ &= \sum F(2^{-k}j) \lim_{x\to x_0+} \varphi(2^k x - j) = B_k(F)(x_0).\end{aligned}$$

For fixed $x^0 = (x_1^0, \dots, x_r^0)$, as a function of x_i, $B_k(F)(x_1^0,\dots,x_{i-1}^0, x_i, x_{i+1}^0,\dots,x_r^0)$ is nondecreasing by (i) and (iii), see [1].

Next, let us show that $B_k(F)$ generates a positive measure on $\mathbb{R}^r$, i.e.,

$$\Delta_{\delta_1,\dots,\delta_r} B_k(F)(x) \ge 0, \quad \text{all } \delta_i \in \mathbb{R}_+;\ i = 1,\dots,r.$$

Observe that $B_k(F)(x) = B_0(F(2^{-k}\cdot))(2^k x)$ and thus it is enough to show the above inequality for $k = 0$. Fix $x^0 = (x_1^0,\dots,x_r^0)$ and note that as a function of x_i, $B_k(F)(x_1^0,\dots,x_{i-1}^0, x_i, x_{i+1}^0,\dots,x_r^0)$ is increasing for $x_i \le \tilde{b}_i = \tilde{b}_i(x_1^0,\dots,x_{i-1}^0, x_{i+1}^0,\dots,x_r^0)$ and decreases for $x_i \ge \tilde{b}_i$. Let $j_i^{(0)}$ $(i = 1,\dots,r)$ be the integer such that

$$x_i^{(0)} + \delta_i - j_i^{(0)} - 1 \le \tilde{b}_i \le x_i^{(0)} + \delta_i - j_i^{(0)}.$$

Remark 1. Observe that it is enough to show that $B_0(F)$ generates a probability measure in case of φ satisfying (i) – (iv). In fact, in the general case $\varphi = \tilde{\varphi} \circ \psi$ we have

$$\begin{aligned}B_0(F)(x) &= \sum_{j\in\mathbb{Z}^r} F(j)(\tilde{\varphi}\circ\psi)(x-j) \\ &= \int \left(\sum_{j\in\mathbb{Z}^r} F(j)(\tilde{\varphi}(x-u-j) \right) \psi(du),\end{aligned}$$

i.e., $B_0(F)$ is a ψ-mixture of probability distributions, and so, it is a probability distribution itself.

With the above remark in hand, invoking condition (i), we have that

$$\sum_{j_i\in\mathbb{Z}} \varphi(x_1,\dots,(x_i - j_i),\dots,x_r) = c_i(x_1,\dots,x_{i-1},x_{i+1},\dots,x_r)$$

and therefore for $\delta = (\delta_1,\dots,\delta_r) \in \mathbb{R}^r_+$

$$\sum_{j_i\in\mathbb{Z}} \Delta_\delta \varphi(x_1,\dots,x_{i-1},x_i^{(0)} - j, x_{i+1},\dots,x_r) = \Delta_\delta c_i(x_1,\dots,x_{i-1},x_{i+1},\dots,x_r) = 0,$$

since c_i does not depend on x_i. Therefore,

$$\sum_{j\in\mathbb{Z}^r} F(j_1,\dots,j_{i-1},j_i^{(0)},j_{i+1},\dots,j_r)\Delta_\delta\varphi(x^{(0)}-j)$$
$$= \sum_{(j_1,\dots,j_{i-1},j_{i+1},\dots,j_r)\in\mathbb{Z}^{r-1}} F(j_1,\dots,j_{i-1},j_i^{(0)},j_{i+1},\dots,j_r) \times \sum_{j_i\in\mathbb{Z}} \Delta_\delta\varphi(x^{(0)}-j) = 0.$$

In the same way, we can change j_k $(k > j)$ with $j_k^{(0)}$ in the above equalities to get

$$\sum_{j\in\mathbb{Z}^r} F(j_1,\dots,j_i^{(0)},\dots,j_k^{(0)},\dots,j_r)\Delta_\delta\varphi(x^{(0)}-j) = 0$$

and so on,

$$\sum_{j\in\mathbb{Z}^r} F(j^{(0)})\Delta_\delta\varphi(x^{(0)}-j)) = 0.$$

Clearly $\Delta_{\delta_1,\dots,\delta_r}F(x) = 0$ if some $\delta_i = 0$. Moreover, if $\delta_i \in \mathbb{R}\setminus\{0\}$ then sign $\Delta_{\delta_1,\dots,\delta_r}F(x) = \prod_{i=1}^r (\text{sign } \delta_i)$, because of $\Delta_{|\delta_1|,\dots,|\delta_r|}F(x) \geq 0$. Using these equalities,

$$\begin{aligned} \Delta_\delta B_0(F)(x^{(0)}) &= \sum_{j\in\mathbb{Z}^r} F(j)\Delta_\delta\varphi(x^{(0)}-j) \\ &= \sum_{j\in\mathbb{Z}^r} \Delta_{j-j^{(0)}}F(j^{(0)})\Delta_\delta\varphi(x^{(0)}-j), \quad \forall\, i, j_i \neq j_i^{(0)}. \end{aligned}$$

where

$$\Delta_{j-j^{(0)}}F(j) = \prod_{i=1}^r \varepsilon_i,$$

where $\varepsilon_i := \text{sign}(j_i - j_i^{(0)})$. From the definition of $j_i^{(0)}$,

$$0 < x_i^{(0)} + \delta_i - j_i^{(0)} - b_i \leq 1.$$

If $\varepsilon_i = -1$, then for δ_i's less than 1,

$$\begin{aligned} 0 &< x_i^{(0)} + \delta_i - j_i^{(0)} - b_i < x_i^{(0)} + 1 = j_i^{(0)} - b_i \\ &\leq x_i^{(0)} - j_i - b_i < x_i^{(0)} + \delta_i - j_i - b_i, \end{aligned}$$

which implies

$$\text{sign}(x_i^{(0)} - j_i - b_i) = \text{sign}(x_i^{(0)} + \delta_i - j_i - b_i) = -\varepsilon_i.$$

If $\varepsilon_i = 1$, then similarly,

$$x_i^{(0)} - j_i - b_i < x_i^{(0)} + \delta_i - j_i - b_i \leq x_i^{(0)} + \delta_i - j_i^{(0)} - 1 - b_i \leq 0,$$

and so, either $\text{sign}(x_i^{(0)} - j_i - b_i) = \text{sign}(x_i^{(0)} + \delta_i - j_i - b_i) = -\varepsilon_i$ or $x_i^{(0)} + \delta_i - j_i = b_i$. Combining all the above information we have that the vectors $x^{(0)} + \delta - j$ and $x^{(0)} - j$ belong to $\overline{T}(-\varepsilon_1,\dots,-\varepsilon_r)$. Invoking condition (iv) we have:

$$\text{sign } \Delta_\delta\varphi(x^{(0)}-j) = \prod_{i=1}^r \varepsilon_i.$$

From this equality and the last representation of $\Delta_\delta B_0(F)(x^{(0)})$, we obtain that the latter is nonnegative as desired.

To show that $B_k(F)$ is a probability distribution function we need to check that

$$B_0(F)(+\infty,\ldots,+\infty) = 1$$

and $\lim_{x \downarrow y_\infty} B_0(F)(x_1,\ldots,x_n) = 0$ if at least one coordinate of y_∞ is $-\infty$. To this end it is enough to show

$$\lim_{x\to(+\infty,\ldots,+\infty)} |B_k(F)(x) - F(x)| = 0$$

and

$$\lim_{x\to y_\infty} |B_k(F)(x) - F(x)| = 0.$$

Choose $N > 0$ such that $F(x) > 1 - \varepsilon$ if all coordinates of x exceed N. Let $x_0 > N\mathbf{1} + a$ where $\mathbf{1} = (1,\ldots,1) \in \mathbb{R}^r$ and recall that supp $\varphi \subset [-a,a]$. If $x_0 - a \le j \le x_0 + a$ then $j \ge x_0 - a > N\mathbf{1}$. Hence,

$$\begin{aligned}
|B_0(F)(x_0) - F(x_0)| &= \left| \sum_{j\in\mathbb{Z}^r, x_0-a\le j\le x_0+j} [F(j) - F(x_0)]\varphi(x_0 - j)\right| \\
&\le \varepsilon \cdot \sum_{j\in\mathbb{Z}^r} \varphi(x_0 - j) = \varepsilon.
\end{aligned}$$

Here we use that φ is nonnegative since it is compactly supported and fulfills (iii). So $lim_{x\to(+\infty,\ldots,+\infty)} B_0(F)(x) = 1$. The same arguments apply to show that $\lim_{x\to y_\infty} B_0(F) = 0$. Therefore, from the definition of distribution function (cf. for example, Shiryayev (1984), p. 158) on $\mathbb{R}^r$, $B_0(F)$ and so $B_k(F)$ for all k are distribution functions in $\mathbb{R}^r$.

Next, let us consider the bound for the uniform distance $\rho(B_k(F), F)$. Using condition (ii) and the boundedness of the support of φ we have for any $x \in \mathbb{R}^r$,

$$\begin{aligned}
|B_k(F)(x) - F(x)| &\le \sum_{j\in\mathbb{Z}^r} |F(2^{-k}j) - F(x)| \cdot \varphi(2^k x - j) \\
&= \sum_{2^k x - a < j < 2^k x + a} |F(2^{-k}j) - F(x)| \cdot \varphi(2^k x - j) \\
&\le \sum_{2^k x - a < j < 2^k x + a} w(F, 2^{-k}d, x) \cdot \varphi(2^k x - j) \\
&= w(F, 2^{-k}d, x) \cdot \int \sum_{2^k x - a < j < 2^k x + a} \tilde{\varphi}(2^k x - u - j)\psi(du)
\end{aligned}$$

(since $\tilde{\varphi}$ satisfying (i) – (iv) is nonnegative)

$$\begin{aligned}
&\le w(F, 2^{-k}d, x) \cdot \int \sum_{j\in\mathbb{Z}^r} \tilde{\varphi}(2^k x - u - j)\psi(du) \\
&= w(F, 2^{-k}d, x).
\end{aligned}$$

Taking the supremum over $x \in \mathbb{R}^d$ we get

$$\rho(F, B_k(F)) \le w(F, 2^{-k}d).$$

To show the sharpness of the above inequality, take φ to be the uniform density φ_0 and F to be the product of univariable distribution functions

$$F_i(x_i) = \begin{cases} 0, & x_i \le -2^{-k-1} \\ 2^{k+1}x_i + 1, & -2^{-k-1} < x_i < 0 \\ 1, & x_i \ge 0. \end{cases}$$

Take $x_k = -2^{-k-1} \cdot \mathbf{1}$. Then

$$\begin{aligned} B_k(F)(x_k) - F(x_k) &= B_k(F)(x_k) \\ &= \sum_{j \in \mathbb{Z}^r} F(2^k j) \cdot \varphi\left(-\frac{1}{2}\mathbf{1} - j\right) \\ &= F(0 \cdot \mathbf{1}) = 1, \end{aligned}$$

since $\varphi(-\frac{1}{2}\mathbf{1} - j) = 0$ for any $j \neq 0\mathbf{1}$. On the other hand we have $d = \|a\|_\infty = \frac{1}{2}$ and $w(F, 2^{-k}d) = w(F, 2^{-k-1}) = F(0\mathbf{1}) - F(-2^{-k-1}\mathbf{1}) = 1$. Therefore the bound for $\rho(F, B_k(F))$ is sharp. This completes the proof of the theorem. □

Corollary 1. *Under the assumptions of Theorem 1, the class of $B_k(F)$ (for any k fixed) is weak *-dense, in the set of distribution functions on $\mathbb{R}^d$.*

Proof. As in Theorem 1, we can show that for any $x \in \mathbb{R}^r$, $k \in \mathbb{Z}$

$$|B_k(F)(x) - F(x)| \leq w(F, 2^{-k}d, x),$$

and moreover the above inequality is sharp. Therefore $B_k(F)$ converges weakly to F as $k \to \infty$. □

Remark 2. Let $\varphi_{n-1}(x)$ be the n-fold ($n = 1, 2, \ldots$) convolution of the uniform density φ_0 on $[-\frac{1}{2}, \frac{1}{2}]^r$, then φ_{n-1} clearly satisfies the assumption of Theorem 1.

Remark 3. (*Approximation of a stochastic process via $B_k(F)$-laws*) Let $X = (X_t)_{t \in T}$ be a stochastic process on a probability space (Ω, A, P) with $t \in T \subset \mathbb{R}$. The probability law P^X of X is completely determined by the set $\{F_{t_1,\ldots,t_n}\}_{t_1 < t_2 < \cdots < t_n}$ of finite-dimensional distribution functions

$$F_{t_1,\ldots,t_n}(x) = P\{X_{t_1} \leq x_1, \ldots, X_{t_n} \leq x_n\}, \qquad x \in \mathbb{R}^n.$$

Suppose now $\varphi^{(n)}(x) = \prod_{i=1}^{n} \varphi(x_i)$, where the univariate function φ satisfies the conditions (i) – (iv) with $r = 1$. Therefore, $\varphi^{(n)}$ possesses the properties (i) – (iv) $r = n$. Applying Theorem 1 we have that for any $k \in \mathbb{Z}^n$

$$B_k(F_{t_1,\ldots,t_n})(x) = \sum_{j \in \mathbb{Z}^n} F_{t_1,\ldots,t_n}(2^{-k}j)\varphi^{(n)}(2^k x - j)$$

is a distribution function on $\mathbb{R}^n$. Observe that the collection

$$\{B_k(F_{t_1,\ldots,t_n})\}_{t_1 < t_2 < \cdots < t_n}$$

possesses the following *consistency property*: for any $i = 1, \ldots, n$, $n \geq 2$,

$$\lim_{x_i \uparrow \infty} B_k(F_{t_1,\ldots,t_n})(x_1, \ldots, x_n) = B_k(F_{t_1,\ldots,\hat{t}_i,\ldots,t_n})(x_1, \ldots, \hat{x}_i, \ldots, x_n),$$

where ˆ indicates an omitted coordinate. From the Kolmogorov theorem on existence of a process, see, for example, Shiryayev, p. 244, the family

$$\{B_k(F_{t_1,\ldots,t_n})\}_{t_1 < \cdots < t_n}$$

determines a probability space (Ω^*, F^*, P^*) and a stochastic process X^* such that,

$$P^*(X^*(t_1) \leq x_1, \ldots, X^*(t_n) \leq x_n) = B_k(F_{t_1,\ldots,t_n})(x_1, \ldots, x_n).$$

If $X = \{X_t\}_{t \in \mathbb{N}}$ is a random sequence with continuous finite-dimensional distributions and let X^* be the corresponding approximating sequence. Define the uniform distance between the laws P^X and P^{X^*},

$$K(P^X, P^{X^*}) = \sum_{n=1}^{\infty} 2^{-n} \min(1, \rho(F_{1,\dots,n}, B_k(F_{1,\dots,n})),$$

where

$$\rho(F_{1,\dots,n}, B_k(F_{1,\dots,n})) = \sup_{x \in \mathbb{R}^n} |F_{1,\dots,n}(x) - B_k(F_{1,\dots,n})(x)|.$$

Applying Theorem 1 we conclude that

$$K(P^*, P^{X^*}) \leq \sum_{n=1}^{\infty} 2^{-n} \min(1, w(F_{1,\dots,n}, 2^{-k} \cdot d)).$$

Letting $k \to \infty$ we have that $K(P^*, P^{X^*}) \to 0$ and therefore the class of P^{X^*} is weak *-dense in the space of probabilities P^X given on $\mathbb{R}^\infty$ with metric

$$d_0(x, y) = \sum_{i=1}^{\infty} 2^{-i} \frac{|x_i - y_i|}{1 + |x_i - y_i|},$$

see Billingsley (1968), Appendix 1.

Remark 4 (*Extensions for bounded and unbounded measures*). From the proof of Theorem 1 it follows that if F is the distribution function of a bounded positive measure then $B_k(F)$ is also a distribution function with the same total mass as F, i.e., $B_k(F)(+\infty, \dots, +\infty) = F(+\infty, \dots, +\infty)$. Moreover $(B_k(F))_{k \in \mathbb{Z}}$ is a weak *-dense set in the space of $\mathcal{M}$ of bounded measures. Let now $\mathcal{N}$ be the space of all Borel positive measures ν on $\mathbb{R}^r$ finite on all bounded Borel subsets of $\mathbb{R}^r$. We endow $\mathcal{N}$ with the vague convergence (see, for example Rachev (1990), see 10.2); for ν_n, $\nu \in \mathcal{N}$, ν_n vaguely converges to ν ($\nu_n \xrightarrow{\nu} \nu$) if

$$\int f \, d\nu_n \to \int f \, d\nu$$

for all $f \in C_b(S_m)$, $S_m = \{x \in \mathbb{R}^r\colon \|x\| \leq m\}$, $m = 1, 2, \dots$. For any $\nu \in \mathcal{N}$ let $\nu^{(m)}$ be the restriction of ν on S_m and $F_\nu^{(m)}$ be the corresponding distribution function of $\nu^{(m)}$. Suppose $F_\nu^{(m)}$ are continuous for all m. Define

$$B_k(F_\nu^{(m)})(x) = \sum_{j \in \mathbb{Z}^r} F_\nu^{(m)}(2^{-k} \cdot j) \varphi(2^k x - j),$$

where φ is defined as in Theorem 1. Then the distance between $(F_\nu^{(m)})_{m \geq 1}$ and $(B_k(F_\nu^{(m)}))_{m \geq 1}$ is given by

$$\sum_{m=1}^{\infty} 2^{-m} \min\{1, \rho(F_\nu^{(m)}, B_k(F_\nu^{(m)}))\} \leq \sum_{m=1}^{\infty} 2^{-m} \cdot \min(1, w(F_\nu^{(m)}, 2^{-k} \cdot d))$$

goes to zero as $k \to \infty$.

REFERENCES

1. G.A. Anastassiou and X.M. Yu, Monotone and probabilistic wavelet approximation, to appear in *J. Stoch. Analysis and Appl.*
2. P. Billingsley, *Convergence of Probability Measures*, Wiley, New York (1968).

3. I. Daubechies, The wavelet transform, time-frequency localization and signal analysis, *IEEE Trans. Information Theory*, 36:961–1005 (1990).
4. S. Mallat, Multiresolution approximations and wavelet orthonormal bases in $L^2(\mathbb{R})$, *Trans. A.M.S.*, 315:69–87 (1989).
5. S.T. Rachev, *Probability Metrics and the Stability of Stochastic Models*, Wiley, New York (1990).
6. A.N. Shiryayev, *Probability*, Springer-Verlag, New York–Berlin (1984).

PROBABILISTIC APPROACH TO THE ROUNDING PROBLEM WITH APPLICATIONS TO FAIR REPRESENTATION

Bessy Athanasopoulos†

University of California
Department of Statistics
and Applied Probability
Santa Barbara, CA 93106, USA

ABSTRACT

Failure to add to 100% occurs frequently for sums of percentages in reported sets of tables. It occurs so frequently, that if many sums of percentages add to exactly 100% in a reported set of tables, one begins to suspect the reporter of forcing the situation. Extending the pioneer works of Mosteller, Youtz and Zahn (1967) and of Diaconis and Freedman (1979) who assess the probability that a table of conventionally (MYZ) rounded proportions adds to 1, Balinski and Rachev (1992) introduced some rules of rounding that can improve the conventional rule. Investigating and developing further the so-called K-stationary divisor rules of rounding we compute, for several of these rules, the limiting probability that the rounded percentages add to 100%. We build up a bridge between the problem of rounding and the problem of apportionment. We apply the theory of apportionment in allocating representation among geographical regions in Greece and among states in U.S.A. We investigate and comment on the methods of apportionment currently being used in the two countries, as well as on other possible options including some K-stationary methods.

INTRODUCTION

In this paper, we deal with the problem of developing and comparing various rules—mainly probabilistic—of rounding percentages reported in statistical tables. Surprisingly enough, the rounded percentages rarely add to 100%.

The importance and frequency of this problem has led to significant interest and research within academia. Fundamental work was done, in 1967, by Mosteller, Youtz and Zahn who investigated how frequently the rounded percentages fail to add up correctly and what the distributions of sums of rounded percentages are for (1) an empirical set of data, (2) the multinomial distribution in small samples, (3) spacings between points dropped on an interval—the broken stick model—and (4) simulation

† Dr. Bessy Athanasopoulos is a postgraduate student under the program of the NATO Scientific Committee in the Greek Ministry of National Economy and under the supervision of Dr. Svetolozar Rachev.

Approximation, Probability, and Related Fields, Edited by
G. Anastassiou and S.T. Rachev, Plenum Press, New York 1994

for several categories. They found that the probability that the sum of rounded percentages adds to exactly 100% is certain for two categories, about three-fourths for three categories, about two-thirds for four categories, and about $\sqrt{\frac{6}{n\pi}}$ for a larger number n of categories.

In 1979, Diaconis and Freedman assessed the probability that a table of rounded percentages adds to 100%. Extending the work of Mosteller, Youtz and Zahn, they gave a mathematical treatment of this phenomenon when the table is drawn from a multinomial distribution or from a mixture of multinomial distributions. Their principal result concerned the Mosteller, Youtz and Zahn broken-stick model.

Balinski and Rachev (1992) continued the work of Diaconis and Freedman by introducing the so-called stationary rules and considering vectors and matrices under varying assumptions concerning the probabilistic structure of the data to be rounded.

In what follows, we investigate a class of rounding rules, called divisor rules of rounding, and in particular, we study the K-stationary divisor rules of rounding.

We first describe the vector problem $(\vec{p}, h)$ and a variety of possible rounding rules, with which we treat a particular case of the vector problem where $\vec{p}$ is uniformly distributed on the simplex S_n.

We compute the limiting probability that the sum of the rounded percentages equals the rounding of the sum of the percentages for different rules of rounding and for various probabilistic models generating the data.

We finally connect the problem of rounding percentages with the problem of apportionment and discuss several solutions to the issue of fair representation among geographical regions.

1. THE VECTOR PROBLEM OF ROUNDING: K-STATIONARY DIVISOR RULES

We start by investigating the so-called K-stationary divisor rules of rounding percentages. Given a vector problem $(\vec{p} = (p_1, \ldots, p_n), h)$ and a K-stationary rule $\vec{x}^{(K)} = \rho_t^{(K)}(\vec{p})$, we first try to evaluate the chance that $x_N^{(K)} := x_1^{(K)} + \ldots + x_n^{(K)} = h$ and then we find the particular rule which maximizes this chance.

This extends the works of Mosteller, Youtz and Zahn (1967), Diaconis and Freedman (1979), and Balinski and Rachev (1992), who assessed the probability that a table of rounded percentages add to 100%.

In Section 1.1, we define the vector problem of rounding and the K-stationary divisor rules of rounding.

In Section 1.2, we show that the maximum of $\lim_{t\to\infty} Pr[1 - \frac{\Delta}{t} \le x_N^{(K)} \le 1 + \frac{\Delta}{t}]$ for every $\Delta = 0, 1, 2, \ldots$ does not change if, instead of rounding the p_i's, $i = 1, \ldots, n$ with the best of 0-stationary divisor rule, we round them with the best of any other K-stationary divisor rule ($K \ge 1$).

In Section 1.3, we display several computer simulations to support our theoretical results.

1.1 Notation and Preliminaries on the Vector Problem of Rounding

A *vector problem* is a pair, $(\vec{p}, h)$, where $\vec{p} = (p_j)$, $j \in N = \{1, \ldots, n\}$ is a vector of real numbers and h is a real number such that $p_N := p_1 + \ldots + p_n = h$. Unless otherwise specified, we assume $p_j \ge 0$, $j \in N$ and set $p_N = 1$, which is not a restriction for h.

Given any positive real number t, a *rule* ρ_t of $\frac{1}{t}$-*rounding* assigns to each vector $\vec{p}$ a set $\{\vec{x} : \vec{x} = \rho_t(\vec{p})\} \subseteq \{\vec{x} = (x_j) : x_j = \frac{k_j}{t}, k_j \text{ integer}, j \in N\}$. Since the rule of $\frac{1}{t}$-rounding does not depend on p_N, it is important, first to evaluate the change that the sum $x_N := x_1 + \ldots + x_n$ is exactly p_N, and then, find the rule that maximizes this chance.

A *divisor rule* $\rho_{t,d}$ of $\frac{1}{t}$*-rounding* assigns to each vector $\vec{p}$ a set of vectors $\{\vec{x}_d : \vec{x}_d = \rho_{t,d}(\vec{p})\} \subseteq \{\vec{x} : \vec{x} = \rho_t(\vec{p})\}$ defined by:

$$(\vec{x}_d)_j := [p_j]_{t,d} := \begin{cases} \frac{k+1}{t} & \text{if } \{k + \frac{1}{2} < tp_j \leq k+1\} \text{ or } \{k + \frac{1}{2} = tp_j,\ k \text{ odd}\}, \\ \frac{k}{t} & \text{if } \{k \leq tp_j < k + \frac{1}{2}\} \text{ or } \{k + \frac{1}{2} = tp_j,\ k \text{ even}\}, \end{cases} \tag{1.1}$$

where for $k \in Z$, $d(k) = k + C \in [k, k+1]$ is said to be the divisor criterion.

Mosteller, Youtz and Zahn (1967) were the first to discuss the conventional rule (for short, MYZ-rule) of $1/t$-rounding $\vec{x} = \rho_t(\vec{p})$ for the problem $(\vec{p}, 1)$. The conventional rule rounds p_j, $j \in N$, to the nearest k/t and, therefore, is the divisor rule with $d(k) = k + 1/2$, that is,

$$x_j \equiv (\vec{x}_d)_j := [p_j]_{t,k+1/2} := \begin{cases} \frac{k+1}{t} & \text{if } \{k + 1/2 < tp_j \leq k+1\} \text{ or } \{k + 1/2 = tp_j,\ k \text{ odd}\}, \\ \frac{k}{t} & \text{if } \{k \leq tp_j < k + 1/2\} \text{ or } \{k + 1/2 = tp_j,\ k \text{ even}\}, \end{cases} \tag{1.2}$$

Mosteller, Youtz and Zahn computed the probability that $x_N := x_1 + \ldots + x_n = 1$ for several probability models generating $\vec{p}$ and found that the probability that $x_N = 1$ is 1 for $n = 2$, about 3/4 for $n = 3$, about 2/3 for $n = 4$ and about $\sqrt{\frac{6}{\pi n}}$ for $n > 4$.

Diaconis and Freedman (1979) assessed the limit probability of $x_N = 1$. They showed that if $\vec{p}$ has an absolutely continuous distribution on the simplex S_n, n large, and $\vec{x}$ is obtained by (1.2), then, as $t \to \infty$, $Pr\{x_N = 1\}$ converges to the probability that $-\frac{1}{2} \leq V_1 + V_2 + \ldots + V_{n-1} \leq \frac{1}{2}$, where the V_j's are independent and uniformly distributed on $[-1/2, 1/2]$. In particular, as $t \to \infty$, $Pr\{x_N = 1\} \to \sqrt{\frac{6}{\pi(n-1)}} + O\left(\frac{1}{\sqrt{n^3}}\right)$.

Balinski and Rachev (1992) slightly extended the above theorem: They stated that if $\vec{p}$ has an absolutely continuous distribution on the simplex S_n, n large, and $\vec{x}'$ is obtained by (1.1) with $d(k) = k + C$, $k \in Z$, $C \in [0,1]$, then as $t \to \infty$, $Pr\{x'_N = 1\}$ converges to the probability that $C - 1 \leq V_1 + \ldots + V_{n-1} \leq C$, where the V_j's are independent and uniformly distributed on $[-C, 1-C]$. Taking $C = 1/2$, the limit is maximized and, in this case, as $t \to \infty$, $Pr\{x'_N = 1\} \to \sqrt{\frac{6}{\pi(n-1)}} + O\left(\frac{1}{\sqrt{n^3}}\right)$.

A *K-stationary divisor rule* $\rho_t^{(K)}$ *of 1/t-rounding* assigns to each vector $\vec{p}$ a set $\{\vec{x}^{(K)} : \vec{x}^{(K)} = \rho_t^{(K)}(\vec{p})\} \subseteq \{\vec{x} : \vec{x} = \rho_t(\vec{p})\}$ defined by (1.1) where, for $k \in Z$,

$$d(k) = \begin{cases} k + C_k\ ,\ C_k \in [0,1] & \text{if } k < K, \\ k + C\ ,\ C \in [0,1] & \text{if } k \geq K. \end{cases} \tag{1.3}$$

Remark: The Mosteller, Youtz and Zahn divisor rule maximizes the $\lim Pr\{x_N = 1\}$ as $t \to \infty$, and therefore it is to be the best among all 0-stationary divisor rules in that it maximizes the limit of $Pr\{x_N = 1\}$. We, further on, study the K-stationary divisor rules with $K \geq 1$. Our primary objective is to enquire if the K-stationary divisor rules can or cannot lead to a better limiting probability of $x_N = 1$.

1.2 K-stationary divisor rules (K $\geq$ 1) for the Vector Problem of Rounding

We define the K-stationary divisor rule $\vec{x}^{(K)} := \rho_t^{(K)}(\vec{p})$ of $1/t$-rounding of a vector $\vec{p}$ by:

$$x_j^{(K)} := [p_j]_{t,d}^K := \begin{cases} \frac{k}{t} & \text{if } \{k \leq p_j t < d(k)\} \text{ or } \{p_j t = d(k),\ k \text{ even}\}, \\ \frac{k+1}{t} & \text{if } \{d(k) < p_j t \leq k+1\} \text{ or } \{p_j t = d(k),\ k \text{ odd}\}, \end{cases} \tag{1.4}$$

where, for $0 \leq k \leq K-1$, $d(k) = k + C_k$, $C_k \in [0,1]$ and, for $k \geq K$, $d(k) = k + C$, $C \in [0,1]$.

Theorem 1.1

Suppose $\vec{p}$ is uniformly distributed on the simplex $S_n (n > 1)$ and $\vec{x}^{(K)}$ is obtained by a K-stationary divisor rule, $\vec{x}^{(K)} = \rho_t^K(\vec{p})$ (see (1.4)). Then

$$\max\left\{\lim_{t\to\infty} Pr\left(1 - \frac{\Delta}{t} \leq x_N^{(K)} \leq 1 + \frac{\Delta}{t}\right) : \vec{x}^{(K)} = \rho_t^{(K)}(\vec{p})\right\}, \quad \Delta = 0, 1, 2, \ldots$$

is attained, for any K-stationary rule ($K \geq 0$) when $C = \frac{1}{2}$ and C_k is any point in $[0,1]$ for every $0 \leq k \leq K-1$. Moreover, if $\vec{x} = \rho_t(\vec{p})$ (see (1.2)) then, $\forall \Delta = 0, 1, \ldots$

$$\max \lim_{t\to\infty} Pr\left(1 - \frac{\Delta}{t} \leq x_N^{(K)} \leq 1 + \frac{\Delta}{t}\right) = \lim_{t\to\infty} Pr\left(1 - \frac{\Delta}{t} \leq x_N \leq 1 + \frac{\Delta}{t}\right),$$

We will later sketch the proof of the above theorem for $K > 1$. Next, we look at the case $K = 1$ and prove theorem 1.2.

According to the definition of K-stationary divisor rule of $1/t$-rounding, a *1-stationary divisor rule* $\vec{x}^{(1)} = \rho_t^{(1)}(\vec{p})$ *of* $1/t$*-rounding* of a vector $\vec{p}$ is defined by

$$x_j^{(1)} := [p_j]_t^1 := \begin{cases} \frac{k+1}{t} & \text{if } k \neq 0 \text{ and } k + C < p_j t \leq k+1, \\ \frac{k}{t} & \text{if } k \neq 0 \text{ and } k \leq p_j t < k + C, \\ \frac{1}{t} & \text{if } C_0 < p_j t \leq 1 + C, \\ 0 & \text{if } 0 \leq p_j t \leq C_0, \end{cases} \tag{1.5}$$

where C_0, $C \in [0,1]$.

Theorem 1.2

Suppose $\vec{p}$ is uniformly distributed on the simplex $S_n (n > 1)$. There is no 1-stationary rule $\vec{x}^{(1)} = \rho_t^{(1)}(\vec{p})$ (see (1.5)) of $1/t$-rounding that is "better" than the Mosteller, Youtz and Zahn rule $\vec{x} = \rho_t(\vec{p})$ (see (1.2)) in the sense that $\vec{x}^{(1)}$ cannot improve the limiting probability $Pr(x_N = 1)$ as $t \to \infty$. In fact,

$$\max\left\{\lim_{t\to\infty} Pr\left(1 - \frac{\Delta}{t} \leq x_N^{(1)} \leq 1 + \frac{\Delta}{t}\right) : \vec{x}^{(1)} = \rho_t^{(1)}(\vec{p})\right\}$$
$$= \lim_{t\to\infty} Pr\left(1 - \frac{\Delta}{t} \leq x_N \leq 1 + \frac{\Delta}{t}\right), \quad \Delta = 0, 1, 2, \ldots, .$$

For proof of Theorem 1.2 we need the following two lemmas:

Lemma 1.3: Let $m_1, m_2, \ldots, m_{n-1}$ be positive integers whose sum is at most $t-n+1$ for t and n fixed, and t large enough. Denote by $A_t(m_1, \ldots, m_{n-1})$ the set

$$A_t(m_1, \ldots, m_{n-1}) = \left\{(p_1, \ldots, p_{n-1}) : \frac{m_i}{t} \leq p_i < \frac{m_i + 1}{t}, \, i = 1, \ldots, n-1\right\}$$

and let A_t be the union of these $A_t(m_1, \ldots, m_{n-1})$ over all choices of $m_1, \ldots, m_{n-1}$. Then:

(i) The probability of A_t tends to 1, as $t \to \infty$.

(ii) Given $A_t(m_1, \ldots, m_{n-1})$, the random variables $\tilde{V}_i := t(x_i^{(1)} - p_i)$ (rounding errors of a 1-stationary divisor rule), $i = 1, \ldots, n-1$ are conditionally independent and uniformly distributed over the $(n-1)$-fold Cartesian product $\otimes_{n-1}[-C, 1-C]$.

Lemma 1.4:

(i) For every real σ_2, the 1-stationary rule of $1/t$-rounding $[\bullet]_t^1$ gives $\left[\frac{\sigma_2}{t}\right]_t^1 = \frac{1}{t}[\sigma_2]_1^1$.

(ii) For every integer $\sigma_1 \neq 0$ and every real σ_2 the 1-stationary rule of $1/t$-rounding $[\bullet]_t^1$ gives the following:

(a) For $C \leq C_0$, assuming that $\sigma_2 \neq -\sigma_1, 1$

$$\left[\frac{\sigma_1+\sigma_2}{t}\right]_t^1 = \frac{\sigma_1}{t} + \frac{1}{t}[\sigma_2]_1^1 - \frac{1}{t}I\{\sigma_1+\sigma_2 \in (C, C_0]\} + \frac{1}{t}I\{\sigma_2 \in (C, C_0)\}.$$

(b) For $C > C_0$, assuming that $\sigma_2 \neq -\sigma_1, 1$,

$$\left[\frac{\sigma_1+\sigma_2}{t}\right]_t^1 = \frac{\sigma_1}{t} + \frac{1}{t}[\sigma_2]_1^1 + \frac{1}{t}I\{\sigma_1+\sigma_2 \in (C_0, C]\} - \frac{1}{t}I\{\sigma_2 \in (C, C_0)\}.$$

Proof of Theorem 1.2

By the definition of the 1-stationary divisor rule of $1/t$-rounding $[\bullet]_t^1$ and the corresponding "rounding errors" $\tilde{V}_i$, we obtain

$$x_i^{(1)} = [p_i]_t^1 = p_i + \frac{1}{t}\tilde{V}_i,\ i = 1, \ldots, n-1 \text{ and, therefore,}$$

$$p_n = 1 - \sum_{i=1}^{n-1} p_i = 1 - \sum_{i=1}^{n-1} x_i^{(1)} + \frac{1}{t}\sum_{i=1}^{n-1}\tilde{V}_i = \frac{1}{t}\left[\left(t - t\sum_{i=1}^{n-1} x_i^{(1)}\right) + \sum_{i=1}^{n-1}\tilde{V}_i\right].$$

Using lemma 1.4, with $\sigma_1 = t - t\sum_{i=1}^{n-1} x_i^{(1)} \neq 0$ and $\sigma_2 = \sum_{i=1}^{n-1}\tilde{V}_i \neq 1, -\sigma_1$, we obtain

$$\begin{aligned} x_n^{(1)} = [p_n]_t^1 &= 1 - \sum_{i=1}^{n-1} x_i^{(1)} + \frac{1}{t}\left[\sum_{i=1}^{n-1}\tilde{V}_i\right]_1^1 + \frac{1}{t}R \\ &= 1 - \sum_{i=1}^{n-1}\left(p_i + \frac{1}{t}\tilde{V}_i\right) + \frac{1}{t}\left[\sum_{i=1}^{n-1}\tilde{V}_i\right]_1^1 + \frac{1}{t}R \\ &= p_n + \frac{1}{t}\left[\sum_{i=1}^{n-1}\tilde{V}_i\right]_1^1 + \frac{1}{t}\sum_{i=1}^{n-1}\tilde{V}_i + \frac{1}{t}R. \end{aligned}$$

According to Lemma 1.4, if $C_0 \geq C$, the remainder $R = R_t$ in the above expression equals

$$R_t = I\{\sigma_1 + \sigma_2 \in (C, C_0]\} + I\{\sigma_2 \in (C, C_0]\}.$$

By Lemma 1.3, $\sigma_2 := \sum_{i=1}^{n-1}\tilde{V}_i$ is, conditionally on A_t, continuous random variable, so without loss of generality we can assume that $\sigma_2 \notin \{-\sigma_1, 1\}$. Consequently, as $t \to \infty$, with probability 1,

$$\begin{aligned} R_t &= -I\left\{t - t\left(\sum_{i=1}^{n-1} x_i^{(1)} - \frac{1}{t}\sum_{i=1}^{n-1}\tilde{V}_1\right) \in (C, C_0]\right\} + I\left\{\sum_{i=1}^{n-1}\tilde{V}_i \in (C, C_0]\right\} \\ &= -I\left\{tp_n \in (C, C_0]\right\} + I\left\{\sum_{i=1}^{n-1}\tilde{V}_i \in (C, C_0]\right\} \underset{t\to\infty}{\longrightarrow} I\left\{\sum_{i=1}^{n-1} V_i \in (C, C_0]\right\} \end{aligned}$$

where, by Lemma 1.3, V_i's are i.i.d. uniforms on $[-C, 1-C]$.

Since $\frac{1}{t}\tilde{V}_i := x_i^{(1)} - p_i$ and, thus, $\frac{1}{t}\sum_{i=1}^{n-1}\tilde{V}_i = \sum_{i=1}^{n-1} x_i^{(1)} - \sum_{i=1}^{n-1} p_i$, we obtain $x_n^{(1)} = p_n + \frac{1}{t}\left[\sum_{i=1}^{n-1}\tilde{V}_i\right]_1^1 - \sum_{i=1}^{n-1} x_i^{(1)} + \sum_{i=1}^{n-1} p_i + \frac{1}{t}R_t$. Since $x_N^{(1)} := \sum_{i=1}^{n} x_i^{(1)}$

and $\sum_{i=1}^{n} p_i = 1$, we, finally, conclude that $x_N^{(1)} = 1 + \frac{1}{t}\left[\sum_{i=1}^{n-1}\tilde{V}_i\right]_1^1 + \frac{1}{t}R_t$, or else, $t\left\{x_N^{(1)} - 1\right\} = \left[\sum_{i=1}^{n-1}\tilde{V}_i\right]_1^1 + R_t$.

By virtue of Lemma 1.3 and since $\sigma_1 \neq 0$ means $\sum_{i=1}^{n-1} x_i^{(1)} \neq 1$, we conclude that

$$t\left\{x_N^{(1)} - 1\right\} I\left\{\sum_{i=1}^{n-1} x_i^{(1)} \neq 1\right\} \xrightarrow{w} \left[\sum_{i=1}^{n-1} V_i\right]_1^1 + I\left\{\sum_{i=1}^{n-1} V_i \in (C, C_0]\right\}$$

and

$$t\left\{x_N^{(1)} - 1\right\} I\left\{\sum_{i=1}^{n-1} x_i^{(1)} = 1\right\} \xrightarrow{w} \left[\sum_{i=1}^{n-1} V_i\right]_1^1$$

where V_i's are i.i.d. uniforms over $[-C, 1-C]$.

In particular, as $t \to \infty$,

$$\begin{aligned}\lim Pr\left(x_N^{(1)} = 1\right) &= \lim Pr\left(x_N^{(1)} = 1, \sum_{i=1}^{n-1} x_i^{(1)} \neq 1\right) + \lim Pr\left(x_N^{(1)} = 1, x_n^{(1)} = 0\right) \\ &= \lim Pr\left(x_N^{(1)} = 1, \sum_{i=1}^{n-1} x_i^{(1)} \neq 1\right).\end{aligned}$$

Therefore,

$$\begin{aligned}Pr\left(x_N^{(1)} = 1\right) \underset{t\to\infty}{\longrightarrow} &Pr\left(\left[\sum_{i=1}^{n-1} V_i\right]_1^1 = 0, \sum_{i=1}^{n-1} V_i \notin (C, C_0]\right), \\ &+Pr\left(\left[\sum_{i=1}^{n-1} V_i\right]_1^1 = -1, \sum_{i=1}^{n-1} V_i \notin (C, C_0]\right),\end{aligned}$$

where, obviously, $Pr\left(\left[\sum_{i=1}^{n-1} V_i\right]_1^1 = -1, \sum_{i=1}^{n-1} V_i \notin (C, C_0]\right) = 0$

Hence,

$$\begin{aligned}Pr\left(x_N^{(1)} = 1\right) \underset{t\to\infty}{\longrightarrow} &Pr\left(-1 + C < \sum_{i=1}^{n-1} V_i < C_0, \sum_{i=1}^{n-1} V_i \notin (C, C_0)\right) \\ =&Pr\left(-1 + C < \sum_{i=1}^{n-1} V_i < C\right)\end{aligned}$$

and, therefore, the limiting probability does not depend on C_0.

Similarly, if $C_0 < C$, with $\sigma_1 = t - t\sum_{i=1}^{n-1} x_i^{(1)} \neq 0$ and $\sigma_2 = \sum_{i=1}^{n-1}\tilde{V}_i$, we obtain

$$\begin{aligned}R_t &= I\{\sigma_1 + \sigma_2 \in (C_0, C]\} - I\{\sigma_2 \in (C_0, C]\} \\ &= I\{tp_n \in (C_0, C]\} - I\left\{\sum_{i=1}^{n-1}\tilde{V}_i \in (C_0, C]\right\} \underset{t\to\infty}{\longrightarrow} -I\left\{\sum_{i=1}^{n-1} V_i \in (C_0, C]\right\}.\end{aligned}$$

Therefore,

$$t\left\{x_N^{(1)}-1\right\} I\left\{\sum_{i=1}^{n-1} x_i^{(1)} \neq 1\right\} \xrightarrow{w} \left[\sum_{i=1}^{n-1} V_i\right]_1^1 - I\left\{\sum_{i=1}^{n-1} V_i \in (C_0, C]\right\}$$

and

$$t\left\{x_N^{(1)}-1\right\} I\left\{\sum_{i=1}^{n-1} x_i^{(1)} = 1\right\} \xrightarrow{w} \left[\sum_{i=1}^{n-1} V_i\right]_1^1,$$

where V_i's are i.i.d. uniforms on $[-C, 1-C]$.

In particular, as $t \to \infty$,

$$\begin{aligned}\lim Pr\left(x_N^{(1)}=1\right) &= \lim Pr\left(x_N^{(1)}=1, \sum_{i=1}^{n-1} x_i^{(1)} \neq 1\right) + \lim Pr\left(x_N^{(1)}=1, x_n^{(1)}=0\right)\\ &= \lim Pr\left(x_N^{(1)}=1, \sum_{i=1}^{n-1} x_n^{(1)} \neq 1\right).\end{aligned}$$

Hence, as $t \to \infty$,

$$\begin{aligned}Pr\left\{x_N^{(1)}=1\right\} &\longrightarrow Pr\left(\left[\sum_{i=1}^{n-1} V_i\right]_1^1 = 1, \sum_{i=1}^{n-1} V_i \in (C_0, C]\right)\\ &\quad + Pr\left(\left[\sum_{i=1}^{n-1} V_i\right]_1^1 = 0, \sum_{i=1}^{n-1} V_i \notin (C_0, C]\right)\\ &= Pr\left(C_0 < \sum_{i=1}^{n-1} V_i < C\right) + Pr\left(-1+C < \sum_{i=1}^{n-1} V_i < C_0\right)\\ &= Pr\left(-1+C < \sum_{i=1}^{n-1} V_i < C\right)\end{aligned}$$

and therefore, once again the limiting probability does not depend on C_0.

Next, we wish to find the optimal C that maximizes

$$\left\{Pr\left(-1+C < \sum_{i=1}^{n-1} V_i < C\right) : 0 \le C \le 1,\ V_i\text{'s are i.i.d. uniforms on } [-C, 1-C]\right\}.$$

Define $U_i := V_i + C - \frac{1}{2}$. Then the above maximum becomes

$$\begin{aligned}&\max_{C\in[0,1]}\left\{Pr\left[-1+C < \sum_{i=1}^{n-1}\left(U_i - C + \tfrac{1}{2}\right) \le C\right],\right.\\ &\qquad\qquad\left. U_i\text{'s are i.i.d. uniforms on } \left[-\tfrac{1}{2}, \tfrac{1}{2}\right]\right\}\\ &= \max_{C\in[0,1]}\left\{Pr\left[-1+C+(n-1)C - \frac{n-1}{2} < \sum_{i=1}^{n-1} U_i \le C+(n-1)C - \frac{n-1}{2}\right],\right.\\ &\qquad\qquad\left. U_i\text{'s are i.i.d. uniforms on } \left[-\tfrac{1}{2}, \tfrac{1}{2}\right]\right\}.\end{aligned}$$

Since $\sum_{i=1}^{n-1} U_i$ has a symmetric distribution around zero, the optimal C is determined by the equation $-\left(-1+C+(n-1)C-\frac{n-1}{2}\right)=C+(n-1)C-\frac{n-1}{2}$, which results in $C=\frac{1}{2}$.

Therefore, the limiting probability of $\left\{x_N^{(1)}=1\right\}$, for a 1-stationary rule, $\vec{x}^{(1)}=\rho_t^{(1)}(\vec{p})$ attains its maximum for the rule with divisor points C_0 and C, where C_0 is any point on $[0,1]$ while $C=\frac{1}{2}$.

Thus, we have proven Theorem 1.2 for $\Delta=0$. Next, we let $\Delta\in\{1,2,\ldots\}$ be fixed and we consider the limit of $Pr\left(1-\frac{\Delta}{t}\leq x_N^{(1)}\leq 1+\frac{\Delta}{t}\right)$, as $t\to\infty$. Assuming $C\leq C_0$, we have seen that

$$t\left\{x_N^{(1)}-1\right\}I\left\{\sum_{i=1}^{n-1}x_i^{(1)}\neq 1\right\}\xrightarrow{w}\left[\sum_{i=1}^{n-1}V_i\right]_1^1+I\left\{\sum_{i=1}^{n-1}V_i\in(C,C_0]\right\}$$

and

$$t\left\{x_N^{(1)}-1\right\}I\left\{\sum_{i=1}^{n-1}x_i^{(1)}\neq 1\right\}\xrightarrow{w}\left[\sum_{i=1}^{n-1}V_i\right]_1^1,$$

where V_i's are i.i.d. uniforms on $[-C,1-C]$.

In particular, as $t\to\infty$,

$$\begin{aligned}&\lim Pr\left(1-\frac{\Delta}{t}\leq x_N^{(1)}\leq 1+\frac{\Delta}{t}\right)=\\&\lim Pr\left(1-\frac{\Delta}{t}\leq x_N^{(1)}\leq 1+\frac{\Delta}{t},\sum_{i=1}^{n-1}x_i^{(1)}\neq 1\right)+\\&\lim Pr\left(1-\frac{\Delta}{t}\leq x_N^{(1)}\leq 1+\frac{\Delta}{t},\sum_{i=1}^{n-1}x_i^{(1)}=1\right)=\\&\lim Pr\left(1-\frac{\Delta}{t}\leq x_N^{(1)}\leq 1+\frac{\Delta}{t},\sum_{i=1}^{n-1}x_i^{(1)}\neq 1\right).\end{aligned}$$

Hence,

$$Pr\left(1-\frac{\Delta}{t}\leq x_N^{(1)}\leq 1+\frac{\Delta}{t}\right)\underset{t\to\infty}{\longrightarrow}Pr(-\Delta\leq L_n\leq\Delta)$$

where

$$L_n:=\left[\sum_{i=1}^{n-1}V_i\right]_1^1+I\left\{\sum_{i=1}^{n-1}V_i\in(C,C_0]\right\}.$$

If $\Delta=1$, then

$$\begin{aligned}&Pr(-\Delta\leq L_n\leq\Delta)=Pr(-1\leq L_n\leq 1)=\\&=Pr\left(\left[\sum_{i=1}^{n-1}V_i\right]_1^1=-1,\sum_{i=1}^{n-1}V_i\notin(C,C_0]\right)+Pr\left(\left[\sum_{i=1}^{n-1}V_i\right]_1^1=0,\sum_{i=1}^{n-1}V_i\notin(C,C_0]\right)\\&+Pr\left(\left[\sum_{i=1}^{n-1}V_i\right]_1^1=-1,\sum_{i=1}^{n-1}V_i\in(C,C_0]\right)+Pr\left(\left[\sum_{i=1}^{n-1}V_i\right]_1^1=1,\sum_{i=1}^{n-1}V_i\notin(C,C_0]\right)\end{aligned}$$

$$+ Pr\left(\left[\sum_{i=1}^{n-1} V_i\right]_1^1 = 0, \sum_{i=1}^{n-1} V_i \in (C, C_0]\right)$$

$$= Pr\left(\left[\sum_{i=1}^{n-1} V_i\right]_1^1 = -1\right) + Pr\left(\left[\sum_{i=1}^{n-1} V_i\right]_1^1 = 0, \sum_{i=1}^{n-1} V_i \notin (C, C_0]\right)$$

$$+ Pr\left(\left[\sum_{i=1}^{n-1} V_i\right]_1^1 = 1\right) + Pr\left(\sum_{i=1}^{n-1} V_i = 0, \sum_{i=1}^{n-1} V_i \in (C, C_0]\right)$$

$$= Pr\left(-2 + C < \sum_{i=1}^{n-1} V_i < -1 + C\right) + Pr\left(-1 + C < \sum_{i=1}^{n-1} V_i < C\right)$$

$$+ Pr\left(C_0 < \sum_{i=1}^{n-1} V_i < 1 + C\right) + Pr\left(-1 + C < \sum_{i=1}^{n-1} V_i < C_0\right)$$

$$= Pr\left(-2 + C < \sum_{i=1}^{n-1} V_i < 1 + C\right).$$

Similarly, for $\Delta \geq 2$, we obtain

$$Pr\left(-1 + \frac{\Delta}{t} \leq x_N^{(1)} \leq 1 + \frac{\Delta}{t}\right) \underset{t\to\infty}{\longrightarrow} Pr(-\Delta \leq L_n \leq \Delta)$$

$$= Pr\left(-\Delta - 1 + C \leq \sum_{i=1}^{n-1} V_i \leq \Delta + C\right).$$

Therefore, $\forall \Delta \geq 1$, the limiting probability of $\left(1 - \frac{\Delta}{t} \leq x_N^{(1)} \leq 1 + \frac{\Delta}{t}\right)$, as $t \to \infty$, does not depend on C_0. Next, we wish to find the C which maximizes the limiting probability.

Recall the definition $U_i := V_i + C - \frac{1}{2}$. Then

$$\max\left\{ Pr\left(-\Delta - 1 + C \leq \sum_{i=1}^{n-1} V_i \leq \Delta + C\right) : C \in [0,1], \right.$$
$$\left. V_i\text{'s are i.i.d. uniforms on } [-C, 1 - C] \right\}$$

$$= \max\left\{ Pr\left(-\Delta - 1 + C \leq \sum_{i=1}^{n-1} \left(U_i - C + \tfrac{1}{2}\right) \leq \Delta + C\right) : C \in [0,1], \right.$$
$$\left. U_i\text{'s are i.i.d. uniforms on } \left[-\tfrac{1}{2}, \tfrac{1}{2}\right] \right\}$$

$$= \max_{C\in[0,1]} \left\{ Pr\left(-\Delta - 1 + C + (n-1)(C - \tfrac{1}{2})\right) \leq \sum_{i=1}^{n-1} U_i \leq \Delta + C + (n-1)(C - \tfrac{1}{2}) : \right.$$
$$\left. U_i\text{'s are i.i.d. uniforms on } \left[-\tfrac{1}{2}, \tfrac{1}{2}\right] \right\}.$$

Since $\sum_{i=1}^{n-1} U_i$ has a symmetric distribution, the optimal C in the above maximum is determined by the equation

$$-\Delta - 1 + C + (n-1)(C - \tfrac{1}{2}) = -\left[\Delta + C + (n-1)(C - \tfrac{1}{2})\right],$$

that is, $C = \frac{1}{2}$.

Therefore,

$$\max \lim_{t\to\infty} Pr\left(1-\frac{\Delta}{t} \le x_N^{(1)} \le 1+\frac{\Delta}{t}\right)$$
$$=Pr\left(-\Delta-1+\tfrac{1}{2} \le \sum_{i=1}^{n-1} V_i \le \Delta+\tfrac{1}{2}\right) = Pr\left(-\Delta-\tfrac{1}{2} \le \sum_{i=1}^{n-1} V_i \le \Delta+\tfrac{1}{2}\right),$$

where V_i's are i.i.d. uniforms on $\left[-\frac{1}{2},\frac{1}{2}\right]$. Hence, if $\vec{x}$ is obtained by the Mosteller-Youtz-Zahn divisor rule $\vec{x}=\rho_t(\vec{p})$,

$$\max \lim_{t\to\infty} Pr\left(1-\frac{\Delta}{t} \le x_N^{(1)} \le 1+\frac{\Delta}{t}\right) \equiv \lim_{t\to\infty} Pr\left(1-\frac{\Delta}{t} \le x_N \le 1+\frac{\Delta}{t}\right),$$

$\forall\Delta = 0,1,2,\ldots$

Lemma 1.5

Let $\vec{m}=(m_1,\ldots,m_{n-1})$ be a vector of integers $m_i \ge K$ whose sum is at most $t-n+1$ for t and n fixed, $n>1$ and t large enough. Let

$$A_t(\vec{m}) = \left\{(p_1,\ldots,p_{n-1}) : \frac{m_i}{t} \le p_i < \frac{m_i+1}{t}, i=1,\ldots,n-1\right\}$$

and $A_t = \bigcup_{\vec{m}} A_t(\vec{m})$. Then $Pr(A_t) \to 1$, as $t\to\infty$. Moreover, given $\vec{p}\in A_t$, the random variable

$$\tilde{V}_i = V_i^{\vec{m},t} := t(x_i^{(K)} - p_i), i=1,\ldots,n-1$$

(rounding errors of a K-stationary rule of rounding $\vec{x}^{(K)} = \rho_t^{(K)}(\vec{p})$), are independent and uniformly distributed on $[-C,1-C]$.

Lemma 1.6

For every integer $\sigma_1 \notin \{0,-1,\ldots,-K+1\}$ and every continuous random variable, the K-stationary rule of $1/t$-rounding $\vec{x}^{(K)} = \rho_t^{(K)}(\vec{p})$ (1.4), gives the following:

$$\left[\frac{\sigma_1+\sigma_2}{t}\right]_t^K = \frac{\sigma_1}{t} + \frac{1}{t}[\sigma_2]_1^K + \frac{1}{t}\sum_{j=0}^{K-1}\left(I\{\sigma_2\in[j,j+C_j]\} - I\{\sigma_2\in[j,j+C]\}\right)$$
$$+\frac{1}{t}\sum_{j=0}^{K-1}\left(I\{\sigma_1+\sigma_2\in[j,j+C]\} - I\{\sigma_1+\sigma_2\in[j,j+C_j]\}\right).$$

In particular, if $C \le C_j$, $j=0,\ldots,K-1$, then

$$\left[\frac{\sigma_1+\sigma_2}{t}\right]_t^K = \frac{\sigma_1}{t} + \frac{1}{t}[\sigma_2]_1^K + \frac{1}{t}\sum_{j=0}^{K-1} I\{\sigma_2\in(j+C,j+C_j]\}$$
$$+\frac{1}{t}\sum_{j=0}^{K-1} I\{\sigma_1+\sigma_2\in(j+C,j+C_j]\}.$$

Sketch of proof of Theorem 1.1

Applying Lemma 1.6, and using the expression $x_i^{(K)} = [p_i]_t^K = p_i + \frac{1}{t}\tilde{V}_i$, $i = 1, \ldots, n-1$ we obtain the following expression for $x_n^{(K)}$:

$$x_n^{(K)} = [p_n]_t^K = \left[1 - \sum_{i=1}^{n-1} x_i^{(K)} + \frac{1}{t}\sum_{i=1}^{n-1} \tilde{V}_i\right]_t^K = \left[\frac{\left(t - t\sum_{i=1}^{n-1} x_i^{(K)}\right) + \sum_{i=1}^{n-1}\tilde{V}_i}{t}\right]_t^K$$

$$= 1 - \sum_{i=1}^{n-1} x_i^{(K)} + \frac{1}{t}\left[\sum_{i=1}^{n-1} V_i\right]_1^K + \frac{1}{t}R_t,$$

where for R_t we have the following:

If $C \le C_j$, $j = 0, 1, \ldots, K-1$ (the general case can be handled in the same way) and $\sigma_1 \notin \{0, -1, \ldots, -K+1\}$,

$$R_t = \sum_{j=0}^{K-1} I\left\{\sigma_2 \in \left(j + C, j + C_j\right]\right\} - \sum_{j=0}^{K-1} I\left\{\sigma_1 + \sigma_2 \in \left(j + C, j + C_j\right]\right\},$$

where $\sigma_1 = t - t\sum_{i=1}^{n-1} x_i^{(K)}$ and $\sigma_2 = \sum_{i=1}^{n-1}\tilde{V}_i$. Consequently, $\sigma_1 + \sigma_2 = tp_n$ and, as $t \to \infty$, $\sum_{j=0}^{K-1} I\left\{\sigma_1 + \sigma_2 \in \left(j + C, j + C_j\right]\right\} \to 0$, with probability 1. On the other hand, by Lemma 1.5,

$$\sum_{j=0}^{K-1} I\left\{\sigma_2 \in \left(j + C, j + C_j\right]\right\} \underset{t\to\infty}{\longrightarrow} \sum_{j=0}^{K-1} I\left\{\sum_{j=0}^{K-1} V_i \in \left(j + C, j + C_j\right]\right\},$$

where V_i's are i.i.d. uniforms on $[-C, 1-C]$. Summing up all expressions of x_i, $i = 1, \ldots, n$, we obtain

$$x_N^{(K)} = \sum_{i=1}^{n-1} x_i^{(K)} = 1 + \frac{1}{t}\left[\sum_{i=1}^{n-1} V_i\right]_1^K + \frac{1}{t}R_t$$

and consequently,

$$t\left\{x_N^{(K)} - 1\right\} = \left[\sum_{i=1}^{n-1} V_i\right]_1^K + R_t.$$

Recall that, to apply Claim 1.5, we must assume $\sigma_1 = t - t\sum_{i=1}^{n-1} x_i^{(K)} \notin \{0, -1, \ldots, -K+1\}$. Then, as $t \to \infty$,

$$\lim Pr(x_N^{(K)} = 1) = \lim Pr\left(x_N^{(K)} = 1, \sum_{i=1}^{n-1} x_i^{(K)} \neq 1 - \frac{\Delta}{t}, \forall\Delta \in \{0, 1, \ldots, K-1\}\right)$$

$$+ \lim Pr\left(x_N^{(K)} = 1, \sum_{i=1}^{n-1} x_i^{(K)} = 1 - \frac{\Delta}{t}, \text{ for some } \Delta \in \{0, 1, \ldots, K-1\}\right).$$

The probability of the second term on the right-hand side is, in fact, equal to the $Pr\left(x_N^{(K)} = 1, x_n^{(K)} = \frac{\Delta}{t} \text{ for some } \Delta \in \{0, 1, \ldots, K-1\}\right)$ and, as $t \to \infty$, it converges to 0, since $\vec{p}$ is uniformly distributed on the simplex S_n. Therefore,

$$\lim_{t\to\infty} Pr\left(x_N^{(K)} = 1\right) = \lim_{t\to\infty} Pr\left(t\left\{x_N^{(K)} - 1\right\} = 0, \sigma_1 \notin \{0, -1, \ldots, -K+1\}\right)$$

$$= Pr\left(\left[\sum_{i=1}^{n-1} V_i\right]_1^K + \sum_{j=0}^{K-1} I\left\{\sum_{i=1}^{n-1} V_i \in (j+C, j+C_j]\right\} = 0\right).$$

The latter probability can be expressed as a sum of K terms, say $T_0, \ldots, T_{K-1}$ where

$$T_0 = Pr\left(\left[\sum_{i=1}^{n-1} V_i\right]_1^K = 0 \text{ and } I\left\{\sum_{i=1}^{n-1} V_i \in (j+C, j+C_j]\right\} = 0 \ \ \forall j \in \{0, \ldots, K-1\}\right)$$

$$T_1 = Pr\left(\left[\sum_{i=1}^{n-1} V_i\right]_1^K = -1, I\left\{\sum_{i=1}^{n-1} V_i \in (j_0+C, j_0+C_{j_0}]\right\} = 1, \right.$$

$$\text{for some } j_0 \in \{0, \ldots, K-1\}$$

$$\left. \text{and } I\left\{\sum_{i=1}^{n-1} V_i \in (j+C, j+C_j]\right\} = 0 \ \ \forall j \neq j_0, j \in \{0, \ldots, K-1\}\right)$$

and so on,

$$T_{K-1} = Pr\left(\left[\sum_{i=1}^{n-1} V_i\right]_1^K = -K \text{ and } I\left\{\sum_{i=1}^{n-1} V_i \in (j+C, j+C_j]\right\} = 1 \right.$$

$$\left. \forall j \in \{0, \ldots, K-1\}\right)$$

Note that $T_1 = T_2 = \ldots = T_{k-1} = 0$. Consequently,

$$\lim Pr\left(x_N^{(K)} = 1\right) = Pr\left(-1 + C < \sum_{i=1}^{n-1} V_i < C_0 \text{ and } \sum_{i=1}^{n-1} V_i \notin (j+C, j+C_j), \right.$$

$$\left. \forall j \in \{0, 1, \ldots, K-1\}\right)$$

$$= Pr\left(-1 + C < \sum_{i=1}^{n-1} V_i < C\right).$$

Hence the limiting probability does not depend on $C_0, \ldots, C_{K-1}$.

The rest of the proof parallels that of Theorem 1.2 and leads us to the following conclusion:

The limiting probability of $\left\{x_N^{(K)} = 1\right\}$ for a K-stationary divisor rule $\vec{x}^{(K)} = \rho_t^{(K)}(\vec{p})$ attains its maximum for the rule with divisor points $C_0, C_1, \ldots, C_{k-1}, C$ where C_j, $0 \leq j \leq K-1$ may be any point on $[0,1]$ while $C = \frac{1}{2}$. Moreover, the maximum of the $\lim_{t\to\infty} Pr\left(1 - \frac{\Delta}{t} \leq x_N^{(K)} \leq 1 + \frac{\Delta}{t}\right)$, $\forall\Delta = 0, 1, 2, \ldots$ is attained by the K-stationary rule with divisor point $C_0, C_1, \ldots, C_{K-1}, C$ as described above.

In addition, if $\vec{x}$ is obtained by the Mosteller-Youtz-Zahn divisor rule $\vec{x} = \rho_t(\vec{p})$, then

$$\max \lim_{t\to\infty} Pr\left(1 - \frac{\Delta}{t} \leq x_N^{(K)} \leq 1 + \frac{\Delta}{t}\right) = \lim_{t\to\infty} Pr\left(1 - \frac{\Delta}{t} \leq x_N \leq 1 + \frac{\Delta}{t}\right).$$

1.3. Simulation Studies

Simulation studies have been conducted to support our theoretical results: Suppose $\vec{p} = (p_1, \ldots, p_n)$ is uniformly distributed over the simplex S_n and $\vec{x}^{(MYZ)} = (x_1, \ldots, x_n)$, $\vec{x}^{(1)} = \left(x_1^{(1)}, \ldots, x_n^{(1)}\right)$ and $\vec{x}^{(2)} = \left(x_1^{(2)}, \ldots, x_n^{(2)}\right)$ are the $1/t$-roundings of $\vec{p}$ obtained by the Mosteller-Youtz-Zahn rule, 1-stationary and 2-stationary rules respectively. Then, if $x_N^{(\bullet)} = x_1^{(\bullet)} + \ldots + x_n^{(\bullet)}$,

$$\lim_{t\to\infty} Pr\left(x_N^{(MYZ)} = 1\right) = \lim_{t\to\infty} Pr\left(x_N^{(1)} = 1\right)$$

$$= \lim_{t\to\infty} Pr\left(x_N^{(2)} = 1\right) = \sqrt{\frac{6}{\pi(n-1)}} + O\left(\frac{1}{\sqrt{n^3}}\right) \quad (1.6)$$

where the second term of the sum is equal to $-\sqrt{\frac{3}{2\pi(n-1)^3}} + O\left(\frac{1}{\sqrt{n^5}}\right)$.

Our simulations (see Tables 1.1–1.5) show that for $n \geq 100$, the numerical results approach the theoretical results of (1.6) when we round in the sixth or seventh decimal point, that is, for $t = 10^6$ or $t = 10^7$. If we wish to obtain precision up to the second term on the right-hand side of (1.6), we need, first, to consider for rounding at least 10^6 vectors $\vec{p}$ and second, to round at least to the 10th decimal point.

In our simulations, the C_0 of the 1-stationary rule and the C_0, C_1 of the 2-stationary rule have been assigned values taken from the interval $[0.35, 0.65]$. The further from 0.5 these values are, the larger the rounding number t should be, in order to obtain the first and second equality in (1.6). In order to get the desired results in the cases where C_0 and C_1 take values outside the interval $[0.35, 0.65]$, we need, once more, to round at least to the 10th decimal.

The expected results in (1.6) change subject to changes on the number n of components that each vector $\vec{p}$ consists of. The following table displays the values of $\sqrt{\frac{6}{\pi(n-1)}}$ and $\sqrt{\frac{3}{2\pi(n-1)^3}}$ for several values of n:

n	$\sqrt{\frac{6}{\pi(n-1)}}$	$\sqrt{\frac{3}{2\pi(n-1)^3}}$
100	0.1389	0.000701662
500	0.0619	0.000062005
1000	0.0437	0.000021889
1500	0.0357	0.000011909
2000	0.0309	0.000007733

In rounding five thousand vectors $\vec{p} = (p_1, \ldots, p_n)$ for each $n \in \{100, 500, 1000, 1500, 2000\}$ we obtain the following Tables:

TABLE 1.1

$n = 100$			
t	$P(x_N^{(MYZ)} = 1)$	$P(x_N^{(1)} = 1)$	$P(x_N^{(2)} = 1)$
10^3	0.1340	0.1332	0.1240
10^4	0.1362	0.1346	0.1322
10^5	0.1378	0.1366	0.1360
10^6	0.1380	0.1380	0.1376
10^7	0.1382	0.1382	0.1380

TABLE 1.2

	$n = 500$		
t	$P(x_N^{(MYZ)} = 1)$	$P(x_N^{(1)} = 1)$	$P(x_N^{(2)} = 1)$
10^3	0.0200	0.0102	0.0224
10^4	0.0536	0.0528	0.0506
10^5	0.0582	0.0590	0.0594
10^6	0.0616	0.0614	0.0618
10^7	0.0618	0.0614	0.0616

TABLE 1.3

	$n = 1000$		
t	$P(x_N^{(MYZ)} = 1)$	$P(x_N^{(1)} = 1)$	$P(x_N^{(2)} = 1)$
10^3	0.0	0.0	0.0034
10^4	0.0405	0.0314	0.0342
10^5	0.0416	0.0414	0.0393
10^6	0.0425	0.0420	0.0418
10^7	0.0436	0.0435	0.0436

TABLE 1.4

	$n = 1500$		
t	$P(x_N^{(MYZ)} = 1)$	$P(x_N^{(1)} = 1)$	$P(x_N^{(2)} = 1)$
10^3	0.00	0.00	0.00
10^4	0.0225	0.0225	0.0128
10^5	0.0325	0.0325	0.0315
10^6	0.0345	0.0340	0.0340
10^7	0.0355	0.0355	0.0355

TABLE 1.5

	$n = 2000$		
t	$P(x_N^{(MYZ)} = 1)$	$P(x_N^{(1)} = 1)$	$P(x_N^{(2)} = 1)$
10^3	0.0	0.0	0.0
10^4	0.0165	0.0165	0.0018
10^5	0.0260	0.0260	0.0268
10^6	0.029	0.029	0.028
10^7	0.030	0.030	0.029

2. THE APPORTIONMENT PROBLEM AND ITS APPLICATION IN DETERMINING POLITICAL REPRESENTATION

The apportionment problem arises every time we are required to round fractions so that their sum is maintained at some given constant value. It appears in many situations, for example, in allocating seats of a legislature according to the populations of regions or to party votes, in assigning faculty to colleges or departments, in allotting service facilities (courts, judges, or hospitals) to areas in proportion to the number of people to be served. In this section we study the problem of fair representation. What is a fair way to determine political representation in democratic institutions?

The aim is that no individual should have a greater voice than another: a region should receive a number of representatives in proportion to its population or a party in proportion to its total vote.

Although proportionality seems to be the solution, it cannot be met in practice. The problem that arises is what to do about the fractions. A representative cannot be cut in pieces!

In section 2.1, we refer to the history of the problem of fair representation and give some necessary definitions and notations. In section 2.2, we comment on the Hamilton method, currently being used in Greece to determine the allocation of the seats of the Greek parliament to the 56 districts. We do the same for the Hill method, currently being used in the United States to determine the allocation of the seats of the House of Representatives to the 50 states. We point out the weak and strong points of each of these methods and we offer alternatives which may be considered for future apportionments.

2.1 History of the Problem of Fair Representation. Notations and Preliminaries

In the United States, there have been many debates over the choice of method to solve the apportionment problem. Beginning at the Constitutional Convention in 1787, the issue came up again in 1791, after the first census was reported, and resurfaces every ten years with the completion of every census. The most recent debate was on March 31, 1992 (New York Times, April 1, 1992) when the Supreme Court upheld the constitutionality of the current method of apportionment (Hill's method or method of Equal Proportions) denying to grant the state of Montana an extra seat.

Next, we define the quota, the fair share and the quotient of a region. The *quota* q_i for each region i is found by dividing the region's population p_i by the total population P and then multiplying by the total number h or seats to be apportioned, i.e., $q_i = \frac{p_i h}{P}$.

The *quotient* Q_i of a region is computed by dividing the region's population by a divisor d, where the value of d depends on the method to be followed for apportioning seats into regions. Specifically, d must be selected in such a way, so that appropriate rounding of the resulting quotients produce number a_i that sum to the given house size h.

In many countries, there are prescribed "floors" and "ceilings" on the regions' permissible allotments. For every region i, $f_i \leq a_i \leq c_i$ where f_i and c_i are the minimum and maximum number of seats that the ith region may receive, and a_i is the number of seats that the ith region receives. The *fair share* s_i of the ith region is defined as $s_i = \text{med}\{f_i, \lambda a_i, c_i\}$, where q_i is the quota of the ith region and λ is such that $\sum_i s_i = h$. Clearly, if there are no floors and ceilings then $s_i \equiv q_i$ for every region i.

A region should receive its fair share but, most of the time, the fair share is not an integer number. The matter has concerned many mathematicians and noted statesmen. This fact attests both to the complexity of the problem and to its profound political consequences. For example, in U.S.A., some of the greatest disputes over method have been over the allocation of a single seat.

The most popular of the methods of apportionment existing around the world are those of Hamilton, Lowndes, Jefferson, Adams, Dean, Hill and Webster (Balinsky and Young, 1982). The last five are referred as *divisor* methods of apportionment.

An apportionment problem is given by a vector of populations $\vec{p} = (p_1, \ldots, p_n)$ of n regions, a total number of seats h to be distributed among the n regions, a vector of floors $\vec{f} = (f_1, \ldots, f_n)$ and a vector of ceilings $\vec{c} = (c_1, \ldots, c_n)$. An apportionment of h is a vector $\vec{a} = (a_1, \ldots, a_n)$ where $\forall i = 1, \ldots, n$, a_i is nonnegative integer, $f_i \leq a_i \leq c_i$

and $\sum_{i=1}^{n} a_i = h$. Given an apportionment problem $(\vec{p}, h, \vec{f}, \vec{c})$, the objective of an apportionment problem is to find a vector $\vec{a}$ of apportionments that most closely approximates the vector of fair share $\vec{s} = \{s_1, \ldots, s_n\}$. But how should we measure the approximation? And, is the optimization of this measure enough to determine the most fair method?

Many "paradoxes" that a fair method should avoid have arisen in the history of the United States. Some well-known paradoxes are the "Alabama paradox," the "Population paradox," and the "New States paradox".

An apportionment method must be *house monotone, population monotone,* and *quota monotone.* A fair apportionment method should stay *within fair share* or at least *near fair share.* It should not favor systematically large states at the expense of the small nor the small at the expense of the large, that is, it should be unbiased. An *unbiased* method is one that sometimes favors the large states and sometimes favors the small, but over many problems these advantages balance out.

Some other properties that a method should enjoy are homogeneity, symmetry, exactness, and uniformity. All the methods to which we referred above are homogeneous, symmetric and exact. A method is uniform and exact if and only if it is a divisor method. A method is population monotone if and only if it is a divisor method. Every divisor method is house monotone and has particular solutions that avoid the Alabama paradox.

There exists no uniform and symmetric method that satisfies fair share. A method is uniform and near the fair share if and only if it is the Webster method. Empirical observation and theory show that Webster's is the unique unbiased divisor method (Balinsky and Young, 1982).

It is clear from the above paragraphs that none of the methods we have mentioned has all the desirable properties. It is evident though, that most of the properties hold for the divisor methods.

Any monotone increasing function $d : Z \to R$ with a $\leq d(a) \leq a + 1$ for every $a \in Z$ is said to be a *divisor criterion.* Different divisor methods are based on different divisor criteria d. Define

$$\left[\frac{p_i}{x}\right]_d := \begin{cases} k & \text{if } k < \frac{p_i}{x} < d(k) \text{ or } \frac{p_i}{x} = d(k) \text{ and } k \text{ odd,} \\ k+1 & \text{if } d(k) < \frac{p_i}{x} \leq k+1 \text{ or } \frac{p_i}{x} = d(k) \text{ and } k \text{ even,} \end{cases}$$

where $k \in Z$.

Then, a divisor method based on d is defined, for each problem of apportionment $(\vec{f}, \vec{c}, \vec{p}, h)$, to be that apportionment $\vec{a}$ that satisfies for all i, $a_i = \text{med}\left\{f_i, \left[\frac{p_i}{x}\right]_d, c_i\right\}$, where x is chosen so that $\sum_{i=1}^{n} a_i = h$. The divisor criteria used by the divisor methods of apportionment we introduced earlier are:

For every $k \in Z$

Adams	:	$d(k) = k,$	
Dean	:	$d(k) = \dfrac{k(k+1)}{k+1/2}$	(Harmonic mean of k and $k+1$),
Hill	:	$d(k) = \sqrt{k(k+1)}$	(Geometric mean of k and $k+1$),
Webster	:	$d(k) = k + 1/2$	(Arithmetic mean of k and $k+1$),
Jefferson	:	$d(k) = k + 1.$	

In this point we introduce the divisor criteria for some of the divisor methods we described in Section 1.

1-Stationary: $d(k) = k + 1/2$ if $k \neq 0,\ k \in Z,\ d(0) \in [0, 1]$,

2-Stationary: $d(k) = k + 1/2$ if $k \neq 0, 1,\ k \in Z,\ d(0) \in [0, 1],\ d(1) \in [0, 1]$.

After applying an apportionment method, between any two regions there will practically be a certain inequality which gives one of the regions a slight advantage over the other. A transfer of one representative from the more favored region to the less favored should be made if the transfer will cause a decrease in the amount of the inequality between the two regions. An apportionment method is called *stable* if no transfer is justified.

The question now that arises is what the measure of inequality should be. The U.S. Constitution expresses two ideals that are suitable for measuring inequalities: Firstly, every representative should have as nearly as possible the same number of constituents, i.e., the sizes of congressional districts should be as nearly as possible equal. Secondly, every inhabitant, no matter in what region he lives, should have as nearly as possible the same representation in the House of Representatives. Note that the size of congressional district in the ith region is given by the ratio $\frac{p_i}{a_i}$, where p_i is the population of the ith region and a_i is the number of seats apportioned to the ith region. On the other hand, the number of representatives per person in the ith region is given by the ratio $\frac{a_i}{p_i}$.

Another approach to select an apportionment method is the optimization of some function. Ideally, one would like to have the a_i "close" to the fair share $s_i = \text{med}\{f_i, c_i, \lambda q_i\}$ for every i. So, naturally, one may select that apportionment method that minimizes $\sum_i |a_i - s_i|$ or $\sum_i (a_i - s_i)^2$.

2.2 Commenting on the Apportionment Methods Used in Greece and in U.S.A. in Determining Political Representation

In Greece, from 1926 until today, very rarely can one find two consecutive elections with exactly the same electoral system. In the last several decades, there is a "tradition" of changing the electoral system. Typically, the changes are taking place a few months before the elections, so the government's chances for reelection are maximized. It is worth mentioning that only in 1954 the government decided to vote for a new electoral law in the middle of its term. This is the only electoral law that, although voted by the parliament, was never used, since the same governmental majority replaced it by another one a few days before the elections of 1956! This bad tradition was made more official, in a way, when the 1974–1978 government rejected a request of the opposition for a stable electoral system. It may be that, the majority of the Greek politicians view the electoral system as the most easily manipulated mechanism of the politics. That is why the electoral system, part of which describes the method of apportionment, is always the center of political discussions, with increasing intensity as the election day approaches. Despite the significance of the electoral system and the fiery discussions about it, there are hardly any scientific monographs, books or papers in Greece that give some analysis of the systems and their characteristics. And, in particular, there is no in-depth study of the methods of apportionment that governments have or could have been using. The limited bibliography is mainly concerned with the effects of the several electoral systems on the operation of the democratic form of government, rather than present a systematic and comparative analysis of the characteristics of the several systems.

Nikolakopoulos, I. (1989) studies the electoral systems that have been applied in some countries, and in particular, analyzes the characteristics of the electoral systems in Greece since 1926. There is no investigation on the methods of apportionment that governments have been using to apportion the seats of the parliament to the districts.

Studying the legislative decrees of Greece concerning the election of the deputees and, in particular, the methods of apportionment from 1929–1990, we find the following:

It was in 1956 when the Greek legislation first introduced the Hamilton method, in the form that is stated in Balinski and Young, 1982. The objective was to apportion the seats of the parliament into the electoral districts. In the years before 1956, the method used was some variation of the Hamilton method.

Currently, the Hamilton method is being used to apportion 288 of the 300 parliamentary seats to the electoral districts (the remaining 12 are given to the parties according to their country-wide percentages received on the election day). Hamilton's approach is simple and direct. It seems reasonable and natural. The major advantage of this method is that it results apportionments that stay within the fair share. Thus minimizing $\sum |a_i - s_i|$, $\sum (a_i - s_i)^2$ and actually, any ℓ_p-norm of $\vec{a} - \vec{s}$ (Birkhoff, 1976). There are some paradoxes of Hamilton's method. One of them may occur when the size of the parliament changes. Another occurs when districts' populations change.

More analytically, one may notice that although a district may be given x seats when the size of the parliament is y, it is given only $x - 1$ when the size becomes $y + 1$. For example, we observe that if 290 seats were to be allotted to the 56 districts (based on the census of 1981) then the district of "Elea" would receive 7 seats while if 291 seats were to be allotted the district of "Elea" would receive only 6 seats (see Table 2.1).

Another problem with Hamilton's approach is that it is not uniform. For example, given the Hamilton-apportionment of the 288 seats to the 56 districts of Greece, say a_i, $i \in I = \{1, \ldots, 56\}$ (Table 2.1), we find a subset of districts with populations p_j, $j \in J \subset I$ that when considered alone for apportionment of its corresponding number of seats $\sum_{i \in J} a_i$, does not receive the preassigned apportionment (Table 2.2). Specifically, we consider a subset of 10 districts whose original allotment was a total of 93 seats. Applying the Hamilton method to apportion these 93 seats to the 10 districts we observe that the districts of "2nd Athenean community" and "Pieria" receive 33 and 3 seats respectively while their original apportionment was 32 and 4 seats respectively.

Thus, the Hamilton method used in Greece to apportion the seats of the parliament to the districts, although stays within the fair share and its apportionments minimize any ℓ_p-norm of $\vec{a} - \vec{s}$, it is not uniform and it falls into paradoxes that may be very critical. Achieving apportionments that accurately reflect relative changes in populations or changes in the size of the parliament seems more important than always staying within the fair share.

Among the five divisor methods, Webster's method is the unique population monotone that is near fair share interpreted absolutely ($|a_i - s_i|$) or relatively $\left(\frac{a_i - s_i}{s_i}\right)$, and it is stable for the test $\frac{a_j}{p_j} - \frac{a_i}{p_i}$ (i.e., no transfer of a seat from one region to another will make this difference smaller).

As one can see in Table 2.1, the Hamilton's and Webster's apportionments on the basis of the 1981 census, happen to coincide. We also investigate the apportionments given by 1-stationary and 2-stationary methods. The fact that each district should receive at least one seat, makes the divisor point $d(0)$ powerless in both methods. In Table 2.4, we display the 1-stationary apportionment. Also displayed are the 2-stationary apportionments for selected values of the divisor point $d(1) \in [0, 1]$.

The 1-stationary apportionment coincides with those of Webster and Hamilton. No district can be brought closer to its fair share without moving another state further from its fair share. On the other hand, the 2-stationary apportionments are slightly different than the 1-stationary one and, consequently, it is possible to take a seat from a state and given it to another and simultaneously bring both of them closer to their fair shares.

In the United States, on the other hand, the method currently being used to apportion the 435 seats of the House of Representatives to the 50 states, is the method of Hill or method of equal proportions. The method of equal proportions minimizes the percentage differences in the proportion (or ratio) of representation in the House among all possible pairs of States, regardless of their size (population). This is true whether representation is calculated on the basis of (a) the number of Representatives per million population, or (b) the population per Representative. This method has been used since 1941 and recently (March 31, 1992), the Supreme Court upheld the constitutionality of the method when Montana asked for an extra seat taken by the State of Washington. As it can be seen in Table 2.3, Montana would maintain both seats it was given in the decade of 1980 if the apportionment was done with one the following methods: Dean, Lowndes or Adams.

The Hill-apportionment, on the basis of the census of 1990, is shown in Table 2.3. In the 1990 census, for only the second time since 1900, the Census Bureau allocated the Department of Defense's overseas employees to particular States for reapportionment purposes, using an allocation method that is determined most closely resembling "usual residence," its standard measure of state affiliation. The 1990 reapportionment, which is based on the populations that include the overseas employees, is shown in Table 2.3. We notice, as Massachusetts did, that by including the overseas employees there was a shift of a seat from the State of Massachusetts to Washington State. Massachusetts appealed to the President and the Secretary of Commerce (April 21, 1992) but they had no luck as their appeal was denied (Decision was taken on June 26, 1992).

In Table 2.3, apart from the Hill apportionment we display the ones due to the Webster, Dean, Lowndes and Adams one. As it can be seem in Table 2.3, Montana would receive two seats if the apportionment was done with one of the following methods: Dean, Lowndes or Adams. Note that only Lowndes method justifies Montana's request for a second seat taken from Washington State.

Hill's method or method of equal proportions does not satisfy fair share, i.e., its apportionments are not always within fair share. Instead it satisfies the principle of pairwise comparisons between regions. Namely, Hill's method is stable for the relative difference between the average district sizes $\left(\frac{p_i/a_i - p_j/a_j}{p_j/a_j} = \frac{p_i a_j}{p_j a_i} - 1\text{, where } \frac{p_i}{a_i} \geq \frac{p_j}{a_j}\right)$. On the other hand, as Balinski and Young observe, the probability that Webster's method will violate fair share is only about 1 in 1,600 apportionments, where Hill's method is five times as likely. Also, even though Webster's method does not stay within the fair share all of the time, it does stay near the fair share all of the time, whether measured in absolute or relative terms. And, it is the only divisor method that does so.

Comparing the 1990 apportionments of Hill and Webster, based on the population that includes the overseas employees, we find that they differ in two states: Oklahoma and Massachusetts. Webster takes a seat from Oklahoma and gives it to Massachusetts. Observe that Massachusetts would receive an extra seat if either the overseas employees were not included in the population or if, instead of the Hill's method, the Webster's method was used for the apportionment.

Note that, if i =Oklahoma and j =Massachusetts, then under Hill's method $\frac{a_i}{p_i} = 0.000002102$ and $\frac{a_j}{p_j} = 0.000001658$, i.e., in Oklahoma there are 2.102 representatives for every million people where, in Massachusetts, there are 1.658 representatives for every million people, a difference of 0.443 per million people in favor of Oklahoma. However, under Webster's method, this difference becomes 0.072 per million people in favor of Massachusetts. So, if our goal is to minimize the difference $\frac{a_i}{p_i} - \frac{a_j}{p_j}$, the Webster's method prevails.

On the other hand, as one can see in Table 2.5, 1-stationary coincides with Webster's apportionment. Probably these two methods would not coincide in the case where a state could receive no seat (and, of course, using an appropriate divisor point $d(0)$ for 1-stationary). The 2-stationary with $0.4 \leq d(1) \leq 0.6$ results exactly the same apportionment with that of Webster's and 1-stationary. With $d(1) \leq 0.3$, 2-stationary helps the very small states, where, with $d(1) \geq 0.7$, it hammers them. Note that, the set of 2-stationary methods contains the set of 1-stationary methods and this, in its turn, contains the Webster method. This fact gives us the opportunity to select the divisor points $d(0)$ and $d(1)$ in $[0, 1]$, and, if possible, to succeed in obtaining a better method. This will depend also on the distribution of the population in study, as well as on the criteria based on which one may choose the best method of apportionment.

TABLE 2.1

Greek Districts	Population	Hamilton's apportionment of 290 seats	Hamilton's apportionment of 291 seats	Hamilton's apportionment of 288 seats	Webster's apportionment of 288 seats
1st Athenean community	715840	21	22	21	21
2nd Athenean community	1083830	32	33	32	32
1st Pirean community	257695	8	8	8	8
2nd Pirean community	273614	8	8	8	8
Etolia & Akarnania	284954	9	9	8	8
Attiki	226527	7	7	7	7
Veotia	127783	4	4	4	4
Evia	211440	6	6	6	6
Evritania	42258	1	1	1	1
Phtiotida	188808	6	6	6	6
Fokida	56674	2	2	2	2
Argolida	98584	3	3	3	3
Arkadia	143782	4	4	4	4
Ahaia	285069	9	9	9	9
Elea	217371	7	6	6	6
Korynthia	135199	4	4	4	4
Lakonia	113042	3	3	3	3
Messinia	218746	7	7	7	7
Zakynthos	37979	1	1	1	1
Kerkyra	110606	3	3	3	3
Kephalenia	46165	1	1	1	1
Leucada	31088	1	1	1	1
Arta	106492	3	3	3	3
Thesprotia	54364	2	2	2	2
Yannina	187460	6	6	6	6
Preveza	71319	2	2	2	2
Karditsa	179148	5	5	5	5
Larissa	263134	8	8	8	8
Magnisia	186771	6	6	6	6
Trikala	171761	5	5	5	5
Grevena	52658	2	2	2	2
Drama	119115	4	4	4	4
Hemathia	139209	4	4	4	4
1st Section of Saloniki	420833	13	13	13	13
2nd Section of Saloniki	218156	7	7	7	7
Kavala	150389	4	4	4	4
Kastoria	52076	2	2	2	2
Kilkis	107786	3	3	3	3
Kozani	167382	5	5	5	5
Pella	154990	5	5	5	5
Pieria	118354	4	4	4	4
Serres	264777	8	8	8	8
Florina	63029	2	2	2	2
Halkidiki	97777	3	3	3	3
Evros	161300	5	5	5	5
Xanthi	97990	3	3	3	3
Rodopi	114545	3	3	3	3
DodeKanisa	134654	4	4	4	4
Cyclades	115369	3	3	3	3
Lesvos	128472	4	4	4	4
Samos	49380	1	1	1	1
Khios	60315	2	2	2	2
Iraklio	249302	7	7	7	7
Lasythi	82222	2	2	2	2
Rethymno	81051	2	2	2	2
Khania	137504	4	4	4	4
TOTAL	**9666138**	**290**	**291**	**288**	**288**

TABLE 2.2

District	Population based on 1981 census	Hamilton's apportionment of 93 seats	Corresponding numbers from Hamilton's apportionment of 288 seats of 56 districts
1st Athenean community	715840	21	21
2nd Athenean community	1083830	33	32
1st Pirean community	257695	8	8
2nd Pirean community	273614	8	8
Kastoria	52076	2	2
Kilkis	107786	3	3
Pella	154990	5	5
Pieria	118354	3	4
Serres	264777	8	8
Florina	63029	2	2
TOTAL	**3091991**	**93**	**93**

TABLE 2.3

States of USA	Population in 1990 including overseas	Hill's app. in 1990	Webster's app. in 1990	Dean's app.	Lownde's app.	Adam's app.
Alabama	4062608	7	7	7	7	7
Alaska	551947	1	1	1	1	1
Arizona	3677985	6	6	7	6	7
Arkansas	2362239	4	4	4	4	4
California	29839250	52	52	50	52	50
Colorado	3307912	6	6	6	6	6
Connecticut	3295669	6	6	6	6	6
Delaware	668696	1	1	2	2	2
Florida	13003362	23	23	22	22	22
Georgia	6508419	11	11	11	11	11
Hawaii	1115274	2	2	2	2	2
Idaho	1011986	2	2	2	2	2
Illinois	11466682	20	20	19	20	19
Indiana	5564228	10	10	10	10	10
Iowa	2787424	5	5	5	5	5
Kansas	2485600	4	4	5	5	5
Kentucky	3698969	6	6	7	7	7
Louisiana	4238216	7	7	8	7	8
Maine	1233223	2	2	3	3	3
Maryland	4798622	8	8	8	8	8
Massachusetts	6029051	10	11	10	10	10
Michigan	9328784	16	16	16	16	16
Minnesota	4387029	8	8	8	8	8
Mississippi	2586443	5	5	5	5	5
Missouri	5137804	9	9	9	9	9
Montana	803655	1	1	2	2	2
Nebraska	1584617	3	3	3	3	3
Nevada	1206152	2	2	2	2	2
New Hampshire	1113915	2	2	2	2	2
New Jersey	7748634	13	14	13	13	13
New Mexico	1521779	3	3	3	3	3
New York	18044505	31	31	30	31	30
North Carolina	6657630	12	12	11	11	11
North Dakota	641364	1	1	2	2	2
Ohio	10887325	19	19	18	19	18
Oklahoma	3157604	6	5	6	6	6
Oregon	2853733	5	5	5	5	5
Pennsylvania	11924710	21	21	20	20	20
Rhode Island	1005984	2	2	2	2	2
South Carolina	3505707	6	6	6	6	6
South Dakota	699999	1	1	2	2	2
Tennessee	4896641	9	9	9	8	9
Texas	17059805	30	30	29	29	29
Utah	1727784	3	3	3	3	3
Vermont	564964	1	1	1	1	1
Virginia	6216568	11	11	11	11	11
Washington	4887941	9	9	9	8	9
West Virginia	1801625	3	3	3	3	3
Wisconsin	4906745	9	9	9	8	9
Wyoming	455975	1	1	1	1	1
TOTAL	**249022783**	**435**	**435**	**435**	**435**	**435**

TABLE 2.4

Greek Districts	Fair Share	1-stationary	2-stationary $d(1) = .2$	$d(1) = .4$	$d(1) = .6$	$d(1) = .8$
1st Athenean community	21.322783	21	21	21	21	21
2nd Athenean community	32.28413	32	32	32	32	33
1st Pirean community	7.675981	8	8	8	8	8
2nd Pirean community	8.150162	8	8	8	8	8
Etolia & Akarnania	8.487947	8	8	8	9	9
Attiki	6.747578	7	7	7	7	7
Veotia	3.806282	4	4	4	4	4
Evia	6.29818	6	6	6	6	6
Evritania	1.258742	1	2	1	1	1
Phtiotida	5.624039	6	6	6	6	6
Fokida	1.688153	2	2	2	2	1
Argolida	2.936529	2	3	3	3	3
Arkadia	4.282846	4	4	4	4	4
Ahaia	8.491373	9	8	8	9	9
Elea	6.474847	6	6	6	7	7
Korynthia	4.027183	4	4	4	4	4
Lakonia	3.367191	3	3	3	3	3
Messinia	6.515804	7	6	7	7	7
Zakynthos	1.131283	1	1	1	1	1
Kerkyra	3.29463	3	3	3	3	3
Kephalenia	1.375121	1	2	1	1	1
Leucada	1.000000	1	1	1	1	1
Arta	3.172086	3	3	3	3	3
Thesprotia	1.619345	2	2	2	2	1
Yannina	5.583886	6	6	6	6	6
Preveza	2.124385	2	2	2	2	2
Karditsa	5.336296	5	5	5	5	5
Larissa	7.837993	8	8	8	8	8
Magnisia	5.563363	6	6	6	6	6
Trikala	5.116259	5	5	5	5	5
Grevena	1.568528	2	2	2	1	1
Drama	3.548088	4	4	4	4	4
Hemathia	4.14663	4	4	4	4	4
1st Section of Saloniki	12.535386	13	13	13	13	13
2nd Section of Saloniki	6.49823	7	6	7	7	7
Kavala	4.479649	4	4	4	4	5
Kastoria	1.551192	2	2	2	1	1
Kilkis	3.21063	3	3	3	3	3
Kozani	4.985821	5	5	5	5	5
Pella	4.616699	5	5	5	5	5
Pieria	3.52542	4	4	4	4	4
Serres	7.886934	8	8	8	8	8
Florina	1.87745	2	2	2	2	2
Halkidiki	2.912491	3	3	3	3	3
Evros	4.804656	5	5	5	5	5
Xanthi	2.918836	3	3	3	3	3
Rodopi	3.411961	3	3	3	3	3
DodeKanisa	4.010494	4	4	4	4	4
Cyclades	3.436506	3	3	3	3	3
Lesvos	3.826806	4	4	4	4	4
Samos	1.470886	1	2	2	1	1
Khios	1.796608	2	2	2	2	2
Iraklio	7.425978	7	7	7	7	7
Lasythi	2.449153	2	2	2	2	2
Rethymno	2.414273	2	2	2	2	2
Khania	4.095843	4	4	4	4	4
TOTAL		**288**	**288**	**288**	**288**	**288**

TABLE 2.5

States of USA	Fair Share	1-stationary	2-stationary $d(1) = .2$	$d(1) = .4$	$d(1) = .6$	$d(1) = .8$
Alabama	7.092533	7	7	7	7	7
Alaska	1.000000	1	1	1	1	1
Arizona	6.421055	6	6	6	6	6
Arkansas	4.124016	4	4	4	4	4
California	52.0936	52	52	52	52	52
Colorado	5.774979	6	6	6	6	6
Connecticut	5.753605	6	6	6	6	6
Delaware	1.167415	1	1	1	1	1
Florida	22.701373	23	23	23	23	23
Georgia	11.36245	11	11	11	11	11
Hawaii	1.947054	2	2	2	2	2
Idaho	1.766733	2	2	2	2	1
Illinois	20.018625	20	20	20	20	20
Indiana	9.714074	10	10	10	10	10
Iowa	4.866307	5	5	5	5	5
Kansas	4.33938	4	4	4	4	4
Kentucky	6.45769	6	6	6	6	6
Louisiana	7.399111	7	7	7	7	7
Maine	2.152971	2	2	2	2	2
Maryland	8.377472	8	8	8	8	8
Massachusetts	10.525565	11	10	11	11	11
Michigan	16.286265	16	16	16	16	16
Minnesota	7.65891	8	8	8	8	8
Mississippi	4.515433	5	4	5	5	5
Missouri	8.969619	9	9	9	9	9
Montana	1.403027	1	2	1	1	1
Nebraska	2.766437	3	3	3	3	3
Nevada	2.10571	2	2	2	2	2
New Hampshire	1.944682	2	2	2	2	2
New Jersey	13.527627	13	13	13	13	14
New Mexico	2.656734	3	3	3	3	3
New York	31.50224	31	31	31	31	31
North Carolina	11.622943	12	12	12	12	12
North Dakota	1.119698	1	1	1	1	1
Ohio	19.007179	19	19	19	19	19
Oklahoma	5.51257	5	5	5	5	6
Oregon	4.98207	5	5	5	5	5
Pennsylvania	20.818254	21	21	21	21	21
Rhode Island	1.756255	2	2	2	2	1
South Carolina	6.120291	6	6	6	6	6
South Dakota	1.222064	1	2	1	1	1
Tennessee	8.548595	9	9	9	9	9
Texas	29.783143	30	30	30	30	30
Utah	3.016379	3	3	3	3	3
Vermont	1.000000	1	1	1	1	1
Virginia	10.852934	11	11	11	11	11
Washington	8.533406	9	9	9	9	9
West Virginia	3.145291	3	3	3	3	3
Wisconsin	8.566234	9	9	9	9	9
Wyoming	1.000000	1	1	1	1	1
TOTAL	**435**	**435**	**435**	**435**	**435**	**435**

To summarize, the Hamilton method used in Greece to apportion the seats of the parliament to the districts, although stays within fair share ($\vec{s}$) and its apportionments ($\vec{a}$) minimize any ℓp-norm of $\vec{a} - \vec{s}$, is neither uniform nor house monotone, and it falls into paradoxes that may be of great importance. On the other hand, the Webster's method and 1-stationary method result apportionments that accurately reflect relative changes in population or changes in the size of the parliament. At the same time, very rarely they violate fair share, and they are always near fair share.

In the United States, Hill's method has been the method of apportionment since 1941. It minimizes the difference between the representation in the House of any two states when measured by the relative difference in the average population per district and also by the relative difference in the individual share in a representative. But if the main purpose of apportionment is to give to any group of individuals as nearly as may be the same weight in choosing representatives in the House whether they happen to live in the large states or the small states, then Webster's method is the one that should prevail. As Balinski and Young observe (1982), Hill's method is five times more likely to violate fair share than Webster's method. The latter, together with the 1-stationary method, are the only divisor methods that stay near the fair share all the time whether measured in absolute or relative terms. Moreover, small shifts in population can lead to large shifts in Hill's apportionment.

The truth is that most of the times Hill's, Webster's and 1-stationary apportionments are identical. But the times where the only difference is one or two transferred seats, the decision to favor one method over another must be based on logic and on constitutional principles, and not be driven by political interests.

ACKNOWLEDGMENTS

Credit must be given to my advisor, Svetlozar T. Rachev for his continuous support, his helpful comments and his guidance on technical matters. I wish to acknowledge Michel Balinski for his hospitality and scientific directions during my stay at the Laboratoire d'Econométrie de l'Ecole Polytechnique, in Paris, France.

REFERENCES

Balinski, M.L. and Young, H.P., 1982, Fair Representation: Meeting the Ideal of One Man, One Vote, Yale University, New Haven.

Balinski, M.A., Demange, G., 1989, Algorithms for Proportional Matrices in Reals and Integers, *Math Programming,* North Holland, 45:193–210.

Balinski, M.L. and Rachev, S.T., 1992, Rounding proportions: rules of rounding, *Technical Report No. 384,* Laboratoire d'Econometrie, École Polytechnique.

Billingsley, P., 1986, *Probability and Measure,* 2nd edition, Wiley.

Birkhoff, G., 1976, Monotone apportionment Schemes, *Proceedings of the National Academy of Sciences,* U.S.A., 684–686.

Diaconis, P. and Freedman, D., 1979, On rounding percentages, *Journal of the American Statistical Association,* 74:359–364.

Feller, W., 1970, *An Introduction to Probability Theory and Its Applications,* John Wiley & Sons, New York, Vol. II, 2nd ed., 504–515.

Hoffman, Mark S., 1992, The World Almanac and Book of Facts, 588:74–75.

Legislative decrees concerning the election of the Greek deputies, 1928–1990, Journals of the governments of the Greek Republic, Athens.

Maejima, M., Rachev, S.T., 1987, An ideal metric and the rate of convergence to a self-similar process, *Annals of Probability,* 15:702–727.

Mosteller, F., Youtz, C. and Zahn, D., 1967, The distribution of sums of rounded percentages, *Demography,* 4:850–858.

Nikolakopoulos, I., 1989, Introduction in the Theory and Practice of Electoral Systems, Sakkoula, A., Athens.

Pyke, R., 1965, Spacings, *The Journal of the Royal Statistical Society,* Series B, 27, No. 3:395–449.

Rachev, S.T., 1991, *Probability Metrics and the Stability of Stochastic Models,* Wiley, New York.

Turing, A.M., 1948, Rounding-off errors in matrix processes, *Quart. J. Mech.,* 1:287–308.

Wilkinson, J.H., 1963, *Rounding errors in algebraic processes,* Prentice-Hall, Englewood Cliffs, NJ.

LIMIT THEOREMS FOR RANDOM MULTINOMIAL FORMS

Alfredas Basalykas

Inst. of Mathem. and Inform.
Akademijos 4, 2600, Vilnius, Lithuania

INTRODUCTION

Let $X_1, X_2, \ldots$ be the sequence of i.i.d. (0,1) random variables with finite $2k$ moments for some integer $k \geqslant 2$. By $Q_n^{(k)}$ we denote the multinomial form of order k

$$Q_n^{(k)} = Q_n^{(k)}(X_1, \ldots, X_n) = n^{-k/2} \sum_{1 \leqslant i_1, \ldots, i_k \leqslant n} a_{i_1 \ldots i_k}^{(n)} X_{i_1} \ldots X_{i_k}, \qquad a_{i_1 \ldots i_k}^{(n)} \in R.$$

When $a_{i_1, \ldots, i_k}^{(n)} = 0$ if two or more indices coincide, then $Q_n^{(k)}$ reduces to the multilinear form $\eta_n^{(k)}$ of order k

$$\eta_n^{(k)} = \eta_n^{(k)}(X_1, \ldots, X_n) = n^{-k/2} \sum_{1 \leqslant i_1 \neq \ldots \neq i_k \leqslant n} a_{i_1 \ldots i_k}^{(n)} X_{i_1} \ldots X_{i_k}.$$

In this paper we show that the multiple stochastic Wiener-Ito integrals (and their sums) are the limit distributions for the multilinear and multinomial forms (Theorems 1.2 and 1.3). Also the speed of this convergence and the distance between the distributions of multinomial forms are given (Theorem 1.1). Theorem 1.4 shows that the distribution of the multiple stochastic integral from standard Wiener process can be approximated by the distribution of the multilinear form in i.i.d. random variables (for example in Bernoulli random variables). And finally we give the speed of convergence in CLT for multilinear forms (Theorem 1.5).
Now we introduce the necessary notations and conditions:
$[m]$ – the whole part of the number m,
λ – Lebesque measure,
$\mathbf{1}(U)$ – indicator of event U,

Approximation, Probability, and Related Fields, Edited by
G. Anastassiou and S.T. Rachev, Plenum Press, New York 1994

$W(t)$ – standard Wiener process on [0,1] with $\mathbf{E}|W(t)|^2 = t$ and $W(0) \overset{\text{a.s.}}{=} 0$,
$L(X,Y)$ – Levy metric between the random variables X and Y,
$K_1,\ K_2,\ K_3,\dots,L_1,L_2,\dots$ – some positive constants.
Also $\xrightarrow{d}$ denotes the weak convergence. Let $\Lambda_s \subseteq \{1,2,\dots,n\}$ and $\#\Lambda_s$ means the power of the set Λ_s.

Condition A1. Assume that for $\Lambda_1,\dots,\Lambda_k$: $\prod_{s=1}^{k} \#\Lambda_s \to \infty$ as $n \to \infty$

$$\sum_{i_1\in\Lambda_1,\dots,i_k\in\Lambda_k} \left(a^{(n)}_{i_1\dots i_k}\right)^2 = O\Bigl(\prod_{s=1}^{k} \#\Lambda_s\Bigr). \tag{1}$$

This condition means that $Q_n^{(k)}$ and $\eta_n^{(k)}$ have finite (and nonzero) variance as $n \to \infty$. Also it is obvious that condition $A1$ is satisfied when for all $i_1,\dots,i_k,n$

$$0 < c_1 \leqslant \left|a^{(n)}_{i_1\dots i_k}\right| \leqslant c_2 < \infty,$$

c_1, c_2 – some positive constants.

Remark 1. It is well known that in $Q_n^{(k)}$ or $\eta_n^{(k)}$ we may instead of $X_1,\dots,X_n$ put another i.i.d. $Z_1,\dots,Z_n$ and this does not seriously effect to the limit distribution of $Q_n^{(k)}$ and $\eta_n^{(k)}$. When coefficients $a^{(n)}_{i_1\dots i_k}$ satisfy (1) then for this it is enough $\mathbf{E}X_1 = \mathbf{E}Z_1$ and $\mathbf{E}X_1^2 = \mathbf{E}Z_1^2$. Firstly it was noticed by Rotar[1]. Theorem 1.1 gives the speed of such approximation. Let $X^{(n)} = \{X_1,\dots,X_n\}$, $Z^{(n)} = \{Z_1,\dots,Z_n\}$.

Theorem 1.1. *1) Let $X_1, X_2,\dots$ is the sequence of i.i.d. random variables with $\mathbf{E}X_1 = 0$, $\mathbf{E}X_1^2 = 1$ and $\mathbf{E}|X_1|^p < \infty$, $p = \max(2k, 3(k-1))$, $k \geqslant 2$. Assume the condition A1 holds. Then there exists a constant $K_1 > 0$ such, that*

$$L(Q_n^{(k)}(X^{(n)}) - \mathbf{E}Q_n^{(k)}(X^{(n)}), Q_n^{(k)}(Z^{(n)}) - \mathbf{E}Q_n^{(k)}(Z^{(n)})) \leqslant K_1 n^{-1/8} \ln n,$$

where $Z_1,\dots,Z_n$ are i.i.d. random variables (independent also from $X_1,X_2,\dots,X_n$) with $\mathbf{E}Z_1 = 0$, $\mathbf{E}Z_1^2 = 1$ and $\mathbf{E}|Z_1|^p < \infty$, $p = \max(2k, 3k-3)$. 2) When $\mathbf{E}X_1 = \mathbf{E}Z_1 = 0$, $\mathbf{E}X_1^s = \mathbf{E}Z_1^s$, $\ s = 2,3,\dots,r-1$, $\max(\mathbf{E}|X_1|^r, \mathbf{E}|Z_1|^r) < \infty$, $\ r \geqslant 3$ and condition A1 holds, then there exists a constant $K_2 > 0$ such, that

$$L(\eta_n^{(k)}(X^{(n)}), \eta_n^{(k)}(Z^{(n)})) \leqslant K_2(\sqrt{n})^{-\frac{(r-2)}{r+1}} \ln n.$$

Theorem 1.2. *Let $X_1, X_2,\dots$ are i.i.d. random variables with zero mean variance 1 and finite $2k$ moment, $k \geqslant 2$. Assume that condition A1 holds. Then as $n \to \infty$*

$$Q_n^{(k)} \xrightarrow{d} Q^{(k)} = C^{(k)}\delta(k) + \sum_{s=0}^{[\frac{k}{2}]-\delta(k)} I_{k-2s}(g),$$

where

$$\delta(k) = \mathbf{1}\{k \textit{ is even number }\},$$

$$I_{k-2s}(g) = \int_0^1 \dots \int_0^1 g(t_1,\dots t_{k-2s}) W(dt_1)\dots W(dt_{k-2s}).$$

Functions $g(t_1,\dots,t_r) \in L_2([0,1]^r,\lambda^r)$ and constants $C^{(k)}$ are known. Moreover, if $\mathbf{E}|X_1|^p < \infty$, $p = \max(2k, 3k-3)$, then there exists a constant $K_3 > 0$ such, that

$$L(Q_n^{(k)}, Q^{(k)}) \leqslant K_3 n^{-1/8} \ln n.$$

Theorem 1.3. *Let $X_1, X_2, \dots$ are i.i.d. random variables with zero mean, variance 1. Assume that condition A1 holds. Then as $n \to \infty$*

$$\eta_n^{(k)} \xrightarrow{d} \eta^{(k)} = I_k(a) = \int_0^1 \dots \int_0^1 a(t_1 \dots t_k) W(dt_1) \dots W(dt_k),$$

where function $a(t_1,\dots,t_k) \in L_2([0,1]^k,\lambda^k)$ and are defined by relation (7). Moreover, if $\mathbf{E}X_1^s = \mathbf{E}\xi^s$, $s = 3,4,\dots,r-1$, $\mathbf{E}|X_1|^r < \infty$, $r \geqslant 3$ (ξ is standart normal random variable), then there exists a constant $K_4 > 0$ such, that

$$L(\eta_n^{(k)}, \eta^{(k)}) \leqslant K_4 (\sqrt{n})^{-\frac{r-2}{r+1}} \ln n.$$

When we have the stochastic integrals we confront the problem of the calculation of such integrals. There are ways how to do it (see Willinger, Taqqu[2]). In our paper we propose to calculate multiple stochastic integrals from standard Wiener measure with the help of the multilinear form in i.i.d. random variables (for example, in Bernoulli random variables). Let

$$I_k(h) = \int_0^1 \dots \int_0^1 h(t_1,\dots,t_k) W(dt_1) \dots W(dt_k),$$

where $h(t_1,\dots,t_k) \in L_2([0,1]^k,\lambda^k)$. Define

$$h_{i_1,\dots,i_k}^{(n)} := h\left(\frac{i_1}{n},\dots,\frac{i_k}{n}\right)$$

for $1 \leqslant i_1 \neq \dots \neq i_k \leqslant n$, $n \geqslant k$ and suppose that $\left\{h_{i_1 \dots i_k}^{(n)}\right\}$ satisfies condition A1.

Theorem 1.4. *For any i.i.d. random variables $Y_1,\dots,Y_n$ with $\mathrm{E}Y_1 = 0$, $\mathrm{E}Y_1^2 = 1$*

$$n^{-k/2} \sum_{1 \leqslant i_1 \neq \dots \neq i_k \leqslant n} h_{i_1,\dots,i_k}^{(n)} Y_{i_1} \dots Y_{i_k} \xrightarrow[n\to\infty]{d} I_k(h).$$

If $\mathbf{E}Y_1^s = \mathbf{E}\xi^s$, $s = 3,4,\dots,r-1$, $\mathbf{E}|Y_1|^r < \infty$, $r \geqslant 3$ (ξ is standart normal random variable), then there exists a constant $K_5 > 0$ such, that

$$L\left(I_k(h), n^{-k/2} \sum_{1 \leqslant i_1 \neq \dots \neq i_k \leqslant n} h_{i_1,\dots,i_k}^{(n)} Y_{i_1} \dots Y_{i_k}\right) \leqslant K_5 (\sqrt{n})^{-\frac{r-2}{r+1}} \ln n. \qquad (2)$$

It is not difficult to see that in Theorem 1.4 we can take $Y_1,\dots,Y_n$ such, that $\mathbf{P}(Y_1 = -1) = \mathbf{P}(Y_1 = 1) = \frac{1}{2}$ and to obtain the rate of convergence in (2) of order $n^{-1/5} \ln n$. Also, for example, if we take $Y_1,\dots,Y_n$ such, that $\mathbf{P}(Y_1 = -\sqrt{3}) = \mathbf{P}(Y_1 = \sqrt{3}) = 1/6$ and $\mathbf{P}(Y_1 = 0) = 2/3$, then we obtain in (2) the rate of convergence of order $n^{-1/4} \ln n$. Therefore we can easy to improve the accuracy of (2) approximation.

Let

$$B_n^2 = n^{-k} \sum_{1 \leqslant i_1 \neq \ldots \neq i_k \leqslant n} \left(a_{i_1 \ldots i_k}^{(n)}\right)^2, \quad \Psi(x) = (2\pi)^{-1/2} \int_{-\infty}^{x} e^{-u^2/2} du,$$

$$\Delta_n^2 = B_n^{-2} \max_{1 \leqslant s \leqslant k-1} \left(\max_{i_1 \neq \ldots \neq i_s} \sum_{i_{s+1} \neq \ldots \neq i_k} \left|a_{i_1 \ldots i_k}^{(n)}\right| \right) \left(\max_{i_{s+1} \neq \ldots \neq i_k} \sum_{i_1 \neq \ldots \neq i_s} \left|a_{i_1 \ldots i_k}^{(n)}\right| \right) n^{-k}.$$

Theorem 1.5. *Let* $\mathbf{E}X_1 = 0$, $\mathbf{E}X_1^2 = 1$, $\mathbf{E}|X_1|^3 < \infty$. *Assume that* $\Delta_n \to 0$ *as* $n \to \infty$. *Then there exists a constant* $K_6 > 0$ *such, that*

$$\sup_{x \in R} \left| \mathbf{P}\left\{ \eta_n^{(k)} / B_n < x \right\} - \Psi(x) \right| \leqslant K_6 \Delta_n \ln(\Delta_n^{-1}).$$

Remark 2. Under stronger moment conditions for random variable X_1 we can get (see Basalykas [3]) the theorems of large deviations and asymptotical expansions for $Q_n^{(k)}$ and $\eta_n^{(k)}$. Δ_n here plays the analogously role as the $n^{-1/2}$ in the same theorems for the sums of i.i.d. random variables.

PROOFS

Lemma 2.1. *1) Assume the conditions of the first part of the Theorem 1.1 hold. Then there exists a constant* $L_1 > 0$ *such, that for all* $t \in R$

$$\left| \mathbf{E}e^{it\left(Q_n^{(k)}(X^{(n)}) - EQ_n^{(k)}(X^{(n)})\right)} - \mathbf{E}e^{it\left(Q_n^{(k)}(Z^{(n)}) - EQ_n^{(k)}(Z^{(n)})\right)} \right| \leqslant L_1\left(|t| + |t|^3\right) n^{-1/2}. \quad (3)$$

2) If conditions of the second part of the Theorem 1.1 hold, then there exists a constant $L_2 > 0$ *such, that for all* $t \in R$

$$\left| \mathbf{E}e^{it\eta_n^{(k)}\left(X^{(n)}\right)} - \mathbf{E}e^{it\eta_n^{(k)}\left(Z^{(n)}\right)} \right| \leqslant L_2 |t|^r n^{-\frac{r-2}{2}}. \quad (4)$$

Proof. We begin with $\eta_n^{(k)}$. In order to estimate the distance between the characteristics functions of $\eta_n^{(n)}(X^{(n)})$ and $\eta_n^{(k)}(Z^{(n)})$ we notice, that

$$\left| \mathbf{E}e^{it\eta_n^{(k)}\left(X^{(n)}\right)} - \mathbf{E}e^{it\eta_n^{(k)}\left(Z^{(n)}\right)} \right| \leqslant \sum_{j=1}^{n} |b_j|, \quad (5)$$

where

$$b_j = \mathbf{E}e^{it\eta_n^{(k)}\left(X_1, \ldots, X_j, Z_{j+1}, \ldots, Z_n\right)} - \mathbf{E}e^{it\eta_n^{(k)}\left(X_1, \ldots, X_j - 1, Z_j, \ldots, Z_n\right)}.$$

It is not difficult to see that

$$b_j = \mathbf{E}e^{itG_0}\left(e^{\frac{itX_j}{\sqrt{n}} G_j} - e^{\frac{itZ_j}{\sqrt{n}} G_j} \right),$$

where $G_j = G_j(X_1, \ldots, X_j - 1, Z_{j+1}, \ldots, Z_n)$ are known multilinear functions of order $k-1$, $G_0 = \eta_n^{(k)}(X_1, \ldots, X_{j-1}, 0, Z_{j+1}, \ldots, Z_n)$. As $\mathbf{E}X_1 = \mathbf{E}Z_1 = 0$ and $\mathbf{E}X_1^s = \mathbf{E}Z_1^s$, $s = 2, 3, \ldots, r-1$, $\mathbf{E}|X_1|^r < \infty$, $\mathbf{E}|Z_1|^r < \infty$, then for $t \in R$

$$|b_j| \leqslant L_3 |t|^r n^{-\frac{r}{2}} \mathbf{E}|G_j|^r.$$

It is known that if $\max\left(\mathbf{E}|X_1|^r, \mathbf{E}|Z_1|^r\right) < \infty$, then $\mathbf{E}|G_j|^r \leqslant L_4 < \infty$, where L_4 does not depend on n (see Basalykas[4]). Therefore, using the estimate for b_j from (5) we obtain (4). The estimate (3) we get analogously. Lemma 2.1 is proved.
Proof of the Theorem 1.1 follows from Lemma 2.1 and well known estimate: for any random variables X and Y and any number $T > 0$

$$L(X, Y) \leqslant \frac{1}{\pi} \int_0^T \left|\mathbf{E}e^{itX} - \mathbf{E}e^{itY}\right| \frac{dt}{|t|} + 5.66 \frac{\ln(T+1)}{T}.$$

Proof of the Theorem 1.3. From Remark 1 it follows that we have to show that

$$\eta_n^{(k)}(\xi_1, \ldots, \xi_n) \xrightarrow{d} I_k(a),$$

where $\xi_1, \ldots, \xi_n$ are i.i.d. $N(0,1)$ random variables. For this we take the interval [0,1] division $\left\{\Delta_j^{(n)}\right\}_{j=0}^n : \Delta_0^{(n)} = \{0\}$, $\Delta_j^{(n)} = \left(\frac{j-1}{n}, \frac{j}{n}\right]$. As $\frac{\xi_j}{\sqrt{n}} = \frac{\xi_1 + \ldots + \xi_j}{\sqrt{n}} - \frac{\xi_1 + \ldots + \xi_{j-1}}{\sqrt{n}} \stackrel{d}{=} W\left(\frac{j}{n}\right) - W\left(\frac{j-1}{n}\right) = W\left(\Delta_j^{(n)}\right)$ we can write $\eta_n^{(k)}(\xi_1, \ldots, \xi_n)$ in the form of multiple Wiener -Ito integral

$$\eta_n^{(k)}(\xi_1, \ldots, \xi_n) = I_k(a_n) = \int_0^1 \cdots \int_0^1 a_n(x_1, \ldots, x_k) W(dx_1) \ldots W(dx_k).$$

Here

$$a_n = a_n(x_1, \ldots, x_k) = \sum a_{t_1 \ldots t_k}^{(n)} \mathbf{1}\left(x_1 \in \Delta_{t_1}^{(n)}\right) \ldots \mathbf{1}\left(x_k \in \Delta_{t_k}^{(n)}\right).$$

The functions $a_n \in L_2([0,1]^k, \lambda^k)$ and $||a||_{L_2}^2 = \frac{1}{n^k} \sum_{1 \leqslant t_1 \neq \ldots \neq t_k \leqslant n} \left(a_{t_1 \ldots t_k}^{(n)}\right)^2$. Moreover, the sequence of functions $\{a_n\}$ is fundamental. This follows from the fact that for every $p = 1, 2, 3, \ldots$

$$\left\|a_n(x_1, \ldots, x_k) - a_{n+p}(x_1, \ldots, x_k)\right\|^2 \leqslant \frac{1}{n^k (n+p)^k} \sum_{1 \leqslant t_1 \neq \ldots \neq t_k \leqslant n} \times$$

$$\times \sum_{\substack{\left[\frac{(t_1-1)(n+p)}{n}\right] - 1 \leqslant j_1 \leqslant \left[\frac{t_1(n+p)}{n}\right] + 1 \\ \cdots\cdots\cdots \\ \left[\frac{(t_k-1)(n+p)}{n}\right] - 1 \leqslant j_k \leqslant \left[\frac{t_k(n+p)}{n}\right] + 1}} \left(a_{t_1, \ldots, t_k}^{(n)} - a_{j_1, \ldots, j_k}^{(n+p)}\right)^2 \leqslant \frac{\text{const}}{n^k} \to 0 \qquad (6)$$

as $n \to \infty$.
This yields that there exists a function $a \in \left(L_2[0,1]^k, \lambda^k\right)$ such that

$$a(x_1, \ldots, x_k) \stackrel{L_2}{=} \lim_{n \to \infty} a_n(x_1, \ldots, x_k) \ . \qquad (7)$$

From the properties of the multiple stochastic integrals we have

$$\mathbf{E}\left|I_k(a_n) - I_k\left(a_{n+p}\right)\right|^2 = \left\|a_n - a_{n+p}\right\|^2 \underset{n,p\to\infty}{\to} 0\,. \tag{8}$$

Therefore there exists a random variable $I_k(a) \in R$ such, that

$$\mathbf{E}\left|I_k(a_n) - I_k(a)\right|^2 \to 0$$

as $n \to \infty$. So $I_k\left(a_n\right) \xrightarrow{d} I_k(a)$.

In order to estimate $L\left(\eta_n^{(k)}(X^{(n)}), \eta^{(k)}\right)$ we notice that

$$\begin{aligned} L\left(\eta_n^{(k)}(X^{(n)}), \eta^{(k)}\right) \leqslant & L\left(\eta_n^{(k)}(X^{(n)}), \eta_n^{(k)}(\xi^{(n)})\right) + \\ & + L\left(\eta_n^{(k)}(\xi^{(n)}), \eta_{n+r}^{(k)}(\xi^{(n+r)})\right) + L\left(\eta_{n+r}^{(k)}(\xi^{(n+r)}), \eta^{(k)}\right) \end{aligned} \tag{9}$$

for every $r \geqslant 1$. Here $\xi^{(s)} = \left\{\xi_1, \ldots, \xi_s\right\}$. From the second part of the Theorem 1.1, relations (6)- (8) and estimate

$$L(X,Y) \leqslant \text{const} \cdot \left(\mathbf{E}|X - Y|^2\right)^{1/3}$$

we get $L\left(\eta_n^{(k)}(X^{(n)}),\ \eta_n^{(k)}(\xi^{(n)})\right) \leqslant K_7 \cdot \left(\sqrt{n}\right)^{-\frac{r'-2}{r'+1}} \ln n$ and $L\left(\eta_n^{(k)}(\xi^{(n)}), \eta_{n+r}^{(k)}(\xi^{(n+r)})\right) \leqslant K_8 \cdot n^{-k/3}$ for every $r \geqslant 1$. Now we put $r \to \infty$ in (9) and obtain finally that $L\left(\eta_n^{(k)},\ \eta^{(k)}\right) \leqslant K_9 \cdot \left(\sqrt{n}\right)^{-\frac{r'-2}{r'+1}} \ln n$. Theorem 1.3 is proved.

Proof of the Theorem 1.2 is analogous to that of Theorem 1.3, only we notice that $Q_n^{(k)}\left(X^{(n)}\right)$ can be written like the sum of multilinear forms of order $k, k-2, k-4, \ldots$,:

$$Q_n^{(k)}(X_1, \ldots, X_n) = \sum_{s=0}^{\left[\frac{k}{2}\right]} n^{-\frac{k-2s}{2}} \sum_{1 \leqslant i_1 \neq \ldots \neq \leqslant i_{k-2s} \leqslant n} \widetilde{a}^{(n)}_{i_1,\ldots,i_{k-2s}} X_{i_1} \ldots X_{i_{k-2s}} + r_n.$$

Here $\mathbf{E}|r_n|^2 \leqslant \text{const} \cdot n^{-1}$ and

$$\widetilde{a}^{(n)}_{i_1,\ldots,i_{k-2s}} := \frac{1}{n^s} \sum_{(s)} a^{(n)}_{j_1 \ldots j_k};$$

where $\sum_{(s)}$ means the summation over $2s$ indices $1 \leqslant p_1, \ldots, p_{2s} \leqslant n,\ p_i \in \left\{j_1, \ldots, j_k\right\}$ and every p_i has multiplicity 2 among $p_1, \ldots, p_{2s}$. The rest $k - 2s$ indices are fixed and are equal to $i_1 \neq \ldots \neq i_{k-2s}$ accordingly. Moreover $\left\{i_1, \ldots, i_{k-2s}\right\} \cap \left\{p_1, \ldots, p_{2s}\right\} = \varnothing$. The functions $g\left(t_1, \ldots, t_{k-2s}\right)$ and constants $C^{(k)}$ in the formulation of Theorem 1.2 are defined with the help of coefficients $\widetilde{a}^{(n)}_{i_1,\ldots,i_{k-2s}}$. The estimate for $L\left(Q_n^{(k)}, Q^{(k)}\right)$ we obtain analogously as in the the proof of Theorem 1.3 . This concludes the proof of Theorem 1.2.

Proof of the Theorem 1.4 follows from Theorem 1.3.

And finally the CLT for $\eta_n^{(k)}$ we prove using the fact that $\eta_n^{(k)}$ and the σ-algebra generated by the random variables $X_1, \ldots, X_n$ is a martingale. This and the verification of the corresponding conditions are the main proving CLT for $\eta_n^{(k)}$ (see Basalykas[4]). The proof of Theorem 1.5 uses the cumulant's estimates for $\eta_n^{(k)}$. From the requirement that cumulants must vanish as $n \to \infty$ follows the condition $\Delta_n \to 0$ as $n \to \infty$.

REFERENCES

1. V.I. Rotar. Some limit theorems for multilinear forms and quasipolynomial functions, Teor. Veroyat. Primen. 20: 527–545 (1975).
2. W. Willinger, M. G.Taqqu. Pathwise stochastic integration and applications to the theory of continous trading, Stoch. Proc. Appl. 32: 253–280 (1989).
3. A. Basalykas. Some asymptotical properties for distributions of the polinomial forms, Litovsk. Matem. Sborn.28: 4, 644–654 (1988).
4. A. Basalykas. Functional central limit theorem for random multilinear forms, Lith. Mathem. J. 32: 175–186 (1992).

MULTIVARIATE BOOLEAN TRAPEZOIDAL RULES

Günter Baszenski[1] and Franz-Jürgen Delvos[2]

[1]FB Nachrichtentechnik
Fachhochschule Dortmund
Postfach 10 50 18
44047 Dortmund, Germany

[2]FB Mathematik I
Universität GH Siegen
Hölderlinstraße 3
57076 Siegen, Germany

ABSTRACT

Boolean methods of interpolation have been applied to the construction of bivariate and trivariate numerical integration formulas[3,4]. These formulas are comparable with lattice rules of multivariate numerical integration[5,6]. In this paper we will construct Boolean trapezoidal rules for multivariate numerical integration in arbitrary dimensions which are based on the ideas of multivariate Boolean interpolation[2] and which extend bi- and trivariate results[3,4]. A detailed error investigation is presented using Boolean remainder formulas[1].

MULTIVARIATE TENSOR PRODUCT TRAPEZOIDAL RULES

Let d be a positive integer. The d-dimensional unit cube is denoted by U^d with $U = [0,1]$. Further we denote by $\mathcal{C}(U^d)$ the algebra of continuous complex-valued functions defined on U^d and by $\mathcal{C}_0(U^d)$ is the subalgebra of functions $f \in \mathcal{C}(U^d)$ having period 1 with respect to each variable. The Korobov spaces $\mathcal{E}^\alpha(U^d)$, $\alpha \in \mathcal{R}$, $\alpha > 1$, are subspaces of $\mathcal{C}_0(U^d)$. By definition, $f \in \mathcal{E}^\alpha(U^d)$ iff its Fourier coefficients which are given for $m_1, \ldots, m_d \in \mathcal{Z}$ by

$$a(m_1,\ldots,m_d)[f]$$
$$= \int_0^1 \cdots \int_0^1 f(x_1,\ldots,x_d) \exp\big(-2\pi i(x_1 m_1 + \ldots + x_d m_d)\big)\, dx_1 \cdots dx_d$$

Approximation, Probability, and Related Fields, Edited by
G. Anastassiou and S.T. Rachev, Plenum Press, New York 1994

satisfy with a positive real constant γ

$$a(m_1,\ldots,m_d)[f] \le \gamma|\overline{m}_1\cdots\overline{m}_d|^{-\alpha} \quad (m_1,\ldots,m_d \in \mathcal{Z})$$

where

$$\overline{m} = \begin{cases} m & \text{if } m \ne 0 \\ 1 & \text{if } m = 0. \end{cases}$$

If $f \in \mathcal{E}^\alpha(U^d)$ with $\alpha > 1$ then f is represented by its absolutely convergent Fourier series:

$$f(x_1,\ldots,x_d) = \sum_{m_1=-\infty}^{\infty} \cdots \sum_{m_d=-\infty}^{\infty} a(m_1,\ldots,m_d)[f]\,\exp\big(2\pi i(m_1x_1+\ldots+m_dx_d)\big)$$
$$\big((x_1,\ldots,x_d) \in U^d\big).$$

For positive integers $k_1,\ldots,k_d$, the d-variate tensor product trapezoidal rule applied to $f \in \mathcal{C}_0(U^d)$ is given by

$$T(k_1,\ldots,k_d)[f] = 2^{-(k_1+\cdots+k_d)} \sum_{j_1=1}^{2^{k_1}} \cdots \sum_{j_d=1}^{2^{k_d}} f(j_1/2^{k_1},\ldots,j_d/2^{k_d}).$$

This formula has an interpolatory background which is important for the following constructions. We consider for $1 \le u \le d$ the parametrically extended periodic linear spline interpolation projectors

$$P_u^{2^{k_u}}(f)(x_1,\ldots,x_d) = \sum_{j_u=1}^{2^{k_u}} f(x_1,\ldots,j_u/2^{k_u},\ldots,x_d)L_{j_u,u}^{2^{k_u}}(x_1,\ldots,x_d).$$

For $1 \le u \le d$, $1 \le j_u \le 2^{k_u}$ the function $L_{j_u,u}^{2^{k_u}}(x_1,\ldots,x_d)$ depends only on the variable x_u and is given by

$$L_{j_u,u}^{2^{k_u}}(x_1,\ldots,x_d) = L^{2^{k_u}}(x_u - j_u/2^{k_u}) \quad (1 \le u \le d, 1 \le j_u \le 2^{k_u})$$

where the univariate function $L^{2^k}(x)$ is the unique periodic linear spline with knots $r/2^k$ satisfying

$$L^{2^k}(r/2^k) = \delta_{0,r}, \quad 0 \le r < 2^k.$$

The univariate periodic linear spline interpolation projector is then given by

$$P^{2^k}(f)(x) = \sum_{j=1}^{2^k} f(2^{-k}j)L^{2^k}(x - j/2^k) \quad (x \in U).$$

The d-variate tensor product periodic linear spline interpolation projector is for $k_1,\ldots,k_d \in \mathcal{N}$ and $(x_1,\ldots,x_d) \in U^d$ given by

$$\begin{aligned} &P_1^{2^{k_1}}\cdots P_d^{2^{k_d}}(f)(x_1,\ldots,x_d) \\ &= \sum_{j_1=1}^{2^{k_1}} \cdots \sum_{j_d=1}^{2^{k_d}} f(j_1/2^{k_1},\ldots,j_d/2^{k_d}) \prod_{u=1}^{d} L_{j_u,u}^{2^{k_u}}(x_1,\ldots,x_d). \end{aligned}$$

The d-variate tensor product trapezoidal rule applied to $f \in \mathcal{C}_0(U^d)$ is obtained by integrating the d-variate tensor product linear periodic spline interpolant over the unit cube U^d:

$$T(k_1,\ldots,k_d)[f] = \int_0^1 \cdots \int_0^1 P_1^{2^{k_1}} \cdots P_d^{2^{k_d}}(f)(x_1,\ldots,x_d)\,dx_1\cdots dx_d.$$

We will derive a remainder formula for the d-variate tensor product trapezoidal rule using Boolean methods of multivariate interpolation[1,2].

Let the remainder projector of $P_u^{2^{k_u}}$ be defined as

$$R_u^{2^{k_u}}(f) = f - P_u^{2^{k_u}}(f).$$

The remainder projector of $P_1^{2^{k_1}} \cdots P_d^{2^{k_d}}$ is then given by

$$f - P_1^{2^{k_1}} \cdots P_d^{2^{k_d}}(f) = f - (I - R_1^{2^{k_1}})\cdots(I - R_d^{2^{k_d}})(f)$$

where I denotes the identity projector on $\mathcal{C}_0(U^d)$. We obtain the explicit formula

$$I - P_1^{2^{k_1}} \cdots P_d^{2^{k_d}} = -\sum_{h=1}^{d}(-1)^h \sum_{1\le v_1<\ldots<v_h\le d} R_{v_1}^{2^{k_{v_1}}} \cdots R_{v_h}^{2^{k_{v_h}}}.$$

We will apply this remainder formula to investigate the approximation power of the tensor product trapezoidal rule in the Korobov spaces $\mathcal{E}^\alpha(U^d)$, $\alpha > 1$.

The discrete orthogonality of the exponentials:

$$\sum_{j=1}^{2^k} \exp(2\pi i m j/2^k) = \begin{cases} 2^k & \text{if } m = r2^k,\ r \in \mathcal{Z} \\ 0 & \text{otherwise} \end{cases}$$

in the univariate case, from which

$$\sum_{j_1=1}^{2^{k_1}} \cdots \sum_{j_d=1}^{2^{k_d}} \exp(2\pi i(m_1 j_1/2^{k_1} + \cdots + m_d j_d/2^{k_d}))$$
$$= \begin{cases} 2^{k_1+\cdots+k_d} & \text{if } m_u = r_u 2^{k_u} \text{ with } r_u \in \mathcal{Z} \text{ for all } u = 1,\ldots,d \\ 0 & \text{otherwise} \end{cases}$$

follows for the multivariate case, and the pointwise convergence of the Fourier series of $f \in \mathcal{E}^\alpha(U^d)$ yields the equality

$$T(k_1,\ldots,k_d)[f] = 2^{-(k_1+\cdots+k_d)} \sum_{j_1=1}^{2^{k_1}} \cdots \sum_{j_d=1}^{2^{k_d}} f(j_1/2^{k_1},\ldots,j_d/2^{k_d})$$
$$= \sum_{r_1=-\infty}^{\infty} \cdots \sum_{r_d=-\infty}^{\infty} a(r_1 2^{k_1},\ldots,r_d 2^{k_d})[f].$$

Note that

$$a(0,\ldots,0)[f] = \int_0^1 \cdots \int_0^1 f(x_1,\ldots,x_d)\,dx_1\cdots dx_d.$$

This gives a series representation of the error for the multivariate tensor product trapezoidal rule. For the univariate case we have

$$\begin{aligned}-\int_0^1 R^{2^k}(f)(x)\,dx &= \int_0^1 \left(P^{2^k}(f)(x) - f(x)\right) dx\\ &= T(k)[f] - a(0)[f]\\ &= \sum_{r\neq 0} a(r2^k)[f]\end{aligned}$$

where R^{2^k} is the univariate remainder projector.

For treating the d-variate case we recall the remainder formula for multivariate tensor product interpolation:

$$I - P_1^{2^{k_1}} \cdots P_d^{2^{k_d}} = -\sum_{h=1}^{d} (-1)^h \sum_{1\leq v_1<\ldots<v_h\leq d} R_{v_1}^{2^{k_{v_1}}} \cdots R_{v_h}^{2^{k_{v_h}}}.$$

Note first that

$$\begin{aligned}&\int_0^1 P_u^{2^{k_u}}(f)(x_1,\ldots,x_d)\,dx_u - \int_0^1 f(x_1,\ldots,x_d)\,dx_u\\ &\quad= -\int_0^1 R_u^{2^{k_u}}(f)(x_1,\ldots,x_d)\,dx_u\\ &\quad= \sum_{m_1=-\infty}^{\infty} \cdots \sum_{r_u\neq 0} \cdots \sum_{m_d=-\infty}^{\infty} a(m_1,\ldots,r_u 2^{k_u},\ldots,m_d)[f]\times\\ &\quad\quad \times \exp\bigl(2\pi i(x_1 m_1 + \cdots + x_u\cdot 0 + \cdots + x_d m_d)\bigr).\end{aligned}$$

The d-variate integral is then given by

$$-\int_0^1 \cdots \int_0^1 R_u^{2^{k_u}}(f)(x_1,\ldots,x_d)\,dx_1\cdots dx_d = \sum_{r_u\neq 0} a(0,\ldots,r_u 2^{k_u},\ldots,0)[f].$$

Similarly we have for the operator $R_{v_1}^{2^{k_{v_1}}} \cdots R_{v_h}^{2^{k_{v_h}}}$:

$$\begin{aligned}&\int_0^1 \cdots \int_0^1 (-R_{v_1}^{2^{k_{v_1}}})\cdots(-R_{v_h}^{2^{k_{v_h}}})(f)(x_1,\ldots,x_d)\,dx_1\cdots dx_d\\ &\quad= \sum_{r_{v_1}\neq 0} \cdots \sum_{r_{v_h}\neq 0} a(0,\ldots,r_{v_1}2^{k_{v_1}},\ldots,r_{v_h}2^{k_{v_h}},\ldots,0)[f]\end{aligned}$$

where the multiindex of the Fourier coefficient in the above series representation is to be understood as

$$(0,\ldots,r_{v_1}2^{k_{v_1}},\ldots,r_{v_h}2^{k_{v_h}},\ldots,0) = (l_1,\ldots,l_d)$$

with

$$l_u = \begin{cases} r_u 2^{k_u} & \text{if } u \in \{v_1,\ldots,v_h\}\\ 0 & \text{if } u \notin \{v_1,\ldots,v_h\}.\end{cases}$$

In particular we have

$$\int_0^1 \cdots \int_0^1 (-R_1^{2^{k_1}}) \cdots (-R_h^{2^{k_h}})(f)(x_1, \ldots, x_d)\, dx_1 \cdots dx_d$$
$$= \sum_{r_1 \neq 0} \cdots \sum_{r_h \neq 0} a(r_1 2^{k_1}, \ldots, r_h 2^{k_h}, 0, \ldots, 0)[f].$$

We introduce the notations

$$I_d[f] = \int_0^1 \cdots \int_0^1 f(x_1, \ldots, x_d)\, dx_1 \cdots dx_d,$$

$$R_{v_1,\ldots,v_h}^{k_{v_1},\ldots,k_{v_h}}[f] = I_d[(-R_{v_1}^{2^{k_{v_1}}}) \cdots (-R_{v_h}^{2^{k_{v_h}}})(f)].$$

Proposition 1. *Let* $1 \leq v_1 < \ldots < v_h \leq d$ *and* $f \in \mathcal{E}^\alpha(U^d)$, $\alpha > 1$. *Then*

$$\left| R_{v_1,\ldots,v_h}^{k_{v_1},\ldots,k_{v_h}}[f] \right| = \mathcal{O}\left(2^{-\alpha(k_{v_1}+\cdots+k_{v_h})}\right) \quad \text{as } k_{v_1}, \ldots, k_{v_h} \to \infty.$$

Proof: We have

$$\left| R_{v_1,\ldots,v_h}^{k_{v_1},\ldots,k_{v_h}}[f] \right| \leq \sum_{r_{v_1} \neq 0} \cdots \sum_{r_{v_h} \neq 0} \left| a(0, \ldots, r_{v_1} 2^{k_{v_1}}, \ldots, r_{v_h} 2^{k_{v_h}}, \ldots, 0)[f] \right|$$
$$\leq \gamma \sum_{r_{v_1} \neq 0} \cdots \sum_{r_{v_h} \neq 0} \left| r_{v_1} 2^{k_{v_1}} \cdots r_{v_h} 2^{k_{v_h}} \right|^{-\alpha}$$
$$= \mathcal{O}\left(2^{-\alpha(k_{v_1}+\cdots+k_{v_h})}\right) \quad (k_{v_1}, \ldots, k_{v_h} \to \infty). \quad \blacksquare$$

Proposition 2. *Let* $f \in \mathcal{E}^\alpha(U^d)$, $\alpha > 1$. *The remainder in the multivariate tensor product trapezoidal rule satisfies the asymptotic relation*

$$\left| T(q, \ldots, q)[f] - I_d[f] \right| = \mathcal{O}(2^{-\alpha q}) \quad \text{as } q \to \infty.$$

Proof: We have

$$P_1^{2^{k_1}} \cdots P_d^{2^{k_d}} - I = \sum_{h=1}^{d} (-1)^h \sum_{1 \leq v_1 < \ldots < v_h \leq d} R_{v_1}^{2^{k_{v_1}}} \cdots R_{v_h}^{2^{k_{v_h}}}$$

with $k_j = q$, $1 \leq j \leq d$. This implies

$$\left| T(k_1, \ldots, k_d)[f] - I_d[f] \right| \leq \sum_{h=1}^{d} \sum_{1 \leq v_1 < \ldots < v_h \leq d} \left| I_d[(-R_{v_1}^{2^{k_{v_1}}}) \cdots (-R_{v_h}^{2^{k_{v_h}}})(f)] \right|$$
$$= \sum_{h=1}^{d} \sum_{1 \leq v_1 < \ldots < v_h \leq d} \left| R_{v_1,\ldots,v_h}^{k_{v_1},\ldots,k_{v_h}}[f] \right| = \mathcal{O}(2^{-\alpha q}), \quad q \to \infty. \quad \blacksquare$$

To obtain an asymptotic error estimate of the form $\mathcal{O}(2^{-\alpha q})$ via the multivariate tensor product trapezoidal rule $T(q,\dots,q)[f]$ it is necessary to evaluate f at 2^{dq} points. Using d-variate Boolean interpolation with linear periodic splines we will construct multivariate Boolean trapezoidal rules which are much more efficient.

Again we consider the parametrically extended linear periodic spline interpolation projectors

$$P_u^{2^{k_u}}, \quad k_u \in \mathcal{N}, \quad 1 \le u \le d.$$

These projectors generate a Boolean algebra of commuting projectors with the operations of composition and Boolean sum:

$$PQ = P \circ Q = \inf\{p, q\}, \quad P \oplus Q = P + Q - PQ = \sup\{P, Q\}$$

where the order relation is given by $P \le Q$ iff $PQ = QP = P$. Special elements of this Boolean algebra of commuting projectors are the tensor product operators $P_1^{2^{k_1}} \cdots P_d^{2^{k_d}}$.

The projector of d-variate Boolean sum interpolation is given by[2]

$$\begin{aligned} B_{q;d} &= \bigoplus_{k_1+\dots+k_d=q} P_1^{2^{k_1}} \cdots P_d^{2^{k_d}} \\ &= \sum_{j=0}^{d-1} (-1)^j \binom{d-1}{j} \sum_{k_1+\dots+k_d=q-j} P_1^{2^{k_1}} \cdots P_d^{2^{k_d}} \\ &= \sup\{P_1^{2^{k_1}} \cdots P_d^{2^{k_d}} : k_1,\dots,k_d \in \mathcal{N},\ k_1 + \dots + k_d = q\}. \end{aligned}$$

The d-variate Boolean trapezoidal rule is now defined as the d-variate integral of the periodic spline interpolant over the unit cube U^d:

$$H^{q;d}[f] = I_d[B_{q;d}(f)] = \int_0^1 \cdots \int_0^1 B_{q;d}(f)(x_1,\dots,x_d)\, dx_1 \cdots dx_d.$$

Thus the d-variate Boolean trapezoidal rule is a linear combination of d-variate tensor product trapezoidal rules:

$$H^{q;d}[f] = \sum_{j=0}^{d-1} (-1)^j \binom{d-1}{j} \sum_{k_1+\dots+k_d=q-j} T(k_1,\dots,k_d)[f].$$

Our objectives are to determine error bounds as well as bounds for the number of point evaluations needed in the computation of $H^{q;d}[f]$.

To derive asymptotic error bounds we use the remainder formula for d-variate Boolean interpolation[1]. As in the tensor product case we will use the remainder projectors

$$R_u^{2^{k_u}} = I - P_u^{2^{k_u}}, \quad k_u \in \mathcal{N}, \quad 1 \le u \le d.$$

Then the remainder projector of $B_{q;d}$ satisfies

$$\begin{aligned} I - B_{q;d} &= I - \sum_{j=0}^{d-1} (-1)^j \binom{d-1}{j} \sum_{k_1+\dots+k_d=q-j} (I - R_1^{2^{k_1}}) \cdots (I - R_d^{2^{k_d}}) \\ &= I - \sum_{j=0}^{d-1} (-1)^j \binom{d-1}{j} \times \\ &\quad \times \sum_{k_1+\dots+k_d=q-j} \left(I - \sum_{h=1}^{d} (-1)^{h-1} \sum_{1 \le v_1 < \dots < v_h \le d} R_{v_1}^{2^{k_{v_1}}} \cdots R_{v_h}^{2^{k_{v_h}}} \right). \end{aligned}$$

Since

$$\left|\{(k_1,\ldots,k_d)\in\mathcal{N}^d : k_1+\cdots+k_d=q-j\}\right| = \binom{q-j-1}{d-1},$$

$$\sum_{j=0}^{d-1}(-1)^j\binom{d-1}{j}\binom{q-j-1}{d-1}=1$$

we obtain

$$I-B_{q;d}=\sum_{j=0}^{d-1}(-1)^j\binom{d-1}{j}\sum_{k_1+\cdots+k_d=q-j}\;\sum_{h=1}^{d}(-1)^{h-1}\sum_{1\le v_1<\ldots<v_h\le d}R_{v_1}^{2^{k_{v_1}}}\cdots R_{v_h}^{2^{k_{v_h}}}.$$

Let

$$S_{v_1,\ldots,v_h}=\sum_{j=0}^{d-1}(-1)^j\binom{d-1}{j}\sum_{k_1+\cdots+k_d=q-j}R_{v_1}^{2^{k_{v_1}}}\cdots R_{v_h}^{2^{k_{v_h}}}.$$

Then we have

$$I-B_{q;d}=\sum_{h=1}^{d}(-1)^{h-1}\sum_{1\le v_1<\ldots<v_h\le d}S_{v_1,\ldots,v_h}.$$

Collecting identical operators in $S_{v_1,\ldots,v_h}$ we get

$$S_{v_1,\ldots,v_h}=\sum_{t=1}^{h}(-1)^{h-t}\binom{h-1}{t-1}\sum_{k_{v_1}+\cdots+k_{v_h}=q-d+t}R_{v_1}^{2^{k_{v_1}}}\cdots R_{v_h}^{2^{k_{v_h}}}.$$

This implies

$$B_{q;d}-I=\sum_{t=1}^{d}\sum_{h=t}^{d}(-1)^t\binom{h-1}{t-1}\sum_{1\le v_1<\ldots<v_h\le d}\;\sum_{k_{v_1}+\cdots+k_{v_h}=q-d+t}R_{v_1}^{2^{k_{v_1}}}\cdots R_{v_h}^{2^{k_{v_h}}}.$$

A detailed proof can be found in another paper of the authors[1].

Recall the notation

$$R_{v_1,\ldots,v_h}^{k_{v_1},\ldots,k_{v_h}}[f]=I_d[(-R_{v_1}^{2^{k_{v_1}}})\cdots(-R_{v_h}^{2^{k_{v_h}}})(f)],$$

and the identities

$$R_{v_1,\ldots,v_h}^{k_{v_1},\ldots,k_{v_h}}[f]=\sum_{r_{v_1}\neq 0}\cdots\sum_{r_{v_h}\neq 0}a(0,\ldots,r_{v_1}2^{k_{v_1}},\ldots,r_{v_h}2^{k_{v_h}},\ldots,0)[f],$$

$$\int_0^1\cdots\int_0^1(-R_1^{2^{k_1}})\cdots(-R_h^{2^{k_h}})(f)(x_1,\ldots,x_d)\,dx_1\cdots dx_d$$
$$=\sum_{r_1\neq 0}\cdots\sum_{r_h\neq 0}a(r_1 2^{k_1},\ldots,r_h 2^{k_h},0,\ldots,0)[f].$$

Integrating the formula for $B_{q;d}-I$ over the unit cube U^d we get the remainder formula of d-variate Boolean trapezoidal rules:

$$H^{q;d}[f]-I_d[f]$$
$$=\sum_{t=1}^{d}\sum_{h=t}^{d}(-1)^{h-t}\binom{h-1}{t-1}\sum_{1\le v_1<\ldots<v_h\le d}\;\sum_{k_{v_1}+\cdots+k_{v_h}=q-d+t}R_{v_1,\ldots,v_h}^{k_{v_1},\ldots,k_{v_h}}[f].$$

Proposition 3. *Let $f \in \mathcal{E}^\alpha(U^d)$, $\alpha > 1$. The remainder in the multivariate Boolean trapezoidal rule satisfies the asymptotic relation*

$$\left|H^{q;d}[f] - I_d[f]\right| = \mathcal{O}(q^{d-1}2^{-\alpha q}) \quad (q \to \infty).$$

Proof: We have for $h \le d$

$$\begin{aligned} &\left|\{(k_1,\ldots,k_h) \in \mathcal{N}^h : k_1 + \cdots + k_h = q-d+t\}\right| \\ &= \binom{q-d+t-1}{h-1} = \mathcal{O}(q^{d-1}) \quad (q \to \infty). \end{aligned}$$

It follows from Proposition 1 and $k_1+\cdots+k_h = q-d+t$ that $\left|R_{1,\ldots,h}^{k_1,\ldots,k_h}[f]\right| = \mathcal{O}(2^{-\alpha q})$ as $q \to \infty$. Similarly we have in view of $k_{v_1} + \cdots + k_{v_h} = q-d+t$ the relation $R_{v_1,\ldots,v_h}^{k_{v_1},\ldots,k_{v_h}}[f] = \mathcal{O}(2^{-\alpha q})$ as $q \to \infty$. Now we can conclude

$$\begin{aligned} &\left|H^{q;d}[f] - I_d[f]\right| \\ &\le \sum_{t=1}^{d}\sum_{h=t}^{d}\binom{h-1}{t-1}\sum_{1\le v_1<\ldots<v_h\le d}\;\sum_{k_{v_1}+\cdots+k_{v_h}=q-d+t}\left|R_{v_1,\ldots,v_h}^{k_{v_1},\ldots,k_{v_h}}[f]\right| \\ &\le \sum_{t=1}^{d}\sum_{h=t}^{d}\binom{h-1}{t-1}\sum_{1\le v_1<\ldots<v_h\le d}\mathcal{O}(q^{d-1}2^{-\alpha q}) \\ &= \mathcal{O}(q^{d-1}2^{-\alpha q}). \quad \blacksquare \end{aligned}$$

Proposition 4. *The number of point evaluations for the d-variate Boolean trapezoidal rule satisfies the asymptotic relation*

$$\left|H^{q;d}\right| = \mathcal{O}(q^{d-1}2^{q}) \quad (q \to \infty).$$

Proof: Recall that

$$H^{q;d}[f] = \sum_{j=0}^{d-1}(-1)^j\binom{d-1}{j}\sum_{k_1+\cdots+k_d=q-j}T(k_1,\ldots,k_d)[f].$$

The number of point evaluations for the d-variate tensor product trapezoidal rule $T(k_1,\ldots,k_d)[f]$ is given by

$$\left|T(k_1,\ldots,k_d)\right| = 2^{k_1+\cdots+k_d} = \mathcal{O}(2^q) \quad (q \to \infty)$$

in view of $k_1 + \cdots + k_d = q-j$, $0 \le j < d$. Since

$$\left|\{(k_1,\ldots,k_d) \in \mathcal{N}^d : k_1 + \cdots + k_d = q-j\}\right| = \binom{q-j-1}{d-1} = \mathcal{O}(q^{d-1})$$

we finally obtain

$$\left|H^{q;d}\right| = \sum_{j=0}^{d-1}\sum_{k_1+\cdots+k_d=q-j}\left|T(k_1,\ldots,k_d)\right| = \mathcal{O}(q^{d-1}2^q) \quad (q \to \infty). \quad \blacksquare$$

Remark. *The asymptotic error bound and the asymptotic bound for the number of point evaluations for the d-variate tensor product trapezoidal rule* $T(q,\ldots,q)[f]$ *satisfy:*

$$T(q,\ldots,q)[f]-I_d[f]=\mathcal{O}(2^{-\alpha q})\quad (q\to\infty),$$

$$\left|T(q,\ldots,q)\right|=\mathcal{O}(2^{qd})\quad (q\to\infty).$$

This contrasts heavily to the d-variate Boolean trapezoidal rule where we have

$$H^{q;d}[f]-I_d[f]=\mathcal{O}(q^{d-1}2^{-\alpha q})\quad (q\to\infty),$$

$$\left|H^{q;d}\right|=\mathcal{O}(q^{d-1}2^{q})\quad (q\to\infty).$$

Thus the d-variate Boolean trapezoidal rules are comparable with good lattice rules[5,6].

REFERENCES

1. G. Baszenski and F.-J. Delvos, Remainders for Boolean interpolation, *in:* "Progress in Approximation Theory," P. Nevai, A. Pinkus, eds., Academic Press, New York (1991).
2. F.-J. Delvos, *d*-variate Boolean interpolation, *J. Approx. Theory* 34 (1982).
3. F.-J. Delvos, Boolean methods for double integration. *Math. Comp.* 55 (1990).
4. F.-J. Delvos and H. Nienhaus, A trivariate Boolean cubature scheme, *in:* "Multivariate Approximation Theory IV," C. K. Chui, W. Schempp, K. Zeller, eds., ISNM 90, Birkhäuser, Basel (1990).
5. I. H. Sloan, Lattice methods for multiple integration, *J. Comput. Appl. Math.* 12–13 (1985).
6. I. H. Sloan, Lattice methods for multiple integration: theory, error analysis and examples, *SIAM J. Numer. Anal.* 24 (1987).

CONVERGENCE RESULTS FOR AN EXTENSION OF THE FOURIER TRANSFORM (*)

Carlo Belingeri[1] and Paolo Emilio Ricci[2]

[1]Dipartimento di Metodi e Modelli Matematici
per le Scienze Applicate
[2]Dipartimento di Matematica "Guido Castelnuovo"
Università degli Studi di Roma "La Sapienza"

1. INTRODUCTION

In a preceding paper (see[2]), for any function f such that: $x^k f(x) \in L(a,b)$, $\forall\, k \in \mathbb{N}_0 := \mathbb{N} \cup \{0\}$, we have considered the integral transform:

$$\hat{f}(y) := \int_a^b F(x,y) f(x) dx \,, \tag{1.1}$$

related to the kernel $F(x,y)$ which is the generating function of a set of polynomials orthogonal in (a,b) with respect to the weight $W(x)$ (shortly O.P.S.). We have proved that the integral transforms of this kind can be considered in some sense as a generalization of the Fourier transform, since they formally verify a property which is analogous to a known property of this classical operator.

The results presented in [2] are obtained under very particular hypotheses, since we consider only classical O.P.S. and we impose some other restrictions on the function f.

Furthermore the results obtained in [2] have only a formal character, since the problem of inverting integral signs and series in the proof of Propositions I-II is not treated in details.

The aim of this paper is to improve the results obtained in [2] considering more general classes of O.P.S. and functions f.

We will be able to show, in the case of classical O.P.S., sufficient conditions in order to guarantee the effective convergence of the power expansions we have considered in [2].

(*) This research was partially supported by 40% funds of the M.U.R.S.T.

2. THE FORMAL RESULT

Consider an O.P.S.: $\{G_k(x)\}_{k\in\mathbb{N}_0}$, generated in (a,b) by a *weight function* $W(x)$. We assume that:

Hp. I: $W(x)\geq 0$ in (a,b), and $W(x)\not\equiv 0$ in any not trivial interval $[\alpha,\beta]\subset(a,b)$;
Hp. II: $\forall\, k\in\mathbb{N}_0$, $x^kW(x)\in L(a,b)$, i.e. all the moments of the measure associated to the weight are finite.

Put

$$\int_a^b G_h(x)G_k(x)W(x)dx = h_k\delta_{h,k}\,,$$

so that: $h_k := \int_a^b G_k^2(x)W(x)dx$.

Let $F(x,y)$ be the Generating Function of the set $\{G_k(x)\}_{k\in\mathbb{N}_0}$, corresponding to the sequence $\{c_k\}_{k\in\mathbb{N}_0}$, i.e.:

$$F(x,y) := \sum_{k=0}^{\infty} c_kG_k(x)y^k\,. \tag{2.1}$$

Suppose further:

Hp. III: $F^2(x,y)W(x)\in L(a,b)$.

Consider for any f such that: $x^kf(x)\in L(a,b)$, $\forall\, k\in\mathbb{N}_0$, the Integral Transform (1.1) (see[5]), related to the kernel: $K(x,y):=F(x,y)$, defined by (2.1). Then the following theorem holds:

THEOREM I. *Suppose the preceding hypotheses* I-II-III *be satisfied, and let* $x^kf(x)\in L(a,b)$, $\forall\, k\in\mathbb{N}_0$. *Denote by* α_k $(k\in\mathbb{N}_0)$ *the Fourier coefficients of the function* $W^{-1}(x)f(x)$*:*

$$\alpha_k = \frac{1}{h_k}\int_a^b f(x)G_k(x)dx\,. \tag{2.2}$$

Then the coefficients of the formal Taylor expansion, in a neighborhood of the origin, of the function $\hat{f}(y)$ *are given by:* $\alpha_kc_kh_k$.

PROOF. Let us prove before the following equality:

$$\int_a^b F(x,y)W(x)\Big(\sum_{k=0}^{\infty}\alpha_kG_k(x)\Big)dx = \int_a^b F(x,y)f(x)dx\,. \tag{2.3}$$

Putting $S_N := \sum_{k=0}^{N}\alpha_kG_k(x)$, we can write the limit relation:

$$\lim_{N\to\infty}\int_a^b\Big(W^{-1}(x)f(x)-S_N\Big)^2W(x)dx = 0\,.$$

Applying Schwarz inequality in $L^2(a,b)$ we obtain:

$$\int_a^b F(x,y)W(x)\Big(W^{-1}(x)f(x) - S_N\Big)dx =$$

$$= \int_a^b F(x,y)W^{1/2}(x)\Big(W^{-1}(x)f(x) - S_N\Big)W^{1/2}(x)dx \leq$$

$$\leq \left\{\int_a^b F^2(x,y)W(x)dx\right\}^{1/2}\left\{\int_a^b \Big(W^{-1}(x)f(x) - S_N\Big)^2 W(x)dx\right\}^{1/2}$$

and consequently, recalling hypothesis III:

$$\lim_{N\to\infty} \int_a^b F(x,y)W(x)S_N dx = \int_a^b F(x,y)f(x)dx\,.$$

We can then repeat the proof of Proposition II in [2] in order to obtain the formal Taylor expansion in a neighborhood of the origin of the function $\hat{f}(y)$:

$$\hat{f}(y) := \int_a^b F(x,y)W(x)W^{-1}(x)f(x)dx = \int_a^b F(x,y)W(x)\left(\sum_{k=0}^{\infty}\alpha_k G_k(x)\right)dx\,.$$

By (2.1) we obtain:

$$\hat{f}(y) := \sum_{k=0}^{\infty}\alpha_k \int_a^b F(x,y)W(x)G_k(x)dx =$$

(2.4)

$$= \sum_{k=0}^{\infty}\alpha_k \int_a^b \sum_{\ell=0}^{\infty} c_\ell W(x)G_\ell(x)G_k(x)y^\ell dx = \sum_{k=0}^{\infty}\alpha_k c_k h_k y^k\,,$$

so that:

(2.5) $$\hat{f}(y) = \sum_{k=0}^{\infty}\alpha_k c_k h_k y^k \overset{(2.2)}{=} \sum_{k=0}^{\infty} c_k\left(\int_a^b f(x)G_k(x)dx\right)y^k\,,$$

which proves the proposition. □

REMARK I. – Note that the result of Theorem I has only a formal character, since no conditions on the function f have been considered in order to guarantee the possibility to invert the symbols of integrals and of series and the convergence (at least for small values of $|y|$) of the series expansion.

Before removing objections exposed in the preceding Remark I, we will show now, for the reader's convenience, that results of Theorem I, considered for the particular case of Hermite polynomials, reduces to a classical property of the Fourier transform.

THEOREM II. *Consider the system of the Hermite polynomials* $\{He_k(x)\}_{k\in\mathbb{N}_0}$, *orthogonal in* $(-\infty,\infty)$ *with respect to the weight function* $e^{-\frac{x^2}{2}}$. *Let* f *be such that* $x^k f(x) \in L(-\infty,\infty)\ \forall\ k \in \mathbb{N}_0$. *If* $\alpha_k := \frac{1}{h_k}\int_{-\infty}^{\infty} f(x)He_k(x)dx$ $(h_k = \sqrt{2\pi}k!)$

denote the Fourier coefficients of the function $e^{\frac{x^2}{2}} f(x)$, *then the Fourier transform* $\mathcal{F}(f)(y) := \frac{1}{\sqrt{2\pi}} \int_{-\infty}^{\infty} e^{ixy} f(x)dx$ *multiplied by* $e^{\frac{y^2}{2}}$ *admits the following formal Taylor expansion in a neighborhood of the origin:*

$$e^{\frac{y^2}{2}} \mathcal{F}(f)(y) = \sum_{k=0}^{\infty} i^k \alpha_k y^k .$$

The proof can be done following the same considerations as before. See also [2], under more particular hypotheses.

3. CONVERGENCE RESULTS FOR THE FORMAL EXPANSIONS

In this paragraph, we will limit ourselves to consider the case of *classical O.P.S.*. We will show that it is possible to find sufficient conditions, consisting in some restrictions on the parameters (α for Laguerre polynomials; α, β for the Jacobi case) and conditions about the asymptotic behavior of the Fourier coefficients α_k of the function $W^{-1}f$, in order to guarantee the convergence of the formal expansions of Theorem I.

3.1. Hermite polynomials $He_k(x)$

We prove the following

THEOREM III. *Let the Fourier coefficients* α_k *of the function* $e^{\frac{x^2}{2}} f(x)$ *satisfy the asymptotic behavior:*

$$|\alpha_k| = O\left((k!)^{-1/2}\right), \tag{3.1}$$

then $\forall\, y \in \mathbb{R}$ *the expansion* (2.5) *has an effective meaning.*

PROOF. We consider the inequality:

$$|He_k(x)| \leq C_1 (k!)^{\frac{1}{2}} e^{\frac{x^2}{4}} \qquad (-\infty < x < \infty), \tag{3.2}$$

that immediately follows for Hermite polynomials $He_k(x)$ from the estimate which is presented in a paper by A. AVANTAGGIATI [1, formula (1.12')].

As a consequence of (3.2), and taking into account hypothesis (3.1), we can write:

$$\left|\sum_{k=0}^{N} e^{-\frac{x^2}{2}} \alpha_k He_k(x)\right| \leq \sum_{k=0}^{\infty} |\alpha_k| e^{-\frac{x^2}{2}} |He_k(x)| \leq$$

$$\leq c_1 \sum_{k=0}^{\infty} |\alpha_k| (k!)^{\frac{1}{2}} e^{-\frac{x^2}{4}} \leq D_1 e^{-\frac{x^2}{4}} .$$

By Lebesgue theorem, it is then possible to perform the following exchange between the limit and the integral sign:

$$\lim_{N\to\infty} \sum_{k=0}^{N} \alpha_k \int_a^b F(x,y) e^{-\frac{x^2}{2}} He_k(x)dx = \int_a^b \lim_{N\to\infty} \sum_{k=0}^{N} \alpha_k e^{-\frac{x^2}{2}+xy-\frac{y^2}{2}} He_k(x)dx ,$$

since

$$\left|\sum_{k=0}^{N} e^{-\frac{x^2}{2}+xy-\frac{y^2}{2}} \alpha_k He_k(x)\right| \leq D_1 e^{-\frac{x^2}{4}+xy-\frac{y^2}{2}} ,$$

and for any fixed $y \in \mathbb{R}$, the second member belongs to $L(-\infty, \infty)$.

REMARK II. – The other inversion of the symbols of series and integral sign, which is necessary to write the last equality of formula (2.4), does not imply any further restriction on the behavior of Fourier coefficients α_k, so that we have proved the thesis of the theorem. □

The result of Theorem III can be summarized in the following way:

HERMITE case.

Assume: $a := -\infty, \quad b := \infty,$

$W(x) := e^{-\frac{x^2}{2}}, \quad G_k(x) = He_k(x)$ (Hermite Polynomials)

Put: $c_k := \frac{1}{k!}, \ \forall\, k \in \mathbb{N}_0; \quad h_k = \sqrt{2\pi}k!.$

Then:

$$F(x,y) = e^{xy - \frac{y^2}{2}},$$

$$\hat{f}(y) := \int_{-\infty}^{\infty} e^{xy - \frac{y^2}{2}} f(x)dx = e^{-\frac{y^2}{2}} \mathcal{F}(f)(y).$$

Suppose: $|\alpha_k| = O\Big((k!)^{-1/2}\Big)$. Then, as a consequence of Theorem III, we have the following expansion formulae:

$$e^{\frac{x^2}{2}} f(x) = \sum_{k=0}^{\infty} \alpha_k He_k(x) \tag{3.3}$$

$$\hat{f}(y) = e^{-\frac{y^2}{2}} \sum_{k=0}^{\infty} i^k \alpha_k y^k = e^{\frac{z^2}{2}} \sum_{k=0}^{\infty} \alpha_k z^k, \qquad (y = iz). \tag{3.4}$$

3.2. Laguerre polynomials $L_k^{(\alpha)}(x)$

We prove the following

THEOREM IV. *If $\alpha \geq -1/2$ and the Fourier coefficients α_k of the function $x^{-\alpha} e^x f(x)$ have the asymptotic behavior:*
$\exists\, \varepsilon > 0$ *s.t.*

$$|\alpha_k| = O\left(\left[\frac{k!}{\Gamma(\alpha+k+1)}\right]^{1/2} \frac{1}{k^{3/4+\varepsilon}}\right), \tag{3.5}$$

then $\forall\, y : |y| < 1$ the expansion (2.5) has an effective meaning.

PROOF. We consider the estimate:

$$|L_k^{(\alpha)}(x)| \leq h_k^{1/2} \left[C_2 k^{-\frac{1}{4}} + C_3 k^{-\frac{1}{2}} x^{\frac{5}{4}}\right] e^{\frac{x}{2}} x^{-\frac{\alpha}{2} - \frac{1}{4}}$$
$$\left(h_k := \frac{\Gamma(\alpha+k+1)}{k!}; \quad \alpha \geq -\frac{1}{2}; \quad 0 \leq x < +\infty\right) \tag{3.6}$$

that is proved for Laguerre polynomials in the book by A.F. NIKIFOROV - V.B. UVAROV[4]. Then we can write:

$$\left|\sum_{k=0}^{N} \alpha_k L_k^{(\alpha)}(x) F(x,y) W(x)\right| \leq (1-y)^{-(\alpha+1)} \sum_{k=0}^{\infty} |\alpha_k| x^{\alpha} e^{\frac{-x}{1-y}} |L_k^{(\alpha)}(x)| \leq$$

$$\leq C_2(\alpha;y) e^{-\frac{x}{2}\left(\frac{1+y}{1-y}\right)} x^{\frac{\alpha}{2}-\frac{1}{4}} \sum_{k=0}^{\infty} |\alpha_k| \left[\frac{\Gamma(\alpha+k+1)}{k!}\right]^{1/2} k^{-\frac{1}{4}} +$$

$$+ C_3(\alpha;y) e^{-\frac{x}{2}\left(\frac{1+y}{1-y}\right)} x^{\frac{\alpha}{2}+1} \sum_{k=0}^{\infty} |\alpha_k| \left[\frac{\Gamma(\alpha+k+1)}{k!}\right]^{1/2} k^{-\frac{1}{2}} .$$

Then, assuming condition (3.5) we have $\forall\, N \in \mathbb{N}_0$:

$$\left|\sum_{k=0}^{N} \alpha_k L_k^{(\alpha)}(x) F(x,y) W(x)\right| \leq \left[D_2(\alpha;y) + D_3(\alpha;y) x^{\frac{5}{4}}\right] e^{-\frac{x}{2}\left(\frac{1+y}{1-y}\right)} x^{\frac{\alpha}{2}-\frac{1}{4}}$$

and it is possible to use Lebesgue theorem since, recalling the condition $|y| < 1$, the second member belongs to $L(0,\infty)$.

A remark similar to the preceding Remark II can be done also in this case, so that the proof is completed. □

The result of Theorem IV can be summarized in the following way:

LAGUERRE case.

Assume: $a := 0$, $b := +\infty$,

$$W(x) := x^{\alpha} e^{-x}, \qquad (\alpha \geq -1/2),$$

$$G_k(x) = L_k^{(\alpha)}(x) \qquad \text{(Laguerre Polynomials)}$$

$$h_k = \frac{\Gamma(\alpha+k+1)}{k!}$$

Put: $c_k := 1$, $\forall k \in \mathbb{N}_0$, then:

$$F(x,y) = (1-y)^{-(\alpha+1)} e^{-\frac{xy}{1-y}}$$

$$\hat{f}(y) := \int_0^{+\infty} (1-y)^{-(\alpha+1)} e^{-\frac{xy}{1-y}} f(x)\, dx .$$

Suppose that $\exists\, \varepsilon > 0$ s.t.

$$|\alpha_k| = O\left(\left[\frac{k!}{\Gamma(\alpha+k+1)}\right]^{1/2} \frac{1}{k^{3/4+\varepsilon}}\right),$$

then, as a consequence of THEOREM III, we have the following expansion formulae:

$$x^{-\alpha} e^{x} f(x) = \sum_{k=0}^{\infty} \alpha_k L_k^{(\alpha)}(x) \tag{3.7}$$

$$\hat{f}(y) = \sum_{k=0}^{\infty} \frac{\Gamma(\alpha+k+1)}{k!} \alpha_k y^k, \qquad (|y| < 1). \tag{3.8}$$

3.3. Jacobi polynomials $P_k^{(\alpha,\beta)}(x)$

We prove the following

THEOREM V. *If $\alpha > -1/2$, $\beta > -1/2$, and the Fourier coefficients α_k of the function $(1-x)^{-\alpha}(1+x)^{-\beta}f(x)$ have the asymptotic behavior:*
$\exists\, \varepsilon > 0$ s.t.

$$|\alpha_k| = O\left(\frac{1}{k^{1/2+\varepsilon}}\right), \tag{3.9}$$

then $\forall\, y : |y| < 1$, the expansion (2.5) has an effective meaning.

PROOF. We consider the estimate:

$$\left|P_k^{(\alpha,\beta)}(x)\right| \le C_4 k^{-\frac{1}{2}}(1-x)^{-\frac{\alpha}{2}-\frac{1}{4}}(1+x)^{-\frac{\beta}{2}-\frac{1}{4}} \tag{3.10}$$
$$\left(\alpha > -\frac{1}{2};\quad \beta > -\frac{1}{2};\quad -1 \le x \le 1\right)$$

that is proved for Jacobi polynomials in the book by A.F. NIKIFOROV - V.B. UVAROV[4]. Then we can write:

$$\left|\sum_{k=0}^{N} \alpha_k P_k^{(\alpha,\beta)}(x) F(x,y) W(x)\right| \le D_4(\alpha|\beta)(1-x)^{\frac{\alpha}{2}-\frac{1}{4}}(1+x)^{\frac{\beta}{2}-\frac{1}{4}} \sum_{k=0}^{\infty} |\alpha_k| k^{-\frac{1}{2}},$$

where $D_4(\alpha|\beta)$ denotes a constant which depends by α if we are studying the problem of inversion near the point $x = -1$, and by β if we are studying the same problem near the point $x = 1$.

Then it is easily seen that condition $|y| < 1$ guarantees the possibility to use Lebesgue theorem also in this case, and a remark similar to the preceding Remark II can be done, so that the proof is completed. □

The result of Theorem V can be summarized in the following way:

JACOBI case.

Assume: $a := -1$, $b := 1$,

$$W(x) := (1-x)^{\alpha}(1+x)^{\beta}, \qquad (\alpha > -1/2, \beta > -1/2),$$

$$G_k(x) = P_k^{(\alpha,\beta)}(x) \qquad \text{(Jacobi Polynomials)}$$

$$h_k = \frac{2^{\alpha+\beta+1}}{\alpha+\beta+2k+1} \frac{\Gamma(\alpha+k+1)\Gamma(\beta+k+1)}{k!\,\Gamma(\alpha+\beta+k+1)}.$$

Put: $c_k := 1$, $\forall\, k \in \mathbb{N}_0$; $R := \sqrt{1-2xy+y^2}$, then:

$$F(x,y) = \frac{2^{\alpha+\beta}}{R(1-y+R)^{\alpha}(1+y+R)^{\beta}},$$

$$\hat{f}(y) := \int_{-1}^{1} \frac{2^{\alpha+\beta}}{R(1-y+R)^{\alpha}(1+y+R)^{\beta}} f(x)\,dx.$$

Suppose that $\exists\, \varepsilon > 0$ s.t.

$$|\alpha_k| = O\left(\frac{1}{k^{1/2+\varepsilon}}\right),$$

then, as a consequence of Theorem III, we have the following expansion formulae:

$$(1-x)^{-\alpha}(1+x)^{-\beta}f(x)=\sum_{k=0}^{\infty}\alpha_k P_k^{(\alpha,\beta)}(x)$$

$$\hat{f}(y)=\sum_{k=0}^{\infty}\frac{2^{\alpha+\beta+1}}{\alpha+\beta+2k+1}\frac{\Gamma(\alpha+k+1)\Gamma(\beta+k+1)}{k!\Gamma(\alpha+\beta+k+1)}\alpha_k y^k\,, \qquad (|y|<1)\,.$$

Acknowledgements

We want to thank Prof. Edward B. Saff of the University of South Florida (TAMPA - FL) for valuable comments in order to improve this paper.

REFERENCES

1. A. Avantaggiati: Sviluppi in serie di Hermite-Fourier e condizioni di analiticità e quasi analiticità, *in* Proc. Intern. Meeting dedicated to the memory of Carlo Miranda, Liguori Editore, Napoli, (1982).
2. C. Belingeri and P.E. Ricci: An extension of a property of the Fourier Transform, *Riv. Matem. Univ. Parma*, 1: (1992).
3. I.S. Gradshteyn and I.M. Ryzhik: "Table of Integrals, Series and Products", Academic Press, New York and London (1965).
4. A.F. Nikiforov - V.B. Uvarov: "Special Functions of Mathematical Physics", Birkhäuser, Basel (1988).
5. D.V. Widder: "An introduction to Transform Theory", Academic Press, New York-London (1971).

EULER FUNCTIONS $E_\alpha(z)$ WITH COMPLEX α AND APPLICATIONS

Paul L. Butzer, Stefan Flocke, and Michael Hauss

Lehrstuhl A für Mathematik
RWTH Aachen
Templergraben 55
52056 Aachen/Germany

Dedicated to Jacob Korevaar on the occasion of his 70th birthday, in friendship and esteem.

1 INTRODUCTION

The aim of this paper is to extend the Euler polynomials $E_n(x)$, which may be defined in terms of their exponential generating function via

$$\frac{2e^{xw}}{e^w+1} = \sum_{n=0}^{\infty} \frac{E_n(x)}{n!} w^n \qquad (x \in \mathbf{R};\ |w| < \pi), \tag{1}$$

to Euler functions $E_\alpha(z)$ with complex indices $\alpha \in \mathbf{C}$; z will also be allowed to belong to $\mathbf{C}$ (see e.g. [26]). The $E_n(x)$ with $\alpha = n \in \mathbf{N}$ are polynomials in x, namely

$$E_n(x) = \sum_{k=0}^{n} \binom{n}{k} E_{n-k} 2^{k-n} (x - \frac{1}{2})^k, \tag{2}$$

$E_k := 2^k E_k(1/2)$ being the classical *Euler numbers*, with $E_{2k-1} = 0$, $k \in \mathbf{N}$.

One of the famous applications of the E_k is the explicit evaluation of the Dirichlet function $\mathcal{L}(s) := \sum_{k=0}^{\infty}(-1)^k(2k+1)^{-s}$, $\Re s > 0$, at the odd integers, namely

$$\mathcal{L}(2m+1) = (-1)^m 2^{-2m-2} \pi^{2m+1} E_{2m}/(2m)! \qquad (m \in \mathbf{N}_0). \tag{3}$$

The evaluation of the $\mathcal{L}$-function at *even* integral values is an open problem, to be considered here. In fact, the $\mathcal{L}(\alpha)$ will be evaluated in closed form for any complex $\alpha \in \mathbf{C}$, if interpreted in a special structural sense.

Already Ramanujan[1] used (3) to *define* "signless" Euler numbers for any complex

[1]Prof. K. Dilcher (Halifax/Canada), who lectured in Aachen on July 6, 1993, kindly made the authors aware of A. Jonquière [19] and especially of the bibliography edited by K. Dilcher et al. [13] listing 1956 publications by 839 authors on Bernoulli and related numbers – e.g. Euler, Genocchi. The designations "Bernoulli functions" or "Euler functions" in many older works, e.g. Raabe, Schlömilch, Glaisher and Bell refer to Bernoulli or Euler (or closely related) polynomials! However, D.M. Sinocov (1890/91) (see e.g. [27], cited in [13]), and Jonquière actually treat true Bernoulli functions.

Approximation, Probability, and Related Fields, Edited by
G. Anastassiou and S.T. Rachev, Plenum Press, New York 1994

α by $E_\alpha^* = 2^{\alpha+1}\Gamma(\alpha)\mathcal{L}(\alpha)/\pi^\alpha$ (see [3, p. 170]), signless in the sense that $E_{2m+1}^* = (-1)^m E_{2m} > 0$; thus he avoided the problem of finding a substitute for $(-1)^m$ with m replaced by α, a major difficulty to be resolved here.

The starting point of this investigation is the difference equation

$$f(z+1) + f(z) = 2z^\alpha, \tag{4}$$

with $z \in \mathbf{C}\backslash\mathbf{R}_0^-$ and $\alpha \in \mathbf{C}$. Here z^α is defined via the principal branch of the logarithm, namely $z^\alpha = \exp(\alpha \log z)$. For $\alpha = n \in \mathbf{N}$ it is well-known that the Euler polynomials $E_n(z)$ satisfy the equation (4). The general solution of this equation is considered, and the Euler functions $E_\alpha(z)$ will be defined as a suitable particular solution of (4). However, it will not be possible to define the $E_\alpha(z)$ for $z \in \mathbf{R}_0^-$ via (4) since the right side z^α is not single-valued in this rather interesting case. This will lead us to a study of the conjugate Euler functions, i.e., the Hilbert transform of $E_\alpha(x)$, denoted by $E_\alpha^\sim(x)$, which can be represented as

$$E_\alpha^\sim(x) = -4\Gamma(\alpha+1)\sum_{k=0}^{\infty} \frac{\cos((2k+1)\pi x - \alpha\pi/2)}{[(2k+1)\pi]^{\alpha+1}} \qquad (\Re\alpha > -1;\ x \in (0,1)).$$

The main properties of the $E_\alpha(x)$ as well as of the $E_\alpha^\sim(x)$ will be studied in this paper; the latter functions will play a basic role here. In particular, an exponential generating function will be deduced for the $E_n^\sim(x)$, together with several representations of these functions. This material is dealt with in Section 3.

As an application to approximation theory there follows in Section 4 a fractional order Jackson inequality for best trigonometric approximation, where the constants are shown to be exact and attainable in a certain way by employing the Euler functions $E_\alpha(x)$.

2 EULER AND CONJUGATE EULER FUNCTIONS

2.1 Euler Functions $E_\alpha(z)$ for $\alpha \in \mathbf{C}$, $\Re z > 0$

Let us first consider the determination of the general solution of the difference equation (4). For this purpose, let us begin with the relation between two arbitrary solutions of (4).

Lemma 2.1 *Let f and g be arbitrary solutions of (4) on $D := \mathbf{C}\backslash\mathbf{R}_0^-$ or on $H_\beta := \{z \in \mathbf{C}; \Re z > \beta\} \cap D$, $\beta \in \mathbf{R}$. Then*

$$\omega(z) := f(z) - g(z) \tag{5}$$

is a one-antiperiodic function on D or H_β, i.e., $\omega(z+1) = -\omega(z)$. On the other hand, every solution of (4) can be obtained via equation (5) when knowing one arbitrary solution.

Proof Given two such solutions f and g, then by (4), $f(z+1) + f(z) = 2z^\alpha = g(z+1) + g(z)$, since with z also $(z+1)$ belongs to the domain of D or H_β. Thus $\omega(z)$, as given by (5), is a 1-antiperiodic function. Concerning the second part, it is obvious since, given any solution $g(z)$ and an arbitrary 1-antiperiodic function $\omega(z)$, then $f(z) := g(z) + \omega(z)$ is a solution of (4), too. ∎

Now we want to find one particular solution of the difference equation (4). The Euler functions will then be defined in terms of this solution. This approach goes back in principle to P. Böhmer[2] [4] and Nörlund [26], particularly in the instance of Bernoulli polynomials. However, they did not work on the "cut" $\mathbf{R}^-$, one basic contribution of this paper. But first to

Lemma 2.2 *The function defined by the contour integral*

$$F(\alpha, z) := \int_C \frac{e^{uz}}{1+e^u} u^{-\alpha-1}\, du \qquad (\alpha \in \mathbf{C};\ \Re z > 0) \tag{6}$$

is analytic in $\alpha \in \mathbf{C}$. Here C denotes a positively oriented loop around the negative real axis, composed of a circle C_2 of radius $0 < c < \pi$ around the origin together with the lower and upper edges C_1 and C_3 of the "cut" in the complex plane along $\mathbf{R}^-$.

Proof Using the parametrization $u = re^{i\pi}$ on C_3 and $u = re^{-i\pi}$ on C_1, $r \in [c, \infty)$, then $u^{-\alpha-1}$ means $r^{-\alpha-1}e^{-i\pi(\alpha+1)}$ on C_3 and $r^{-\alpha-1}e^{i\pi(\alpha+1)}$ on C_1. Consider an arbitrary compact disk $|\alpha| \leq M$, $M > 1$, and prove that the integrals along C_3 and C_1 converge uniformly on every such disk. As in [2, p. 253] one has along C_3 and C_1 the estimate, for $r \geq 1$, $|u^{-\alpha-1}| \leq r^{M-1}e^{\pi M}$.

Hence on either C_3 or C_1 there holds for $r > \log 2$,

$$\left|\frac{e^{uz}}{1+e^u} u^{-\alpha-1}\right| \leq \frac{e^{-r\Re z}}{1-e^{-r}} r^{M-1} e^{\pi M} < A e^{-r\Re z} r^{M-1}$$

with $A := 2e^{\pi M}$. Since $\int_c^\infty r^{M-1}e^{-r\Re z}\, dr$ converges if $c > 0$ and $\Re z > 0$, the integrals along C_3 and C_1 converge uniformly on every disk $|\alpha| \leq M$, establishing Lemma 2.2. ∎

Definition 2.1 *The Euler functions $E_\alpha(z)$ are defined for $\alpha \in \mathbf{C}$ and $\Re z > 0$ by*

$$E_\alpha(z) := \frac{\Gamma(\alpha+1)}{\pi i} \int_C \frac{e^{uz}}{1+e^u} u^{-\alpha-1}\, du, \tag{7}$$

C being the loop of Lemma 2.2.

Observe that these Euler functions have removable discontinuities at $\alpha \in \mathbf{Z}^-$ since the Gamma function has simple poles at these points, and the integral vanishes by Cauchy's theorem, since then the integrand is analytic. Thus we have by Lemma 2.2

Corollary 2.1 *$E_\alpha(z)$ is an analytic function of $\alpha \in \mathbf{C}$ for $\Re z > 0$.*

It remains to show that these Euler functions are indeed solutions of (4).

Proposition 2.1 *For $\alpha \in \mathbf{C}$, $\Re z > 0$ one has*

$$E_\alpha(z+1) + E_\alpha(z) = 2z^\alpha. \tag{8}$$

Proof By Definition 2.1 there follows

$$E_\alpha(z+1) + E_\alpha(z) = \frac{\Gamma(\alpha+1)}{\pi i} \int_C e^{uz} u^{-\alpha-1}\, du =: \frac{\Gamma(\alpha+1)}{\pi i} I_C(\alpha).$$

Now it can be shown after some calculations that $I_C(\alpha) = 2\pi i z^\alpha / \Gamma(\alpha+1)$ provided $\Re \alpha < 0$ and $\Re z > 0$. This yields (8) under these restrictions. However, since for $\Re z > 0$ the functions $E_\alpha(z)$, $E_\alpha(z+1)$ and z^α are analytic for all $\alpha \in \mathbf{C}$ the identity theorem for holomorphic functions gives that (8) is valid for all $\alpha \in \mathbf{C}$ and $\Re z > 0$. ∎

[2]The authors would like to thank Priv. Doz. Dr. D. Klusch (Rendsburg/Germany) for pointing out the important paper by Paul E. Böhmer (1939 Professor in Dresden/Germany) to them; it is hardly ever cited (one exception being Nörlund [26]).

The nomenclature "Euler functions"[3] of Definition 2.1 is justified since the $E_\alpha(z)$ coincide with the classical Euler polynomials $E_n(z)$ in the case $\alpha = n \in \mathbf{N}_0$. In fact, according to the Cauchy integral formula for derivatives, and (1), noting that $C_1 = -C_3$, there holds for $\Re z > 0$

$$\frac{\Gamma(n+1)}{\pi i}\int_C \frac{e^{uz}u^{-n-1}}{1+e^u}\,du = \frac{n!}{\pi i}\int_{C_2}\frac{e^{uz}u^{-n-1}}{1+e^u}\,du = 2\Big(\frac{d}{du}\Big)^n\Big(\frac{e^{uz}}{1+e^u}\Big)\Big|_{u=0} = E_n(z).$$

2.2 $E_\alpha(z)$ for $\alpha \in \mathbf{C}, z \in \mathbf{C}\backslash\mathbf{R}_0^-$

Now Lemma 2.1 together with Proposition 2.1 yields all solutions of (4) in the half-plane $\Re z > 0$. But the Euler functions can also be defined for $\Re z \leq 0$ provided, to begin with, $z \notin \mathbf{R}_0^-$. For this purpose one proceeds from the strip $\Re z \in (0,1]$ and defines $E_\alpha(z)$ in the strip $\Re z \in (-m, -m+1]$, $m \in \mathbf{N}$, as follows.

Definition 2.2 *The Euler functions $E_\alpha(z-m)$ with argument $(z-m)$ are defined for $\alpha \in \mathbf{C}$, $z \in \mathbf{C}\backslash\mathbf{R}$ with $\Re z \in (0,1]$ and $m \in \mathbf{N}$ by*

$$E_\alpha(z-m) := (-1)^m E_\alpha(z) + 2\sum_{j=0}^{m-1}(-1)^{m+j+1}(z-j-1)^\alpha.$$

By this definition it is clear that the Euler functions, now defined on $\mathbf{C}\backslash\mathbf{R}_0^-$, are solutions of (4) on the whole $\mathbf{C}\backslash\mathbf{R}_0^-$-plane. In fact, for arbitrary $m \in \mathbf{N}$, $E_\alpha(z-m+1) + E_\alpha(z-m) = 2(z-m)^\alpha$. For $\alpha \in \mathbf{C}$ the functions $E_\alpha(z)$ are analytic with respect to $z \in \mathbf{C}\backslash\mathbf{R}_0^-$ (see Corollary 3.1).

Corollary 2.2 *$E_\alpha(z)$ is an analytic function of $\alpha \in \mathbf{C}$ for $z \in \mathbf{C}\backslash\mathbf{R}_0^-$.*

2.3 $E_\alpha(x)$ represented as a Fourier Series

Let us now define the Euler functions in the remaining area, namely $\mathbf{R}_0^-$. This is surely the most interesting case in the study of fractional Euler numbers and functions. The definition in this particular instance has as its model the authors' [6] extension of Bernoulli numbers and polynomials to the fractional instance. But first to a representation of the Euler functions on $\mathbf{R}^+$ in terms of Fourier series.

Proposition 2.2 *There holds for $\Re\alpha > -1$, $x \in (0,1]$ ($x \neq 1$ if $\Re\alpha \in (-1,0]$), $m \in \mathbf{N}_0$,*

$$E_\alpha(x+m) = (-1)^m\mathcal{E}_\alpha(x) + 2\sum_{k=0}^{m-1}(-1)^{m+k+1}(x+k)^\alpha,$$

where $\mathcal{E}_\alpha(x)$ are the so-called α-th one-antiperiodic Euler functions, given by

$$\mathcal{E}_\alpha(x) := 4\Gamma(\alpha+1)\sum_{k=0}^{\infty}\frac{\sin((2k+1)\pi x - \alpha\pi/2)}{((2k+1)\pi)^{\alpha+1}} \qquad (x \in \mathbf{R}; x \notin \mathbf{Z} \text{ if } \Re\alpha \in (-1,0]). \quad (9)$$

In the particular case $m = 0$, $E_\alpha(x) = \mathcal{E}_\alpha(x)$ for $x \in (0,1]$ ($x \neq 1$ if $\Re\alpha \in (-1,0]$).

For the proof we need

[3]Our Euler functions are (also) not to be confused with the well-known Euler (indicator) function $\varphi(n)$, defined by the number of positive integers k which do not exceed n, and are relatively prime to n (see [12, pp. 193, 162]).

Lemma 2.3 *Let $S(r)$ denote the region that remains when from the* **C***-plane all open circular disks are removed which have radius r, $0 < r < 1$, with centers at $u = (2n+1)\pi i$, $n \in \mathbf{Z}$. Then if $0 < x \leq 1$ the function $g(u) := e^{ux}/(1+e^u)$ is bounded in $S(r)$ (the bound depending on r).*

Proof Write $u = u_1 + iu_2$ and, as in [2, pp. 256ff], consider the punctured rectangle

$$Q(r) := \{u \in \mathbf{C}; |u_1| \leq 1, |u_2 - \pi| \leq \pi, |u - \pi i| \geq r\}.$$

Because this set is compact, g is bounded on $Q(r)$. Also, since $|g(u+2\pi i)| = |g(u)|$, g is bounded on the punctured infinite strip $\{u \in \mathbf{C}; |u_1| \leq 1, |u-(2n+1)\pi i| \geq r, n \in \mathbf{Z}\}$. But g is also bounded outside this strip. Indeed, let $|u_1| \geq 1$. Then one has $|g(u)| \leq e^{xu_1}/|1-e^{u_1}|$. Similarly as in [2, p. 257] one then has $|g(u)| \leq e/(e-1)$ for $|u_1| \geq 1$. ∎

Proof of Prop. 2.2 Let us evaluate the integral (7) defining $E_\alpha(z)$. To this end, consider the integral

$$I_N(\alpha, x) := \frac{1}{2\pi i}\int_{C(N)} \frac{e^{ux}u^{-\alpha-1}}{1+e^u}\,du, \tag{10}$$

where $C(N)$, $N \in \mathbf{N}$, is the contour that results when the upper and lower edges of the contour C are cut off at an arbitrary point $-2N\pi$, $N \in \mathbf{N}$, and these endpoints are connected by a negatively oriented circle of radius $2N\pi$ around 0 (cf. [2, p. 258]). First we show that $\lim_{N\to\infty} I_N(\alpha,x) = E_\alpha(x)/(2\Gamma(\alpha+1))$ for $\Re\alpha > 0$. It suffices to show that the integral along the outer circle tends to zero as $N \to \infty$. Here $u = Re^{i\vartheta}$, $-\pi \leq \vartheta \leq \pi$, so that $|u^{-\alpha-1}| \leq R^{-\Re\alpha-1}e^{\pi|\Im\alpha|}$. Since the outer circle lies in $S(r)$ (cf. Lemma 2.3), the integrand of (10) is bounded by $MR^{-\Re\alpha-1}e^{\pi|\Im\alpha|}$, M being the bound of $|g(u)|$. Thus, the integral tends to zero as $N \to \infty$ if $\Re\alpha > 0$. Now let us compute $I_N(\alpha,x)$ by Cauchy's residue theorem. Indeed,

$$I_N(\alpha,x) = -\sum_{k=-N}^{N-1} R(k) = -\sum_{k=0}^{N-1} R(k) - \sum_{k=1}^{N} R(-k),$$

where $R(k) := Res_{u=(2k+1)\pi i}\{u^{-\alpha-1}e^{ux}/(1+e^u)\}$. But for $k \in \mathbf{Z}$,

$$R(k) = \lim_{u\to(2k+1)\pi i}(u-(2k+1)\pi i)\frac{u^{-\alpha-1}e^{ux}}{1+e^u} = -\frac{e^{(2k+1)\pi i x}}{[(2k+1)\pi i]^{\alpha+1}},$$

which gives, noting $i^{\alpha+1} = \exp(i\pi(\alpha+1)/2)$, and $(-i)^{\alpha+1} = \exp(-i\pi(\alpha+1)/2)$, after some calculations,

$$I_N(\alpha,x) = 2\sum_{k=0}^{N-1}\frac{\sin((2k+1)\pi x - \alpha\pi/2)}{((2k+1)\pi)^{\alpha+1}}. \tag{11}$$

Hence one obtains for $x \in (0,1]$, $\Re\alpha > 0$, $E_\alpha(x) = \lim_{N\to\infty} 2\Gamma(\alpha+1)I_N(\alpha,x)$, which in turn is equal to the right side of (9). Thus $E_\alpha(x) = \mathcal{E}_\alpha(x)$ for $x \in (0,1]$, so that Proposition 2.2 readily follows.

Finally, let us consider the case $\Re(\alpha) \in (-1,0]$, $x \in (0,1)$. In view of Abel's lemma of partial summation (9) is well-defined and converges uniformly on each compact subset of $\Re(\alpha) > -1$. Therefore, for each $x \in (0,1)$, $\mathcal{E}_\alpha(x)$ is a holomorphic function of α, satisfying $\mathcal{E}_\alpha(x) = E_\alpha(x)$ for $\Re(\alpha) > 0$. This equality holds true for $\Re(\alpha) > -1$ by the interior uniqueness theorem, because $E_\alpha(x)$ is holomorphic in $\alpha \in \mathbf{C}$ ($x \in (0,1)$), which proves Proposition 2.2. ∎

There exist two further representations of $\mathcal{E}_\alpha(x)$ which are of interest and will be needed.

Proposition 2.3 *There hold for $\alpha \in \mathbf{C}$, $\Re\alpha > -1$, $x \in \mathbf{R}$ ($x \notin \mathbf{Z}$ if $\Re\alpha \in (-1,0]$),*

$$\mathcal{E}_\alpha(x) = 2\Gamma(\alpha+1)\Big\{\cos\frac{\pi(\alpha+1)}{2}\,\phi_\alpha(x) + \sin\frac{\pi(\alpha+1)}{2}\,\psi_\alpha(x)\Big\}, \tag{12}$$

$$\mathcal{E}_\alpha(x) = 2\Gamma(\alpha+1)\sum_{k=-\infty}^{\infty}\frac{e^{(2k+1)\pi i x}}{[(2k+1)\pi i]^{\alpha+1}}, \tag{13}$$

where the functions $\phi_\alpha(x)$ and $\psi_\alpha(x)$ are defined and given by

$$\phi_\alpha(x) := \frac{2}{\pi^{\alpha+1}}\sum_{k=0}^{\infty}\frac{\cos((2k+1)\pi x)}{(2k+1)^{\alpha+1}} = \sum_{k=-\infty}^{\infty}\frac{e^{(2k+1)\pi i x}}{[|2k+1|\pi]^{\alpha+1}}, \tag{14}$$

$$\psi_\alpha(x) := \frac{2}{\pi^{\alpha+1}}\sum_{k=0}^{\infty}\frac{\sin((2k+1)\pi x)}{(2k+1)^{\alpha+1}} = \sum_{k=-\infty}^{\infty}\{-i\,\mathrm{sgn}(2k+1)\}\frac{e^{(2k+1)\pi i x}}{[|2k+1|\pi]^{\alpha+1}}. \tag{15}$$

Proof The representation (12) is a direct consequence of (9) by using some trigonometric identities. Formula (13) then follows observing that $i^\alpha = e^{i\alpha\pi/2}$ and $(-i)^\alpha = e^{-i\alpha\pi/2}$. The representations in (14) and (15) are readily obtained in view of $e^{ix} = \cos x + i\sin x$. ∎

2.4 $E_\alpha(x)$ for $x \in \mathbf{R}_0^-$; Conjugate Euler Functions

Now let us define the Euler functions for the argument $x \in \mathbf{R}_0^-$.

Definition 2.3 *For $\Re\alpha > -1$, $x \in (0,1]$ ($x \neq 1$ if $\Re\alpha \in (-1,0]$) the Euler functions with argument $(-x-m)$, $m \in \mathbf{N}_0$, are defined by $E_\alpha(0) := \mathcal{E}_\alpha(0)$ ($\Re\alpha > 0$) and*

$$E_\alpha(-x-m) := \cos\pi\alpha\Big\{(-1)^{m+1}E_\alpha(x) + 2\sum_{k=0}^{m}(-1)^{m-k}(x+k)^\alpha\Big\}$$
$$-\sin\pi\alpha\Big\{(-1)^{m+1}E_\alpha^\sim(x) + 2\sum_{k=0}^{m}(-1)^{m-k}(x+k)^\alpha\Big\},$$

where $E_\alpha^\sim(x) := \mathcal{E}_\alpha^\sim(x)$ on $[0,1]$ (on $(0,1)$), $\mathcal{E}_\alpha^\sim(x) := (H_2\mathcal{E}_\alpha(\cdot))(x) = (H_{2\pi}\mathcal{E}_\alpha(\cdot/\pi))(\pi x)$ being the Hilbert transform (or conjugate function) of the 2-periodic Euler function $\mathcal{E}_\alpha(x)$ (cf. Lemma 2.4). Its definition for p-periodic functions reads ([9, p. 334], [6])

$$H_pf(x) := \mathrm{PV}\Big(\frac{1}{p}\int_{-p/2}^{p/2} f(x-u)\cot\Big(\frac{u\pi}{p}\Big)\,du\Big).$$

Lemma 2.4 *There hold for $\Re\alpha > -1$, $x \in \mathbf{R}$ ($x \notin \mathbf{Z}$ if $\Re\alpha \in (-1,0]$),*

$$\mathcal{E}_\alpha^\sim(x) = 2\Gamma(\alpha+1)\Big\{\cos\frac{\pi(\alpha+1)}{2}\psi_\alpha(x) - \sin\frac{\pi(\alpha+1)}{2}\phi_\alpha(x)\Big\}, \tag{16}$$

$$\mathcal{E}_\alpha^\sim(x) = -4\Gamma(\alpha+1)\sum_{k=0}^{\infty}\frac{\cos((2k+1)\pi x - \alpha\pi/2)}{[(2k+1)\pi]^{\alpha+1}}. \tag{17}$$

Proof The representation (16) follows from (12) by observing that $\phi_\alpha^\sim(x) = \psi_\alpha(x)$, $\psi_\alpha^\sim(x) = -\phi_\alpha(x)$, and (17) after some calculations. ∎

In view of Proposition 2.3 and Lemma 2.4 it follows that for $\Re\alpha \in (-1,0]$ there holds $\mathcal{E}_\alpha(\cdot/\pi), \mathcal{E}_\alpha^\sim(\cdot/\pi) \in L^1_{2\pi}$ (see Theorem 6.3.7 of [9] if α is real; if α is complex, the proof in [18, p. 33] can be modified accordingly).

The Hilbert transform $\mathcal{E}_\alpha^\sim(x)$ is a 1-antiperiodic function of x. Now $E_\alpha^\sim(x)$, as defined (on $[0,1]$) in Definition 2.3, is equal to this $\mathcal{E}_\alpha^\sim(x)$ on $[0,1]$. Let us now see how $E_\alpha^\sim(x)$ can be extended to the whole real axis $\mathbf{R}$ in a similar way as $E_\alpha(x)$ (or $B_\alpha(x)$ and $B_\alpha^\sim(x)$ in [6]). The complete definition for $E_\alpha^\sim(x)$, $x \in \mathbf{R}$, $\alpha \in \mathbf{C}$, $\Re\alpha > -1$, reads (if $\Re\alpha \in (-1,0]$, then $x \in (0,1)!$):

$$E_\alpha^\sim(x) := \mathcal{E}_\alpha^\sim(x) \quad (x \in [0,1]), \tag{18}$$

$$E_\alpha^\sim(x+m) := (-1)^m \mathcal{E}_\alpha^\sim(x) + 2\sum_{k=0}^{m-1}(-1)^{m+k+1}(x+k)^\alpha \quad (x \in (0,1],\ m \in \mathbf{N}), \tag{19}$$

$$\begin{aligned} E_\alpha^\sim(-x-m) :=& -\cos(\pi\alpha)\Big\{(-1)^{m+1}\mathcal{E}_\alpha^\sim(x) + 2\sum_{k=0}^{m}(-1)^{m-k}(x+k)^\alpha\Big\} - \\ & -\sin(\pi\alpha)\Big\{(-1)^{m+1}\mathcal{E}_\alpha(x) + 2\sum_{k=0}^{m}(-1)^{m-k}(x+k)^\alpha\Big\} \quad (x \in (0,1],\ m \in \mathbf{N}_0). \end{aligned} \tag{20}$$

It can easily be shown that Proposition 2.2, Definition 2.3, (19) and (20) are also valid on the compact interval $[0,1]$ in the case $\Re(\alpha) > 0$.

Finally let us define the *fractional Euler numbers* E_α by $E_\alpha := 2^\alpha E_\alpha(1/2)$, $\alpha \in \mathbf{C}$, generalizing the classical Euler numbers $E_n := 2^n E_n(1/2)$, $n \in \mathbf{N}_0$, as well as the *conjugate fractional Euler numbers* by $E_\alpha^\sim := 2^\alpha E_\alpha^\sim(1/2)$, $\Re\alpha > -1$.

3 PROPERTIES: CONNECTIONS WITH KNOWN FUNCTIONS

3.1 Basic Properties

Let us first return to the difference equation already considered in Section 2.

Theorem 3.1 *(Difference Property) There hold ($x \notin \mathbf{Z}$ if $\Re\alpha \in (-1,0]$)*

$$E_\alpha(z+1) + E_\alpha(z) = \begin{cases} 2z^\alpha &, \text{ if } z \in \mathbf{C}\backslash\mathbf{R}_0^-,\ \alpha \in \mathbf{C} \\ 2|x|^\alpha(\cos(\pi\alpha) - \sin(\pi\alpha)), & \text{ if } z = x \in \mathbf{R}_0^-,\ \Re\alpha > -1, \end{cases} \tag{21}$$

$$E_\alpha^\sim(x+1) + E_\alpha^\sim(x) = \begin{cases} 2x^\alpha &, \text{ if } x \in \mathbf{R}^+,\ \Re\alpha > -1 \\ -2|x|^\alpha(\cos(\pi\alpha) + \sin(\pi\alpha)), & \text{ if } x \in \mathbf{R}_0^-,\ \Re\alpha > -1. \end{cases} \tag{22}$$

Proof In view of Proposition 2.1 and the remark following Definition 2.2 it remains to show the second half of (21) and formula (22). Firstly, let $x \in (-\infty,-1]$. Then there is $y \in [0,1)$ with $x = -y-m$, $m \in \mathbf{N}$. Thus, by Definition 2.3,

$$\begin{aligned} E_\alpha(x+1) + E_\alpha(x) &= E_\alpha(-y-(m-1)) + E_\alpha(-y-m) \\ &= 2\cos(\pi\alpha)(y+m)^\alpha - 2\sin(\pi\alpha)(y+m)^\alpha = 2|x|^\alpha(\cos(\pi\alpha) - \sin(\pi\alpha)). \end{aligned}$$

Secondly, let $x \in (-1,0]$. Then $y := x+1 \in (0,1]$, so that by Proposition 2.2 and Definition 2.3,

$$E_\alpha(x+1) + E_\alpha(x) = E_\alpha(y) + E_\alpha(-(1-y)) = 2|x|^\alpha(\cos(\pi\alpha) - \sin(\pi\alpha)) + G_\alpha(y),$$

where, it readily follows by (9) and (17) that for $\Re\alpha > -1$,

$$G_\alpha(y) := \mathcal{E}_\alpha(y) - \cos(\pi\alpha)\mathcal{E}_\alpha(1-y) + \sin(\pi\alpha)\mathcal{E}^\sim_\alpha(1-y) = 0.$$

Formula (22) follows in a similar way. ∎

In the case $\alpha = n \in \mathbf{N}_0$, $x \in \mathbf{R}$, Theorem 3.1 reduces to the classical result $E_n(x+1) + E_n(x) = 2x^n$.

The extension of $E_n'(x) = nE_{n-1}(x)$ to the fractional instance reads as follows.

Theorem 3.2 *(Derivative Property) There hold ($x \notin \mathbf{Z}$ if $\Re\alpha \in (0,1]$)*

$$\frac{d}{dz}E_\alpha(z) = \alpha E_{\alpha-1}(z) \qquad (z \in \mathbf{C}\backslash\mathbf{R}_0^-,\ \alpha \in \mathbf{C}), \tag{23}$$

$$\frac{d}{dx}E_\alpha(x) = \alpha E_{\alpha-1}(x) \qquad (x \in \mathbf{R}_0^-,\ \Re\alpha > 0),$$

$$\frac{d}{dx}E^\sim_\alpha(x) = \alpha E^\sim_{\alpha-1}(x) \qquad (x \in \mathbf{R},\ \Re\alpha > 0). \tag{24}$$

Proof Firstly, let $\alpha \in \mathbf{C}$, $\Re z > 0$. Differentiating (7), the interchange of the order of integration and differentiation being permissible (after tedious calculations) by Theorems 17.19, 17.20 of [25, pp. 418ff],

$$\frac{d}{dz}E_\alpha(z) = \frac{\Gamma(\alpha+1)}{\pi i}\int_C \frac{u^{-\alpha-1}}{1+e^u}\Big(\frac{d}{dz}e^{uz}\Big)\,du = \alpha E_{\alpha-1}(z).$$

Secondly, let $\alpha \in \mathbf{C}$, $z \in \mathbf{C}\backslash\mathbf{R}_0^-$ with $\Re z \le 0$. Then there is $w \in \mathbf{C}$, $\Re w \in (0,1]$, $m \in \mathbf{N}$ with $z = w - m$. Thus, by Definition 2.2, and the first step,

$$\frac{d}{dz}E_\alpha(z) = \frac{d}{dz}\Big\{(-1)^m E_\alpha(z+m) + 2\sum_{j=0}^{m-1}(-1)^{m+j+1}(z+m-j-1)^\alpha\Big\}$$

$$= (-1)^m\alpha E_{\alpha-1}(z+m) + 2\sum_{j=0}^{m-1}(-1)^{m+j+1}\alpha(z+m-j-1)^{\alpha-1} = \alpha E_{\alpha-1}(z).$$

Thirdly, let $x \in \mathbf{R}_0^-$. Then there is $u \in [0,1)$, $m \in \mathbf{N}_0$ with $x = -u-m$. Thus, by Definition 2.3, and (23),

$$\frac{d}{dx}E_\alpha(x) = \cos(\pi\alpha)\Big\{(-1)^{m+1}\frac{d}{dx}\mathcal{E}_\alpha(-x-m) + 2\sum_{k=0}^{m}(-1)^{m+k}\frac{d}{dx}(-x-m+k)^\alpha\Big\} -$$

$$-\sin(\pi\alpha)\Big\{(-1)^{m+1}\frac{d}{dx}\mathcal{E}^\sim_\alpha(-x-m) + 2\sum_{k=0}^{m}(-1)^{m+k}\frac{d}{dx}(-x-m+k)^\alpha\Big\} = \alpha E_{\alpha-1}(x).$$

The proof of (24) follows in a similar way. ∎

Corollary 3.1 *$E_\alpha(z)$ is an analytic function of $z \in \mathbf{C}\backslash\mathbf{R}_0^-$ for arbitrary $\alpha \in \mathbf{C}$.*

Proof Since, by (23), $E_\alpha(\cdot)$ is differentiable in the open set $\mathbf{C}\backslash\mathbf{R}_0^-$, it is analytic there. ∎

Observe that the $E_n(z)$ for $\alpha = n \in N$ are actually polynomials in z (see (2)), which are of course analytic functions.

The counterpart of the classical symmetry property $E_n(1-x) = (-1)^n E_n(x)$, $n \in \mathbf{N}_0$, reads as follows.

Theorem 3.3 *For $x \in \mathbf{R}$ and $\Re\alpha > -1$ ($x \notin \mathbf{Z}$ for $\Re(\alpha) \in (-1,0]$) there holds*

$$E_\alpha(1-x) = \cos(\pi\alpha)E_\alpha(x) - \sin(\pi\alpha)E_\alpha^\sim(x). \tag{25}$$

Proof Let $x \in [0,1]$. By Proposition 2.2 there follows

$$E_\alpha(1-x) = 4\Gamma(\alpha+1)\sum_{k=0}^{\infty}\Big\{\frac{\sin(-(2k+1)\pi x + \alpha\pi/2)\cos((2k+1)\pi - \pi\alpha)}{[(2k+1)\pi]^{\alpha+1}} +$$

$$+\frac{\cos(-(2k+1)\pi x + \alpha\pi/2)\sin((2k+1)\pi - \pi\alpha)}{[(2k+1)\pi]^{\alpha+1}}\Big\},$$

which yields (25). Now let $x > 1$, so that there is $y \in (0,1]$, $m \in \mathbf{N}$ with $x = y+m$. Then, by Definition of the $E_\alpha^\sim(x)$ and Proposition 2.2

$$\cos(\pi\alpha)E_\alpha(x) - \sin(\pi\alpha)E_\alpha^\sim(x) = \cos(\pi\alpha)\Big\{(-1)^m\mathcal{E}_\alpha(y) + 2\sum_{k=0}^{m-1}(-1)^{m+k+1}(y+k)^\alpha\Big\} -$$

$$-\sin(\pi\alpha)\Big\{(-1)^m\mathcal{E}_\alpha^\sim(y) + 2\sum_{k=0}^{m-1}(-1)^{m+k+1}(y+k)^\alpha\Big\} = E_\alpha(-y-(m-1)) = E_\alpha(1-x).$$

In the case $x < 0$ there is $y \in (0,1]$, $m \in \mathbf{N}_0$ with $x = -y-m$, and the proof proceeds similarly. ∎

Observe that the right side of (25) is well-defined for $x > 0$ by Definition 2.1 (for $E_\alpha(x)$ is a solution of the difference equation (4)) and by (18) and (19) (for $E_\alpha^\sim(x)$ extended in a similar way). Because (25) is the only formula in the classical case $\alpha = n$ connecting the Euler polynomials with positive and negative arguments, this equation was the motivation for our Definition 2.3 (see also the remark after Proposition 3.2), which then yields a unique extension of the Euler functions to $\mathbf{R}^-$ (see Definition 2.3).

Theorem 3.4 *(Integration Property) For $\alpha \in \mathbf{C}$, $a, z \in \mathbf{C}\backslash\mathbf{R}_0^-$ there holds*

$$\int_a^z E_\alpha(t)\,dt = \frac{E_{\alpha+1}(z) - E_{\alpha+1}(a)}{\alpha+1}, \tag{26}$$

the integration being along an arbitrary rectifiable curve in $\mathbf{C}\backslash\mathbf{R}_0^-$ connecting a and z.

Proof Since $E_\alpha(\cdot)$ is analytic in the simply connected domain $\mathbf{C}\backslash\mathbf{R}_0^-$ (Corollary 3.1), integration is independent of the curve joining $a, z \in \mathbf{C}\backslash\mathbf{R}_0^-$. On account of Theorem 3.2, $E_{\alpha+1}(z)$ is a primitive of $(\alpha+1)E_\alpha(z)$. Thus the result follows by the fundamental theorem of calculus. ∎

Theorem 3.5 *(Sum of Alternating Fractional Powers) For $\alpha \in \mathbf{C}$, $z \in \mathbf{C}\backslash\mathbf{R}_0^-$ and $m \in \mathbf{N}_0$ there holds*

$$\sum_{k=0}^{m}(-1)^k(k+z)^\alpha = \frac{1}{2}\{E_\alpha(z) + (-1)^m E_\alpha(z+m+1)\}.$$

Proof The result follows easily from the relation

$$2\sum_{k=0}^{m}(-1)^k(k+z)^\alpha = \sum_{k=0}^{m}(-1)^k\{E_\alpha(z+k+1) + E_\alpha(z+k)\}$$

valid for $k+z \in \mathbf{C}\backslash\mathbf{R}_0^-$ by Theorem 3.1. ∎

The results of Theorems 3.4 and 3.5 in the particular case $\alpha = n \in \mathbf{N}$ are well-known (see [26, p. 24] and [23, p. 30f]).

Extensions of (2) to the fractional case are the following power series expansions.

Theorem 3.6 *There hold for* $\alpha \in \mathbf{C}$, $w \in \mathbf{C}\backslash\mathbf{R}_0^-$,

$$a)\quad E_\alpha(z+w) = \sum_{k=0}^{\infty}\binom{\alpha}{k}E_{\alpha-k}(w)z^k \qquad \left(|z| < \begin{cases} |\Im w|, & \text{if } \Re w \le 0, \\ |w|\,, & \text{if } \Re w > 0 \end{cases}\right),$$

$$b)\quad E_\alpha(z) = \sum_{k=0}^{\infty}\binom{\alpha}{k}2^{k-\alpha}E_{\alpha-k}\cdot(z-\frac{1}{2})^k \qquad (|z-\frac{1}{2}| < \frac{1}{2}).$$

Proof The Taylor Theorem, applicable in view of Corollary 3.1, yields

$$E_\alpha(z+w) = \sum_{k=0}^{\infty}\frac{(d/dz)^k E_\alpha(z+w)|_{z=0}}{k!}z^k,$$

valid for those $z \in \mathbf{C}$ satisfying the restrictions of part a). This yields a), since, by Theorem 3.2, $(d/dz)^k E_\alpha(z+w)|_{z=0}/k! = \binom{\alpha}{k}E_{\alpha-k}(w)$. Part b) follows from a) by setting $w = 1/2$, and $z - 1/2$ instead of z, and by Definition of the fractional Euler numbers, i.e. $E_\alpha := 2^\alpha E_\alpha(1/2)$. ∎

The connection between Definition 2.2 and Definition 2.3 reads as follows.

Proposition 3.1 *For* $x_0 \in \mathbf{R}^-$, $\Re\alpha > -1$ *(*$x_0 \notin \mathbf{Z}$ *if* $\Re\alpha \in (-1,0]$*) there holds*

$$\lim_{\substack{z\to x_0\\ \Im(z)>0}} E_\alpha(z) + \lim_{\substack{z\to x_0\\ \Im(z)<0}} E_\alpha(z) = 2E_\alpha(x_0) + 4\sin(\pi\alpha)\sum_{k=0}^{m-1}(-1)^{m+k+1}(-x_0-m+1+k)^\alpha. \quad (27)$$

Proof Definition 2.2 yields

$$E_\alpha(z) = (-1)^m E_\alpha(z+m) + 2\sum_{k=0}^{m-1}(-1)^{k+m+1}(z+m-k-1)^\alpha.$$

By taking the limits $\lim_{z\to x_0}$ for $\Im(z) > 0$ and $\Im(z) < 0$, respectively, and adding the results, the proof follows by Definition 2.3. ∎

Of course it would be preferable if formula (27) would be somewhat more elegant, e.g. $\lim_{z\to x_0, \Im(z)>0} E_\alpha(z) = E_\alpha(x_0)$, but then one would have to alter Definition 2.3 for $E_\alpha(x_0)$, $x_0 \in \mathbf{R}^-$, and accordingly the symmetry property (25) – even in the classical case the only connection between $E_\alpha(\mathbf{R}^+)$ and $E_\alpha(\mathbf{R}^-)$ – would have a strange appearance. But by demanding this symmetry property, the values $E_\alpha(x)$, $x \in \mathbf{R}^-$, are uniquely defined.

3.2 Connections with the Bernoulli Functions

The Euler functions $E_\alpha(z)$ are also connected with the Bernoulli functions $B_\alpha(z)$ studied by the authors in [6]. In fact, they defined the Bernoulli functions $B_\alpha(x)$ with index $\alpha \in \mathbf{C}$, $\Re\alpha > 1$ (not via the solution of a difference equation as a contour integral, but in a converse way) first for $0 \le x < 1$ by $B_\alpha(x) := \mathcal{B}_\alpha(x)$, where the $\mathcal{B}_\alpha(x)$[4] are the one-periodic functions

$$\mathcal{B}_\alpha(x) := -2\Gamma(\alpha+1)\sum_{k=1}^{\infty}\frac{\cos(2\pi kx - \alpha\pi/2)}{(2\pi k)^\alpha} \qquad (x \in \mathbf{R}), \qquad (28)$$

[4] $\mathcal{B}_\alpha(x)$ also is well-defined in the case $\Re(\alpha) \in (0,1]$ and $x \in \mathbf{R}\backslash\mathbf{Z}$.

extended $B_\alpha(x)$ to $x \in \mathbf{R}_0^+$ and $x \in \mathbf{R}^-$ by procedures similar to those of Proposition 2.2 and Definition 2.3, then extended $B_\alpha(x)$, $x \in \mathbf{R}$, beyond the line $\Re\alpha = 1$ to the entire α-plane $\mathbf{C}$ via the contour integral given in Definition 3.1 a) for *real* $z \in (0,1]$.

Moreover, $B_\alpha(x)$ can even be extended to the instance of complex z as follows (see M. Leclerc [21]). We shall formulate this extension in terms of

Definition 3.1 *(Bernoulli Functions) The Bernoulli functions* $B_\alpha(z)$ *for* $\alpha \in \mathbf{C}$, $z \in \mathbf{C}\backslash\mathbf{R}_0^-$ *are defined first for* $\Re z > 0$ *by*

$$a) \quad B_\alpha(z) := \frac{\Gamma(\alpha+1)}{2\pi i}\int_C \frac{e^{wz}}{e^w - 1} w^{-\alpha}\, dw,$$

where C *is the above curve, and the* $B_\alpha(z)$ *for* $z \in \mathbf{C}\backslash\mathbf{R}_0^-$ *by*

$$b) \quad B_\alpha(z-m) := B_\alpha(z) - \alpha\sum_{k=0}^{m-1}(z-k-1)^{\alpha-1},$$

where $\Re z \in (0,1]$, $z \notin \mathbf{R}$, $m \in \mathbf{N}$.

It can be shown that the $B_\alpha(z)$ for $\alpha \in \mathbf{C}$ and $z \in \mathbf{C}\backslash\mathbf{R}_0^-$ satisfy the difference property

$$B_\alpha(z+1) - B_\alpha(z) = \alpha z^{\alpha-1} \tag{29}$$

as well as the multiplication formula, an extension of the Raabe formula (see [6], [21]),

$$B_\alpha(mz) = m^{\alpha-1}\sum_{k=0}^{m-1} B_\alpha(z+k/m) \qquad (m \in \mathbf{N};\, z \in \mathbf{C}\backslash\mathbf{R}_0^-). \tag{30}$$

These results will enable us to establish the following connections, which in the case $\alpha = n \in \mathbf{N}$ can be found in e.g. [26, p. 24], [24, p. 41].

Theorem 3.7 *a) For* $\alpha \in \mathbf{C}\backslash\mathbf{Z}^-$, $z \in \mathbf{C}\backslash\mathbf{R}_0^-$ *there hold*

$$E_\alpha(z) = \frac{2^{\alpha+1}}{\alpha+1}\left\{B_{\alpha+1}(\frac{z+1}{2}) - B_{\alpha+1}(\frac{z}{2})\right\}, \tag{31}$$

$$E_\alpha(z) = \frac{2}{\alpha+1}\left\{B_{\alpha+1}(z) - 2^{\alpha+1}B_{\alpha+1}(\frac{z}{2})\right\}. \tag{32}$$

b) For $\Re(\alpha) > -1$, $x \in \mathbf{R}$ *(*$x \notin \mathbf{Z}$ *if* $\Re(\alpha) \in (-1,0]$*) there holds*

$$\mathcal{E}_\alpha(x) = \frac{2^{\alpha+1}}{\alpha+1}\{\mathcal{B}_{\alpha+1}(\frac{x+1}{2}) - \mathcal{B}_{\alpha+1}(\frac{x}{2})\}. \tag{33}$$

Proof First to the proof of (31) for $\Re z > 0$. Then, by Definition 3.1 a),

$$B_{\alpha+1}(\frac{z+1}{2}) - B_{\alpha+1}(\frac{z}{2}) = \frac{\Gamma(\alpha+2)}{2\pi i}\int_C \frac{e^{wz/2}(e^{w/2}-1)w^{-\alpha-1}}{(e^{w/2}-1)(e^{w/2}+1)}\, dw = \frac{(\alpha+1)}{2^{\alpha+1}}E_\alpha(z).$$

Next to the instance $\Re z \le 0$, $z \notin \mathbf{R}$. Firstly, let there be $m \in \mathbf{N}$ with $\Re(z/2+m) \in (0,1/2]$. Then by Definition 3.1 b),

$$B_{\alpha+1}(\frac{z+1}{2}) - B_{\alpha+1}(\frac{z}{2}) = B_{\alpha+1}(\frac{z+1}{2}+m) - (\alpha+1)\sum_{k=0}^{m-1}(\frac{z+1}{2}+m-k-1)^\alpha -$$

$$-B_{\alpha+1}(\frac{z}{2}+m) + (\alpha+1)\sum_{k=0}^{m-1}(\frac{z}{2}+m-k-1)^\alpha$$

$$= (\alpha+1)2^{-\alpha-1}\left\{E_\alpha(z+2m)+2\sum_{k=1}^{2m}(-1)^k(z+2m-k)^\alpha\right\} = (\alpha+1)2^{-\alpha-1}E_\alpha(z),$$

the latter being valid by Theorem 3.5.

Secondly, let there be $m \in \mathbf{N}$ with $\Re(z/2+m) \in (1/2,1]$. Then by Definition 3.1 b)

$$B_{\alpha+1}(\frac{z+1}{2}) - B_{\alpha+1}(\frac{z}{2}) = B_{\alpha+1}(\frac{z+1}{2}+m-1) - (\alpha+1)\sum_{k=0}^{m-2}(\frac{z+1}{2}+m-k-2)^\alpha$$

$$-B_{\alpha+1}(\frac{z}{2}+m) + (\alpha+1)\sum_{k=0}^{m-1}(\frac{z}{2}+m-k-1)^\alpha,$$

the proof now proceeding similarly as above.

Regarding (32), noting formulae (30), and (31), there follows

$$2\left\{B_{\alpha+1}(2\frac{z}{2}) - 2^{\alpha+1}B_{\alpha+1}(\frac{z}{2})\right\} = 2\left\{2^\alpha\left[B_{\alpha+1}(\frac{z}{2}) + B_{\alpha+1}(\frac{z}{2}+\frac{1}{2})\right] - 2^{\alpha+1}B_{\alpha+1}(\frac{z}{2})\right\}$$

$$= 2^{\alpha+1}\left\{B_{\alpha+1}(\frac{z+1}{2}) - B_{\alpha+1}(\frac{z}{2})\right\} = (\alpha+1)E_\alpha(z).$$

Part b) follows immediately by using the Fourier series representations of $\mathcal{E}_\alpha(x)$ and $\mathcal{B}_{\alpha+1}(x)$. ∎

3.3 Raabe-Type Multiplication Formulae

Now we come to multiplication formulae which are the fractional counterparts of classical results, e.g. [26, p. 24], [23, p. 30].

Theorem 3.8 *There hold for* $\alpha \in \mathbf{C}\backslash\mathbf{Z}^-$, $m \in \mathbf{N}$, $z \in \mathbf{C}\backslash\mathbf{R}_0^-$,

$$E_\alpha(mz) = -\frac{2}{\alpha+1}m^\alpha\sum_{k=0}^{m-1}(-1)^k B_{\alpha+1}(z+\frac{k}{m}) \qquad (m \text{ even}), \tag{34}$$

$$E_\alpha(mz) = m^\alpha\sum_{k=0}^{m-1}(-1)^k E_\alpha(z+\frac{k}{m}) \qquad (m \text{ odd}). \tag{35}$$

Proof Let m be even, $m=2n$. Noting (31) and applying (30) twice,

$$E_\alpha(mz) = E_\alpha(2nz) = \frac{2^{\alpha+1}}{\alpha+1}\left\{B_{\alpha+1}\left(n(z+\frac{1}{2n})\right) - B_{\alpha+1}(nz)\right\}$$

$$= \frac{2^{\alpha+1}}{\alpha+1}(\frac{m}{2})^\alpha\sum_{j=0}^{2n-1}(-1)^{j+1}B_{\alpha+1}(z+\frac{j}{2n}),$$

yielding (34). If m is odd, i.e., $m=2n-1$, one has by (32) and (30) again,

$$E_\alpha(mz) = E_\alpha((2n-1)z) = \frac{2m^\alpha}{\alpha+1}\{\sum_{k=0}^{n-1}[B_{\alpha+1}(2(\frac{z}{2}+\frac{k}{2n-1})) - 2^{\alpha+1}B_{\alpha+1}(\frac{z}{2}+\frac{k}{2n-1})]$$

$$+\sum_{k=0}^{n-2}B_{\alpha+1}(z+\frac{2k+1}{2n-1}) - 2^{\alpha+1}\sum_{k=n}^{2n-2}B_{\alpha+1}(\frac{z}{2}+\frac{k}{2n-1})\} = m^\alpha\sum_{k=0}^{n-1}E_\alpha(z+\frac{2k}{2n-1}) + H(\alpha,n),$$

observing (32), where

$$H(\alpha,n) := \frac{2m^\alpha}{\alpha+1}\Big\{\sum_{k=0}^{n-2} B_{\alpha+1}(z+\frac{2k+1}{2n-1}) - 2^{\alpha+1}\sum_{k=n}^{2n-2} B_{\alpha+1}(\frac{z}{2}+\frac{k}{2n-1})\Big\}.$$

Now again by (30) for $m=2$, noting (31),

$$H(\alpha,n) = \frac{2^{\alpha+1}m^\alpha}{\alpha+1}\Big\{\sum_{k=0}^{n-2} B_{\alpha+1}(\frac{z+(2k+1)/(2n-1)}{2})-$$

$$-\sum_{k=0}^{n-2} B_{\alpha+1}(\frac{z+(2k+1)/(2n-1)+1}{2})\Big\} = -m^\alpha \sum_{k=0}^{n-2} E_\alpha(z+\frac{2k+1}{2n-1}).$$

Combining the results finally establishes (35). ∎

3.4 Connections with the "alternating" Hurwitz Function

Whereas the Bernoulli functions $B_\alpha(z)$ are connected with the Hurwitz Zeta function $\zeta(\alpha,z) := \sum_{k=0}^{\infty}(k+z)^{-\alpha}, \Re\alpha>1, z\in \mathbf{C}\backslash\mathbf{R}_0^-$, the Euler functions $E_\alpha(z)$ are connected with the function

$$\zeta^*(\alpha,z) := \sum_{k=0}^{\infty}\frac{(-1)^k}{(k+z)^\alpha} \qquad (\Re\alpha>1,\ z\in\mathbf{C}\backslash\mathbf{R}_0^-),$$

for which it is well-known that (cf. [23, p. 25])

$$\zeta^*(\alpha,z) = 2^{-\alpha}\{\zeta(\alpha,z/2)-\zeta(\alpha,(z+1)/2)\} \qquad (\Re\alpha>1,\ z\in\mathbf{C}\backslash\mathbf{R}_0^-). \tag{36}$$

Now it is again well-known that (cf. [31, p. 266])

$$\zeta(\alpha,z) = \frac{\Gamma(1-\alpha)}{2\pi i}\int_C \frac{e^{zw}}{1-e^w}w^{\alpha-1}dw \qquad (\Re z>0;\ \alpha\in\mathbf{C}\backslash\mathbf{N}) \tag{37}$$

(C being the contour of Lemma 2.2), which is simultaneously the analytic continuation of $\zeta(\alpha,z)$ to $\alpha\in\mathbf{C}\backslash\mathbf{N}$ (removable singularities at $\alpha\in\mathbf{N}$). Hence also $\zeta^*(\alpha,z)$ has an analytic extension beyond the line $\Re\alpha=1$ to the cut-plane $\mathbf{C}\backslash\mathbf{N}$ in the case $\Re z>0$.

Theorem 3.9 *One has for* $\alpha\in\mathbf{C}\backslash\mathbf{Z}^-$ *and* $\Re z>0$: $E_\alpha(z) = 2\zeta^*(-\alpha,z)$.

Proof The proof follows immediately by (36), (37) as well as Definition 2.1. ∎

Notice that Definition 2.3 would enable one to define $\zeta^*(-\alpha,z)$ via Theorem 3.9 also for $z\in\mathbf{R}_0^-$ (a fact which has not been considered as yet, according to a communication by Dr. D. Klusch). All in all, the Euler functions $E_\alpha(z)$ are thus connected with the functions $\phi_\alpha(x)$ and $\psi_\alpha(x)$ of Fourier analysis (recall (12)), with the Bernoulli functions $B_\alpha(z)$ (recall (31) - (34)), as well as with the alternating Hurwitz zeta function $\zeta^*(\alpha,z)$. These connections are just as close as those of the classical Euler polynomials with these functions. Indeed, for $\alpha=n$ these connections reduce precisely to the classical ones. These close connections with these classical functions confirm that our definition of the $E_\alpha(z)$ is a meaningful generalization of the $E_n(z)$. In this respect, first of all, $\zeta^*(\alpha,z)$ can now be extended to $z\in\mathbf{R}_0^-$ - the critical cut - which is one of the innovations of this paper (and not dealt with e.g. by Sinocov [27], Jonquière [19] and Böhmer [4]). Secondly, an independent theory of Euler functions $E_\alpha(z)$ has been built up, encompassing essentially the whole classical theory, with the hope that it will also yield a good part of its applications as well as new ones. Probably, Theorem 3.5 is already a new application and result. An application to best approximation is to follow below.

Theorem 3.10 *(Weyl-Integral Representation) For $\Re z > 0$ one has*

$$E_\alpha(z) = \frac{2}{\Gamma(-\alpha)} \int_0^\infty \frac{e^{-zt}}{1+e^{-t}} t^{-\alpha-1} dt \qquad (\Re\alpha < 0), \tag{38}$$

$$E_\alpha = (\frac{2}{\pi})^{\alpha+1} \cos(\frac{\pi\alpha}{2}) \int_0^\infty t^\alpha \operatorname{sech}(t) dt \qquad (\Re\alpha > -1).$$

The proof is a consequence of the corresponding known integral representation of $\zeta^*(-\alpha, z)$ (cf. [31, p. 266], [23, p. 24], [24, p. 35]) and Theorem 3.9. It was brought to our attention that formula (38) can also be proved by using the contour integral in the definition of the Euler functions.

Observe that the integral (38) representing $E_\alpha(z)$ can be regarded as the fractional Weyl integral of order $(-\alpha)$, $\Re\alpha < 0$ (see e.g. [28, p. 286f]) of the generating function $f(t) := 2e^{-zt}/(1+e^{-t})$ at $t = 0$, i.e., $E_\alpha(z) = {}_\infty\mathcal{D}_t^\alpha(f(t))|_{t=0}$, which is the counterpart of formula (1) for $\alpha = n \in \mathbf{N}$.

3.5 Functional Equation for the Euler Numbers E_α

Already Euler showed in 1742 ([30, pp. 264, 272]) that the values of the Dirichlet function $\mathcal{L}(\alpha) := \sum_{k=0}^\infty (-1)^k (2k+1)^{-\alpha}$, $\Re\alpha > 0$, at the odd integers $\alpha = 2m+1, m \in \mathbf{N}_0$, can be written in the closed form

$$\mathcal{L}(2m+1) = (-1)^m 2^{-2m-2} \pi^{2m+1} E_{2m}/(2m)!,$$

where E_{2m} are the classical Euler numbers. It will be shown below that this relation is valid more generally for complex $\alpha \in \mathbf{C}$.

Riemann's greatest discovery in mathematics is often regarded as his functional equation for the Riemann Zeta function. The counterpart of this equation for the $\mathcal{L}$-function reads (see [22], [3, p. 171]; [30, p. 185]), for general $\alpha \in \mathbf{C}$,

$$\mathcal{L}(-\alpha) = 2^{1+\alpha} \pi^{-\alpha-1} \cos(\pi\alpha/2) \Gamma(1+\alpha) \mathcal{L}(1+\alpha).$$

Noting that by Theorem 3.9, $E_\alpha \equiv 2^\alpha E_\alpha(1/2) = 2^{\alpha+1}\zeta^*(-\alpha, 1/2) = 2\mathcal{L}(-\alpha)$ for $\alpha \in \mathbf{C}\backslash\mathbf{Z}^-$, the latter being regarded as the extension of $\mathcal{L}(\alpha)$ beyond the line $\Re\alpha = 0$, we have

Theorem 3.11 *(Riemann-Type Functional Equation) For $\alpha \in \mathbf{C}$,*

$$E_\alpha = (2/\pi)^{\alpha+1} \cos(\pi\alpha/2) \Gamma(1+\alpha) E_{-\alpha-1}. \tag{39}$$

Theorem 3.12 *Assume that $\alpha \in \mathbf{C}$: a) There holds for $\alpha \neq 2m$, $m \in \mathbf{Z}$,*

$$\mathcal{L}(\alpha) = \operatorname{cosec}(\pi\alpha/2) 2^{-\alpha-1} \pi^\alpha E_{\alpha-1}/\Gamma(\alpha). \tag{40}$$

b) There holds for $\alpha = 2m$, $m \in \mathbf{N}$,

$$\mathcal{L}(2m) = (-1)^m 2^{-2m-1} \pi^{2m} E^\sim_{2m-1}/(2m-1)!. \tag{41}$$

Proof As to part a), in view of (39), and the relations $\pi(\Gamma(1-\alpha)\Gamma(\alpha))^{-1} = \sin(\pi\alpha)$ and $\sin(\pi\alpha) = 2\sin(\pi\alpha/2)\cos(\pi\alpha/2)$,

$$\mathcal{L}(\alpha) = (1/2)E_{-\alpha} = (1/2)(2/\pi)^{1-\alpha} \cos(\pi\alpha/2)\Gamma(1-\alpha)E_{\alpha-1}\Gamma(\alpha)/\Gamma(\alpha),$$

which immediately yields (40). Concerning b), according to the representation of $\mathcal{E}^\sim_\alpha(x)$ (i.e., Lemma 2.4) and (18),

$$E^\sim_{2m-1} \equiv 2^{2m-1}\mathcal{E}^\sim_{2m-1}(1/2) = -2^{2m+1}(2m-1)! \sum_{k=0}^\infty \frac{\cos((k-m)\pi + \pi)}{[(2k+1)\pi]^{2m}},$$

which readily yields (41) by the definition of $\mathcal{L}(2m)$. ∎

Lemma 3.1 *For $m \in \mathbf{N}$ there holds the connection*

$$\lim_{\alpha \to 2m} \operatorname{cosec}(\pi\alpha/2) E_{\alpha-1} = (-1)^m E^{\sim}_{2m-1}.$$

The proof of Lemma 3.1 follows by Proposition 2.2, noting L'Hospital's rule. This is one way of studying the $E^{\sim}_{2m-1}$, knowing $E_{\alpha-1}$ (e.g. by (38)).

Part b) of Theorem 3.12 resolves a conjecture open since the days of Euler ([30]) in the following sense. Comparing (3) and (41), essentially the numbers E_{2m} of (3) are replaced in (41) by their conjugates $E^{\sim}_{2m-1}$. But the complete solution of Euler's problem depends upon a possible evaluation of the Hilbert transform of E_{2m-1}, to be considered below.

Part a) handles the conjecture for complex $\alpha \in \mathbf{C}$, $\alpha \neq 2m$, $m \in \mathbf{Z}$, in the sense that the classical Euler numbers E_{2m} of (3) are replaced in (40) by the *fractional* Euler numbers $E_{\alpha-1}$, which can be represented via their Fourier series (9), the contour integral (7), or the Weyl integral (38), all taken at $z = 1/2$, or simply via Theorem 3.9.

On the other hand, in view of (3) and (41) it would be possible to calculate the values of E_α and $E^{\sim}_\alpha$ for $\alpha = n \in \mathbf{N}$ by bringing into play the rapidly converging series representations of $\mathcal{L}(n)$, $n \in \mathbf{N}$, established by the authors ([7], [8]). Thus, one obtains the numerical approximations $E^{\sim}_1 = -0.7424537453$, $E^{\sim}_2 = 0$, $E^{\sim}_3 = 1.949277546$, which are exact to 10 significant figures when just summing up 12 terms of the series.

According to Theorem 3.12 a) the "signless" Euler numbers E^*_α of the introduction are thus connected with our "true" (not signless) Euler numbers E_α in terms of

$$E_\alpha = \cos(\pi\alpha/2) E^*_{\alpha+1} \qquad (\alpha \in \mathbf{C}).$$

Observe that the rôle of $(-1)^m$ in $E_{2m} = (-1)^m E^*_{2m+1}$ is here replaced for $m = \alpha/2$ by $\cos(\pi\alpha/2)$. This is the simplest substitute for $(-1)^m$ to be found in this paper; e.g. in Theorem 3.1 this substitute reads $(\cos(\pi\alpha) - \sin(\pi\alpha))$. The substitute in Theorem 3.3 is even more complicated in view of the required introduction of the conjugate Euler functions.

3.6 $E^{\sim}_n(x)$ and Generating Functions

In this subsection representations of $E^{\sim}_n(x)$ as well as an exponential generating function of these functions are derived. In all of the representations to follow one special function is involved, a function to be called the *omega* function, defined by

Definition 3.2 *For $w \in \mathbf{C}$ and $-1/2 \leq a < b \leq 1/2$ $(a, b \neq 0)$ the incomplete Ω-function is defined by*

$$\Omega(w; a, b) := \int_a^b e^{uw} \cot(\pi u)\, du. \tag{42}$$

If $0 \in (a, b)$, the integral is to be understood as the Cauchy principal value at zero. In the case $a = -1/2$ and $b = 1/2$ we write $\Omega(w) := \Omega(w; -1/2, 1/2)$ in abbreviation, called the (complete) Ω-function.

This function as well as its νth derivatives at $w = 0$, $\Omega_\nu(a, b) := D^\nu_w \Omega(w; a, b)|_{w=0}$ (with $\Omega_\nu := \Omega_\nu(-1/2, 1/2)$), play an important rôle in deriving representations for the conjugate Euler functions. In this respect, $\Omega(w; a, b)$ is the exponential generating function of the $\Omega_\nu(a, b)$:

$$\Omega(w; a, b) = \sum_{\nu=0}^{\infty} \frac{\Omega_\nu(a, b)}{\nu!} w^\nu \qquad (w \in \mathbf{C}). \tag{43}$$

First of all, we recall $E^{\sim}_{2m-1}$ may be represented by the Fourier series of Lemma 2.4 at $z = 1/2$. But the following representation of the $E^{\sim}_n(x)$ is of greater interest.

Theorem 3.13 *For $x \in (0,1)$ and $n \in \mathbf{N}_0$ there holds*

$$E_n^{\sim}(x) = \sum_{j=0}^{n}\sum_{\nu=0}^{j}(-1)^{\nu}\binom{j}{\nu}\binom{n}{j}E_{n-j}2^{j+\nu-n}\Big\{-[(x+\frac{1}{2})^{j-\nu}+(x-\frac{1}{2})^{j-\nu}]\Omega_{\nu}(\frac{x}{2},\frac{1}{2}) \quad (44)$$
$$-[(x-\frac{1}{2})^{j-\nu}+(x-\frac{3}{2})^{j-\nu}]\Omega_{\nu}(-\frac{1}{2},\frac{x-1}{2})+(x-\frac{1}{2})^{j-\nu}\Omega_{\nu}\Big\},$$

where for $\nu \in \mathbf{N}_0$ and $0 < a < b \leq 1/2$ (or $-1/2 \leq a < b < 0$) there hold the representations

$$\Omega_{\nu} = \mathrm{PV}\int_{-1/2}^{1/2} u^{\nu}\cot(\pi u)\,du, \qquad \Omega_{\nu}(a,b) = \int_{a}^{b} u^{\nu}\cot(\pi u)\,du.$$

Proof The proof consists of a long but straight-forward evaluation for $x \in (0,1)$ of

$$E_n^{\sim}(x) = \mathcal{E}_n^{\sim}(x) = \frac{1}{2}\mathrm{PV}\int_{x-1}^{x+1}\mathcal{E}_n(u)\cot(\frac{\pi(x-u)}{2})\,du$$

by considering the anti-periodicity of $\mathcal{E}_n(x)$, using formula (2), and noting the formulae

$$(a-u)^j-(a-1-u)^j = \sum_{\mu=1}^{j}\binom{j}{\mu}(a-1)^{j-\mu}\sum_{\nu=0}^{\mu-1}\binom{\mu}{\nu}(-1)^{\nu}u^{\nu},$$
$$\sum_{\mu=\nu+1}^{j}\binom{j}{\mu}\binom{\mu}{\nu}x^{j-\mu} = \binom{j}{\nu}[(1+x)^{j-\nu}-x^{j-\nu}]. \quad ■$$

Note that further $\Omega_{2\nu} = 0$, $\nu \in \mathbf{N}_0$, and that (cf. [16, p. 465])

$$\Omega_{2\nu+1} = \frac{1}{\pi 4^{\nu}}\Big(\frac{1}{2\nu+1}+\sum_{j=1}^{\infty}\frac{(-1)^j B_{2j}\pi^{2j}}{(2j)!(2\nu+2j+1)}\Big), \quad (45)$$

a rapidly converging series, the terms in the sum being of order $j^{-1}4^{-j}$.

In the case $x = 1/2$ Theorem 3.13 can, noting formula (41), simply be rewritten as (cf. [1, p. 807])

Lemma 3.2 *For $m \in \mathbf{N}$ there holds*

$$\mathcal{L}(2m) = \frac{(-1)^m\pi^{2m}}{4(2m-1)!}\int_0^1 E_{2m-1}(u)\sec(\pi u)\,du. \quad (46)$$

As a counterpart to the exponential generating function (1) there is

Theorem 3.14 *For $|w| < \pi$ and $x \in (0,1)$ one has*

$$\sum_{n=0}^{\infty}\frac{E_n^{\sim}(x)}{n!}w^n = \frac{2e^{xw}}{e^w+1}\{-\Omega(2w)+(e^w+1)\Omega(2w;-\frac{1}{2},-\frac{x}{2})+ \quad (47)$$
$$+(e^{-w}+1)\Omega(2w;\frac{1-x}{2},\frac{1}{2})\},$$

together with the further representation for the omega function

$$\Omega(w) = 2\int_{0+}^{1/2}\sinh(uw)\cot(\pi u)\,du.$$

Proof For $|w| < \pi$ and $x \in (0,1)$ one obtains by the Cauchy product formula, noting (1) and (43),

$$\frac{2e^{xw}}{e^w+1}\Omega(2w) = \sum_{k=0}^{\infty}\{\sum_{j=0}^{k}\binom{k}{j}E_{k-j}(x)2^j\Omega_j\}\frac{w^k}{k!}. \tag{48}$$

Using Theorem 3.6 a) in the case $\alpha = k$, the expression in brackets now can be written as

$$\frac{1}{2}\mathrm{PV}\int_{-1}^{1} E_k(x+u)\cot\frac{\pi u}{2}\,du = -\mathcal{E}_k^{\sim}(x) - \int_x^1 (x-u)^k \cot\frac{\pi u}{2}\,du - \\ - \int_{-1}^{x-1}(x-u-1)^k\cot\frac{\pi u}{2}\,du.$$

Therefore, substituting this expression into (48), Theorem 3.14 follows, because, e.g.,

$$\sum_{k=0}^{\infty}\{\int_x^1 (x-u)^k\cot\frac{\pi u}{2}\,du\}\frac{w^k}{k!} = \int_x^1 e^{(x-u)w}\cot\frac{\pi u}{2}\,du = -2e^{xw}\Omega(2w;-\frac{1}{2},-\frac{x}{2}). \quad ∎$$

Observe that the generating function (47) differs from that for $E_n(x)$ given in (1) by the factor in the curly brackets.

Now one can apply the theory of generating functions to establish further properties of $E_n^{\sim}(x)$. In particular, a recurrence relation for the $E_n^{\sim}(x)$ can be deduced:

Proposition 3.2 *For $k \in \mathbf{N}$, $x \in (0,1)$, there holds*

$$\sum_{j=0}^{k}\binom{k}{j}(-1)^{k-j}E_j^{\sim}(x)x^{k-j} = -\sum_{j=0}^{k}\binom{k}{j}E_{k-j}(0)2^j\Omega_j + 2^{k+1}\Omega_k(-\frac{1}{2},-\frac{x}{2}) + \\ +\sum_{j=0}^{k}\binom{k}{j}(-1)^{k-j}2^{j+1}\Omega_j(\frac{1-x}{2},\frac{1}{2}), \tag{49}$$

$$E_0^{\sim}(x) = -\frac{2}{\pi}\log|\cot\frac{\pi x}{2}|.$$

Proof Proposition 3.2 follows from Theorem 3.14, observing the Cauchy product

$$e^{-xw}\sum_{k=0}^{\infty}\frac{E_k^{\sim}(x)}{k!}w^k = \sum_{k=0}^{\infty}\{\sum_{j=0}^{k}\binom{k}{j}(-1)^{k-j}E_j^{\sim}(x)x^{k-j}\}\frac{w^k}{k!}. \quad ∎$$

According to Theorem 3.13, Theorem 3.14 and Proposition 3.2, the study of $E_{2m+1}^{\sim}$ has essentially been reduced to the study of the complete and incomplete omega functions $\Omega(t)$ and $\Omega(t;a,b)$.

In regard to the conjugate Euler numbers $E_n^{\sim} := 2^n E_n^{\sim}(1/2)$ there is

Theorem 3.15 *For $|w| < \pi/2$ there holds*

$$\sum_{n=0}^{\infty}\frac{E_n^{\sim}+iE_n}{n!}w^n = \frac{4i}{\pi}\int_0^{\infty}\frac{e^{2iwu/\pi}}{e^u+e^{-u}}\,du = -\frac{i}{\pi}\{\psi(\frac{1}{4}-\frac{iw}{2\pi}) - \psi(\frac{3}{4}-\frac{iw}{2\pi})\} \\ = \frac{4i}{\pi-2iw}\,{}_2F_1(1,\frac{1}{2}-\frac{iw}{\pi};\frac{3}{2}-\frac{iw}{\pi};-1).$$

Proof Concerning the integral representation, one has by Theorem 3.12 and some elementary calculations for $|t| < \pi/2$,

$$\sum_{n=0}^{\infty} \frac{E_n^{\sim} + iE_n}{n!} t^n = 2\sum_{n=0}^{\infty} \mathcal{L}(n+1)(\frac{2i}{\pi})^{n+1} t^n. \tag{50}$$

Noting ([24], p. 24)

$$\mathcal{L}(n) = 2^{-2n}(\zeta(n,1/4) - \zeta(n,3/4)) = \frac{1}{(n-1)!}\int_0^{\infty} \frac{t^{n-1}e^{-t}}{1+e^{-2t}}\,dt,$$

one readily deduces the first part of Theorem 3.15. In view of formula (54.10.1) in Hansen [17], (50) can be rewritten in terms of the ψ-function $\psi(x) := \Gamma'(x)/\Gamma(x)$. Finally, one has

$$\sum_{n=0}^{N}\{\frac{1}{n+1} - \frac{1}{1/4 - z + n}\} - \sum_{n=0}^{N}\{\frac{1}{n+1} - \frac{1}{3/4 - z + n}\} = -2\sum_{n=0}^{2N+1} \frac{(-1)^n}{n + 1/2 - 2z}.$$

Letting $N \to \infty$, this sum tends to ${}_2F_1(1, 1/2 - 2z; 3/2 - 2z; -1)/(z - 1/4)$, the desired result. ∎

Finally, Theorem 3.14 is a motivation for an important conjecture, namely the following possible representation of $E_\alpha^{\sim}(x)$ as a contour integral:

$$E_\alpha^{\sim}(x) = \frac{\Gamma(\alpha+1)}{2\pi i}\int_C \frac{2e^{wt}}{e^w+1}\{-\Omega(2w) + (e^w+1)\Omega(2w; -\frac{1}{2}, -\frac{x}{2}) + $$
$$+(e^{-w}+1)\Omega(2w; \frac{1-x}{2}, \frac{1}{2})\}\frac{dw}{w^{\alpha+1}},$$

where C is the contour of Lemma 2.2. Again observe that this representation differs from that for $E_\alpha(x)$ in (7) by the expression in the curly brackets. This integral does not seem to be comparable to a known integral as is $\zeta^*(\alpha, x)$ for $E_\alpha(x)$.

4 EXACT CONSTANTS IN BEST APPROXIMATION

A famous result on exact constants in the theory of best trigonometric approximation is the theorem of Favard ([15], [20, p. 166]) stating that in case of the space $L_{2\pi}^{\infty}$,

$$\sup_{f\in W_{L_{2\pi}^{\infty}}^{r}} \frac{E_{n-1}[f]_{L_{2\pi}^{\infty}}}{\|f^{(r)}\|_{\infty}} = \frac{K_r}{n^r} := \frac{1}{n^r}\frac{4}{\pi}\sum_{j=0}^{\infty}\frac{(-1)^{j(r+1)}}{(2j+1)^{r+1}} \qquad (r \in \mathbf{N}),$$

where $W_{X_{2\pi}}^r = \{f \in X_{2\pi}; f^{(j)} \in X_{2\pi} \cap AC_{2\pi}, 1 \le j \le r-1, f^{(r)} \in X_{2\pi}\}$, $X_{2\pi}$ standing for one of the spaces $C_{2\pi}$, $L_{2\pi}^p$, $1 \le p \le \infty$, and the best approximation of $f \in X_{2\pi}$ of degree $n-1$ being

$$E_{n-1}[f]_{X_{2\pi}} := \min_{t_{n-1}\in\Pi_{n-1}} \|f - t_{n-1}\|. \tag{51}$$

Here Π_{n-1} is the set of all trigonometric polynomials of degree $\le n-1$. The Favard constants K_r, which occur in practically every book on approximation theory, are nothing but $K_{2r} = 4/\pi\mathcal{L}(2r+1)$ and $K_{2r-1} = 4/\pi(1 - 2^{-2r})\zeta(2r)$, respectively. These facts never seem to be mentioned anywhere.

The purpose of this section is to apply the theory of Euler functions in a systematic way to the study of exact constants whereby the classical derivatives $f^{(j)}$ here are replaced by fractional order derivatives $D^{[\alpha]}f$ with $\alpha \in \mathbf{R}^+$. It will turn out that the Euler functions will occur both in the exact constants of the fractional order Jackson inequality to be established as well as in the extremal sequences yielding the sharpness of the constants.

First let us introduce the fractional order Liouville-Grünwald derivative ([5], [11]).

Definition 4.1 *(Liouville-Grünwald derivative) If for $f \in L^1_{2\pi}$ (or $C_{2\pi}$, $C_{2\pi}$) there exists $g \in L^1_{2\pi}$ ($C_{2\pi}$, $L^\infty_{2\pi}$) such that in the norm $\|\cdot\|_1$ ($\|\cdot\|_C$, $\|\cdot\|_\infty$)*

$$\lim_{t\to 0+} \|t^{-\alpha}\nabla_t^\alpha f - g\| = 0, \tag{52}$$

then g (=: $D^{[\alpha]}f$) will be called the αth Liouville-Grünwald derivative ($\alpha > 0$) of f in the norm of $L^1_{2\pi}$ ($C_{2\pi}$, $L^\infty_{2\pi}$). In short, let us write $D^{[\alpha]}f \in L^1_{2\pi}$ ($C_{2\pi}$, $L^\infty_{2\pi}$). Here

$$\nabla_t^\alpha f(x) := \sum_{j=0}^{\infty}(-1)^j\binom{\alpha}{j}f(x-tj)$$

is the fractional receding difference of f of order $\alpha > 0$ with increment t. For $\alpha = 1$ it is the classical receding difference $\nabla_t f(x) = f(x) - f(x-t)$. The norm $\|\cdot\|_1$ is defined by $\|f\|_1 := \int_{-\pi}^{\pi}|f(u)|\,du$.

Definition 4.2 *For $\alpha > 0$ define $W^\alpha_{L^1_{2\pi}} := \{f \in L^1_{2\pi}; D^{[\alpha]}f \in L^1_{2\pi}\}$, $W^\alpha_{C_{2\pi}} := \{f \in C_{2\pi}; D^{[\alpha]}f \in C_{2\pi}\}$, and $W^\alpha_{L^\infty_{2\pi}} := \{f \in C_{2\pi}; D^{[\alpha]}f \in L^\infty_{2\pi}\}$.*

Using this fractional derivative, Theorem 3.2 (in case $z = x \in (0,1]$) can be generalized as follows:

Proposition 4.1 *For $\beta > 0$, $n \in \mathbf{N}$, $x \in \mathbf{R}$ ($nx/\pi \notin \mathbf{Z}$, if $(\alpha - \beta) \in (-1, 0]$)*

$$\left(D^{[\beta]}\mathcal{E}_\alpha(\frac{n}{\pi}\cdot)\right)(x) = \frac{\Gamma(\alpha+1)}{\Gamma(\alpha-\beta+1)}\left(\frac{n}{\pi}\right)^\beta \mathcal{E}_{\alpha-\beta}(\frac{n}{\pi}x) \tag{53}$$

in $L^1_{2\pi}$ if $\alpha > \beta - 1$ and in $C_{2\pi}$ if $\alpha > \beta$.

Proof Noting the Fourier series representation (13), one has for all $\nu \in \mathbf{Z}$

$$(i\nu)^\beta(\mathcal{E}_\alpha(\frac{n}{\pi}\cdot))^\wedge(\nu) = \begin{cases} 0 & ,\nu \neq (2k+1)n, \quad \text{for all } k \in \mathbf{Z} \\ (\frac{n}{\pi})^\beta \dfrac{2\Gamma(\alpha+1)}{[(2k+1)\pi i]^{\alpha-\beta+1}} & ,\nu = (2k+1)n, \quad \text{for one } k \in \mathbf{Z} \end{cases}$$
$$= \frac{\Gamma(\alpha+1)}{\Gamma(\alpha-\beta+1)}(\frac{n}{\pi})^\beta(\mathcal{E}_{\alpha-\beta}(\frac{n}{\pi}\cdot))^\wedge(\nu).$$

Applying Theorem 4.1 of [11] (for $L^1_{2\pi}$ and $C_{2\pi}$), the proof of Proposition 4.1 is complete. ∎

Lemma 4.1 *There holds for $\alpha > 0$*

$$\max_{x\in[0,1]}|E_\alpha(x)| = |E_\alpha(\xi_\alpha)|, \tag{54}$$

where in case $\alpha \in (0,1]$ the number $\xi_\alpha = 0$, and for $\alpha > 1$ $\xi_\alpha \in [0,1)$ satisfies $d/dx\, E_\alpha(\xi_\alpha) = 0$. Further, for $\alpha > 0$, the $\mathcal{E}_\alpha(x)$ are alternately increasing and decreasing on $[k + \xi_\alpha, k + 1 + \xi_\alpha]$, $k \in \mathbf{Z}$.

Proof For $\alpha = 1$ Lemma 4.1 is a direct consequence of $E_1(x) = x - 1/2$. For $\alpha \in (0,1)$ one has in view of Theorem 3.2, Equation (33) and Formula (11.5.20) of [9] (together with the analog of Theorem 9 of [6] in case of $\alpha \in (0,1)$) for $x \in (0,1)$

$$\begin{aligned}\frac{d}{dx}\mathcal{E}_\alpha(x) &= 2^\alpha\{\mathcal{B}_\alpha(\frac{x+1}{2}) - \mathcal{B}_\alpha(\frac{x}{2})\}\\ &= -2\alpha\Big(\frac{1}{(x+1)^{1-\alpha}} - \frac{1}{x^{1-\alpha}} + \sum_{j=1}^{\infty}\big(\frac{1}{(x+1+2j)^{1-\alpha}} - \frac{1}{(x+2j)^{1-\alpha}}\big)\Big);\end{aligned}$$

this yields $d/dx\,\mathcal{E}_\alpha(x) > 0$ for all $x \in (0,1)$ and $\alpha \in (0,1)$. Observing that $\mathcal{E}_\alpha(0) = -\mathcal{E}_\alpha(1)$ and $\mathcal{E}_\alpha(\cdot)$ being continuous, the proof is complete. For $\alpha > 1$ with the notation $\alpha = 2m + \alpha'$, $\alpha' \in [0,2)$, $m \in \mathbf{N}_0$, one has

$$\frac{d}{dx}\mathcal{E}_\alpha(x) = \frac{4\Gamma(\alpha+1)(-1)^m}{\pi^\alpha}\sum_{k=0}^{\infty}\frac{\cos((2k+1)\pi x - \alpha'\pi/2)}{(2k+1)^\alpha}.$$

Then, by Lemma 2 of [29], there exists $\xi_\alpha \in [0,1)$ with $d/dx\,\mathcal{E}_\alpha(\xi_\alpha) = 0$. This ξ_α is unique, since in view of the corollary following Lemma 2 of [29], $\operatorname{sgn}(d/dx\,\mathcal{E}_\alpha(x)) = \pm\operatorname{sgn}(\sin\pi(x-\xi_\alpha))$.

This yields further that for $\alpha > 1$ the $\mathcal{E}_\alpha(x)$ are increasing (or decreasing) on $[0,\xi_\alpha)$ and decreasing (or increasing) on $(\xi_\alpha, 1]$. By the one-antiperiodicity of $\mathcal{E}_\alpha(x)$ the proof is complete. ∎

Now to the main object of this section, namely the Jackson inequality for fractional order derivatives. It gives an estimate between the best approximation of $f \in W^\alpha_{X_{2\pi}}$ ($X_{2\pi}$ being $L^1_{2\pi}$, $C_{2\pi}$ or $L^\infty_{2\pi}$), see (51), and the fractional Liouville-Grünwald derivative of the function f, $D^{[\alpha]}f$.

Theorem 4.1 *For $n \in \mathbf{N}$, $\alpha > 0$ and $f \in W^\alpha_{X_{2\pi}}$ ($X_{2\pi} = L^1_{2\pi}$, $C_{2\pi}$ or $L^\infty_{2\pi}$) there holds the Jackson inequality*

$$E_{n-1}[f]_{X_{2\pi}} \le \frac{1}{\Gamma(\alpha+1)}\max_{x\in[0,1]}|E_\alpha(x)|\left(\frac{\pi}{n}\right)^\alpha\|D^{[\alpha]}f\|_{X_{2\pi}}. \tag{55}$$

The constant in (55) is sharp in the sense that there exist extremal sequences (see also below)

$$m_h(\cdot) := \nabla_h\mathcal{E}_{\alpha+1}(\frac{n}{\pi}\cdot)/h, \quad h \to 0+ \quad \text{in } W^\alpha_{C_{2\pi}} \text{ and } W^\alpha_{L^\infty_{2\pi}}$$

and

$$m_h(\cdot) := \nabla_h\mathcal{E}_\alpha(\frac{n}{\pi}\cdot)/h, \quad h \to 0+ \quad \text{in } W^\alpha_{L^1_{2\pi}},$$

such that by inserting $m_h(\cdot)$ in (55) and then taking the limit $h \to 0+$, one obtains equality in (55), i.e.,

$$\sup_{f\in W^\alpha_{X_{2\pi}}}\frac{E_{n-1}[f]_{X_{2\pi}}}{\|D^{[\alpha]}f\|_{X_{2\pi}}} = \lim_{h\to0+}\frac{E_{n-1}[m_h(\cdot)]_{X_{2\pi}}}{\|D^{[\alpha]}m_h(\cdot)\|_{X_{2\pi}}} = \frac{1}{\Gamma(\alpha+1)}\max_{x\in[0,1]}|E_\alpha(x)|\left(\frac{\pi}{n}\right)^\alpha.$$

Proof Concerning the periodic Bernoulli functions $\mathcal{B}_\alpha(x)$ of [6], one readily obtains by [14] the best approximation

$$\frac{1}{2\pi}E_{n-1}[\mathcal{B}_\alpha(\frac{\cdot}{2\pi})]_{L^1_{2\pi}} = \frac{|E_\alpha(\xi_\alpha)|}{(2n)^\alpha}, \tag{56}$$

ξ_α given as in Lemma 4.1. Using the convolution formula (see [11] with $(f*g)(x) := \frac{1}{2\pi}\int_0^{2\pi} f(x-u)g(u)\,du$) for $\alpha > 0$,

$$f(x) = f^\wedge(0) - \frac{(2\pi)^\alpha}{\Gamma(\alpha+1)}(\mathcal{B}_\alpha(\frac{\cdot}{2\pi}) * D^{[\alpha]}f)(x) \quad \text{(a.e. if } X_{2\pi} = L^1_{2\pi}),$$

and

$$U_{n-1}f(x) := f^\wedge(0) - \frac{(2\pi)^\alpha}{\Gamma(\alpha+1)}(t^*_{n-1,\alpha} * D^{[\alpha]}f)(x) \in \Pi_{n-1},$$

$t^*_{n-1,\alpha} \in \Pi_{n-1}$ being the polynomial of best $L^1_{2\pi}$-approximation to $\mathcal{B}_\alpha(\cdot/2\pi)$, one has (55) by observing (56) and Lemma 4.1,

$$\begin{aligned} E_{n-1}[f]_{X_{2\pi}} &\le \|f - U_{n-1}f\|_{X_{2\pi}} \\ &= \frac{(2\pi)^\alpha}{\Gamma(\alpha+1)}\|(t^*_{n-1,\alpha} - \mathcal{B}_\alpha(\frac{\cdot}{2\pi})) * D^{[\alpha]}f\|_{X_{2\pi}} \\ &\le \frac{(2\pi)^{\alpha-1}}{\Gamma(\alpha+1)}\|t^*_{n-1,\alpha} - \mathcal{B}_\alpha(\frac{\cdot}{2\pi})\|_1\|D^{[\alpha]}f\|_{X_{2\pi}} \\ &= \frac{1}{\Gamma(\alpha+1)}\max_{x\in[0,1]}|E_\alpha(x)|(\frac{\pi}{n})^\alpha\|D^{[\alpha]}f\|_{X_{2\pi}}. \end{aligned}$$

Now let us show that the constant in the estimate (55) is indeed sharp. As in [20, p. 166f], one has for $\alpha > -1$ that $0 \in \Pi_{n-1}$ is the polynomial of best approximation to $\mathcal{E}_\alpha(n\cdot/\pi)$,

$$E_{n-1}[\mathcal{E}_\alpha(\frac{n}{\pi}\cdot)]_{X_{2\pi}} = \|\mathcal{E}_\alpha(\frac{n}{\pi}\cdot)\|_{X_{2\pi}}, \tag{57}$$

$X_{2\pi}$ being $L^1_{2\pi}$ in case $\alpha \in (-1,0]$ and $L^1_{2\pi}$, $C_{2\pi}$ or $L^\infty_{2\pi}$ in case $\alpha > 0$. Further, for all $\alpha > 0$ there holds

$$\|\mathcal{E}_{\alpha-1}(\frac{n}{\pi}\cdot)\|_1 = \frac{\pi}{n\alpha}\mathrm{Var}[\mathcal{E}_\alpha(\frac{n}{\pi}\cdot)]_0^{2\pi} = \frac{4\pi}{\alpha}\max_{x\in[0,1]}|E_\alpha(x)|, \tag{58}$$

the last equation being justified in view of the monotonicity of $\mathcal{E}_\alpha(x)$ (see Lemma 4.1).

If $X_{2\pi} = C_{2\pi}$, $L^\infty_{2\pi}$, one has together with Definition 4.1, (53), the continuity of the functional of best approximation and (57),

$$\lim_{h\to 0+} E_{n-1}[\frac{\nabla_h\mathcal{E}_{\alpha+1}(n\cdot/\pi)}{h}]_{X_{2\pi}} = (\alpha+1)\frac{n}{\pi}\|\mathcal{E}_\alpha(\frac{n}{\pi}\cdot)\|_{X_{2\pi}} = (\alpha+1)\frac{n}{\pi}\max_{x\in[0,1]}|E_\alpha(x)|.$$

On the other hand, for $0 < h < \pi/n$ there holds by (53)

$$\begin{aligned} \|D^{[\alpha]}(\frac{\nabla_h\mathcal{E}_{\alpha+1}(n\cdot/\pi)}{h})\|_{X_{2\pi}} &= (\alpha+1)\Gamma(\alpha+1)(\frac{n}{\pi})^\alpha\|\frac{\nabla_h\mathcal{E}_1(n\cdot/\pi)}{h}\|_{X_{2\pi}} \\ &= (\alpha+1)\Gamma(\alpha+1)(\frac{n}{\pi})^{\alpha+1}. \end{aligned}$$

Thus, inserting the extremal sequence $\nabla_h\mathcal{E}_{\alpha+1}(n\cdot/\pi)/h$ into (55) and taking the limit $\lim_{h\to 0+}$, the desired equality in (55) is given.

If $X_{2\pi} = L^1_{2\pi}$, one has similarly by (58),

$$\lim_{h\to 0+} E_{n-1}[\frac{\nabla_h\mathcal{E}_\alpha(n\cdot/\pi)}{h}]_{L^1_{2\pi}} = 4n\max_{x\in[0,1]}|E_\alpha(x)|,$$

and for $0 < h < \pi/n$,

$$\|D^{[\alpha]}(\frac{\nabla_h\mathcal{E}_\alpha(n\cdot/\pi)}{h})\|_1 = \Gamma(\alpha+1)(\frac{n}{\pi})^\alpha\|\frac{\nabla_h\mathcal{E}_0(n\cdot/\pi)}{h}\|_1 = 4n\Gamma(\alpha+1)(\frac{n}{\pi})^\alpha,$$

in view of the fact that

$$(\nabla_h \mathcal{E}_0(n \cdot /\pi))(x) = \text{sgn}(\sin(nx)) - \text{sgn}(\sin n(x-h)).$$

Therefore, again an extremal sequence is found, and the proof is complete. ∎

Note that in the particular case $\alpha = 2r$, $r \in \mathbf{N}$, the constant in (55) can be given in terms of the Euler numbers (see also [15]),

$$\max_{x\in[0,1]} |E_{2r}(x)| = (-1)^r 2^{-2r} E_{2r}.$$

Observe that in the literature on exact constants instead of the Liouville-Grünwald derivative the more classical Weyl derivative is employed (cf. [14], [29]).

What one really would have liked to find is an element $f \in W^{\alpha}_{C_{2\pi}}$ for which equality in (55) is actually attained. This would formally be the function $f = \mathcal{E}_\alpha(n \cdot /\pi)$, since then

$$E_{n-1}[\mathcal{E}_\alpha(n \cdot /\pi)]_{C_{2\pi}} = \max_{x\in[0,1]} |E_\alpha(x)| \; \|\mathcal{E}_0(n \cdot /\pi\|_C.$$

However, this is unfortunately not possible, since the function $\mathcal{E}_\alpha(n \cdot /\pi)$ does not belong to $W^{\alpha}_{C_{2\pi}}$ for $\alpha > 0$. Let us see this for the case $\alpha = 1$. Here for $\mathcal{E}_1(nx/\pi)$ one has

$$\frac{\nabla_h \mathcal{E}_1(nx/\pi)}{h} = \begin{cases} -\frac{n}{\pi} + \frac{2nx}{\pi h}, & 0 \le x < h \\ \frac{n}{\pi}, & h \le x \le \frac{n}{\pi} \end{cases} \tag{59}$$

(continued anti-periodically on $\mathbf{R}$), which yields

$$\left\| \frac{\nabla_{h/2} \mathcal{E}_1(n \cdot /\pi)}{h/2} - \frac{\nabla_h \mathcal{E}_1(n \cdot /\pi)}{h} \right\|_C \ge \frac{n}{\pi}.$$

Since this expression does not tend to zero for $h \to 0+$, the limit of (59) for $h \to 0+$, which would be $D^{[1]}\mathcal{E}_1(n \cdot /\pi)$, does not exist in $C_{2\pi}$.

For this reason one cannot avoid using the extremal sequence $m_h(\cdot)$ which smoothens the $\mathcal{E}_\alpha(n \cdot /\pi)$. In case of $X_{2\pi} = L^1_{2\pi}$ this extremal sequence is different to those used in the literature. Here, instead of the Maurer-Steklov means (cf. e.g. [20, p. 167], [10]), the receding differences of the definition of the fractional order derivatives are used effectively as a smoothing process.

Finally observe that in the instance of the space $L^{\infty}_{2\pi}$, when the Weyl derivative would be used instead of the Liouville derivative, then equality in (55) would actually be obtained for the extremal element $f(x) = \mathcal{E}_\alpha(nx/\pi)$, the αth Weyl derivative of which belongs to $L^{\infty}_{2\pi}$[5].

REFERENCES

[1] M. Abramowitz and I.A. Stegun. "Handbook of Mathematical Functions," Dover Publications, Inc., New York 1964.

[2] T.M. Apostol. "Introduction to Analytic Number Theory," Springer–Verlag, New York 1976.

[3] B.C. Berndt. "Ramanujan's Notebooks, Part I," Springer–Verlag, New York 1985.

[5]Concerning Theorem 7.1 in [11], note that of the six assertions given there the implication (2) ⇒ (1) is invalid. A counterexample is the function $\mathcal{E}_\alpha(n \cdot /\pi)$ considered above; there $D^{[\alpha]}\mathcal{E}_\alpha(n \cdot /\pi) \notin L^{\infty}_{2\pi}$.

[4] P. Böhmer, Über die Bernoullischen Funktionen. Math. Ann. **68** (1910), 338-360.

[5] P.L. Butzer, H. Dyckhoff, E. Görlich and R.L. Stens, Best trigonometric approximation, fractional order derivatives and Lipschitz classes. Canad. J. Math. **29** (1977), 781-793.

[6] P.L. Butzer, M. Hauss and M. Leclerc, Bernoulli numbers and polynomials of arbitrary complex indices. Appl. Math. Lett. **5** (1992), No. 6, 83–88.

[7] P.L. Butzer and M. Hauss, Integral and rapidly converging series representations of the Dirichlet L-functions $L_1(s)$ and $L_{-4}(s)$. Atti Sem. Mat. Fis. Univ. Modena, **XL** (1992), 329-359.

[8] P.L. Butzer and M. Hauss, Riemann zeta function: Rapidly converging series and integral representations. Appl. Math. Lett. **5**, No. 2 (1992), 83-88.

[9] P.L. Butzer and R.J. Nessel. "Fourier Analysis and Approximation; Vol. I," Birkhäuser Verlag, Basel; Academic Press, New York 1971.

[10] P.L. Butzer, M. Schmidt and E.L. Stark, Observations on the history of central B-splines. Arch. Hist. Exact Sci. **39** (1988), 137-156.

[11] P.L. Butzer and U. Westphal, An access to fractional differentiation via fractional difference quotients, in: Fractional Calculus and its Applications [Lecture Notes in Mathematics, No. 457]. Springer-Verlag, Berlin 1975, 116-145.

[12] L. Comtet. "Advanced Combinatorics," D. Reidel Publishing Company, Dordrecht 1974 (Revised Edition).

[13] K. Dilcher, L. Skula and I.Sh. Slavutskiĭ. "Bernoulli Numbers: Bibliography (1713-1990)," Queen's Papers in Pure and Appl. Math., No. 87, Queen's University, Kingston 1991.

[14] V.K. Dzyadyk, Best approximation on classes of periodic functions defined by kernels which are integrals of absolutely monotonic functions (Russ.). Izv. Akad. Nauk SSSR Ser. Mat. **23** (1959), 933-950.

[15] J. Favard, Sur les meilleurs procédés d'approximation de certaines classes de fonctions par des polynomes trigonométriques. Bull. Sci. Math. Ser. 2 **61** (1937), 209-224.

[16] I.S. Gradstein and I.M. Ryshik. "Tables of Series, Products, and Integrals; Vol. 1," Verlag Harri Deutsch, Thun 1981.

[17] E.R. Hansen. "A Table of Series and Products," Prentice-Hall, Englewood Cliffs, N.J. 1975.

[18] G.H. Hardy and W.W. Rogosinski. "Fourier Series," Cambridge University Press, Cambridge 1962 (1. Edition 1944).

[19] A. Jonquière, Über eine Verallgemeinerung der Bernoulli'schen Functionen und ihren Zusammenhang mit der verallgemeinerten Riemann'schen Reihe. Stockh. Akad. Bihang. XVI. I. 6., 1-28 (see Jahrbuch über d. Fortschr. Math. **23** (1891), p. 432).

[20] N.P. Korneichuk. "Exact Constants in Approximation Theory," Cambridge University Press, Cambridge 1991.

[21] M. Leclerc, Fraktionierte Bernoulli und Euler Polynome. Diplomarbeit, Lehrstuhl A für Mathematik, RWTH Aachen (in preparation).

[22] P. Lichtenbaum, Über die Funktion $\chi(s) = \sum(-1)^k(2k+1)^{-s}$. Math. Z. **33** (1931), 641–647.

[23] W. Magnus, F. Oberhettinger and R.P. Soni, "Formulas and Theorems for the Special Functions of Mathematical Physics," Springer–Verlag, Berlin 1966.

[24] W. Magnus, F. Oberhettinger and F.G. Tricomi. "Higher Transcendental Functions, Vol. I," McGraw–Hill, New York 1953.

[25] A.I. Markushevich. "Theory of Functions of a Complex Variable, Vol. I," Prentice–Hall, Englewood Cliffs, N.J. 1965 (Revised English Edition).

[26] N.E. Nörlund. "Vorlesung über Differenzenrechnung," Chelsea Publishing Company, New York 1954.

[27] D.M. Sinocov, Bernoulli functions with arbitrary indices. Kasan Soc. (2), **1** (1891), 234-256 (Russ.) (see Jahrbuch über d. Fortschr. Math. **24** (1892), p. 404).

[28] H.M. Srivastava and H.L. Manocha. "A Treatise on Generating Functions," Cambridge University Press 1966.

[29] Sun Yong-sheng, On the best approximation of periodic differentiable functions by trigonometric polynomials (Russ.). Izv. Akad. Nauk SSSR Ser. Mat. **23** (1959), 67-92.

[30] A. Weil. "Number Theory: An Approach through History; From Hammurapi to Legendre," Birkhäuser Verlag, Boston 1984.

[31] E.T. Whittaker and G.N. Watson. "A Course of Modern Analysis," Cambridge University Press 1952 (4. Edition).

ON THE ROLE OF ℓ_∞ IN APPROXIMATION THEORY

B. L. Chalmers
Department of Mathematics
University of California
Riverside, CA 92521

B. Shekhtman
Department of Mathematics
University of South Florida
Tampa, FL 33620

1. Motivation: Given a family of finite-dimensional subspaces V_n in $C[0,1]$, it is important to develop an algorithm that to a given function $f \in C[0,1]$ assigns an approximation $P_n f \in V_n$. This algorithm should be "simple" and "good". That translates into P_n being a continuous linear map from $C[0,1]$ in V_n such that $\|f - P_n f\| \le C \operatorname{dist}(f, V_n)$ where the constant C does not depend on n. Using the principle of uniform boundedness it is easy to show that the above-mentioned conditions are equivalent to the existence of projections (linear, idempotent operators) P_n from $C[0,1]$ onto V_n such that $\|P_n\| \le C$.

Hence one would want to know:

I. For what subspaces V_n of $C[0,1]$ there exists a sequence of uniformly bounded projections?

II. What are those projections?

Question I is not just a matter of idle curiosity. Indeed, taking V_n to be subspaces of polynomials on $[0,1]$ of degree n we know (from the Faber Theorem) that there is no sequence of uniformly bounded projections onto V_n. On the other hand, if V_n are subspaces of splines of fixed degree with n nodes, then there exist projections P_n onto V_n that are uniformly bounded. The proof of it (due to de Boor) hinges upon the existence of bases $v_1, \ldots, v_n \in V_n$ (B-splines) such that

$$\max |\alpha_j| \le \|\sum \alpha_j v_j\| \le C \max |\alpha_j| \tag{1}$$

where the constant C again does not depend on n.

The existence of such a constant C is easily seen to be sufficient for the existence of a constant C_1 and projections P_n with

$$\|P_n\| \le C_1. \tag{2}$$

Is it necessary? This is a basic question in Approximation Theory. It turns out to be equivalent to a long-standing open problem in Banach Space Theory: the P_λ-problem.

In order to analyze this question, we will need some definitions presented in the next section. For simplicity we will restrict the field to be the reals ($\mathbb{R}$), although all the results of Sections 2 and 3 go over without change to the complex field $\mathbb{C}$.

In Section 3 we present some relationships between the inequality (1) and the actual form of the projections. In Section 4 we will establish an exact relationship between constants C and C_1 in (1) and (2) for two-dimensional real spaces.

Approximation, Probability, and Related Fields, Edited by
G. Anastassiou and S.T. Rachev, Plenum Press, New York 1994

2. Definitions and interrelations

2.1. Given a pair of n-dimensional spaces U and V define the Banach-Mazur distance $d(U,V)$ by

$$d(U,V) := \inf\{\|T\|\|T^{-1}\| : T \text{ isomorphism from } U \text{ to } V\}.$$

Given $d_0 \geq 1$ then $d(U,V) \leq d_0$ iff there exists a pair of basis $v_1, \dots, v_n \in V$ and $u_1, \dots, u_n \in U$ such that $c_2^{-1}\|\sum \alpha_j u_j\| \leq \|\sum_1^n \alpha_j v_j\| \leq c_1\|\sum_1^n \alpha_j u_j\|$ for all $(\alpha_j) \in \mathbb{R}^n$ and

$$c_1 c_2 \leq d_0.$$

In this case $T : u_j \to v_j$ gives the desired isomorphism.

Geometrically, $d(U,V) \leq d_0$ means that we can position the unit ball at U, denoted by $B(U) := \{(\alpha_j) \in \mathbb{R}^n : \|\sum \alpha_j u_j\| \leq 1\}$ inside the unit ball $B(V) = \{(\alpha_j) \in \mathbb{R}^n : \|\sum \alpha_j v_j\| \leq 1\}$ so that $d_0 B(U) \subset B(V)$. Clearly, for all n-dimensional spaces U, V, W,

$$1 \leq d(U,V) = d(V,U) < \infty;$$
$$1 = d(U,U);$$
$$d(U,V) \leq d(U,W)\, d(W,V).$$

Hence $\log d$ defines a metric on the Grassman manifold of all n-dimensional spaces.

It is possible to show that the diameter of a metric space so defined is $\leq \log n$, and hence $d(U,V) \leq n$ for all U and V with $\dim U = \dim V = n$.

2.2. Given a Banach space X and an n-dimensional subspace $V \subset X$ define its relative projection constant.

$$\lambda(V,X) := \inf\{\|P\| : P \text{ is a projection from } X \text{ onto } V\}.$$

and its absolute projection constant

$$\lambda(V) = \sup\{\lambda(V,X) : X \supset V\}.$$

Let us mention a few properties.

For every pair of n-dimensional spaces U, V,

$$1 \leq \lambda(U) < \sqrt{n} \quad (n \geq 2; \lambda(U) = 1 \text{ if } n = 1);$$
$$\lambda(U) \leq d(U,V)\lambda(V).$$

If X is one of the $\mathcal{L}_\infty$-spaces, like $C(K), L_\infty, \ell_\infty$, then $\lambda(V) = \lambda(V,X)$.

In the last section of this paper we will need some additional information about projection constants for two-dimensional spaces. If the field is $\mathbb{R}$ and $\dim V = 2$ then there exists an isometric copy of V which is a subspace of L_1. If furthermore $V \subset L_1$ or $V \subset \ell_1$ then (see[4])

$$\lambda(V, L_1) = \lambda(V).$$

The following result is well-known and easy to prove.

Theorem 1. If $\dim V = n$, then

$$\lambda(V) \leq d(V, \ell_\infty^n).$$

Proof. We assume that $V \subset X$. Let $T : V \to \ell_\infty^n$ such that $\|T\|\|T^{-1}\| = d(V, \ell_\infty^n)$ and let $e_1, \dots, e_n$ be the canonical basis in ℓ_∞^n and $e_1^*, \dots, e_n^*$ be the canonical basis in ℓ_1^n. Denote $T^{-1}e_j =: v_j$ and let $f_j \in X^*$ be the Hahn-Banach extensions of $T^*e_j^* \in V^*$. Then $\|f_j\| = \|T^*e_j^*\| \leq \|T\|$. Consider the operator P

$$Px := \sum_{j=1}^n f_j(x) v_j : X \to V.$$

Clearly $f_j(v_k) = (T^*e_j^*)(T^{-1}e_k) = e_j^*(e_k) = \delta_{jk}$ and hence P is a projection onto V. Now

$$\begin{aligned}\|Px\| = \|\sum f_j(x)T^{-1}e_j\| \leq \|T^{-1}\|\|\sum f_j(x)e_j\|_\infty = \\ \|T^{-1}\| \max |f_j(x)| \leq \|T^{-1}\|(\max \|f_j\|)\|x\| \\ \leq \|T^{-1}\|\|T\|\|x\|. \quad \square\end{aligned}$$

The P_λ-problem asks for some sort of a converse inequality.

Is it true that for every constant C_1 there exists a constant $C_2 = C_2(C_1)$ such that if $\lambda(V) \leq C_1$ then $d(V, \ell_\infty^{\dim V}) \leq C_2$.

Known partial results are:

a) $\lambda(V) = 1$ iff $d(V, \ell_\infty^{\dim V}) = 1$. In fact in this case there exists an interpolating projection onto V with norm 1 (L. Nachbin).

b) If V has an unconditional basis then such a C_2 exists (Lindenstrauss and Pelczynski).

c) A generalization of a) says that there exists $\varepsilon > 0$ such that if $\lambda(V) \leq 1 + \varepsilon$ then there exists C_2 with $d(V, \ell_\infty^{\dim V}) \leq C_2$ (Zippin).

d) For every C_1 there exists C_2 such that if $\lambda(V) \leq C_1$ then there exists $V_1 \subset V$ such that $\dim V_1 \geq \frac{1}{2} \dim V$ and $d(V_1, \ell_\infty^{\dim V_1}) \leq C_2$ (Bourgain).

3. The P_λ-problem and the form of the projection

While it is difficult to understand the relationship between $d(V, \ell_\infty^n)$ and $\lambda(V)$, the problem simplifies if there is additional knowledge about the projection P onto V with $\|P\|$ being close to $\lambda(V)$.

Since every projection is uniquely determined by $\operatorname{Im} P^*$ we intend to utilize $\operatorname{Im} P^*$ to estimate $d(V, \ell_\infty^n)$. We start with a very simple proposition.

Proposition 1. Let K be a compact Hausdorff space and $C(K)$ the space of continuous functions on K. Let V be an n-dimensional subspace of $C(K)$ and let P be an interpolating projection from $C(K)$ onto V, i.e., there exist points $t_1, \dots, t_n \in K$ such that $(Pf)(t_j) = f(t_j)$ for $j = 1, \dots, n$. Then

$$d(V, \ell_\infty^n) \leq \|P\|.$$

Proof. We need to find a basis $v_1, \dots, v_n$ in V so that

$$\max |\alpha_j| \leq \|\sum_{j=1}^n \alpha_j v_j\| \leq \|P\| \max |\alpha_j| , \quad \text{for all } (\alpha_j) \in \mathbb{R}^n .$$

We represent the projection P in the form

$$Pf = \sum_{j=1}^{n} f(t_j)v_j$$

where $v_j(t_k) = \delta_{jk}$. Now

$$\|\sum_{j=1}^{n} \alpha_j v_j\| \geq \max |(\sum_{j=1}^{n} \alpha_j v_j)(t_k)| = \max |\alpha_k| \ .$$

For the right-hand side of the inequality we start with an arbitrary vector $(\alpha_1, \ldots, \alpha_n) \in \mathbb{R}^n$ and pick a function $f \in C(K)$ with $\|f\| = \max |\alpha_k|$ and $f(t_k) = \alpha_k$.
Then

$$\|\sum \alpha_j v_j\| = \|Pf\| \leq \|P\|\|f\| = \|P\| \max |\alpha_j|. \quad \square$$

Corollary. Let $\mathcal{P}_n \subset C[0,1]$ be the space of polynomials of degree $n-1$, $n = 1, 2, \ldots$. Then there exist $C_1, C_2 > 0$ such that $C_1 \log n \leq d(\mathcal{P}_n, \ell_\infty^n) \leq C_2 \log n$.

Proof. By the well known Faber theorem $\lambda(\mathcal{P}_n) \geq C_1 \log n$ and hence we get the first inequality by Theorem 1.

If P is an interpolation projection at the Chebyshev points, it is also well-known that $\|P\| \leq C_2 \log n$. Hence Proposition 1 supplies the second inequality. $\square$

Observe that Range $P^* = \operatorname{span}[\delta_{t_1}, \ldots, \delta_{t_n}] \subset \mathcal{M}(K)$. Since

$$\|\sum \alpha_j \delta_{t_j}\| = \sum |\alpha_j|$$

we conclude that

$$d(\operatorname{Im} P^*, \ell_1^n) = 1 \ .$$

We are now ready to generalize Proposition 1.

Proposition 2. Let V be an n-dimensional subspace of a Banach space X and let P be a projection from X onto V. Then

$$d(V, \ell_\infty^n) \leq \|P\| \, d(\operatorname{Im} P^*, \ell_1^n);$$
$$d(V, \ell_1^n) \leq \|P\| \, d(\operatorname{Im} P^*, \ell_\infty^n).$$

Proof. Pick a basis $u_1, \ldots, u_n \in \operatorname{Im} P^* \subset X^*$ so that

$$[d(\operatorname{Im} P^*, \ell_1^n)]^{-1} \sum |\alpha_j| \leq \|\sum \alpha_j u_j\|_{X^*} \leq \sum |\alpha_j| \ .$$

Hence $\|u_j\| \leq 1$.

Let $Px = \sum_{j=1}^{n} u_j(x)v_j$ where $u_j(v_k) = \delta_{jk}$. Take $(\alpha_1, \ldots, \alpha_n) \in \mathbb{R}^n$ and $\varepsilon > 0$. By Helly's Theorem (cf [1]) there exists an element $x \in X$ such that $u_j(x) = \alpha_j$ and $\|x\| \leq [d(\operatorname{Im} P^*, V) + \varepsilon] \max |\alpha_j|$.

Hence we get

$$\|\sum \alpha_j v_j\| = \|Px\| \leq \|P\|\|x\| \leq \|P\|(d(\operatorname{Im} P^*, V) + \varepsilon) \ .$$

The proof of the second inequality is similar. We again pick a basis $u_1, \dots, u_n \in \operatorname{Im} P^* \subset X^*$ so that

$$\max |\alpha_j| \leq \|\sum \alpha_j u_j\| \leq d(\operatorname{Im} P^*, \ell_\infty^n) \max |\alpha_j|.$$

Let P be a projection from X onto V given by

$$Px = \sum u_j(x) v_j \ .$$

We have

$$\begin{aligned} \|P\| &= \sup\{|\sum u_j(x) f(v_j)| : \|x\| = 1 \ ; \ \|f\| = 1; \ f \in X^*\} \\ &= \sup\{\|\sum f(v_j) u_j\| \ : \ f \in X^*; \|f\| = 1\} \geq \max \|v_j\|. \end{aligned}$$

Hence, for an arbitrary $(\alpha_1, \dots, \alpha_n) \in \mathbb{R}^n$, we have

$$\|\sum \alpha_j v_j\| \leq (\sum |\alpha_j|) \max \|v_j\| \leq \|P\| \sum |\alpha_j|.$$

On the other hand

$$\begin{aligned} \|\sum \alpha_j v_j\| &\geq |(\sum (\operatorname{sign} \alpha_j) u_j)(\sum \alpha_j v_j)| / \ \|\sum (\operatorname{sign} \alpha_j) u_j\| \\ &= \frac{\sum |\alpha_j|}{\|\sum (\operatorname{sign} \alpha_j) u_j\|} \geq \frac{1}{d(\operatorname{Im} P^*, \ell_\infty^n)} \sum |\alpha_j|. \end{aligned}$$

Combining these two inequalities, we have

$$d(V, \ell_1^n) \leq \|P\| \, d(\operatorname{Im} P^*, \ell_\infty^n). \quad \square$$

In view of this proposition it is to our advantage to find a projection P from X onto V so that $\|P\| = \lambda(V, X)$ and $\operatorname{Im} P^*$ is close to ℓ_1^n. That is not always possible. Yet it is often possible to pick an isometric copy of V say $\tilde{V}$ and a projection $\tilde{P}$ from X onto $\tilde{V}$ so that $\|\tilde{P}\| = \|P\|$ and yet

$$d(\operatorname{Im} P^*, \ell_1^n) > d(\operatorname{Im} \tilde{P}^*, \ell_1^n)$$

and

$$\lambda(\operatorname{Im} P^*, X^*) > \lambda(\operatorname{Im} \tilde{P}^*, X^*).$$

If two n-dimensional spaces are isometric, the projection constants of these spaces are the same; yet the minimal projection onto these spaces need not have isometric kernels.

Example 1. Let $v_1, v_2, v_3 \in \ell_\infty^4$ be given by

$$v_1 = (1,1,1,1), v_2 = (1,1,-1,-1), v_3 = (1,-1,1,-1)$$

and u_j be the same vectors as v_j considered as elements of ℓ_1^4. Let $V = \operatorname{span}[v_1, v_2, v_3] \subset \ell_\infty^4$. The space V is an isometric copy of the span of the first three Rademacher functions in $L_\infty[0,1]$. Hence $d(V, \ell_1^3) = 1$ and $\lambda(V) = \frac{3}{2}$. The unique minimal projection from ℓ_∞^4 onto V is given by $Px = \frac{1}{4}[u_1(x)v_1 + u_2(x)v_2 + u_3(x)v_3]$.

P^* is the unique minimal projection from ℓ_1^4 onto $\operatorname{span}[u_1, u_2, u_3] =: U$ and hence $\lambda(U, \ell_1^4) = \frac{3}{2}$.

We now introduce the following vectors in ℓ_∞^8:

$$\tilde{v}_1 = (1,1,1,1,1,1,1,1),\ \tilde{v}_2 = (1,1,-1,-1,1,1,-1,-1),$$
$$\tilde{v}_3 = (1,-1,1,-1,1,-1,1,-1).$$

Clearly $\tilde{V} := \operatorname{span}[\tilde{v}_1, \tilde{v}_2, \tilde{v}_3] \subset \ell_\infty^8$ is isometric to V. Let

$$\tilde{u}_{11} = (1,1,1,1,0,0,0,0),\ \tilde{u}_{21} = (1,1,-1,-1,0,0,0,0),$$

$$\tilde{u}_{31} = (1,-1,1,-1,0,0,0,0),\ \tilde{u}_{12} = (0,0,0,0,1,1,1,1),$$

$$\tilde{u}_{22} = (0,0,0,0,1,1,-1,-1),\ \tilde{u}_{32} = (0,0,0,0,1,-1,1,-1),$$

$$\tilde{u}_1 = \frac{1}{3}\tilde{u}_{11} + \frac{2}{3}\tilde{u}_{12},\ \tilde{u}_2 = \frac{1}{2}\tilde{u}_{21} + \frac{1}{2}\tilde{u}_{22},\ \tilde{u}_3 = \frac{2}{3}\tilde{u}_{31} + \frac{1}{3}\tilde{u}_{32} \in \ell_1^8.$$

Consider a projection $\tilde{P}$ from ℓ_∞^8 onto $\tilde{V}$ given by

$$\tilde{P}x = \frac{1}{4}[\tilde{u}_1(x)\tilde{v}_1 + \tilde{u}_2(x)\tilde{v}_2 + \tilde{u}_3(x)\tilde{v}_3].$$

It is easy to see that

$$\|\tilde{P}\| = \frac{1}{4} \max_{k=1,\dots,8} \sum_{j=1}^{8} |\tilde{u}_1(j)\tilde{v}_1(k) + \tilde{u}_2(j)\tilde{v}_2(k) + \tilde{u}_3(j)\tilde{v}_3(k)|\ .$$

(Here $z(j)$ denotes the jth coordinate of the vector z). By direct calculation we have $\|\tilde{P}\| = \frac{3}{2}$ and thus $\tilde{P}$ is a minimal projection from ℓ_∞^8 onto $\tilde{V}$.

We now introduce the following vectors in ℓ_∞^8:

$$\tilde{\omega}_1 = (\frac{3}{5}, \frac{3}{5}, \frac{3}{5}, \frac{3}{5}, \frac{6}{5}, \frac{6}{5}, \frac{6}{5}, \frac{6}{5}),\ \tilde{\omega}_2 = (\frac{1}{2}, \frac{1}{2}, -\frac{1}{2}, -\frac{1}{2}, \frac{1}{2}, \frac{1}{2}, -\frac{1}{2}, -\frac{1}{2}),$$
$$\tilde{\omega}_3 = (\frac{6}{5}, -\frac{6}{5}, \frac{6}{5}, -\frac{6}{5}, \frac{3}{5}, -\frac{3}{5}, \frac{3}{5}, -\frac{3}{5}).$$

It is easy to check that $\tilde{\omega}_k(\tilde{u}_j) = 4\delta_{jk}$ and Q, defined by

$$Qx := \frac{1}{4}(\tilde{\omega}_1(x)\tilde{u}_1 + \tilde{\omega}_2(x)\tilde{u}_2 + \tilde{\omega}_3(x)\tilde{u}_3),$$

is a projection from ℓ_1^8 onto $\tilde{U} := \operatorname{span}[\tilde{u_1}, \tilde{u_2}, \tilde{u_3}]$. Its norm is

$$\|Q\| = \frac{1}{4} \max_{k=1,\dots,8} \sum_{j=1}^{8} |\tilde{\omega}_1(k)\tilde{u}_1(j) + \tilde{\omega}_2(k)\tilde{u}_2(j) + \tilde{\omega}_3(k)\tilde{u}_3(j)|.$$

Again, by direct calculation $\|Q\| = \frac{29}{20} < \frac{3}{2}$. Hence

$$\lambda(\tilde{U}, \ell_1^8) < \lambda(U, \ell_1^8).\quad \square$$

Comment. Further relationships between minimal projections and their adjoints can be found in [3] and [5].

4. The evaluation of the "P_λ-ratio"

In this section we restrict ourselves to the real field $\mathbb{R}$ and compute the *exact* value of the supremum of

$$\frac{d(V, \ell_\infty^2)}{\lambda(V)},$$

which we call the P_λ-ratio, over all two-dimensional subspaces $V \subset \ell_1^3$. To do this we will view every two-dimensional space $V \subset \ell_1^3$ (up to isometry) as spanned by vectors $(1, a, 0)$ and $(0, b, 1)$, $a, b \geq 0$. We then compute the values of $d(V, \ell_\infty^2)$ and $\lambda(V)$ in terms of a, b. Since ℓ_∞^2 is isometric to ℓ_1^2 and $\lambda(V, \ell_1^3) = \lambda(V)$, it suffices to compute $d(V, \ell_1^2)$ and $\lambda(V, \ell_1^3)$. We will start with $d(V, \ell_1^2)$ and a small lemma.

Lemma 1. Let T be an isomorphism from V onto ℓ_1^2 such that $\|T\|\|T^{-1}\| = d(V, \ell_1^2)$. Let e_1, e_2 be the canonical basis in ℓ_1^2. Then

$$\|T^{-1}\| = \|T^{-1}e_1\| = \|T^{-1}e_2\|.$$

Proof. Without loss of generality let $\|T\| = 1$. Then $\|T^{-1}\| = \max\{\|T^{-1}e_1\|, \|T^{-1}e_2\|\}$.

If $\|T^{-1}\| = 1$, then T is an isometry and

$$1 = \|T^{-1}\| = \|T^{-1}e_1\| = \|T^{-1}e_2\|,$$

and there is nothing to prove. Thus

$$1 < \|T^{-1}\| := \alpha = \|T^{-1}e_1\| = d(v, \ell_1^2) > \|T^{-1}e_2\|.$$

Then consider an operator $S : V \to \ell_1^2$ such that

$$Sx = Tx + \varepsilon f(x)e_1 \text{ where } f \in V^*, \|f\| = 1, \text{ and } f(T^{-1}e_1) = \alpha.$$

We have $\|S\| \leq \|T\| + \varepsilon\|f\|$ and

$$S^{-1}y = T^{-1}y - \frac{\varepsilon f(T^{-1}y)}{1 + \varepsilon f(T^{-1}e_1)}T^{-1}e_1.$$

Hence

$$S^{-1}e_1 = T^{-1}e_1 - \frac{\varepsilon f(T^{-1}e_1)}{1 + \varepsilon f(T^{-1}e_1)}T^{-1}e_1 = \frac{1}{1 + \varepsilon f(T^{-1}e_1)}T^{-1}e_1$$

and

$$S^{-1}e_2 = T^{-1}e_2 - \frac{\varepsilon f(T^{-1}e_2)}{1 + \varepsilon f(T^{-1}e_2)}T^{-1}e_1.$$

Then for sufficiently small ε we have $\|S^{-1}e_1\| > \|S^{-1}e_2\|$ and

$$\|S\|\|S^{-1}\| \leq \frac{(1+\varepsilon)\alpha}{1 + \varepsilon f(T^{-1}e_1)} = \frac{(1+\varepsilon)\alpha}{1 + \varepsilon\alpha} < \alpha.$$

Hence $\|S\|\|S^{-1}\| < d(V, \ell_1^2)$ and we have a contradiction. □

Geometrically this lemma means that the position of a parallelogram optimally circumscribing the unit ball of V must be such that the ratio of the (euclidean) length of a vertex vector of the parallelogram to the length of this same vector projected onto the boundary of the ball is the same for all vertices.

Lemma 2. An arbitrary two-dimensional subspace of ℓ_1^3 is isometric to a space $V = [v_1, v_2] \subset \ell_1^3$, where $v_1 = (1, a, 0)$ and $v_2 = (0, b, 1)$, with $0 < a$, $0 \leq b$. Let $\rho = a + b + a^2(1+b) + b^2(1+a)$ and

$$\begin{pmatrix} c_1 & d_1 \\ c_2 & d_2 \\ c_3 & d_3 \end{pmatrix} = \frac{1}{\rho} \begin{pmatrix} a+b+b^2(1+a) & -ab(1+a) \\ a(1+b) & b(1+a) \\ -ab(1+b) & a+b+a^2(1+b) \end{pmatrix}.$$

For $u_1 = (c_1, c_2, c_3)$ and $u_2 = (d_1, d_2, d_3)$, the operator $P = u_1 \otimes v_1 + u_2 \otimes v_2$ is a *minimal* projection from ℓ_1^3 onto V with $\|P\| = 1/(1-\epsilon)$, where $\epsilon = 2ab/(1+a)(1+b)(a+b)$. Further, $\|P\| \leq 4/3$, with equality occurring if and only if $a = b = 1$.

Proof. First, by applying a linear transformation, it is immediate that any two-dimensional subspace $\widetilde{V}$ of ℓ_1^3 is isometric to a subspace with a basis of the form $(\tilde{v}_1, \tilde{v}_2)$, where $\tilde{v}_1 = (1, a', 0)$ and $\tilde{v}_2 = (0, b', \pm 1)$, with $0 < |a'|$ and $0 \leq a'b'$. Thus, for any real α and β,

$$\|\alpha\tilde{v}_1 + \beta\tilde{v}_2\| = |\alpha| + |\alpha a' + \beta b'| + |\beta| = \|\alpha v_1 + \beta v_2\|,$$

where $v_1 = (1, a, 0)$ $v_2 = (0, b, 1)$, $a = |a'|$, $b = |b'|$. This establishes the isometry of Lemma 1.

Secondly, a simple direct check shows that P is a projection (i.e., $\langle v_i, u_j \rangle = \delta_{ij}$, $1 \leq i, j \leq 2$).

For the sequel, introduce the notation $\vec{\alpha} = (\alpha_1, \alpha_2) \in \mathbb{R}^2$ and let $\vec{\alpha} \cdot \vec{\beta} = \alpha_1\beta_1 + \alpha_2\beta_2$ denote the usual dot product.

To calculate $\|P\|$, form the projection matrix $P = (p_{ij})_{3\times 3} = (\vec{u}(j) \cdot \vec{v}(i))_{3\times 3}$:

$$P = \frac{1}{\rho} \begin{pmatrix} a+b+b^2(1+a) & a(1+b) & -ab(1+b) \\ a(a+b) & a^2(1+b)+b^2(1+a) & b(a+b) \\ -ab(1+a) & b(1+a) & a+b+a^2(1+b) \end{pmatrix}.$$

By noting that the sign configuration of P is

$$\begin{pmatrix} + & + & - \\ + & + & + \\ - & + & + \end{pmatrix}$$

it is straightforward to verify that all three absolute column sums $(\sum_{i=1}^3 |p_{ij}|$, $j = 1, 2, 3)$ are equal to $1/(1-\epsilon)$. But, $2/\epsilon$ has the form

$$(x+y)(y+z)(x+z)/xyz = \left(\sqrt{x/y} + \sqrt{y/x}\right)\left(\sqrt{y/z} + \sqrt{z/y}\right)\left(\sqrt{x/z} + \sqrt{z/x}\right)$$

≥ 8, with equality holding iff $x = y = z$ (since, for any $0 < s$, $2 \leq s + 1/s$ with equality iff $s = 1$).

To show that P is minimal we use the theory of [2] as follows. Consider the operator

$$E_P = \frac{b(1+b)}{\rho} \begin{pmatrix} 1 & 1 & -1 \\ \mu_2 & \mu_2 & \mu_2 \\ -\mu_3 & \mu_3 & \mu_3 \end{pmatrix},$$

with $\mu_2 = a(a+b)/(1+b)$, $\mu_3 = a(1+a)/b(1+b)$. It can then be checked directly that E_P is an extremal operator for P, i.e., E_P is a norm one integral operator such that $\langle P, E_P \rangle = \mathrm{tr}(E_P \circ P) = \|P\|$. Furthermore it is easily checked that $E_P(V) \subset V$, and thus P is minimal by [2]. $\square$

Lemma 3. If $V = [(1,a,0),(0,b,1)] \subset \ell_1^3$, with $a > 0$ and $b \geq 0$, then

$$d(V, \ell_1^2) = \min\{\delta_1, \delta_2, \delta_3\}, \text{ where}$$

$$\delta_1 := \frac{(1+a)(1+a+b)}{1+b+ab+a^2}, \; \delta_2 := \frac{(1+b)(1+a+b)}{1+a+ab+b^2}, \; \delta_3 := \frac{(a+b+1)(a+b)}{a^2+a+b+b^2}.$$

(Note: $b > 1 \Rightarrow \delta_2 < \delta_1$; $b < 1 \Rightarrow \delta_1 < \delta_2$; $a > 1 \Rightarrow \delta_3 < \delta_1$; $a < 1 \Rightarrow \delta_1 < \delta_3$; $b < a \Rightarrow \delta_3 < \delta_2$; $a < b \Rightarrow \delta_2 < \delta_3$.)

Proof. The proof follows easily by first computing that the unit ball of V is a hexagon (symmetric through the origin) in $\mathbb{R}^2$ with (half of its) coordinates given by $(\frac{1}{1+a}, 0)$, $(0, \frac{1}{1+b})$, $\frac{(-b,a)}{a+b}$ and then applying Lemma 1 to determine the appropriate parallelograms. □

Theorem 2.

$$\frac{d(V, \ell_\infty^2)}{\lambda(V)} \leq \frac{9}{8}$$

for all two-dimensional subspaces $V \subset \ell_1^3$, with equality holding precisely when the unit ball of V is the regular hexagon.

Proof. ¿From the above, it suffices to consider the ratio $\frac{d(V,\ell_1^2)}{\lambda(V)}$, where $V = [(1,a,0),(0,b,1)] \subset \ell_1^3$, with $a > 0$ and $b \geq 0$. I.e., we must examine the following ratios (determined from Lemmas 2 and 3):

$$r_1 := \frac{\delta_1}{\lambda(V)} = \frac{N}{(1+b)(1+a)(a^2+a+b+b^2)} \text{ if } a \leq b \leq 1 \text{ or } b \leq a \leq 1,$$

$$r_2 := \frac{\delta_2}{\lambda(V)} = \frac{N}{(1+b)(a+b)(a^2+1+b+ab)} \text{ if } 1 \leq a \leq b \text{ or } a \leq 1 \leq b,$$

$$r_3 := \frac{\delta_3}{\lambda(V)} = \frac{N}{(1+a)(a+b)(b^2+a+1+ab)} \text{ if } 1 \leq b \leq a \text{ or } b \leq 1 \leq a,$$

where $N = (a + a^2 + a^2b + b + b^2 + ab^2)(1+a+b)$.

But now the partial derivative pairs $(\partial r_i/\partial a, \partial r_i/\partial b)$ can be computed and determined never to have a zero $((0,0))$ in the admissible regions. For example, defining $\gamma := 2ab[(1+a)(1+b)(a^2+a+b+b^2)^2]^{-1}$, we have $(\partial r_1/\partial a, \partial r_1/\partial b) =$

$$\gamma\left(\frac{b(2b+a+ab+2b^2+ab^2-a^3)}{1+a}, \frac{a(2a+b+ab+2a^2+a^2b-b^3)}{1+b}\right),$$

which cannot $= (0,0)$ if $a > 0$ and $b \geq 0$, as is easily determined. We therefore conclude that a maximum for the ratio ($= \frac{3/2}{4/3}$) must occur at $a = 1$, $b = 1$. □

Conjecture.

$$\frac{d(V, \ell_\infty^2)}{\lambda(V)} \leq \frac{9}{8}$$

for all two-dimensional subspaces V, with equality holding precisely when the unit ball of V is the regular hexagon.

Acknowledgment. We would like to express our gratitude to Bill Johnson whose insightful questions motivated Section 3 of this paper.

5. References

1. B. Beauzamy, *Introduction to Banach Spaces and their Geometry*, No. Holland, 1985.
2. B. L. Chalmers and F. T. Metcalf, *A characterization and equations for minimal projections and extensions*, J. Oper. Theory, to appear.
3. B. L. Chalmers, K. C. Pan, and B. Shekhtman, When is the adjoint of a minimal projection also minimal, Proc. of Memphis Conf., Lect. Notes in Pure and Applied Math., **138** (1991), 217-226.
4. J. Lindenstrauss, On the extension of operators with a finite-dimensional range, Illinois J. Math. **8** (1964), 488-499.
5. K. C. Pan and B. Shekhtman, *On minimal interpolating projections and trace duality*, J. of Approx. Theory, **65** (1991), 216-230.

THE ACTION CONSTANTS

B. L. Chalmers
Department of Mathematics
University of California
Riverside, CA 92521

B. Shekhtman
Department of Mathematics
University of South Florida
Tampa, FL 33620

I. Introduction. In this paper we introduce isometric invariants of a given n-dimensional Banach space (cf [1]), conjecture (cf [1]) that these invariants characterize (up to isometry) the Banach space, and obtain several results and examples illustrating their usefulness. To start with, for a given space V, we look at the absolute projection constant and choose to view it as an extension constant of an identity operator on V.

As was mentioned in the previous article [4], the projection constant is an isometric invariant; i.e., if V_1 is another n-dimensional space isometric to V, then the extension constant of the identity on V_1 is the same as the extension constant of the identity on V. This constant does not characterize the Banach space V. It is well-known (for instance) that $\lambda(\ell_1^3) = \lambda(\ell_2^3)$, yet ℓ_1^3 and ℓ_2^3 are not isometric. Hence it is necessary to consider extension constants of operators on V other than the identity. But there is a problem. While the meaning of the identity is well understood independently of the space on which it acts (in the language of category theory, the identity is a universal notion), an arbitrary operator T is defined only on a fixed Banach space V. That is, given two Banach spaces V and V_1 and an operator T on V, it is not clear what operator T_1 should be considered on V_1 to compare V and V_1. One way to insure that T_1 relates to T is to assume that there are bases $v_1, \cdots, v_n$ in V and $v_1', \dots, v_n'$ in V_1 so that the matrix of T with respect to $\{v_j\}$ is the same as the matrix of T_1 with respect to $\{v_j'\}$. To avoid depending on the choices of the basis we start with a matrix A and define the action constant on V to be the infimum of the extension constants of operators T where the infimum is taken over all operators T that have a matrix A with respect to *some* basis in V. This definition does not depend on the space V and the value of this constant depends (up to similarity) only on the matrix A and the space V, and not on the choice of basis for V.

II. Definitions. We will use the notions of Banach-Mazur distance and projection constant defined in the previous article [4].

Definition 1. Given a pair of Banach spaces $X, \mathcal{F}$ a subspace $V \subset X$ and an operator $T : V \to \mathcal{F}$ define the *relative extension constant*

$$e(T, X) := \inf\{\|\tilde{T}\| : \tilde{T} : X \to \mathcal{F},\ \tilde{T}|_V = T|_V\}$$

and *absolute extension constant*

$$e(T) := \sup\{e(T, X) : X \supset V\}.$$

Approximation, Probability, and Related Fields, Edited by
G. Anastassiou and S.T. Rachev, Plenum Press, New York 1994

It is convenient to express this extension by using a commuting diagram:

$$\begin{array}{ccc} X & & \\ \uparrow i & \searrow \tilde{T} & \\ V & \underset{T}{\rightarrow} & \mathcal{F} \end{array}$$

Here $\tilde{T}$ is an extension of T and hence $\|\tilde{T}\| = e(T, X)$, and i is the identical embedding of V into X.

Definition 2. Given a Banach space X, an n-dimensional subspace $V \subset X$, and an $n \times n$ matrix A define a *relative action constant*

$$\lambda(A, V, X) := \inf\{e(T, X)\}$$

where the infimum is taken over all operators $T : V \to V$ such that the matrix of an operator T with respect to some basis in V is equal to A.

Define an *absolute action constant*

$$\lambda(A, V) := \sup\{\lambda(A, V, X),\ X \supset V\}.$$

Conjecture 1([1]). $\lambda(A, V) \le (\rho + \sqrt{(N-1)(N\beta - \rho^2)})/N$, where $N = n(n+1)/2$ or $N = n^2$ if the field is $\mathbb{R}$ or $\mathbb{C}$, respectively, $\rho := |tr(A)|$, $\beta := \sum_{i=1}^{n} |d_i|^2$, and d_i are the eigenvalues of A.

The relationship between the two notions of extension constant and action constant is pretty clear. Let A be a given matrix and $v_1, \dots, v_n$ be a fixed basis in V. Let $\hat{A}$ be an operator on V defined by the matrix A with respect to $v_1, \dots, v_n$. Then

$$\lambda(A, V) = \inf\{e(\widehat{UAU^{-1}}) : \ U \text{ is an invertible } n \times n \text{ matrix}\}$$

Conversely, given an operator T on V, let A be a matrix given by the representation of T with respect to some basis $v_1, \dots, v_n$. Then

$$\lambda(A, V) = \inf\{e(UTU^{-1}) : \ U : V \to V \text{ is invertible}\}.$$

Finally, we need the notion of an L_∞-factorization constant.

Definition 3. Let T be an operator from V into V. Its L_∞-factorization constant is defined by

$$\gamma_\infty(T) = \inf\{\|A\|\ \|B\|\ ,\ AB = T\}$$

where the infimum is taken over all pairs of operators $B : V \to L_\infty(\mu)$, $A : L_\infty(\mu) \to V$ and all $L_\infty(\mu)$ spaces.

Again we can use a commuting diagram:

$$\begin{array}{ccc} V & \xrightarrow{T} & V \\ B \searrow & & \nearrow A \\ & L_\infty(\mu) & \end{array}$$

By use of standard theory of local structure (cf [5]) it is easy to replace $L_\infty(\mu)$ by $\ell_\infty(\Gamma), \ell_\infty^N$ or $C(K)$ space in the above definition when dim $V < \infty$. In fact any $\mathcal{L}_\infty$-space will do (cf [5]).

III. Properties of the Action Constant: In this section we establish some isomorphic properties of the action constants that correspond to the classical properties of the projection constant. Again the fundamental role is played by the well-known
Proposition 1(cf[5]). Let T be an operator from a Banach space V into $L_\infty(\mu)$. Then $e(T) = \|T\|$. In other words, any operator $T : V \to L_\infty(\mu)$ can be extended to a bigger space with the same norm.

Proposition 2. Let T be an operator on V. Then a) $e(T) = \gamma_\infty(T)$; b) $e(T) = e(T, L_\infty(\mu))$.
Proof. We assume without loss of generality that V is already a subspace of some $L_\infty(\mu)$ space. Hence there exists a natural embedding $i : V \hookrightarrow L_\infty(\mu)$. Given $\varepsilon > 0$, let $\tilde{T}$ be an arbitrary extension of T onto $L_\infty(\mu)$ with $\|\tilde{T}\| \le e(T) + \varepsilon$. Then the diagram

$$\begin{array}{ccc} V & \xrightarrow{T} & V \\ i\searrow & & \nearrow \tilde{T} \\ & L_\infty(\mu) & \end{array}$$

commutes and hence $\gamma_\infty(T) \le \|i\|\,\|\tilde{T}\| \le e(T)+\varepsilon$. To establish the converse inequality let X be a Banach space containing V. For $\varepsilon > 0$, let $A : L_\infty(\mu) \to V$ and $B : V \to L_\infty(\mu)$ be a factorization of T such that $\|A\|\,\|B\| \le \gamma_\infty(T) + \varepsilon$:

$$\begin{array}{ccccc} X & \hookleftarrow & V & \xrightarrow{T} & V \\ \tilde{B} & & B\searrow & & \nearrow A \\ & & & L_\infty(\mu) & \end{array}$$

Then by Proposition 1 there exists an extension $\tilde{B}$ of B to X so that $\|\tilde{B}\| = \|B\|$. Observe that the operator $A\tilde{B} : X \to V$ is an extension of T to X. Hence

$$e(T, X) \le \|A\tilde{B}\| \le \|A\|\|B\| \le \gamma_\infty(T) + \varepsilon$$

Now sup'ing over all X and inf'ing over all ε we obtain a converse inequality. Hence $e(T) = \gamma_\infty(T)$. In particular, this inequality together with the first part of the proof of this proposition implies b). □

Theorem. Let V and V_1 be two n-dimensional Banach spaces. Let $v_1, \dots, v_n$ be a fixed basis of V. Let A be an arbitrary $n \times n$ matrix and let $\hat{A}$ be an operator on V that corresponds to A with respect to $v_1, \dots, v_n$. Then

$$a)\quad \lambda(A, V) = \inf\{\gamma_\infty(U\hat{A}U^{-1}) : \ U : V \to V \text{ is invertible}\}$$
$$b)\quad \lambda(A, V_1) \le \lambda(A, V) d(V, V_1).$$

Proof. a) Since all operators on V that correspond to the matrix A have a form $U\hat{A}U^{-1}$, Proposition 2 completes the proof.
b) Let $T : V \to V_1$ so that $\|T\|\|T^{-1}\| = d(V, V_1)$. We have

$$V_1 \xrightarrow{T^{-1}} V \xrightarrow{U\hat{A}U^{-1}} V \xrightarrow{T} V_1$$

and $\gamma_\infty(TU\hat{A}U^{-1}T^{-1}) \le \|T^{-1}\|\gamma_\infty(U\hat{A}U^{-1})\|T\| \le \gamma_\infty(U\hat{A}U^{-1})d(V, V_1)$. Observe that the operator $TU\hat{A}U^{-1}T^{-1}$ corresponds to the matrix A with respect to some basis in V_1. Hence

$$\lambda(A, V_1) \le \gamma_\infty(U\hat{A}U^{-1})d(V, V_1).$$

Inf'ing over all U we obtain the desired conclusion. □

Corollary. The quantity $\lambda(A,V)$ is an isometric invariant; i.e., if V and V_1 are isometric, then

$$\lambda(A,V)=\lambda(A,V_1).$$

Conjecture 2([1]). Let $C_{n\times n}$ be the set of all $n\times n$ matrices. Let V and V_1 be two n-dimensional Banach spaces such that $\lambda(A,V)=\lambda(A,V_1)$ for all $A\in C_{n\times n}$. Then V and V_1 are isometric.

Remark. It follows from Theorem 1 that $\lambda(A,V)$ is a continuous function with respect to A. Hence it suffices to consider only diagonal matrices in the Conjecture.

As we mentioned in the introduction $\lambda(\ell_1^3)=\lambda(\ell_2^3)$. We now show that the action constants distinguish ℓ_1^3 and ℓ_2^3.

Proposition 3. Let

$$A=\begin{bmatrix}1&0&0\\0&1&0\\0&0&0\end{bmatrix}$$

Then $\lambda(A,\ell_1^3)=1\quad;\quad\lambda(A,\ell_2^3)=\frac{4}{\pi}>1.$

Proof. We first embed ℓ_1^3 into $L_\infty[0,1]$. That is done by considering a space V spanned by the first three Rademacher functions

$$r_1(t)\equiv 1\ ,\ r_2(t)=\begin{cases}1 & \text{if } t\in[0,\frac{1}{2}]\\ -1 & \text{if } t\in(\frac{1}{2},1]\end{cases}\quad,\quad r_3(t)=\begin{cases}1 & \text{if } t\in[0,\frac{1}{4}]\cup[\frac{1}{2},\frac{3}{4}]\\ -1 & \text{otherwise}\end{cases}.$$

Clearly

$$\|a_1r_1+a_2r_2+a_3r_3\|=|a_1|+|a_2|+|a_3|$$

and thus $V=\operatorname{span}[r_1,r_2,r_3]$ is isometric to ℓ_1^3. The two-dimensional subspace U of V spanned by the functions $\frac{r_1+r_2}{2}$, $\frac{r_1-r_2}{2}$ is isometric to ℓ_∞^2 since

$$\|a_1\frac{r_1+r_2}{2}+a_2\frac{r_1-r_2}{2}\|=\frac{1}{2}(|a_1+a_2|+|a_1-a_2|)=\ \max\{|a_1|,|a_2|\}.$$

Hence there exists a projection $P:L_\infty\to U$ such that $\|P\|=1$. Consider P as a map from $L_\infty\to V$. Then the restriction of this map onto V has the matrix A with respect to the basis r_1,r_2,r_3. Hence

$$\lambda(A,V)\le 1.$$

On the other hand any map from L_∞ into V that corresponds to the matrix A has to fix at least one element of V (actually a two-dimensional subspace of V) and hence the norm at any operator $\hat{A}$ corresponding to A has a norm at least 1. Hence $\lambda(A,\ell_1^3)=1$.

Considering the second equality, observe that any map $\hat{A}$ that corresponds to A on ℓ_2^3 is a projection from ℓ_2^3 onto a two dimensional subspace $U\subset\ell_2^3$. Let T be an extension of this map to an L_∞ space and let P be a projection from ℓ_2^3 onto U of norm 1. (Such projection exists since ℓ_2^3 is a Hilbert space.) Observe then that the map PT maps L_∞ onto U and is in fact a projection. U being a subspace of a Hilbert space is itself a two dimensional Hilbert space. Hence

$$\|PT\|\ge\lambda(U)=\lambda(\ell_2^2)=\frac{4}{\pi}$$

and thus

$$\frac{4}{\pi}\le\|P\|\|T\|=\|T\|.$$

Hence $\lambda(A,\ell_2^3)\ge\frac{4}{\pi}$.

On the other hand, picking any two-dimensional subspace $U \subset \ell_2^3$, let $P : L_\infty \to U$ be a projection such that $\|P\| = \frac{4}{\pi}$. Then $P|_{\ell_2^3}$ has a matrix similar to A and hence

$$\lambda(A, \ell_2^3) = \frac{4}{\pi}. \quad \square$$

IV. Application to Projection Constants: It is very easy to construct examples of a space (say of dimension $2n$) such that the projection constant is big $\approx \sqrt{n}$ yet it has a large subspace (or dimension n) with a very small projection constant. An easy example is $\ell_\infty^n \oplus \ell_2^n$. The projection constant of this space is $\sim \sqrt{n}$. Yet $\ell_\infty^n \subset \ell_\infty^n \oplus \ell_2^n$ has projection constant 1.

There is a special class of spaces (such as ℓ_p^n) called symmetric spaces for which we will prove that the situation described above cannot happen.

A space V is called symmetric if there exists a basis $v_1, \ldots, v_n \subset V$ such that

$$\|\sum \alpha_k v_k\| = \|\sum \pm\alpha_{\sigma(k)}\, v_k\|$$

for all choices of $\pm$ signs and all $(\alpha_k) \in \mathbb{R}^n$ and for all permutations σ.

We will prove that for any subspace E of a finite-dimensional symmetric space V

$$\lambda(E) \geq \frac{\dim E}{\dim V} \lambda(V).$$

Here are a few needed definitions.

For a finite-dimensional operator $T : X \to Y$ we define its absolutely summing norm $\pi_1(T)$ as the least constant C such that the inequality

$$\sum_{j=1}^{N} \|Tx_j\| \leq C \sum_{j=1}^{N} |f(x_j)|$$

holds for all N, all choices of vectors $x_1, \ldots, x_N \in X$ and all choices of functionals $f \in X^*$ with $\|f\| = 1$.

The π_1 norm and γ_∞ norm are in trace-duality:

$$\gamma_\infty(A) = \sup_{B \neq 0} \frac{tr(AB)}{\pi_1(B)}.$$

Hence for a finite-dimensional space V

$$\lambda(V) = \gamma_\infty(I) = \sup_B \frac{tr(B)}{\pi_1(B)}.$$

For a symmetric space V the "sup" is attained for $B = I$ (see[5]). Hence $\lambda(V) = \dim V / \pi_1(I)$.

Theorem. Let V be a symmetric n-dimensional space. Let A be an $n \times n$ matrix given by

$$A = \begin{bmatrix} I_{m\times m} & 0 \\ 0 & 0 \end{bmatrix} \qquad \text{with } m \leq \dim V.$$

Then

$$\lambda(A, V) \geq \frac{m}{\dim V} \lambda(V).$$

In particular for any m-dimensional subspace $E \subset V$

$$\lambda(E) \geq \frac{m}{\dim V}\lambda(V).$$

Proof. For any map $\hat{A}$ corresponding to A we have

$$\begin{aligned}\lambda(A,V) &= \inf\{\gamma_\infty(U\hat{A}U^{-1}) : \ U : V \to V \text{ is invertible}\} \\ &= \inf\{\sup_{0\neq B:V\to V} \frac{tr(U\hat{A}U^{-1}B)}{\pi_1(B)}\} \geq \inf\{tr(U\hat{A}U^{-1})/\pi_1(I)\} \\ &= \frac{m}{\pi_1(I)} = \frac{m}{\dim V}\lambda(V). \quad \square\end{aligned}$$

Note: Further examples and discussion of computing action constants can be found in [2] and [3].

References

1. B. L. Chalmers and K. C. Pan, *Finite-dimensional action constants*, submitted.
2. B. L. Chalmers, K. C. Pan, and B. Shekhtman *A strategy for proving extensions of the 4/3 conjecture*, Proc. of Memphis Conf., Lect. Notes in Pure and Applied Math., **138** (1991), 207-215.
3. B. L. Chalmers and B. Shekhtman, *Extension constants of unconditional two-dimensional operators*, submitted.
4. B. L. Chalmers and B. Shekhtman, *On the role of ℓ_∞ in approximation theory*, this volume.
5. N. Tomczak-Jaegermann, *Banach-Mazur Distances and Finite-dimensional Operator Ideals*, John Wiley and Sons, New York, 1989.

BIVARIATE PROBABILITY DISTRIBUTIONS SIMILAR TO EXPONENTIAL

B. Dimitrov[1], S. Chukova[2] and Z. Khalil[3]

[1,2] GMI Engineering and Management Institute

Flint, Michigan U.S.A.

[3] Dept. of Mathematics & Statistics

Concordia University

Montreal, Quebec H4B 1R6

ABSTRACT

The class of bivariate distribution having the almost-lack-of-memory property (ALM-distribution) is introduced and the exact form of these distributions in a sub-class with almost independent components is derived. Some of the corresponding probability properties are discussed.

INTRODUCTION

This study has been stimulated by the univariate concept of probability distributions having the almost-lack-of-memory (ALM) property, introduced by Chukova (1992) and developed by the authors in 1993. Thus we start with a definition.

Let $Z = (X, Y), P\{X \geq 0, Y \geq 0\} = 1$, be a nonnegative two-dimensional random vector with components X and Y, defined on a probability space $(\Omega, \mathcal{F}, \mathcal{B})$.

Definition 1. We say, a random vector $Z = (X, Y)$ has the almost-lack-of-memory property iff

$$P\{X \geq a + x,\, Y \geq b + y \mid X \geq a,\, Y \geq b\} = P\{X \geq x,\, Y \geq y\} \tag{1}$$

Approximation, Probability, and Related Fields, Edited by
G. Anastassiou and S.T. Rachev, Plenum Press, New York 1994

holds for any $x \geq 0$, $y \geq 0$ and for infinitely many different nonnegative values of a and b. The class of distributions having the above property will be denoted by $K_0(a,b)$.

It is well known from Marshall and Olkin (1967), that if (1) is true for all nonnegative a, b, it implies that Z has the "trivial" bivariate exponential distribution, which is the product of its marginals. Thus, the class of distributions having the ALM property appears to be an extension of the bivariate exponential one.

THE CLASS OF BIVARIATE ALM DISTRIBUTIONS

Definition 2. The distribution function of a nonnegative random vector $Z = (X, Y)$ belongs to the class $K_0(a,b)$, $a > 0$, $b > 0$, iff equation (1) is true for the given values of a and B, and for any $x \geq 0$ and $y \geq 0$. We denote this fact simply by $Z \in K_0(a,b)$.

Remark, that if the probability $P\{X \geq a,\ y \geq b\}$ equals to zero, then it immediately follows that some of the components of the random vector Z is restricted with probability 1 on the corresponding interval $[0, a)$ or $[0, b)$. Thus we consider only those $Z \in K_0(a,b)$ for which is true that

$$\delta := P\{X \geq a,\ Y \geq b\} > 0,$$

as well as

$$\alpha = P\{X \geq a\} < 1; \ \beta = P\{Y \geq b\} < 1 \tag{2}$$

are satisfied. The condition $\delta > 0$ is necessary in order that the probability in the left side of (1) to be defined. We will see later in Lemma 1 that $\delta = 1$ implies that Z degenerates at infinity, with probability 1, and then $P\{X \geq na,\ Y \geq nb\} = 1$ for any positive integer n. Thus we would like to avoid this case. Now either of the equalities $\alpha = 1$ or $\beta = 1$ also imply that the corresponding component degenerates at infinity with probability 1 (see Corollary 1 below), and we would like to avoid such cases too. The assumption $\delta > 0$ and (2) imply that $\alpha > 0$ and $\beta > 0$ too.

First we prove:

Lemma 1. If $Z \in K_0(a,b)$, then Z has the ALM property.

Proof. It is sufficient to show that for any integer $n \geq 0$ and any $x \geq 0$, $y \geq 0$ it is true that

$$P\{X \geq na + x,\ Y \geq nb + y \mid X \geq na,\ Y \geq nb\} = P\{X \geq x,\ Y \geq y\}. \tag{3}$$

Equation (1) can be rewritten in the following equivalent forms

$$\frac{P\{X \geq a + x,\ Y \geq b + y\}}{P\{X \geq a,\ Y \geq b\}} = P\{X \geq x,\ Y \geq y\}, \tag{4}$$

or

$$P\{X \geq a+x,\, Y \geq b+y\} = P\{X \geq a,\, Y \geq b\} \cdot P\{X \geq x,\, Y \geq y\}. \tag{5}$$

Setting in (5) $x = a$, and $y = b$ we obtain by induction

$$P\{X \geq na,\, Y \geq nb\} = [P\{X \geq a,\, Y \geq b\}]^n. \tag{6}$$

Hence (3) is true for $n = 0$ (trivial), and for $n = 1$ (coincides with (5)). Suppose (3) is true for given integer $n > 1$, we show that it holds for the next integer $n+1$. Taking into account (5) and (6), we get

$$P\{X \geq (n+1)a+x,\, Y \geq (n+1)b+y \mid X \geq (n+1)a,\, Y \geq (n+1)b)$$

$$= \frac{P\{X \geq na+(a+x),\, Y \geq nb+(b+y)\}}{P\{X \geq (n+1)a,\, Y \geq (n+1)b\}} = \frac{P\{X \geq na+(a+x),\, Y \geq nb+(b+y)\}}{P\{X \geq na,\, Y \geq nb\} \cdot P\{X \geq a,\, Y \geq b\}},$$

which equals by assumption to

$$\frac{P\{X \geq a+x,\, Y \geq b+y\}}{P\{X \geq a,\, Y \geq b\}} = P\{X \geq x,\, Y \geq y\}.$$

□

Corollary 1. If in (2) either $\alpha = 1$ or $\beta = 1$ then either $P\{X = \infty\} = 1$ or $P\{Y = \infty\} = 1$.

Proof. Let $\alpha = 1$. Then $\delta = \beta$ and from (6) we obtain

$$P\{X \geq na,\, Y \geq nb\} = \beta^n = P\{Y \geq nb\},$$

i.e. $P\{X \geq na\} = 1$ for any $n \geq 1$. Analogously we consider $\beta = 1$. □

The class $K_0(a, b)$ appears to be sufficiently large. We narrow it to those distributions which have the ALM marginals. Moreover, an assumption for "almost independence" in the sense of the next definition implies that the required marginals have the ALM property. This corresponds in some sense to the characterization of the bivariate exponential with independent components (Galambos and Kotz, 1978). Also we introduce the next concretization. We further specify following:

Let X and Y be defined on a probability space $(\Omega, \mathcal{F}, \mathcal{P})$.

Definition 3. We say that the r.v.'s X and Y are almost independent if there exist infinitely many different constants $a > 0$, $b > 0$ such that for any $x \geq 0$, $y \geq 0$ the following is true:

$$P\{X \geq a,\, Y \geq y\} = P\{X \geq a\} \cdot P\{Y \geq y\}$$

and

$$P\{X \geq x,\, Y \geq b\} = P\{X \geq x\} \cdot P\{Y \geq b\}.$$

We consider further, the subclass $Z \in K_0(a,b)$, introduced by the following definition:

Definition 4. The random vector $Z \in K_0(a,b)$ belongs to the subclass $K(a,b)$ iff for any $x, y (0 \le x \le a,\ 0 \le y \le b)$ and arbitrary integers $n \ge 0$, $m \ge 0$ the equations

$$P\{X \ge na + x,\ Y \ge y\} = P\{X \ge na\} \cdot P\{X \ge x,\ Y \ge y\}, \tag{7}$$

and

$$P\{X \ge x,\ Y \ge mb + y\} = P\{X \ge x,\ Y \ge y\} \cdot P\{Y \ge mb\} \tag{8}$$

hold.

In particular for $Z \in K(a,b)$ it is true that:

$$\delta = P\{X \ge a,\ Y \ge b\} = P\{X \ge a\} \cdot P\{Y \ge b\} = \alpha\beta. \tag{9}$$

Lemma 2. If $Z \in K(a,b)$ then X and Y are one dimensional ALM random variables and their survival functions have either of the form

$$G_X(x) = \begin{cases} 1, & x < 0; \\ \alpha^{n+1} + \alpha^n(1-\alpha)G_1(x), & na \le x < (n+1)a;\ n = 0,1,2,\ldots, \end{cases} \tag{10}$$

and

$$G_Y(y) = \begin{cases} 1, & y < 0; \\ \beta^{m+1} + \beta^m(1-\beta)G_2(y), & mb \le y < (m+1)b;\ m = 0,1,2,\ldots, \end{cases} \tag{11}$$

where α and β are defined by (2) and

$$G_1(x) = P\{X \ge x \mid X < a\}; \qquad G_2(y) = P\{Y \ge y \mid Y < b\}. \tag{12}$$

Proof. We will prove this assertion for the component X. Using (8), (5) and (9) we obtain

$$\begin{aligned} P\{X \ge a + x,\ Y \ge b\} &= P\{X \ge a + x\} \cdot P\{Y \ge b\} \\ &= P\{X \ge a,\ Y \ge b\} \cdot P\{X \ge x\} = P\{X \ge a\}P\{Y \ge b\}P\{X \ge x\}, \end{aligned}$$

and after reduction

$$P\{X \ge a + x\} = P\{X \ge a\} \cdot P\{X \ge x\}.$$

This is equivalent to

$$P\{X \ge a + x \mid X \ge a\} = P\{X \ge x\} \tag{13}$$

for given $a > 0$ and for any $x \ge 0$. Equation (13) is equivalent to the definition of the one-dimensional ALM property and the representation (10) follows as it is shown by

Chukova and Dimitrov (1992). The statements concerning Y are obtained analogously from (7), (5) and (9). □

Obviously, in (12) we have

$$G_1(0) = G_2(0) = 1; \qquad G_1(a) = G_2(b) = 0. \tag{14}$$

Lemma 3. If $Z \in K(a,b)$, then for any $x \in [0,b)$ and $y \in [0,b)$ and for any integers $n \geq 0$, $m \geq 0$ the equalities

$$P\{X \geq na + x, Y \geq mb\} = \alpha^n \beta^m [\alpha + (1-\alpha) G_1(x)], \tag{15}$$

and

$$P\{X \geq na, Y \geq mb + y\} = \alpha^n \beta^m [\beta + (1-\beta) G_2(y)] \tag{16}$$

hold.

Proof. Relations (15) and (16) follow immediately from (8) and (7), together with (10) and (11) respectively. □

For $x = y = 0$ we obtain:

Corollary 2. It is true that

$$P\{X \geq na, Y \geq mb\} = P\{X \geq na\} \cdot P\{Y \geq mb\} = \alpha^n \beta^m \quad \text{for } n \geq 0,\ m \geq 0.$$

Remark: It is worthy to mention that Corollary 2 is true if (7) and (8) are fulfilled only for $y = b$, $x = a$, and $n = m = 1$, i.e. for vectors Z in a larger class than $K(a,b)$. But then (15) is true only for $na + x \geq ma$ as well as (16) is true only for $mb + y \geq nb$.

Lemma 4. If $Z \in K(a,b)$, then for any $x, y (0 \leq x < a,\ 0 \leq y < b)$ and for arbitrary integers $n \geq 0$, $m \geq 0$ the equations

$$P\{X \geq na + x, Y \geq mb + y \mid X \geq na, Y \geq mb\} = P\{X \geq x, Y \geq y\} \tag{17}$$

take place.

Proof. Let $n \leq m$. By Lemma 1, equation (7), Lemma 2 and Corollary 2 we obtain

$$\frac{P\{X \geq na + x, Y \geq nb + (m-n)b + y\}}{P\{X \geq na, Y \geq nb + (m-n)b\}} = \frac{\alpha^n \beta^n P\{X \geq x, Y \geq (m-n)b + y\}}{\alpha^n \beta^n P\{Y \geq (m-n)b\}}$$

$$= \frac{P\{X \geq x, Y \geq y\} \beta^{m-n}}{\beta^{m-n}} = P\{X \geq x, Y \geq y\}.$$

Analogously we can consider the case $n \geq m$. □

Lemma 5. If $Z \in K(a,b)$, then

$$P\{X < a, Y < b\} = (1-\alpha)(1-\beta), \tag{18}$$

where α and β are defined by (2).

Proof. Using the obvious decomposition

$$\{X < a,\, Y < b\} + \{X \geq a,\, Y \geq b\} + \{X < a,\, Y \geq b\} + \{X \geq a,\, Y < b\} = \Omega$$

we obtain

$$P\{X < a,\, Y < b\} = 1 - [P\{X \geq a,\, Y \geq b\} + P\{X < a,\, Y \geq b\} + P\{X \geq a,\, Y < b\}]. \tag{19}$$

Because of (7), Lemma 2 and Corollary 2, we get

$$\begin{aligned} P\{X < a,\, Y \geq b\} &= P\{X < a \mid Y \geq b\} \cdot P\{Y \geq b\} \\ &= [1 - P\{X \geq a \mid Y \geq b\}] \cdot \beta = (1-\alpha)\beta, \end{aligned}$$

and analogously

$$P\{X \geq a,\, Y < b\} = \alpha(1-\beta).$$

Corollary 1 implies that $P\{X \geq a,\, Y \geq b\} = \alpha\beta$. Applying these results in (19) we get

$$P\{X < a,\, Y < b\} = 1 - [\alpha\beta + \beta(1-\alpha) + \alpha(1-\beta)],$$

which proves (18). □

Let $G(x, y)$ denote the conditional survival function

$$G(x,y) = P\{X \geq x,\, Y \geq y \mid X < a,\, Y < b\} \tag{20}$$

of the random vector Z. Obviously we have:

$$G(x,0) = G_X(x); \qquad G(0,y) = G_Y(y), \tag{21}$$

where G_X and G_Y are given by (10) and (11) respectively, and

$$G(x,b) = G(a,y) = G(a,b) = 0; \quad G(0,0) = 1. \tag{22}$$

Thus, the conditional survival function $G(x, y)$ defines a random vector $Z_1 = (X_1, Y_1)$, which takes values in the set $D_{a,b} := \{(x,y) : 0 \leq x < a,\, 0 \leq y < b\}$ with probability 1.

Now we formulate the main result of this section.

Theorem 1. If $Z \in K(a, b)$ then its survival function has the form

$$\begin{aligned} G_Z(x,y) &= P\{X \geq x,\, Y \geq y\} \\ &= \alpha^n\beta^m[(1-\alpha)(1-\beta)G(x-na,\, y-mb) + \beta(1-\alpha)G_X(x-na) \\ &+ \alpha(1-\beta)G_Y(y-mb) + \alpha\beta], \\ &\text{for } na \leq x < (n+1)a,\ mb \leq y < (m+1)b;\ n,m = 1,2,\ldots, \end{aligned} \tag{23}$$

where α, β being defined by (2), G_1 and G_2 by (12) and $G(x, y)$ by (20).

Proof. Let $t \geq 0$, $s \geq 0$ be arbitrary real numbers. Since $a > 0$, $b > 0$, there exist integers $n \geq 0$, $m \geq 0$ such that

$$t = na + x, \; s = mb + y \text{ and } 0 \leq x < a, \; 0 \leq y < b. \tag{24}$$

Thus $P\{X \geq s, Y \geq t\} = P\{X \geq na + x, Y \geq mb + y\}$.

The random event $\{X \geq na + x, Y \geq mb + y\}$ can be disjointly decomposed in the following way:

$$\begin{aligned} &\{(X,Y) : X \geq na + x, \; Y \geq mb + y\} \\ = &\{(X,Y) : na + x \leq X < (n+1)a, \; mb + y \leq Y < (m+1)b\} \\ +&\{(X,Y) : X \geq (n+1)a, \; mb + y \leq Y < (m+1)b\} \\ +&\{(X,Y) : na + x \leq X < (n+1)a Y \geq (m+1)b\} \\ +&\{(X,Y) : X \geq (n+1)a, \; Y \geq (m+1)b\}. \end{aligned} \tag{25}$$

1) By the total probability rule, the ALM property (Lemma 4), Corollary 1 and (20) we obtain:

$$\begin{aligned} &P\{na + x \leq X < (n+1)a, \; mb + y \leq Y < (m+1)b\} \\ &= P\{na + x \leq X < (n+1)a, \; mb + y \leq Y \\ &< (m+1)b \mid X \geq na, Y \geq mb\} \cdot P\{X \geq na, Y \geq mb\} \\ &= \alpha^n \beta^m P\{X \in [x,a), \, Y \in [y,b)\} \\ &= \alpha^n \beta^m P\{X \in [x,a), \, Y \in [y,b) \mid X < a, \, Y < b\} \cdot P\{X < a, \, Y < b\} \\ &= \alpha^n \beta^m (1-\alpha)(1-\beta) G(x,y). \end{aligned}$$

Here $G(x,y)$ is defined by (20).

2) Using Lemma 3 we obtain

$$\begin{aligned} P\{na+x \leq X < (n+1)a, \, Y \geq (m+1)b\} &= \beta^{m+1}[P\{X < (n+1)a\} - P\{X < na+x\}] \\ &= \beta^{m+1}\alpha^n(1-\alpha)G_1(x). \end{aligned}$$

3) Analogously we get

$$P\{X \geq na \cdot mb + y \leq Y < (m+1)b\} = \alpha^{n+1}\beta^m(1-\beta)G_2(y).$$

4) By Corollary 2 we get

$$P\{X \geq (n+1)a, \, Y \geq (m+1)b\} = \alpha^{n+1}\beta^{m+1}.$$

By simple algebraic manipulations it is easily seen that these results imply (23). □

By Theorem 1 the survival function of any random vector $Z \in K(a,b)$ is given. Next we show that $K(a,b)$ contains all possible solutions of equations (1), (7) and (8), when $a > 0$ and $b > 0$ are given and fixed.

Let $Z_1 = (X_1, Y_1)$ be a random vector with survival function

$$G(x,y) = P\{X_1 \geq x,\ Y_1 \geq y\}, \tag{26}$$

and satisfying the condition

$$P\{0 \leq X_1 < a,\ 0 \leq Y_1 < b\} = 1. \tag{27}$$

Then we obtain the marginal survival functions

$$G_1(x) = G(x,0);\ G_2(y) = G(0,y) \tag{28}$$

Definition 5. We say that the distribution of a random vector $Z = (X,Y)$ belongs to the class $S(a,b)$ iff the survival function $G_Z(x,y)$ is given by (23) with $G_X(x) = G_1(x)$, $G_Y(y) = G_2(y)$, where $G(x,y)$, $G_1(x), G_2(y)$, are introduced by (26) and (28) respectively and α, β are real numbers such that $0 < \alpha < 1$, $0 < \beta < 1$.

We have:

Theorem 2. If $Z \in S(a,b)$, then Z satisfies (1), (7) and (8) with the given a and b.

Proof. First of all we see from (23), that it is true

$$P\{X \geq na,\ nb\} = \alpha^n \beta^m \tag{29}$$

for any pair of integers $n \geq 0$, $m \geq 0$. Moreover, for any $x \geq 0$, $y \geq 0$ there exist two integers $n \geq 0$, $m \geq 0$ such that

$$x = na + t, \quad y = mb + s \quad \text{with} \quad t \in [0,a), \quad x \in [0,b).$$

Relation (4) is equivalent to (1), and the numerator on the left-hand side of (4) can be rewritten, using (23), in the form:

$$P\{X \geq a + x,\ Y \geq b + y\} = P\{X \geq (n+1)a + t,\ Y \geq (m+1)b + s\}$$

$$= \alpha^{n+1}\beta^{m+1}[(1-\alpha)(1-\beta)G(t,s) + \beta(1-\alpha)G_1(t) + \alpha(1-\beta)G_2(s) + \alpha\beta]. \tag{30}$$

Thus with (29) and (30) the left-hand side of (4), can finally be represented using again (23), in the following way:

$$P\{X \geq a+x,\, Y \geq b+y \mid X \geq a,\, Y \geq b\}$$

$$= \alpha^n \beta^m [(1-\alpha)(1-\beta)G(t,s) + \beta(1-\alpha)G_1(t) + \alpha(1-\beta)G_2(s) + \alpha\beta]$$

$$= P\{X \geq na+t,\, Y \geq mb+s\},$$

which proves that (1) is true. To prove that the conditions (7) and (8) of Definition 4 are satisfied we put in (23) $m = 0$ and obtain

$$G_Z(na+x, y) = P\{X \geq na+x,\, Y \geq y\}$$

$$= \alpha^n [(1-\alpha)(1-\beta)G(x,y) + \beta(1-\alpha)G_1(x) + \alpha(1-\beta)G_2(y) + \alpha\beta]$$

$$= P\{X \geq na\} \cdot P\{X \geq x,\, Y \geq y\},$$

since $0 \leq x \leq a$, $0 \leq y \leq b$, and this is (7). Analogously we get (8). □

Corollary 3. For any given $a > 0$, $b > 0$ the classes $K(a,b)$ and $S(a,b)$ coincide.

Proof. According to Theorem 1 the survival function of any $Z \in K(a,b)$ is given by (23) with α and β defined by (2) and $G(x,y)$ defined by (20) implying $Z \in S(a,b)$. Theorem 2 states that if $Z \in S(a,b)$ then according to Definition 4 we conclude $Z \in K(a,b)$. □

Remark. Using Theorem 2 it is possible to construct any distribution from $K(a,b)$. Since the ALM property

$$P\{X \geq na+x,\, Y \geq mb+y \mid X \geq na,\, Y \geq mb\} = P\{X \geq x,\, Y \geq y\} \tag{31}$$

holds for pair of any integers $n \geq 0$, $m \geq 0$, the following inclusions are true for $n \geq 1$, $m \geq 1$:

$$K\left(\frac{a}{n}, \frac{b}{m}\right) \subset K(a,b) \subset K(na, mb). \tag{32}$$

Relation (32) means that if a random vector Z has the ALM property with constants a and b, it possesses also the ALM property with constants na and mb. Thus, the representation (23) of the survival function $G_Z(x,y)$ of Z is not unique but it is one of the possible representations.

From the representation (23) we immediately obtain:

Corollary 3. (i) If $Z \in K(a,b)$ has a continuous distribution, then its probability density function is defined by:

$$f_Z(x,y) = \begin{cases} \alpha^n \beta^m (1-\alpha)(1-\beta) f_1(x-na, y-mb), & \text{for } na \leq x < (n+1)a, \\ \quad mb \leq y < (m+1)b,\ n,m = 0,1,2,\ldots & \\ 0, & \text{elsewhere.} \end{cases} \tag{33}$$

where $f_1(x,y)$ is a probability density function with support $[0,a)\times[0,b)$, satisfying

$$f_1(x,y)\begin{cases}\geq 0, & \text{for } (x,y)\in[0,a)\times[0,b);\\ =0, & \text{elsewhere,}\end{cases}\qquad \text{and}\quad \int_0^a\int_0^b f_1(x,y)\,dxdy=1;$$

(ii) If $a\geq 1$, $b\geq 1$ are integers and $Z\in S(a,b)$ has integer valued components, then the probabilities of Z are defined by

$$P\{X=k,\,Y=j\}=:p_Z(k,j)=\alpha^n\beta^m(1-\alpha)(1-\beta)p_1(k-na,\,j-mb)$$

for $na\leq k<(n+1)a$; $\quad mb\leq j<(m+1)b$,

with $p_1(k,j)\geq 0$; $\quad \sum_{k=0}^{a-1}\sum_{j=0}^{b-1}p_1(k,j)=1$ and $0<\alpha<1$; $0<\beta<1$. (34)

Thus the construction of a random vector $Z\in K(a,b)$ is possible by means of a probability distribution on $[0,a)\times[0,b)$ and two arbitrary real numbers α,β with $0<\alpha<1$ and $0<\beta<1$ respectively.

Theorem 3. Let $Z=(X,Y)\in K(a,b)$ then there exist two independent random vectors $Z_1=(X_1,Y_1)$ and $Z_{a,b}=(X_a,Y_b)$ so that the equation

$$Z=Z_1+Z_{a,b}\tag{35}$$

holds. Here Z_1 is given by its survival function $G(x,y)$, defined by (26) and (27) and $Z_{a,b}$ has also two independent components X_a and Y_b, being geometrically distributed over the sets $\{0,a,2a,\ldots\}$ and $\{0,b,2b,\ldots\}$, with parameters α and β respectively.

Proof. Since we have the form (23) of the survival function for Z, it is easy to calculate its bivariate Laplace-Stieltjes transform:

$$\varphi_Z(\mu,\nu)=E[\exp\{-(\mu,\nu)(X,Y)\}]=\int_0^\infty\int_0^\infty e^{-\mu x-\nu y}dG_Z(x,y)$$

$$=\sum_{n=0}^\infty\sum_{m=0}^\infty\alpha^n\beta^m(1-\alpha)(1-\beta)\cdot\int_{na}^{(n+1)a}\int_{mb}^{(m+1)b}e^{-\mu x-\nu y}dG_Z(x,y)$$

$$=(1-\alpha)(1-\beta)\sum_{n=0}^\infty\sum_{m=0}^\infty(\alpha e^{-\mu a})^n(\beta e^{-\nu b})^m\int_0^{a-0}\int_0^{b-0}e^{-\mu x-\nu y}dG(x,y)$$

$$=\varphi_{Z_1}(\mu,\nu)\cdot\frac{1-\alpha}{1-\alpha\cdot\exp(-\mu a)}\frac{1-\beta}{1-\beta\cdot\exp(-\nu b)}=\varphi_{Z_1}(\mu,\nu)\cdot\varphi_{Z_{a,b}}(\mu,\nu).$$

Hence, $\varphi_Z(\mu,\nu)$ is a product of the Laplace-Stiltjes transforms of the random vectors Z_1 and $Z_{a,b}$, described in the theorem. Thus, Z_1 and $Z_{a,b}$ are independent and the decomposition (35) holds. □

The above results explain the probability structure of bivariate vectors in the class $K(a,b)$ having the ALM property. Possible applications of bivariate and multivariate random vectors having the lack-of-memory property are given by Marshall and Olkin (1967), or by Galambos and Kotz (1978).

4.1 Contorted bivariate exponential distribution. The bivariate exponential distribution with independent exponential components of parameters λ_1 and λ_2 has the ALM property (1) and satisfies the requirements of Definition 4 for any pair of constants $a > 0$ and $b > 0$. It is defined by a survival function of the form (23) with

$$G_Z(x,y) = \frac{(e^{-\lambda_1 x} - e^{-\lambda_1 a})(e^{-\lambda_2 y} - e^{-\lambda_2 b})}{(1-e^{-\lambda_1 a})(1-e^{-\lambda_2 b})}, \text{ for } x \in [0,a),\ y \in [0,b),$$

and with $\alpha = \exp(-\lambda_1 a)$; $\beta = \exp(-\lambda_2 b)$.

Referring to Theorem 3, Z is representable in the form (35) as a sum of two independent random vectors $Z_1 = (X_1, Y_1)$ concentrated on $[0,a) \times [0,b)$ and $Z_{a,b} = (X_a, Y_b)$ with independent geometrically distributed components with parameters α and β on the sets $\{0, a, 2a, \ldots\}$ and $\{0, b, 2b, \ldots\}$ respectively. Any other choice of α and β (say $\alpha \neq \exp(-\lambda_1 a)$ or $\beta \neq \exp(-\lambda_2 b)$) yields continuous bivariate distributions from the class $K(a,b)$ whch are not exponential. We call these the contorted bivariate exponential distributions which can be suitable for some shock reliability models.

4.2 Contorted bivariate geometric random vectors. Let $Z = (X,Y)$ be an integer valued random vector with probability distribution determined by (34) where

$$P_1\{X = k,\ Y = 1\} = \frac{(1-\gamma)\gamma^k(1-\delta)\delta^j}{(1-\gamma^a)(1-\delta^b)}; \quad k = 0,1,\ldots,a-1; \quad j = 0,1,\ldots,b-1,$$

with γ and δ chosen arbitrary but fixed numbers from the interval $(0,1)$; a and b chosen also arbitrary but fixed positive integers. If one puts $\alpha = \gamma^a$, $\beta = \delta^b$ in (34), one gets that the distribution of Z is

$$P_Z(na+k,\ mb+j) = \gamma^{na+k}\delta^{mb+j}, \text{ where } k = 0,1,\ldots,a-1;\ j = 0,1,\ldots,b-1$$
$$\text{and } n,m = 0,1,2,\ldots,$$

i.e. Z has bivariate geometric distribution with independent components. According to Theorem 3, the random vector Z is decomposable into the form (35) as a sum of two independent random vectors $Z_1 = (X_1, Y_1)$ and $Z_{a,b} = (X_a, Y_b)$. Any other choice of α and β distinct from γ^a and δ^b yields a contorted bivariate geometric distribution.

4.3 Use of the uniform distribution. Let $a > 0$, $b > 0$ be given and fixed numbers. Let $Z_1 = (X_1, Y_1)$ have bivariate uniform distribution defined by the survival function

$$G(x,y) = (a-x)(b-y)/(ab); \ 0 \le x \le a,\ 0 \le y \le b,$$

Then for any $\alpha \in (0,1)$ and $\beta \in (0,1)$ the density function

$$f_Z(x,y) = \alpha^n \beta^m (1-\alpha)(1-\beta)/(ab); \quad x \in [na,(n+1)a),\ y \in [mb,(m+1)b)$$

$n, m = 0, 1, 2, \ldots$ yields a continuous bivariate distribution, which has the ALM property and is not exponential.

Remark: In the same way we can use other known distribution, to construct bivariate random vectors having the ALM property and with almost independent components in the sense of Definition 4, according to Theorem 1.

ACKNOWLEDGEMENTS

We acknowledge the support of the Natural Sciences and Engineering Research Council of Canada, Grant 9095.

REFERENCES

Chukova, S., and Dimitrov, B., 1992, On the distributions having almost lack-of-memory property, *J. Appl. Prob.*, **29**, No. 3, 691-698.

Chukova, S., Dimitrov, B., and Khalil, Z., 1993, Probability distributions similar to the exponential, *Canadian J. of Statistics*(to appear).

Galambos, J., and Kotz, S., 1978, *"Characterization of Probability Distributions"*, Lecture Notes in Math. **675**, Springer-Verlag, Berlin, Heidelberg, New York.

Marshall, A.W., and Olkin, I., 1967, A multivariate exponential distribution, *J. Amer. Statist. Assoc.*, **62**, 30-44.

PROBABILITY, WAITING-TIME RESULTS FOR PATTERN AND FREQUENCY QUOTAS IN THE SAME INVERSE SAMPLING PROBLEM VIA THE DIRICHLET*

M. Ebneshahrashoob[1] and Milton Sobel[2]

[1]Department of Mathematics
California State University
Long Beach, CA 90840, USA
[2]Department of Statistics and Applied Probability
University of California
Santa Barbara, CA 93106-3110, USA

ABSTRACT

Several papers have been written recently on the study of probabilities and waiting-time questions associated with inverse sampling schemes in a multinomial setting. The stopping rule used has been either reaching some frequency (or quota) in a particular cell and/or reaching some run of length r in some other cell disjoint from the former cell. In some cases we considered frequency quotas for some cells and run quotas for other cells in the same problem and stopped (for example) with the soonest of these events. In the present paper we replace the run quota by its natural (but by no means trivial) generalization: the attainment of some preassigned pattern. The pattern problem has been studied by different authors, most notably by Feller [2], but none of these authors have combined different patterns and frequency quotas in the same problem or have given explicit formulas for either case. Hence our general results can be checked against published results for special cases that are already in the literature. In this paper, we allow several different pattern quotas and/or several frequency quotas in the same problem, provided they each involve disjoint cells; one (so-called) slack cell without any quota is also included. Explicit, simple, exact formulas are given in most cases. The stopping rule of principal interest is to wait for the soonest of all quotas; the Dirichlet integral is the principal tool used in the solution. Our formulation also allows several quota-free cells to be present and our results include the derivation of the probability generating function (pgf) and joint moments for the frequencies of quota-free cells at stopping time. It is interesting to note that in the Case 3 (described below) the correlation of frequencies is positive for all pairs of cell frequencies at stopping time, unlike the usual multinomial structure (with fixed sample size) where the answer is always negative.

Approximation, Probability, and Related Fields, Edited by
G. Anastassiou and S.T. Rachev, Plenum Press, New York 1994

1. INTRODUCTION

In this paper we consider independent trials in a multinomial setting and wish to generalize the idea of a run of observations in some particular cell and replace the "run" by a preassigned "pattern" (or by a collection of patterns) with given structure and length. These patterns each apply to disjoint subsets of a fixed total collection of cells; let $q_1^{(1)}, q_2^{(1)}, \ldots, q_\beta^{(1)}$, denote the cell probabilities for the cells associated with the 1st pattern, ..., and let $q_1^{(h)}, q_2^{(h)}, \ldots, q_{\beta_h}^{(h)}$ denote the cell probabilities for the cells associated with the last (or hth) pattern. Each pattern is regarded as a type of quota. We also wish simultaneously to have frequency quotas for a subset of cells disjoint from those considered for patterns. In addition we allow one or more quota-free (QF) cells and if there is only one we also refer to it as a slack cell. The name QF will also indicate that we wish to study the frequency at stopping time as well as the correlation between two such frequencies at stopping time. Our principal interest is in the "soonest" stopping rule, i.e., to stop as soon as any one quota (frequency or pattern) is satisfied.

The concept of a "pattern" is somewhat subtle and will be understood better by giving a few examples. Suppose that in a multinomial setting the cells numbered 5, 6, 7, and 8 are set aside for pattern quotas. We could ask for the pattern 5,5,6,6,5,5, for example, which uses only two of these four cells. Simultaneously we could have a run quota of length 3 for cell #7 and no quota for cell #8. Any pattern is satisfied if it occurs once and hence we are not concerned with the question of overlapping patterns. Moreover, with two patterns, each one has to be prespecified and they must apply to disjoint subsets of cells. It should be noted that the requirement of having at least 3 heads in any "window" of length 4 tosses is not a specified pattern (although it could be regarded as a generalized pattern). In a sequence of Bernoulli trials there are only two possibilities for each observation and hence in any one problem only one pattern can be considered that uses both H and T, or else they would not be disjoint; thus the pattern HTHTH is a possible pattern for the Bernoulli case. In order to have more than one pattern in this Bernoulli case, we must specify 2 run (pattern) quotas; for example, we could stop if we observe a run or r_1 heads or a run of r_2 tails, whichever comes sooner.

In some cases we can recombine cells and a "non-pattern" problem may become a pattern problem. Thus if we wait for a run of length 4 of either cell #2 or cell #3, viz. 2,2,3,2 or 3,3,2,2 or etc., then by combining cells 2 and 3 this becomes a run pattern of length 4.

The Dirichlet integral which was used in previous work on this subject is especially useful for frequency quotas [4], but was also found to be applicable to inverse sampling when we used frequency and run quotas in the same problem. Here again we bring in the Dirichlet integral to help us get simple attainable results (i.e., results that can be calculated) since tables and algorithms for these integrals have been made available in [5] and [6]. Whenever we can obtain a simple result without Dirichlet integrals, we do so.

2. BREAKDOWN OF THE GENERAL CASE INTO THREE GENERAL SUBCASES

Although we start with a general formulation at the outset, with both pattern and frequency quotas present (but no slack cell), we wish to break this down fairly soon into several special cases in which the results take on a fairly simple form. These special cases are not numerical examples but are again general subcases. Thus for Case 1 we allow (only) 1 pattern quota, 1 slack cell and an arbitrary number α of frequency quotas, all in disjoint cells. In Case 2 we consider the soonest of h disjoint patterns (without frequency quotas and with one slack cell); here $h = 2$ is a special subcase which checks with previous results for the sooner of two runs. Case 3 deals with γ quota-free cells and considers one pattern to be present but no frequency quotas; here the interest is to study the effect of stopping rules on the moments for frequencies at stopping time of quota-free cells. Many modifications of these cases are possible. The effect of any modification on the resulting formulas is straightforward but the details have to be worked out.

The concept of a generalized probability generating function (gpgf) is that of a pgf which introduces markers x, y such that $x = 1$, $y = 1$ gives the usual pgf for the sooner problem (x goes with frequency and y with pattern), the values $x = 1$, $y = 0$ gives the conditional pgf given that we stop with the frequency quota and $x = 0$, $y = 1$ gives the conditional pgf given that we stop with the pattern. For several frequency quotas we use a vector marker $\underset{\sim}{x}$ and similarly $\underset{\sim}{y}$ for several pattern quotas; this was used extensively in [4]; these concepts and notations are also used in the present paper.

One way of writing the results for our general problem is in the form of a multiple summation formula similar to that given on p. 160 of [4]. This is a basic long form that is difficult and tedious to use for calculations; the alternative is to use Dirichlet integrals and simultaneously reduce the multiple sums to single summations. Since these integrals are available as tables and algorithms (cf. [5] and [6]), the reduction in complexity is much more than just notational simplification. The formal definition of a Dirichlet integral was given in [5], [6] and also in [4] and need not be repeated again. Without the Dirichlet methodology many of these problems would be extremely difficult, if not impossible, to solve. In those (simple) cases where the multiple sums are computable, we use the two methods as a check, one upon the other. In general we wish to omit the result as a lengthy multiple summation.

2.1 General Case Model

Consider a sequence of independent trials in a multinomial setting with the following structure: there are α cells each associated with the frequency quotas with cell probabilities $\underset{\sim}{p} = (p_1, p_2, \ldots, p_\alpha)$ and quota vector $\underset{\sim}{f} = (f_1, f_2, \ldots, f_\alpha)$; there are h patterns, each associated with disjoint subsets of cells, with a prespecified pattern quota for the ith subset ($i = 1, 2, \ldots, h$) (all disjoint from the α frequency quota cells) with cell probabilities q_{ij} for the jth cell in the ith subset ($j = 1, 2, \ldots, \beta_i$; $i = 1, 2, \ldots, h$). Thus (without the slack cell) we have

$$\sum_{i=1}^{\alpha} p_i + \sum_{i=1}^{h} \sum_{j=1}^{\beta_i} q_{ij} = 1. \tag{2.1}$$

For any one pattern (disregarding the other patterns and frequency quotas), the expected waiting time $E(WT)$ is denoted by $\frac{u}{\ell}$ where u (for numerator or upper) and ℓ (for denominator or lower) are both functions of the q's for that particular pattern. For this we follow a method outlined by S.R. Li in [3]. For example, in Bernoulli trials, if we are waiting for the pattern HHTTHH with q_H and q_T for the probabilities of H and T, respectively, then by the matching method in [3] we have

$$E(WT|\text{for HHTTHH}) = \frac{1}{q_H^4 q_T^2} + \frac{1}{q_H^2}. \tag{2.2}$$

and writing this as a single fraction in simplest form we obtain

$$u = 1 + q_H^2 q_T^2, \qquad \ell = q_H^4 q_T^2. \tag{2.3}$$

Whenever we replace each q by qt (where t is the pgf variable) in the analysis below, then we will use the same notation as above with capitals, i.e., U for u and L for ℓ. In the general case where there are h patterns we will use U_i and L_i for the ith pattern

$(i = 1, 2, \ldots, h)$. In addition we need the notation

$$U^* = \prod_{i=1}^{h} U_i; \; L_i^* = \frac{L_i}{U_i} U^*, \; i = 1, 2, \ldots, h; \; L^* = \sum_{i=1}^{h} L_i^*;$$
$$\underset{\sim i}{U} = (U_1, U_2, \ldots, U_{i-1}, U_{i+1}, \ldots, U_h); \; \underset{\sim i}{p} = (p_1, p_2, \ldots, p_{i-1}, p_{i+1}, \ldots, p_\alpha); \tag{2.4}$$

for $t = 1$, all capital letters in (2.4) are replaced with the respective small letters. We also use the Dirichlet integral notation

$$D_{\underset{\sim}{a}}^{(b)}(\underset{\sim}{r}, m) = \frac{\Gamma(m+R)}{\Gamma(m)\prod_{i=1}^{b}\Gamma(r_i)} \int_{a_1}^{\infty} \cdots \int_{a_b}^{\infty} \frac{\prod_{i=1}^{b} x_i^{r_i - 1} dx_i}{\left(1 + \sum_{i=1}^{b} x_i\right)^{m+R}}, \; \left(R = \sum_{i=1}^{b} r_i\right) \tag{2.5}$$

as in (4.1b) with $j = 0$ on page 38 of [6], but we have to substitute for the arguments, $\underset{\sim}{r}$ and m.

For the soonest problem with h patterns and α frequency quotas, all associated with disjoint cells, the gpgf can now be written as

$$\varphi(t; \underset{\sim}{x}, \underset{\sim}{y}) = \sum_{i=1}^{\alpha} x_i \left[\frac{p_i t U^*}{L^* + (1 - t + p_i t) U^*}\right]^{f_i} D^{(\alpha-1)}_{\left(\frac{\underset{\sim i}{p}\, tU^*}{L^* + (1-t+p_i t)U^*}\right)} \left(\underset{\sim i}{L}; f_i\right)$$
$$+ \left[\sum_{i=1}^{h} y_i \left(\frac{L_i^*}{L^* + (1-t)U^*}\right)\right] D^{(\alpha)}_{\left(\frac{\underset{\sim}{p}\, tU^*}{L^* + (1-t)U^*}\right)} \left(\underset{\sim}{L}; 1\right). \tag{2.6}$$

With the slack cell (with cell probability p_0) present the only change in (2.6) is to replace $1 - t$ wherever it appears (in four places) by $(1 - t)(1 - p_0)$. For $t = 1$ and $\underset{\sim}{x} = \underset{\sim}{1}$, $\underset{\sim}{y} = \underset{\sim}{1}$ we note that (2.6) reduces to one, i.e.,

$$\varphi(1; \underset{\sim}{1}, \underset{\sim}{1}) = \sum_{i=1}^{\alpha} \left(\frac{p_i u^*}{\ell^* + p_i u^*}\right)^{f_i} D^{(\alpha-1)}_{\left(\frac{\underset{\sim i}{p}\, u^*}{\ell^* + p_i u^*}\right)} \left(\underset{\sim i}{L}; f_i\right) + D^{(\alpha)}_{\frac{\underset{\sim}{p}\, u^*}{\ell^*}} \left(\underset{\sim}{L}; 1\right) = 1. \tag{2.7}$$

It is easily seen that the initial sum in (2.7) is the probability that some frequency quota is satisfied before any pattern quota and the second Dirichlet term is the probability that one of the patterns occurs before any frequency quota; hence the total has to be one. We shall check this identity in several special cases below; this also checks our analysis. Analytic methods for Dirichlet integrals help us to differentiate the gpgf in (2.6) and hence we can obtain all moments of the WT from (2.6).

Case 1: <u>One Pattern, Any number of Frequency Quotas and 1 Slack Cell</u>

In this case $h = 1$ pattern and we use β (the size of the subset for this one pattern) in place of β_1 and q_j instead of q_{ij}; we denote the probability of the slack cell by p_0, so that (with α frequency quotas and with $\beta + 1$ additional cells) we have

$$\sum_{i=1}^{\alpha} p_i + p_0 + \sum_{j=1}^{\beta} q_j = 1. \tag{2.8}$$

Here the result (2.6), with slack cell included, reduces to

$$\varphi(t; \underset{\sim}{x}, y) = \sum_{i=1}^{\alpha} x_i \left(\frac{p_i t U}{L + [(1-p_0)(1-t) + p_i t]U}\right)^{f_i} D^{(\alpha-1)}_{\left(\frac{\underset{\sim i}{p}\, tU}{L + [(1-p_0)(1-t) + p_i t]U}\right)} \left(\underset{\sim i}{L}; f_i\right)$$

$$+y\left(\frac{L}{L+(1-p_0)(1-t)U}\right)D^{(\alpha)}_{\left(\frac{\underset{\sim}{p}\, tU}{L+(1-p_0)(1-t)U}\right)}\left(\underset{\sim}{f};1\right). \tag{2.9}$$

Illustration 1.1 As a numerical illustration for Case 1, consider the one pattern 1212 from the 4 cells marked 1, 2, 3 and 4 on a single, fair 6-sided die. Simultaneously we have a frequency quota $f_1 = 3$ for cell #5 and cell #6 is a slack cell. We want the mean, variance and gpgf for the waiting time for the sooner of the two quotas.

From (2.9) with $\alpha = 1$ the matching method of [3] yields

$$U = 1+\frac{t^2}{6^2},\ L = \frac{t^4}{6^4},\ u = \frac{37}{36},\ \ell = \frac{1}{1296} \text{ and}$$

$$\varphi(t;x,y) = x\left(\frac{216t+6t^3}{1080-864t+30t^2-24t^3+t^4}\right)^3 \tag{2.10}$$

$$+y\left(\frac{t^4}{1080-1080t+30t^2-30t^3+t^4}\right)D^{(1)}_{\left(\frac{216t+6t^3}{1080-1080t+30t^2-30t^3+t^4}\right)}(3;1)$$

and as a check we find that $\varphi(1^*,1,1) = \left(\frac{222}{223}\right)^3 + D^{(1)}_{222}(3;1) = 1$. From (2.10) by differentiation with respect to t and setting $x = y = t = 1$, we obtain

$$E(WT) = 14.86587. \tag{2.11}$$

The second derivative with $x = y = t = 1$ gives $E\{WT(WT-1)\}$ which yields

$$\sigma^2(WT) = 58.89636 \tag{2.12}$$

Actually if we go back to (2.9) with $\alpha = 1$ we can differentiate in (2.9) for any illustration of this type ($\alpha = h = 1$) and obtain in terms of u, ℓ, f, p_0 and p the simple result

$$E(WT) = (1-p_0)\frac{u}{\ell}D^{(1)}_{pu/\ell}(f;1) \tag{2.13}$$

and, letting $Q = \ell\left(\frac{\partial U}{\partial t}\Big|_{t=1}\right) - u\left(\frac{\partial L}{\partial t}\Big|_{t=1}\right)$, and using (2.13),

$$\sigma^2(WT) = \left(\frac{2Q+\ell u+2(1-p_0)u^2}{\ell u}\right)E(WT) - [E(WT)]^2$$

$$-\frac{2f(1-p_0)}{\ell(\ell+pu)}\left(\frac{pu}{\ell+pu}\right)^f\{Q+u[\ell+(1-p_0)u]\}. \tag{2.14}$$

The numerical values (2.11) and (2.12) can be obtained from (2.13) and (2.14), respectively, by setting $f = 3$, $p_0 = p = 1/6$ and using the results in (2.10). If there is no slack cell present, we simply set $p_0 = 0$.

Another illustration is to compare two different problems of the same type as above each with one frequency quota 4 for cell #6 (one fair 6-sided die) and each with one pattern: For problem 1 we use the pattern 121 from cells #1 through 5 and for problem 2 we use the pattern 123 from the same five cells. Using (2.13) and (2.14) with $U = 1+t^2/6^2$, $L = t^3/6^3$ for problem 1 and with $U = 1$, $L = t^3/6^3$ for problem 2, we obtain

$$E(WT|\text{Problem 1}) = 22.46206,\ \text{Var}(WT|\text{Problem 1} = 114.70062, \tag{2.15}$$
$$E(WT|\text{Problem 2}) = 22.42162,\ \text{Var}(WT|\text{Problem 2} = 114.37742. \tag{2.16}$$

These numerical results are very close and it is difficult to obtain intuitive results for them.

Case 2: The Number of Patterns is h with One Slack Cell and No Frequency Quotas

Using the same notation U, L, u, ℓ introduced above, we define

$$S = \sum_{i=1}^{h} \frac{L_i}{U_i} \quad \text{and} \quad s = \sum_{i=1}^{h} \frac{\ell_i}{u_i}, \tag{2.17}$$

the gpgf takes the simple form

$$\varphi(t; \underset{\sim}{y}) = \frac{\sum_{i=1}^{h} \frac{L_i}{U_i} y_i}{S + (1-p_0)(1-t)} \tag{2.18}$$

which is clearly equal to 1 for $t = 1$ and $\underset{\sim}{y} = \underset{\sim}{1}$. For the expectation and variance we differentiate with respect to t and set $\underset{\sim}{y} = \underset{\sim}{1}$ and $t = 1$ obtaining

$$E(WT) = \frac{1-p_0}{s} \tag{2.19}$$

$$\sigma^2(WT) = 2(1-p_0)\left[\frac{\partial}{\partial t}\left(\frac{1}{S}\right)\Big|_{t=1}\right] + [E(WT)]^2 + E(WT). \tag{2.20}$$

For $h = 2$ these results coincide with results (eq.'s (3) and (8) in [1]) in a previous paper for the sooner of 2 runs on disjoint cells. For $h = 1$ they also agree with the results of Case 1 above if we set $\alpha = 0$ therein.

Illustration 2.1 With a single fair 6-sided die and $h = 6$ and $p_0 = 0$ we consider stopping as soon as we have a run of prescribed length ℓ on any one of the six faces. Our results will be given as a table with ℓ varying ($\ell = 1, 2, \ldots$).

The values of U_i, L_i, S and s for this example are

$$U_i = \frac{1-(t/6)^\ell}{1-t/6},\ L_i = (t/6)^\ell\ (i = 1, 2, \ldots, 6),\ S = \frac{6(t/6)^\ell(1-t/6)}{1-(t/6)^\ell},\ s = \frac{5}{6^\ell - 1}. \tag{2.21}$$

The gpgf in (2.18) takes the form

$$\varphi(t; \underset{\sim}{y}) = \frac{\left(\frac{t}{6}\right)^\ell \left(1-\frac{t}{6}\right) \sum_{i=1}^{6} y_i}{6\left(\frac{t}{6}\right)^\ell \left(1-\frac{t}{6}\right) + (1-t)\left[1-\left(\frac{t}{6}\right)^\ell\right]}, \tag{2.22}$$

and the mean and variance in (2.19) and (2.20) respectively become

$$E(WT) = \frac{6^\ell - 1}{5}, \quad \sigma^2(WT) = \frac{6^{2\ell} - 5(2\ell-1)6^\ell - 6}{25}. \tag{2.23}$$

In table for these results are

ℓ	$E(WT)$	$\sigma^2(WT)$
1	1	0
2	7	30
3	43	1,650
4	259	65,370
5	1,555	2,404,650
6	9,331	86,968,650

Algebraically it can be seen that both $E(WT)$ and $\sigma^2(WT)$ are integers for all ℓ.

Case 3. The Number of Quota-Free Cells is γ, One Pattern ($h = 1$), No Frequency Quota.

Let $p_{01}, p_{02}, \ldots, p_{0\gamma}$ denote the cell probabilities of the γ quota-free (QF) cells and let $q_1, q_2, \ldots, q_\beta$ denote the cell probabilities associated with the cells from which our single pattern is specified, so that

$$\sum_{i=1}^{\gamma} p_{0i} + \sum_{i=1}^{\beta} q_j = 1. \tag{2.24}$$

Let X_i denote the frequency of the ith QF cell at stopping time determined only by the pattern ($i = 1, 2, \ldots, \gamma$). We are concerned with moments, covariances and correlations for the frequencies of the QF cells at stopping time.

From the reference [3] we determine u and ℓ as before. Let $\underset{\sim}{t} = (t_1, t_2, \ldots, t_\gamma)$. The pgf (without markers) is given by

$$\varphi(\underset{\sim}{t}) = \frac{1}{1 + \frac{u}{\ell}\sum_{i=1}^{\gamma} p_{0i}(1 - t_i)} \tag{2.25}$$

By differentiating with respect to t_i and setting $\underset{\sim}{t} = \underset{\sim}{1}$, we obtain

$$E(X_i) = p_{0i}\frac{u}{\ell}, \ \sigma^2(X_i) = [E(X_i)][1 + E(X_i)] \quad (i = 1, 2, \ldots, \gamma) \tag{2.26}$$

and the covariances and correlations are given by

$$\mathrm{Cov}(X_i, X_j) = [E(X_i)][E(X_j)] > 0 \ \text{ for all pairs } (i, j, i \neq j), \tag{2.27}$$

$$\rho(X_i, X_j) = \left\{\frac{E(X_i)E(X_j)}{[1 + E(X_i)][1 + E(X_j)]}\right\}^{1/2} > 0 \ \text{ for all pairs } (i, j, i \neq j) \tag{2.28}$$

For the case of h patterns with $h \geq 1$ the results are quite similar. In fact, all the equations above remain the same and only the pgf in (2.25) and the expectation in (2.26) have to be generalized. The only change is to replace $\frac{u}{\ell}$ by $\frac{1}{s}$ where s is defined in (2.17).

It is quite remarkable that the structure of the QF cells is so different from that in the usual multinomial case; in the fixed sample size multinomial the pairwise covariances and correlations are all negative but here they are all positive!

Illustration 3.1 With a single fair 6-sided die and a single ($h = 1$) pattern of the form 1223334444, we let cells #5 and #6 be the $\gamma = 2$ quota-free (QF) cells and wish to find their moments, covariance and correlation at stopping time.

From (2.25) the value of the pgf is given by

$$\varphi(t_1, t_2) = \frac{6^{11}}{6^{11} + 2 - t_1 - t_2}. \tag{2.29}$$

By differentiation with respect to t_i, we obtain from (2.26), (2.27) and (2.28)

$$E(X_i) = 6^9, \ \sigma^2(X_i) = 6^9(1 + 6^9) \ (i = 1, 2) \tag{2.30}$$

$$\mathrm{Cov}(X_i, X_j) = 6^{18}, \ \rho(X_i, X_j) = \frac{6^9}{1 + 6^9} = .9999999. \tag{2.31}$$

For any value of the fixed sample size N in the usual multinomial problem the correlation would be exactly $\rho = -.2$ for all N, but in our problem $E(WT) = \frac{2}{3}(6^{10})$ is very large and the correlation is extremely close to one.

REFERENCES

[1] Ebneshahrashoob, M. and Milton Sobel (1990). Sooner and later waiting time problems for Bernoulli trials: frequency and run quotas. *Statist. Probab. Lett.* 9, 5–11.

[2] Feller, W. (1957). An Introduction to Probability Theory and its Applications Vol. 1, John Wiley, New York [pp. 303–304].

[3] Li, Shuo-Yen R. (1980). A martingale approach to the study of occurrence of sequence patterns in repeated experiments. *Ann. of Probab.* 8, 1171–1176.

[4] Sobel, Milton and M. Ebneshahrashoob (1992). Quota sampling for multinomial via Dirichlet. *Jour. Statist. Planning & Inference* 33, 154–164.

[5] Sobel, Milton, V.R.R. Uppuluri and K. Frankowski (1977). Dirichlet distributions—type 1. *Selected Tables in Mathematical Statistics* Vol. 4, IMS and AMS, Providence, R.I.

[6] Sobel, Milton, V.R.R. Uppuluri and K. Frankowski (1985). Dirichlet integrals of type 2 and their application. *Selected Tables in Mathematical Statistics* Vol. 9, IMS & AMS, Providence, R.I.

ASYMPTOTIC NORMALITY OF THE COLLISION RESOLUTION INTERVAL FOR A MULTIPLE ACCESS PROTOCOL

Phillip M. Feldman

Illgen Simulation Technologies, Inc.
250 Storke Rd., Suite 10
Goleta, CA. 93117
and
Dept. of Computer Science
University of California at Santa Barbara
Santa Barbara, CA. 93106

ABSTRACT

The CTM protocol permits an unlimited number of users to share a single slotted communications channel. A collision occurs when two or more users transmit messages in the same slot. Users who collide re-transmit their messages by randomly splitting into subgroups until each subgroup contains at most one message. L(n), the number of slots required to resolve a collision among n users, is a random variable whose asymptotic properties are of considerable interest. Under fairly general conditions, the asymptotic distribution of L(n) is normal. We briefly review the early research on this problem, and then present some newer results.

INTRODUCTION

Multiple Access Communications

In a multiple access system, users transmit messages to one another using a shared communications medium. This shared medium could take any of several forms. Here are three examples of multiple access systems: (1) In a mobile satellite system, ground terminals transmit to a satellite using one frequency band; the satellite re-broadcasts the signal (with some delay) using a different frequency band. All terminals (including the sender) can hear the re-broadcast. (2) In a Local Area Network (LAN), two or more computers are connected to a common bus (coaxial cable or fiber optic cable). If one computer transmits a message, all other computers hear the transmission after a delay τ. (3)

In a packet radio system for ground vehicles, a user's transmission can be received by other users within some range which depends on the transmit power level and other factors.

Although the shared medium is quite different in the above examples, a common feature of all three systems is that when any single user transmits, many other users can receive the message. In the mobile satellite system, users receive all transmissions, including their own transmissions. In the LAN, users receive all transmissions except their own. In the packet radio system, in general, each user can receive the transmissions of some subset of the user population (this subset is different for each receiver, and "A can hear B" does not imply "B can hear A"). For any of these systems, suppose that user A can receive transmissions from B and C, and that B and C transmit messages during overlapping time intervals; this event, which is called a *collision*, results in the loss of both messages, i.e., nothing intelligible is received.[1] Over the last few decades, a variety of multiple access communications protocols have been proposed. A multiple access protocol is a set of rules which specifies when users are allowed to transmit, what action is to be taken when a collision occurs, and so forth.

One of the earliest and simplest protocols, called Frequency Division Multiple Access (FDMA), divides the available frequency bandwidth into several channels, such that each user who needs to transmit is assigned a unique channel. Simultaneous transmissions in different channels do not collide. A somewhat more flexible scheme, called Time Division Multiple Access, divides time into intervals called *frames*. Each frame is further divided into N *slots*, where N is greater than or equal to the number of users in the system. A user who needs to transmit is assigned (via some scheduling procedure) one or more slots within each frame, and is permitted to transmit only during these slots. More slots can be allocated to a user who generates messages at a higher average rate. FDMA and TDMA are simple schemes which prevent all collisions. If the number of users in the system is small, and all users generate messages at a fairly steady and predictable rate, these schemes can be very efficient.

Because "efficient" is a rather vague term, we define two specific measures of protocol performance: *Throughput* is the long-term average fraction of slots which contain exactly one transmission, and *delay* is the difference between the time when a message is received and the time when it is was generated. For TDMA, throughput can approach unity under heavy-load conditions, i.e., where all users constantly have messages to send. Many multiple access communications systems are characterized by the following conditions:

(1) The number of users is large, and the user population varies (users enter and leave the system).
(2) Individual users generate message in a bursty (unpredictable and sporadic) fashion, such that the average message generation rate for any user is small compared to the capacity (data rate) of the medium.

Protocols like FDMA and TDMA, which allocate a fixed fraction of the capacity to each user, are not well suited here. To see why, consider the operation of TDMA under these conditions. Each user must be allocated at least one slot in each frame (assume for simplicity that all frames have the same structure). Thus, frames will be long, and most of the slots in each frame will be empty. Furthermore, users will experience a mean delay of

[1] In some systems, the more powerful of two signals will "capture" the receiver and be correctly received; we ignore this possibility here.

at least 1/2 frame under even the lightest traffic load. Another disadvantage of FDMA and TDMA is that they require centralized control, i.e., a master node which allocates resources. For systems with large numbers of users and/or bursty message generation, it is desirable to have a protocol which does not allocate resources to specific users.

In what follows, we consider only protocols in which all users share a single frequency band and use the entire band for the duration of each transmission. For such protocols, any simultaneous (or overlapping) transmissions will collide and be lost. This is not true for FDMA.. Also excluded are schemes such as Frequency Hop Code Division Multiple Access (FH-CDMA). Some FH-CDMA systems permit a limited amount of interference between users; this interference tends to produce random errors in the received messages, but does not necessarily cause the loss of messages which are transmitted simultaneously.

Before discussing alternatives to FDMA and TDMA, we present a specific model for message generation and channel behavior; this model, although quite simple, provides a convenient basis for comparing different protocols.

The Classical Multiple Access Model

In the classical multiple access model (Ref. Bertsekas), a large (potentially infinite) population of users share a single slotted communications channel. New messages are generated according to a (stationary) Poisson process with rate λ. All messages have equal length, and the time required to transmit a message is less than the duration of a slot. All messages generated during slot k are transmitted at the beginning of slot $k+1$. During each slot, one of the following events occurs:

1. No one transmits. In this case, the slot is wasted.
2. Exactly one user transmits, in which case the message is successfully received.
3. Two or more users transmit, interfering with each other. In this case, there is a collision and the slot is wasted.

At end of each slot, every user knows which of these three events occurred; this is sometimes called *trinary feedback.*

Consider the three examples of multiple access systems mentioned at the beginning of the paper. The classical model applies to the first two of these (a user's ability to detect a collision does not depend on being able to hear his own transmission). This model does not apply to the packet radio system for the following reason: Suppose that A can receive from B and C, but that B and C cannot receive from each other. Overlapping transmissions from B and C will collide at A, i.e., A does not receive either message, but neither B nor C is aware that a collision occurred.

Although this model specifies that new messages are transmitted immediately, the handling of collisions is not specified. As we shall see, different protocols employ different mechanisms for resolving collisions.

There are many possible variations on this classical model. We mention a few examples:

(1) In some real systems (e.g., the packet radio system mentioned above), there is a small but significant probability that a user detects a collision for a slot which actually contained

a successful transmission. Thus, one might wish to include imperfect feedback in the model.
(2) Feedback might be delayed, such that users receive information about the outcome of slot k in slot $k+d$.
(3) In some situations, it might be possible to estimate the number of users who collided and thus obtain more information than is provided by trinary feedback.
(4) Suppose that the user population is finite, and that a user who has generated a message does not generate new messages (is *backlogged*) until that message is successfully transmitted. In this system, one might wish to model the message generation process as a non-stationary Poisson process whose rate is proportional to the number of non-backlogged users.

THE CAPETANAKIS-TSYBAKOV-MIKHAILOV (CTM) MULTIPLE ACCESS PROTOCOL

Background

Over the last few decades, a variety of random access protocols have been devised to overcome the limitations of FDMA and TDMA; most permit collisions. The first of these, known as the Aloha protocol, was implemented in 1970. The slotted version of this protocol (the original Aloha protocol did not use time slots) resolves collisions by requiring that all users who participate in a collision wait a random number of slots and then re-transmit; the random delay is generated according to a discrete distribution such as uniform or geometric. The maximum throughput of this protocol is $1/e \approx 0.368$. The simple slotted Aloha protocol has undesirable delay characteristics; however, a protocol refinement called "stabilized Aloha" solves this problem (see Bertsekas (1992)).

Definition of the CTM Protocol

An alternative approach to collision resolution was devised independently by J. Capetanakis and by B. S. Tsybakov and V. A. Mikhailov. The Capetanakis-Tsybakov-Mikhailov (CTM) protocol provides an elegant solution to the classical multiple access problem, and is the simplest of a family of related "tree splitting protocols". In the CTM protocol, all users involved in a collision divide randomly into two groups. Each user performs the equivalent of an independent coin toss, selecting group 1 with probability p and group 2 with probability $q = 1-p$. Users in group 1 re-transmit immediately, i.e., during the slot following the one in which the collision occurred. Users in group 2 defer re-transmissions until all messages in group 1 have been successfully transmitted. If group 1 or group 2 contains more than one user, these users will collide again, and must split as before. Even if a group contains only a single message, that message may still collide with a new message. All collisions are resolved on LCFS basis. The state of the system is equivalent to a stack. Each user who has been involved in a collision keeps track of their own level within the stack, incrementing it by 1 whenever another collision occurs, and decrementing it whenever a slot is empty or contains a successful transmission. During each slot, all messages having level zero are transmitted.

Figure 1A demonstrates the operation of the CTM protocol via a tree diagram. Each circle corresponds to one time slot; T is the set of users who transmitted during that slot. An initial collision in slot S=1 involves 3 users. Users 2 and 3 fall into group 1, which re-transmits in slot S=2, producing another collision. When group 1 splits, the sub-groups,

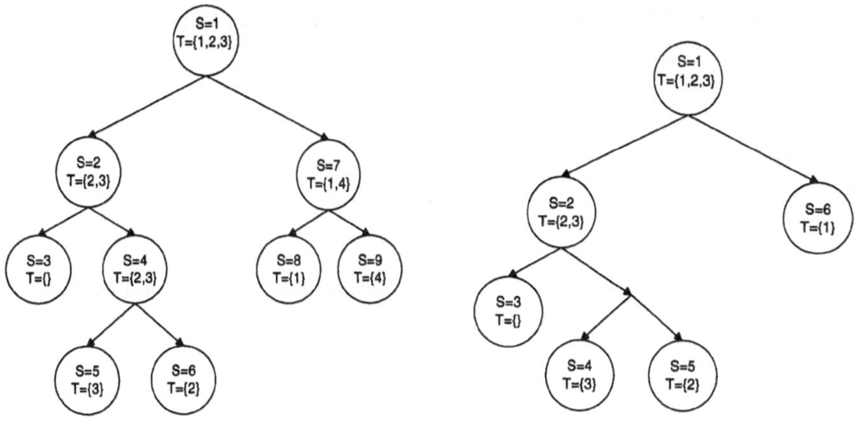

A. Original CTM Protocol B. Massey's Improvement

Figure 1. Example of Protocol Operation

which we denote as 11 and 12, re-transmit in slots S=3 and S=4, respectively (group 11 is empty). Note that the set of users associated with any group is not the union of the subgroups because new users can transmit at any time (user 4, who did not participate in the original collision, transmits a message in slot 7).

The interval of time from an initial collision until its resolution is called a Collision Resolution Interval, or CRI. The CRI includes the slot in which the initial collision occurred, and all slots required to resolve collisions involving those users and any new users who transmit while the original collision is being resolved. In the example of Figure 1A, the length of the CRI is 9.

The CTM protocol provides low mean delay under light-load conditions. The maximum throughput is approximately 0.43, close to that of slotted Aloha, and the delay characteristics are better than those of slotted Aloha. Several improvements to the original CTM protocol have been devised. We mention three of these (see Bertsekas (1992) for a more complete discussion):

(1) In the example of the protocol operation, the collision in slot 4 is unnecessary. The fact that slot 3 is empty indicates that all users who collided in slot 2 have selected group 12, i.e., are deferring their transmissions. This means that another collision is guaranteed to occur. Rather than waste a slot, the users can divide again immediately. This improvement, which is due to J. Massey, increases the maximum throughput to approximately 0.46. Operation of the improved protocol, which we denote by CTM2, is illustrated in Figure 1B.

(2) In practice, there is always some non-zero probability of falsely detecting a collision. If this happens, CTM2 will result in unlimited splitting, and the CRI will be infinite, i.e., the protocol fails catastrophically. One can prevent this by imposing a maximum limit s_{max} on the number of consecutive empty slots allowed before group 2 re-transmits (without

splitting). The original CTM corresponds to $s_{max}=1$, while CTM2 with unlimited splitting corresponds to $s_{max}=\infty$.

(3) Further improvement can be obtained by requiring that all newly-generated messages wait for resolution of collisions. (This is slightly more difficult to implement, because it requires that all users must continually monitor the channel, even if they have nothing to transmit).

Recursions for the Length of the Collision Resolution Interval (CRI) and for Delay

Let $L(n)$ denote the number of slots required to resolve a collision involving n users. The distribution of $L(n)$ depends not only on n, but also on the rate λ at which new messages are generated. Let $I(n)$ denote the number of users who select group 1, i.e., who re-transmit immediately ($n-I(n)$ select group 2). Since each user selects a group on the basis of an independent coin flip, $I(n)$ is binomial with parameters n and p. Let $\tilde{X}$ and $\tilde{\tilde{X}}$ denote the numbers of new arrivals during the slots in which groups 1 and 2, respectively, re-transmit.

$$P\{\tilde{X}=k\}=P\{\tilde{\tilde{X}}=k\}=\frac{(\lambda s)^k e^{-\lambda s}}{k!}$$

where s is the duration of a slot. We can now write a recursion for $L(n)$ which follows directly from the definition of the CTM protocol:

$$L(n)\overset{d}{=}\begin{cases}1+\tilde{L}(I(n)+\tilde{X})+\tilde{\tilde{L}}(n-I(n)+\tilde{\tilde{X}}), & n\geq 2\\ 1, & n=0,1\end{cases} \tag{1}$$

$L(n)\overset{d}{=}\tilde{L}(n)\overset{d}{=}\tilde{\tilde{L}}(n)$. $\tilde{L}(i)$ and $\tilde{\tilde{L}}(j)$ are independent for all fixed i, j. Since $\tilde{X}$ and $\tilde{\tilde{X}}$ are independent, it follows that $\tilde{L}(I(n)+\tilde{X})$ and $\tilde{\tilde{L}}(n-I(n)+\tilde{\tilde{X}})$ are conditionally independent given $I(n)$. For $n\geq 2$, $L(n)$ is decomposed into three terms, corresponding to the slot in which the initial collision occurs, the CRI for group 1, and the CRI for group 2.

The recursion for CTM2 with unlimited splitting ($s_{max}=\infty$) is slightly more complex:

$$L(n)\overset{d}{=}\begin{cases}1+\tilde{L}(I(n)+\tilde{X})+\tilde{\tilde{L}}(n-I(n)+\tilde{\tilde{X}}), & n\geq 2, I(n)+\tilde{X}>0\\ 1+\tilde{\tilde{\tilde{L}}}(n+\tilde{\tilde{\tilde{X}}}), & n\geq 2, I(n)+\tilde{X}=0\\ 1, & n=0,1\end{cases} \tag{2}$$

It is straightforward to write a recursion for CTM2 with general s_{max}. Because the notations become cumbersome, we present the recursion only for the case $s_{max}=2$:

$$
L(n) \overset{d}{=} \begin{cases} 1+\tilde{L}(I(n)+\tilde{X})+\tilde{\tilde{L}}(n-I(n)+\tilde{\tilde{X}}), & n \geq 2, I(n)+\tilde{X}>0 \\ 2+\tilde{\tilde{\tilde{L}}}(\tilde{I}(n)+\tilde{\tilde{\tilde{X}}})+\tilde{\tilde{\tilde{\tilde{L}}}}(n-\tilde{I}(n)+\tilde{\tilde{\tilde{\tilde{X}}}}), & n \geq 2, I(n)+\tilde{X}=0 \\ 1, & n=0,1 \end{cases} \tag{3}
$$

As mentioned before, throughput and delay are the two most important measures of performance for practical applications of multiple access protocols. There is a close connection between delay and length of the CRI. For convenience, let us treat all messages as being generated exactly at the beginning of the slot in which they are first transmitted, so that the delay is one slot if the message is successfully transmitted on the initial attempt. Clearly, $L(n)$ is an upper bound on the maximum delay for all users who transmit during the CRI. One can also write a recursion in terms of $L(n)$ for the delay $D(n)$ of a randomly selected user among n colliding users:

$$
D(n) \overset{d}{=} \begin{cases} B\left(\frac{I(n)}{n}\right)\left[1+\tilde{D}(I(n)+\tilde{X})\right]+\left[1-B\left(\frac{I(n)}{n}\right)\right]\left[1+L(I(n)+\tilde{X})+\tilde{D}(n-I(n)+\tilde{\tilde{X}})\right], & n \geq 2 \\ 1, & n=1 \end{cases} \tag{4}
$$

where $L(n)$ is defined by (1), (2), or (3), depending on the version of the protocol ((4) applies for all variants of the CTM protocol discussed above). $B(r)$ denotes a Bernoulli r.v. with parameter r. In this case, the parameter r is itself a r.v.: $\frac{I(n)}{n}$ is the conditional probability that a randomly selected user falls into group 1 given that $I(n)$ users select group 1. One cannot replace $B\left(\frac{I(n)}{n}\right)$ by $B(p)$ because a user's decision to select the first group is not independent of the delay for that group (if more users select the first group, the delay for that group will increase).

THE ASYMPTOTIC DISTRIBUTION OF L(n)

Background and Previous Results

According to the classical model, the distribution of the number n of users who initiate a CRI is Poisson. This distribution goes to zero faster than e^{-n}, i.e., large values of n are extremely unlikely. Thus, it might seem that the asymptotic distribution of $L(n)$ as $n \to \infty$ is of limited practical interest. However, because there is no closed-form analytical solution for the distribution of $L(n)$ for finite n, the asymptotic behavior is of practical interest to the extent that it gives insight into the behavior of $L(n)$ for reasonable values of n. From the standpoint of protocol performance, it is desirable that the weight in the tail of the distribution be as small as possible. From the standpoint of mathematical analysis, the asymptotic behavior of such stochastic recursions poses some difficult and intriguing problems which are interesting in their own right.

In what follows, we briefly review some of the most important previous results in this area; all of these concern the orignal CTM algorithm. Fayolle et. al. (1985) established that $\lim_{n \to \infty} E\{L(n)\}/n$ exists iff

$$\frac{\log p}{\log(1-p)} \text{ is irrational} \tag{5}$$

Otherwise, $E\{L(n)\}/n$ exhibits an oscillatory behavior. It is intuitively reasonable that the length of the CRI should be asymptotically proportional to n, but rather disquieting that an infinitesimal change in the splitting probability p has a radical affect on the behavior of the protocol. (We later address the question of whether this instability w.r.t. p is a legitimate cause for concern about the practical utility of such protocols). For any integers k and m, any real root $p \in [0.5, 1)$ of $p^{k/m} + p - 1 = 0$ violates (5). For m=1, the table below shows the root of interest for several values of k:

k	1	2	3	4	5	20	∞
root p	0.5	0.618	0.6823	0.7245	0.7549	0.8939	1

It is apparent that values of p which violate (5) cover the interval (0,1) densely.

Fayolle et. al. (1986) showed that under the same condition (5), $\lim_{n\to\infty} \text{var}\{L(n)\}/n$ exists and all moments of $L(n)$ are finite. In more recent work, Jacquet and Regnier (1988, 1989) showed that a standardized version of $L(n)$ tends to normal:

$$Y(n) = \frac{L(n) - E\{L(n)\}}{\sqrt{\text{var}\{L(n)\}}} \overset{d}{\rightarrow} N(0,1) \tag{6}$$

New Results

We now examine the asymptotic normality of the distribution of $L(n)$ under more general conditions. Specifically, we remove the assumptions on the distribution types of $I(n)$, $\tilde{X}$, and $\tilde{\tilde{X}}$, and instead rely solely on the asymptotic linearity of $\text{var}\{L(n)\}/n$.

Theorem 1. Let $Y(n) = (L(n) - l(n))/\sqrt{n}$, where $l(n) = E\{L(n)\}$. $Y(n) \overset{d}{\rightarrow} N(0, \sigma^2)$ if the following three conditions are satisfied for some $r \in (2,3]$:

(a) $E\{(\tilde{X})^{r/2}\} + E\{(\tilde{\tilde{X}})^{r/2}\} < \infty$ and $\frac{I(n)}{n} \overset{L^r}{\rightarrow} p \in (0,1)$,

(b) $\sigma^2(n) = \frac{\text{var}\, L(n)}{n} \rightarrow \sigma^2$, and

(c) $\sup_n E|Y(n)|^r < \infty$.

Sketch of the Proof. We now sketch the proof of Theorem 1; a complete proof will be given in Feldman et. al. (to appear). From the definition of $Y(n)$ and from (1),

$$Y(n) \overset{d}{=} \left(\frac{I(n)+\tilde{X}}{n}\right)^{1/2} Y(I(n)+\tilde{X}) + \left(\frac{n-I(n)+\tilde{\tilde{X}}}{n}\right)^{1/2} \tilde{Y}(n-I(n)+\tilde{\tilde{X}}) + C(n, I(n), \tilde{X}, \tilde{\tilde{X}})$$

where $Y(n)$ and $\tilde{Y}(n)$ are i.i.d., and $C(n,k,m,\tilde{m}):=n^{-1/2}(1+l(k+m)+l(n-k+\tilde{m})-l(n))$. Define a sequence of normal $N(0,\sigma^2(n))$-distributed independent r.v.'s $Z(n)$ which are independent of $\{I(n)\}$, X, and Y, and let

$$Z^*(n)=\left(\frac{I(n)+\tilde{X}}{n}\right)^{1/2} Z(I(n)+\tilde{X})+\left(\frac{n-I(n)+\tilde{\tilde{X}}}{n}\right)^{1/2} \tilde{Z}(n-I(n)+\tilde{\tilde{X}})+C(n,I(n),\tilde{X},\tilde{\tilde{X}})$$

where $\tilde{Z}(n)$ is an independent version of $Z(n)$. Let μ_r be the following ideal metric of order $r>0$:

$$\mu_r(X,Y)=\sup\left\{|E\{f(X)-f(Y)\}|:\left\|f^{(s)}\right\|_{\tilde{q}}\leq 1\right\}$$

where $r=s+1/\tilde{p}$, $s\in\mathbb{N}$, $\tilde{p}\in[1,\infty]$, and $\frac{1}{\tilde{p}}+\frac{1}{\tilde{q}}=1$ (see Rachev (1991), Sect. 14.2, and Rachev and Rüschendorf (1991)).

Claim 1. (μ_r-closeness of $Z^*(n)$ and $Y(n)$). Let $a(n)=\mu_r(Z(n),Y(n))$ and assume that

$$a:=\sup_n a(n)<\infty \tag{7}$$

Then $\sup_n \mu_r(Z^*(n),Y(n))\leq a\left[p^{r/2}+(1-p)^{r/2}\right]$. The proof of this claim follows from the ideality of μ_r.

Claim 2. (7) holds. To show this, note that $\mu_r(X,Y)\leq C\sup_n\left(E|X|^r+E|Y|^r\right)<\infty$ provided that $E\{X^i-Y^j\}=0$ for $j=1,2$, (see, for example, Rachev (1991), chapters 14-15). Thus, $a\leq C\sup_n\left(E|Y(n)|^r+E|Z(n)|^r\right)<\infty$, where C is an absolute constant.

Claim 3. (Asymptotic normality of $Z^*(n)$). $\mu_r(Z(n),Z^*(n))\to 0$ as $n\to\infty$. To show this, let κ_r be the r-th pseudomoment, i.e.,

$$\kappa_r(X,Y)=r\int_{\mathbf{R}}|x|^{r-1}|F_X(x)-F_Y(x)|dx$$

Then, since the mean and variance of $Z(n)$ coincide with those of $Z^*(n)$, it follows that $\mu_r(Z(n),Z^*(n))\leq C\kappa_r(Z(n),Z^*(n))$. Recall that $\{Z(n)\}_{n\geq 1}$ is independent of $\{I(n)\}_{n\geq 1}$ and of $\tilde{X}$ and $\tilde{\tilde{X}}$. Let N_0 denote a standard normal r.v. independent of $\{I(n)\}_{n\geq 1}$ and of $\tilde{X}$ and $\tilde{\tilde{X}}$. Then,

$$Z^*(n)\overset{d}{=}\eta(n)N_0+C(n,I(n),\tilde{X},\tilde{\tilde{X}})$$

The convergence of $\eta(n)$ follows from assumptions (a) and (b): $\eta(n)\overset{P}{\to}\left(p\sigma^2+(1-p)\sigma^2\right)^{1/2}$.

Since $Z^*(n)$ has the same mean and variance as $Z(n)\overset{d}{=}\sigma(n)N_0$, $\sigma^2(n) = E\{\eta^2(n)\} + E\{C(n, I(n), \tilde{X}, \tilde{X})^2\}$. As $\eta(n)\overset{L^2}{\to}\sigma$, we conclude that $C(n, I(n), \tilde{X}, \tilde{X})\overset{p}{\to}0$. This implies that $\mu_r(Z^*(n), Z(n)) \to 0$.

Claim 4. With $a(n) = \mu_r(Z(n), Y(n))$ as before, and $\bar{a} := \limsup_n\{a(n)\}$, $\bar{a} = 0$. This claim is readily checked. From Claim 2, it follows that $\bar{a} \le (\bar{a} + \varepsilon)(p^{r/2} + (1-p)^{r/2})$. Since $r > 2$, $p^{r/2} + (1-p)^{r/2} < 1$. This implies that $\bar{a} = 0$.

The proof of the theorem is complete since μ_r convergence implies weak convergence.

In more recent work, the result of Theorem 1 has been extended to CTM2 with limited splitting, and to the multinomial case. In the multinomial version of the protocol, users respond to a collision by splitting into r groups, where r is a constant greater than 2. (The multinomial version of the protocol is of mathematical interest, but it is not a practical protocol because its maximum throughput is very low).

NUMERICAL AND SIMULATION RESULTS

The analyses discussed in the previous section indicate that the asymptotic behavior of the CTM protocol for large n is unstable with respect to the parameter p, i.e., an infinitesimal change in the value of p has a large effect on the value of $L(n)$. However, these analyses do not tell us how large n must be before this phenomenon is observed. In a practical application of a multiple access protocol, at most a few thousand users would share a single channel. Under reasonable traffic conditions, not more than 1 or 2 percent of these would be expected to collide in any slot. This means that the interesting regime of n for practical applications is roughly [1,50].

We have investigated the distribution of the $L(n)$ for finite n for the case $\lambda = 0$ ($\tilde{X} = \tilde{\tilde{X}} = 0$) using two completely different methods:

(1) We converted the stochastic recursions (1), (2), and (3) into deterministic recursions for $l(n) = E\{L(n)\}$ and $s(n) = E\{(L(n))^2\}$. (See Feldman et. al. (to appear)). These recursions were then evaluated numerically for values of n up to 100,000. In order to prevent numerical roundoff from affecting the solution, we found that it was necessary to use extended-precision arithmetic (128 bits for mantissa and exponent). Ordinary double-precision arithmetic (64 bits) was found to be insufficient for values of n greater than about 20,000. In order to speed up the calculations, the binomial distribution was truncated to include only the center portion; tests were performed to insure that this did not have any significant effect on the solution.

(2) We generated samples of $L(n)$ for values of n up to about 5,000 via direct simulation of the protocols. The protocols were simulated using a recursive function which implements (1), (2), and (3). This is an extremely simple way to simulate the protocol operation. (For some purposes, e.g., estimating mean delay, a more complex simulation is required which stores the full state of the stack, including the times at which all messages were generated).

Each of these methods has advantages and disadvantages. Numerical calculation of the moments allows one to investigate the non-linearity of $l(n)$, e.g., to plot the increments of

$l(n)/n$. This is impossible with simulation, because for practical sample sizes the variances of the sample means are too large. On the other hand, numerical calculation of moments does not permit one to investigate the rate of convergence to the normal except in a very indirect fashion. For this, one needs samples of $L(n)$ which can be subjected to statistical tests of distribution; such samples can be generated only via simulation.

Figure 2 shows the dependence, for p=0.5, of the absolute increment $\left|\frac{l(n)}{n}-\frac{l(n-1)}{n-1}\right|$ of the normalized mean CRI length $l(n)/n$ on the number of users n who initially collide. (Recall that since p=0.5 violates the condition (5), the limit of $l(n)/n$ does not exist). Both axes on the graph are logarithmic. Note that the curve is almost perfectly linear for $n<5{,}000$. A straight line on a log-log plot indicates power law dependence, i.e., $\frac{l(n)}{n}=cn^r$. Because the slope of the linear region is less than -1, a numerical study based on $n<5{,}000$ might eroneously conclude that $\lim_{n\to\infty} l(n)/n$ exists (recall that a summation of the form $\sum_{n=1}^{\infty} cn^r$ exists iff $r<-1$). For $n>38{,}000$, large fluctuations and changes of sign appear. Changes of sign tend to speed convergence, but the increments are no longer bounded from above by a line (on the log-log scale) having slope less than -1; thus, the numerical study does not permit one to conclude anything about the existence of the limit.

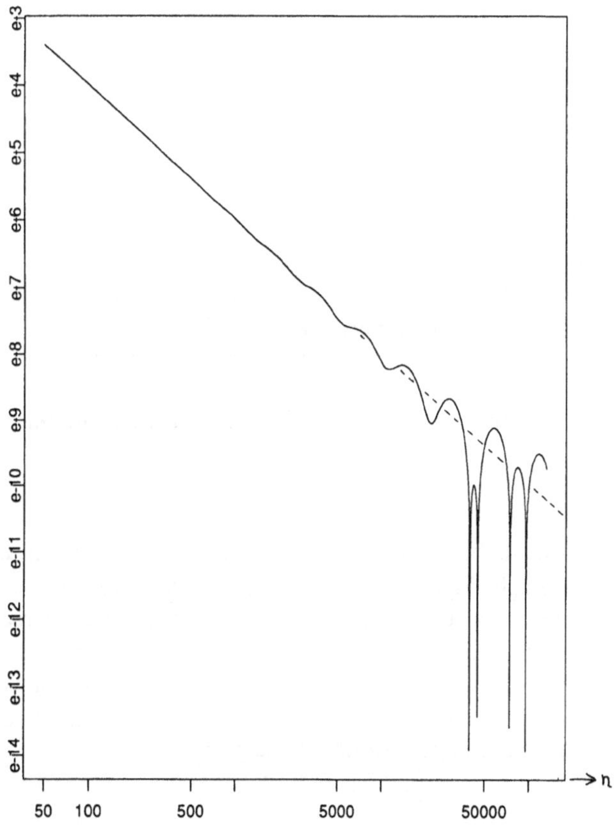

Figure 2. Dependence of increments of $l(n)/n$ on n for p= 0.5

Because the non-linearity of $l(n)/n$ is observed only for values of n much greater than 50, we can safely conclude that this behavior is of no concern for practical applications of the CTM protocol.

Simulation results were used to compare the behavior of the standard CTM (binary splitting) with $p_1 = p_2 = 1/2$ against the trinary splitting version with $p_1 = p_2 = p_3 = 1/3$. Behaviors are similar ($E\{L(n)\}$ and $\text{var}\{L(n)\}$ were found to be almost identical except for very small values on n). However, the third moment goes to zero more rapidly for the trinary version. For each of several values of n, a sample of 5000 values of L(n) was generated for each of these two protocols. Figure 3 shows the results of a two-sided Kolmogorov-Smirnov test of the hypothesis that the samples come from a normally distributed population. As can be seen from the graph, with increasing n, the value of the test statistic (D) rapidly diminishes and eventually crosses the level corresponding to 5 percent significance. Although the standard CTM is initially closer to normality, the trinary splitting version eventually overtakes it for large n; this is in agreement with theoretical predictions. (The irregular appearance of the graph is a consequence of using a single sample for each value of n; a better-looking curve could have been produced by averaging results from 5 to 10 samples, but we did not have time to do this).

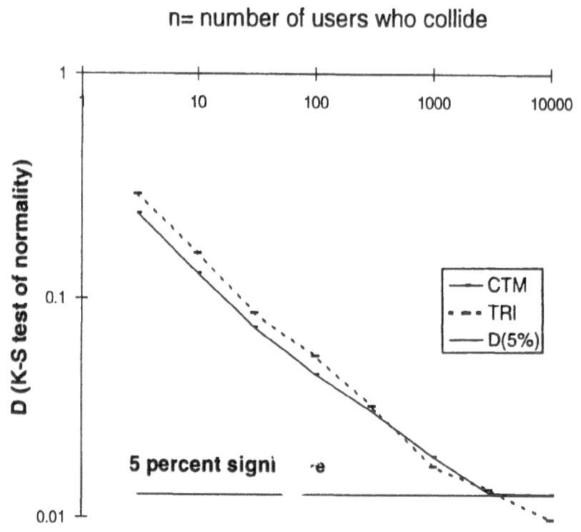

Figure 3. Kolmogorov-Smirnov test of normality (D statistic vs. number who collide)

In order to explain how the bizarre asymptotic behavior of $l(n)/n$ might arise, we have included an additional graph. Figure 4 shows the dependence of the absolute increments of $l(n)/n$ on the parameter p for four values of n. It is seen that there is a regime of values (from approximately 0.55 to 0.75) for which the increments show practically no variation. For $p > 0.75$, we observe oscillations in the increments, with the number of extrema increasing as n increases. It seems reasonable to assume that this phenomenon would be observed over the entire range of p for sufficiently large n.

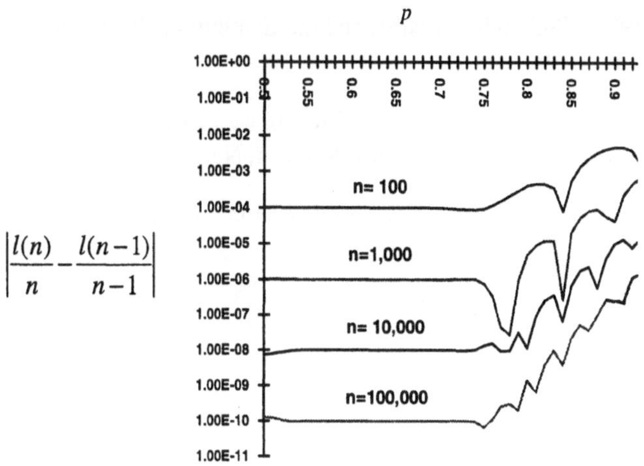

Figure 4. Dependence of increments of $l(n)/n$ on p for four values of n.

ACKNOWLEDGEMENT

I would like to thank Dr. Svetlozar T. Rachev for persistently encouraging me to write this paper, and for many helpful discussions.

REFERENCES

Aldous, D. J., 1987, Ultimate instability of exponential backoff protocol for acknowledgement-based transmission control of random access communication channels, *IEEE Trans. Inform. Theory*, vol. IT-33, 219-223.

Bertsekas, D., and Gallager, R., 1992, "Data Networks", Second Edition, Prentice-Hall, New Jersey.

Capetanakis, J. I., 1979, Tree algorithms for packet broadcast channels, *IEEE Trans. Inform. Theory*, vol. IT-25, No. 5, 505-515.

Fayolle, G., Flajolet, P., Hofri, M., and Jacquet, P., 1985, Analysis of a stack algorithm for random multiple-access communication, *IEEE Trans. Inform. Theory*, vol. IT-31, 244-254.

Fayolle, G., Flajolet, P., and Hofri, M., 1986, On a functional equation arising in the analysis of a protocol for a multi-access broadcast channel, *Adv. Appl. Prob.*, vol. 18, 441-472.

Feldman, P., Rachev, S. T., and Rüschendorf, L, to appear, Limit theorems for recursive algorithms, *J. of Computational and Applied Math.*

Jacquet, P. and Regnier, M., 1988, Normal limiting distribution of the size of tries, *Performance '87*, P. J. Courtois and G. Latouche (edt.), 209-221. Elsevier Science Publ.

Rachev, S. T., 1991, "Probability Metrics and the Stability of Stochastic Models", Wiley, Chichester, New York, 1991.

Rachev, S. T. and Rüschendorf, L., 1991, Probability metrics and recursive algorithms. Technical Report, Inst. fur Math. Stat., Univ. Munster, No. 1.

Regnier, M., and Jacquet, P., 1989, New results on the size of tries, *IEEE Trans. Inform. Theory*, vol. 35, No. 1, 203-205.

AN EXAMPLE OF NON-GAUSSIAN DENSITY PROCESS

Raisa Epstein Feldman[1, 2, 3]

Department of Statistics and Applied Probability
University of California at Santa Barbara, CA 93106, U.S.A

ABSTRACT

The Brownian density process is a distribution-valued process which can be defined either as a solution of a stochastic differential equation or as a limit of a functional over an infinite system of Brownian particles. We show that with an appropriate time change the process solves a particular stochastic differential equation with non-Gaussian measure, possibly having infinite second moment.

1. INTRODUCTION

The Brownian density process has been studied by many authors, starting with the work of Martin-Löf (1976) and Itô (1983). More recent references are Walsh (1986), Adler, Feldman and Lewin (1991), Adler (1993), and Adler and Rosen (1993).

Preliminaries. Since we will deal with distribution valued processes, let us start by introducing the space $\mathcal{S}(R^d)$, which is the Schwartz space of infinitely differentiable functions on R^d decreasing rapidly at infinity. A process $\eta(\phi), \phi \in \mathcal{S}(R^d)$ is called a random linear functional if for any real a and b and for any test functions $\phi, \psi \in \mathcal{S}(R^d)$

---

[1] *AMS 1991 subject classifications*: Primary, 60G60, 60F17; secondary, 60F05, 60E07

[2] *Key words and phrases*. Brownian density process, infinite particle system, invariance principle, non-Gaussian martingale measures, stochastic partial differential equation

[3] This research was supported by ONR Grant No. N00014/89/J/1870

Approximation, Probability, and Related Fields, Edited by
G. Anastassiou and S.T. Rachev, Plenum Press, New York 1994

holds

$$\eta(a\phi + b\psi) = a\eta(\phi) + b\eta(\psi) \quad \text{a. s.}$$

A random linear functional η which is continuous in probability on $\mathcal{S}(R^d)$ has a version in the dual space $\mathcal{S}'(R^d)$ (Corollary 4.2 of Walsh 1986).

First Definition. The shortest possible way to define the Brownian density process $\{\eta_t(\phi), t \in [0,1]\}$, is as the mean zero Gaussian $\mathcal{S}'(R^d)$-valued process with covariance given by

$$E\{\eta_t(\phi_1) \cdot \eta_s(\phi_2)\} = \int\int \phi_1(x) p_{t-s}(x,y) \phi_2(y) dx dy, \tag{1}$$

when $\phi_1, \phi_2 \in \mathcal{S}(R^d)$ and $t > s$, and for $t = s$

$$E\eta_t(\phi_1) \cdot \eta_t(\phi_2) = \int \phi_1(x)\phi_2(x) dx. \tag{2}$$

Although this definition is short and neat, it does not suggest why this process might be interesting. For this, one has to see how this process appears as a certain limit of the net charge density of an infinite system of particles. We do this next.

Particle Approximation to the Density Process. Here we closely follow Adler, Feldman and Lewin (1991). Let π^λ be a Poisson point process on R^d of intensity λ, i.e., the number of points of π^λ in a Borel set $A \subset R^d$ is a Poisson random variable with parameter $\lambda|A|$, ($|\cdot|$ denotes Lebesgue measure), and the numbers in disjoint sets are independent. Since the probability that any two points of π^λ lie exactly the same distance from the origin is 0, we can order them by magnitude, and shall denote them by $X_0^1, X_0^2, \ldots$.

Let $X_t^1, X_t^2, \ldots, t \in [0,1]$, be a sequence of independent, R^d-valued Brownian motions, with initial values given by $X_0^1, X_0^2, \ldots$, and let $\sigma^1, \sigma^2, \ldots$ denote a sequence of independent Rademacher random variables taking values $+1$ and -1 with equal probability. The two sequences and π^λ are assumed independent of one another except for the fact that π^λ determines the initial values of the X^i's. We can think of the system as consisting of signed (positive and negative) particles moving in R^d according to Brownian motion.

Define $\{\eta_t^\lambda, t \in [0,1]\}$ as the $\mathcal{S}'(R^d)$-valued random process

$$\eta_t^\lambda(\phi) = \lambda^{-1/2} \sum_{i=1}^{\infty} \sigma^i \phi(X_t^i), \quad \phi \in \mathcal{S}(R^d). \tag{3}$$

The reason that we are interested in this process (or rather its limit, when $\lambda \to \infty$) is that when ϕ is an approximation to the indicator function I_A of a set A divided by the volume $|A|$, then $\sqrt{\lambda}\eta_t^\lambda$ describes the average net charge density in A at time t. We are interested in a limit of the average charge density when the initial density of the particles goes to infinity.

Second Definition. The Brownian density process $\{\eta_t, t \in [0,1]\}$, is the $\mathcal{S}'(R^d)$-valued Gaussian process which appears as a weak limit in $D([0,1], \mathcal{S}'(R^d))$ of η_t^λ, as $\lambda \to \infty$. (Here $D([0,1], \mathcal{S}'(R^d))$ means the Skorohod space.)

This weak convergence result is due to Walsh and was derived using the Itô formula. We repeat the main steps of his proof in Section 2.

Stochastic Differential Equation. The Brownian density process can also be defined as a Gaussian distribution valued process which is a solution of the following stochastic differential equation (SPDE):

$$\begin{aligned}\frac{\partial \eta}{\partial t} &= \frac{1}{2}\Delta \eta + \nabla \cdot W \\ \eta_0 &= V^0,\end{aligned} \tag{4}$$

where V^0 is a Gaussian white noise on $\Re^d$: for Borel $A, B \in R^d$

$$V^0(A) \sim N(0, |A|), \;\; EV^0(A)V^0(B) = |A \cap B|$$

and W is an $\Re^d$-valued Gaussian white noise on $\Re^d \times \Re_+$ such that for all components

$$W_i([s,t], A) \sim N(0, (t-s)|A|), \;\; EW_i([s,t], A)W_j([u,v], B) = \delta_{ij}|A \cap B||[s,t] \cap [u,v]|.$$

The equation (4) is understood in the weak form developed in Walsh (1986), i.e., its solution is the solution of the following integral equation:

$$\eta_t(\phi) = V^0(\phi) + \frac{1}{2}\int_0^t \eta_s(\Delta\phi)ds + \int_0^t \int_{R^d} < \nabla\phi(x), W(ds, dx) > \tag{5}$$

(Here and below, for a measure μ, $\mu(\phi) \equiv \int \phi d\mu$.) We will show below how to get this equation using Itô's formula for $\phi(X_t^i)$ and passage to the limit as $\lambda \to \infty$.

Note that the last integral in (5) with respect to the martingale measure W, as was introduced in Walsh (1986):

A process $\{M_t(A), \mathcal{F}_t, t \geq 0, A \in \mathcal{A}\}$ is a martingale measure if (i) $M_0(A) = 0$; (ii) if $t > 0$ is fixed, M_t is a σ-finite L^2-valued measure; (iii) if $A \in \mathcal{A}$ is fixed, the process $\{M_t(A), t \geq 0\}$ is a martingale.

Under some additional conditions Walsh constructs a stochastic integral with respect to the martingale measures and considers stochastic partial differential equations involving this type of an integral. Walsh's stochastic calculus is based heavily on L^2 theory.

Question and Motivation. We are interested to study equations similar to (4) when the last integral is taken with respect to a non-Gaussian martingale measure which might not have a finite second moment, for example, stable measure.

Stable processes were studied intensively in the recent years. In Feldman and Rachev (1993) integrals with respect to stable measures were constructed using limit theorems on infinite systems of signed Markovian particles. Rosinski and Woyczynsky (1986) consider stochastic Itô integrals with respect to stable motions. S. Cambanis, J. Rosinski, G. Samorodnitsky, J. Szulga, M. Taqqu, and W. Woyczynsky, are a few of the many names associated with the topic; see extensive literature lists on the subject in the forthcoming book, Samorodnitsky and Taqqu (1993), or review article Weron (1984). (See also Feldman and Krishnakumar (1993), this volume)

Although it would be interesting to develop a theory similar to that of Walsh in the case of integration with respect to stable martingale measures, here we will only provide an easy example of the stable distribution valued process $\tilde{\eta}_t$ which solves an equation similar to (4) when V^0 remains Gaussian, but W is changed to a non-Gaussian (for example, stable) measure $\tilde{W}$.

2. TOOLS

Walsh's Convergence Result. Let us see how the equation (5) was derived in Walsh (1986). To do this, apply Itô's formula for Brownian motion:

$$\phi(X_t^i) = \phi(X_0^i) + \int_0^t \nabla\phi(X_s^i)dX_s^i + \frac{1}{2}\int_0^t \Delta\phi(X_s^i)ds.$$

We then multiply the previous equation by σ^i, sum over all i's and divide by $\sqrt{\lambda}$. Using notations (3), and

$$V_0^\lambda(\phi) := \lambda^{-1/2}\sum_{i=1}^{\infty}\sigma^i\phi(X_0^i), \quad \phi \in \mathcal{S}(R^d); \tag{6}$$

$$W_t^\lambda(A) \equiv W^\lambda(t,A) := \lambda^{-1/2}\sum_{i=1}^{\infty}\sigma^i\int_0^t I_A(X_s^i)dX_s^i, \tag{7}$$

we obtain:

$$\eta_t^\lambda(\phi) = V_0^\lambda(\phi) + \frac{1}{2}\int_0^t \eta_s^\lambda(\Delta\phi)ds + W_t^\lambda(\nabla\phi). \tag{8}$$

The consequence of the Proposition 8.16 in Walsh (1986) is the fact:

Result (A) *As $\lambda \to \infty$, $(V_0^\lambda, W^\lambda, \eta^\lambda) \Rightarrow (V^0, W, \eta)$, where V^0, W, η were defined above and the convergence is on the Skorohod space $D([0,1], \mathcal{S}'(R^d)\times\mathcal{S}'(R^d)\times\mathcal{S}'(R^d))$.*

We will also use the fact that convergence of martingale measures implies convergence of integrals with respect to those measures. (c.f., Proposition 7.8 of Walsh, 1986)

Convergence of Compositions. The second tool which we will need is the following Exercise 4.4.2.2 from Resnick (1987):

Result (B) *Suppose $x_n, n \geq 0$, are real functions in $D[0,\infty)$ and $\tau_n \in D[0,\infty)$ is non-decreasing, non-negative for $n \geq 0$, $x_n \Rightarrow x_0$, $\tau_n \Rightarrow \tau_0$ (converge in $D[0,\infty)$). If τ_0 is continuous, then $\tau_n \circ x_n \Rightarrow \tau_0 \circ x_0$.*

Whitt (1980) allows application of the above result to functions x_n with values in a complete separable metric space.

3. RESULTS AND PROOFS

Let us start with some notations. For $\lambda > 0$, let $N(\lambda)$ be a positive random variable taking integer values independently of the X^i's and σ^i's and such that

$$N(\lambda)/\lambda \xrightarrow{\mathcal{D}} Y \text{ as } \lambda \to \infty \tag{9}$$

for some positive random variable Y independent of X's and σ's.

Take a positive sequence $\lambda_n, n \geq 0$, such that $\lambda_n \to \infty$ as $n \to \infty$. Set $\Lambda_n := N(\lambda_n)/\lambda_n$.

Make random time change in the processes:

$$\tilde{\eta}_t^n(\phi) := \eta_{\Lambda_n t}^{\lambda_n}(\phi), \ \tilde{W}_t^n(\phi) := W_{\Lambda_n t}^{\lambda_n}(\phi) \tag{10}$$

Lemma 1. *When $n \to \infty, \tilde{W}^n \Rightarrow \tilde{W}$ in $D([0,1], \mathcal{S}'(R^d))$, where $\tilde{W}_t(\phi) \overset{D}{=} \sqrt{Y} W_t(\phi)$, the later being defined on the product of probability spaces.*

Note that $\tilde{W}$, defined in Lemma 1, is non-Gaussian: if $\mathcal{L}_Y$ denotes the Laplace transform of Y, then

$$E \exp\{i\theta_1 \tilde{W}_{t_1}(A_1) + \ldots + i\theta_k \tilde{W}_{t_k}(A_k)\} = \mathcal{L}_Y\{(1/2) \sum_{i,j=1}^{k} \theta_i \theta_j t_i t_j |A_i \cap A_j|\}. \tag{11}$$

Note also, that $\tilde{W}_t(\phi) \overset{D}{=} W_{Yt}(\phi)$. A similar method for obtaining non-Gaussian distributions was considered in Feldman and Rachev (1993). We learn from there that the distribution of $\tilde{W}$ will be Laplace if $N(\lambda)$ is a geometric random variable with mean λ; it will be symmetric stable (sub-Gaussian) if the distribution of $N(\lambda)$ is a discretized version of symmetric stable random variable, multiplied by λ.

Proof of Lemma 1. Fix ϕ. In notations of the previous Section, take $x_n(t) = W_t^{\lambda_n}(\phi)$. By Result (A) $x_n \Rightarrow x_0$ is $D[0,1]$, where $x_0(t) := W_t(\phi)$. Take also $\tau_n(t) = \Lambda_n t$ and $\tau_0(t) = Yt$. From (9) and the independence of $N(\lambda)$ and W^{λ_n}, it follows that $(x_n, \tau_n) \equiv (W_t^{\lambda_n}, \Lambda_n t)$ converges weakly to $(x_0, \tau_0) \equiv (W_t, Yt)$. The Skorohod-Dudley theorem (cf., Dudley 1989, Theorem 11.7.2) allows us to put the sequence $(W_t^n, \Lambda_n t)$ and the limit (W_t, Yt) into the same probability space, and so, using the Result (B) we get:

$$\tilde{W}_t^n(\phi) \Rightarrow W_{Yt}(\phi) \overset{D}{=} \tilde{W}_t, \text{ in } D[0,1]. \tag{12}$$

We used the fact that $W_{Yt}(\phi) \overset{D}{=} \sqrt{Y} W_t(\phi)$. Thus, we proved the desired convergence for one fixed test function ϕ. Now note that by Mitoma theorem (see, for example, Chapter 6 of Walsh, 1986) tightness of the distribution-valued sequence of processes $\tilde{W}^n$ follows from the tightness of the real-valued sequence of processes $\tilde{W}^n(\phi)$, which of course follows from (12). The convergence of the finite dimensional distributions also follows from (12) because of the linearity of $\tilde{W}^n$ and $\tilde{W}$ in ϕ. This completes the proof of Lemma 1.

Similarly,

Lemma 2. *When $n \to \infty, \tilde{\eta}^n \Rightarrow \tilde{\eta}$ in $D([0,1], \mathcal{S}'(R^d))$, where $\tilde{\eta}_t(\phi) \overset{D}{=} \sqrt{Y} \eta_t(\phi)$, the later being defined on the product of probability spaces.*

Proof. Identical to the one of Lemma 1 with η instead of W.

Note that $\tilde{\eta}_t(\phi) \overset{D}{=} \eta_{Yt}(\phi)$.

Now the main result of the paper follows easily:

Theorem *If $n \to \infty$, then $\tilde{\eta}_t^n$ converges in $D([0,1], \mathcal{S}'(R^d))$ to a solution $\tilde{\eta}_t$ of the SPDE*

$$\begin{aligned} \frac{\partial \tilde{\eta}}{\partial t} &= \frac{1}{2} Y \Delta \tilde{\eta} + \nabla \cdot \tilde{W} \\ \tilde{\eta}_0 &= V^0. \end{aligned} \tag{13}$$

The last equation is understood as

$$\tilde{\eta}_t(\phi) = V^0(\phi) + \frac{1}{2}Y\int_0^t \tilde{\eta}_s(\Delta\phi)ds + \int_0^t \int_{R^d} < \nabla\phi(x), \tilde{W}(ds,dx) >$$

Proof. Using exactly the same arguments as in the proof of Lemma 1, we show that as $n \to \infty$, the vector $(V_0^{\lambda_n}, \tilde{W}^n, \tilde{\eta}^n) \Rightarrow (V^0, \tilde{W}, \tilde{\eta})$, where the convergence is on the Skorohod space $D([0,1], S'(R^d) \times S'(R^d) \times S'(R^d))$.
Now make a time change $t \to \Lambda_n t$ in the equation (8):

$$\eta^{\lambda_n}_{\Lambda_n t}(\phi) = V_0^{\lambda_n}(\phi) + \frac{1}{2}\int_0^{\Lambda_n t} \eta_s^{\lambda_n}(\Delta\phi)ds + W^{\lambda_n}_{\Lambda_n t}(\nabla\phi). \tag{14}$$

Clearly, we need only to check convergence of the second term:

$$\begin{aligned} \frac{1}{2}\int_0^{\Lambda_n t} \eta_s^{\lambda_n}(\Delta\phi)ds &= \frac{1}{2}\int_0^{\Lambda_n t} \frac{1}{\sqrt{\lambda_n}}\sum_{i=1}^{\infty} \sigma^i \Delta\phi(X_s^i)ds \\ &= \frac{1}{2}\int_0^{t} \frac{1}{\sqrt{\lambda_n}}\sum_{i=1}^{\infty} \sigma^i \Delta\phi(X^i_{\Lambda_n u})\Lambda_n du \quad (\text{by } s = \Lambda_n u) \\ &= \frac{1}{2}\Lambda_n \int_0^t \eta^{\lambda_n}_{\Lambda_n s}(\Delta\phi)ds \end{aligned}$$

The proof is completed.

REFERENCES

Adler, R. J. (1993), Superprocess local and intersection local times and their corresponding particle pictures, *Seminar on Stochastic Processes 1992*. Birkhäuser, Boston.

Adler, R. J., Feldman, R. and Lewin, M. (1991), Intersection local times for infinite systems of planar Brownian motions and the Brownian density process, *Ann. Probab.* **19**, 192-220.

Adler, R. J. and Rosen, J. S. (1993), Intersection local times of all orders for Brownian and stable density processes–construction, renormalisation and limit laws, *Ann. Probab.* **21**, 1073-1123.

Dudley, R. M. (1989), *Real Analysis and Probability*, Wadsworth & Brooks/Cole, Pacific Grove, California.

Feldman, R. E. and Rachev, S. T. (1993), U-statistics of Random-Size samples and limit theorems for systems of Markovian particles with non-Poisson initial distributions, *Ann. Probab.* **21**.

Feldman, R. E. and Krishnakumar, N. (1993), Limit theorems for functionals of Markov processes and renormalizable stable fields, This volume.

Itô, K. (1983), Distribution valued processes arising from independent Brownian motions, *Math. Z.* **182**, 17-33.

Martin-Löf, A. (1976), Limit theorems for the motion of a Poisson system of independent Markovian particles with high density, *Z. Wahrsch. verw. Gebiete* **34**, 205-223.

Resnick, S. I. (1987), *Extreme Values, Regular Variation, and Point Processes*, Springer-Verlag, New York.

Rosinski, J., Woyczynski, W. A. (1986) On Itô stochastic integration with respect to p-stable motion: inner clock, integrability of sample paths, double and multiple integrals, *Ann. Probab.* **14**, 271-286.

Samorodnitsky, G. and Taqqu, M. S. (1993), *Stable Random Processes*, in press.

Walsh, J. B. (1986), An introduction to stochastic partial differential equations, *Springer Lecture Notes in Math.* **1180**, 265-439.

Weron, A. (1984), Stable processes and measures: A survey, in *Springer Lecture Notes in Math.* **1080**, 306-364.

LIMIT THEOREMS FOR FUNCTIONALS OF MARKOV PROCESSES AND RENORMALIZABLE STABLE FIELDS

Raisa E. Feldman and Nagamani Krishnakumar[1, 2, 3]

Department of Statistics and Applied Probability
University of California at Santa Barbara, CA 93106, U.S.A.

ABSTRACT

We study the limiting distribution of the amount of charge left in some set by an infinite system of charged Markovian particles, when the charge distribution belongs to the domain of attraction of a symmetric α-stable law. The limits are symmetric α-stable generalized random fields. Their multiple integrals are built in a similar manner. We also study the renormalizability of these families of random fields and use the construction to simulate stable fields on R^1 and R^2.

1. INTRODUCTION

Stable processes have been studied intensively in recent years. Samorodnitsky and Taqqu (1990), Weron (1984), Rosinski and Woyczynski (198j), Rosinski (1986), Kallenberg and Szulga (1989), Szulga (1992), Maejima (1990), Janicki and Weron (1991), are some of the references related to this work. Because all these authors work on an abstract measurable space, or with specific models on R^1, we were motivated to define families of stable random fields on $R^d, d \geq 1$, and of generalized stable fields on the Sobolev space $\mathcal{S}_d$. We build the fields as limits of sums of functionals on paths of Markov processes and show that using this construction one can easily analyze some of the properties of the fields in the limit.

Motivated by work of Maejima (1990) on self-similar stable processes, we choose to study the renormalizability of the families of stable fields and their functionals. The property of renormalizability of families of random fields was defined in Adler and Epstein (=Feldman) (1987) and Epstein (1989), as an extension of the property of self-similarity to families of fields. These authors discuss the renormalizability of families of Gaussian fields. Here we show that stable fields built from self-similar Markov processes are renormalizable.

Our method of construction of the stable fields also allows us to simulate these fields on a computer and display the surfaces to see what they look like. Our moti-

---

[1] AMS 1991 subject classifications: Primary, 60G60, 60G20, 60E07; secondary, 60G18, 60F05.

[2] Keywords and phrases. Stable random fields, generalized fields, infinite particle system, renormalizability, simulations of stable processes

[3] This research was supported by ONR Grant No. N00014/89/J/1870

vation in this direction comes from work of Janicki and Weron (1991) on simulation of stable processes (all on R^1).

We now look at our results from a somewhat different point of view. In recent years, much attention has been given to the description of infinite systems of particles moving according to some law (usually Markovian). Some of these papers e.g., Adler and Epstein (1987) and Adler (1989), deal with particle systems which behave as follows: Initially (at time zero) a number of independent particles pop into existence at locations within the space R^d, according to a Poisson point process with intensity λ. (Actually the initial distribution of the particles was created differently in the above papers, but one could use Poisson point processes instead.) The particles then move according to some Markov law. A positive or negative charge is assigned to each particle initially, according to a Rademacher random variable. The charge of each decays exponentially with time. The number of particles in the system is then set to infinity and the limiting distribution of the charge left by the system in a set in R^d, after all the particles have lost their charge, is studied. The limiting field, which is indexed by sets, or more generally by functions, was shown to have a Gaussian distribution in Adler and Epstein (1987). In Feldman and Rachev (1993), the authors obtain limiting fields that have sub-Gaussian, Laplace and other distributions, by changing the initial distribution of the particles appropriately.

The second aim of this work is to answer the question, "What happens to the limiting charge distribution if the initial charge of each particle follows a symmetric α-stable law?"

As in the above-cited work, our main tools are limit theorems for U-statistics. Here we use results found in Szulga (1992) on the convergence of resampled U-statistics to multiple stable integrals.

This paper is organized as follows: In Section 2, we present our construction of generalized stable random fields and their multiple integrals. The proofs are given in Section 3. In Section 4, we show how to construct stable fields on R^d. In Section 5 we study the renormalizability of the families of stable fields and their multiple integrals. Section 6 is devoted to simulation results for some stable processes and fields on R^1 and R^2.

2. STABLE RANDOM FIELDS AND THEIR FUNCTIONALS

Let us now define the particle system of the Introduction precisely. On an arbitrary probability space $(\Omega, \mathcal{F}, \mathrm{P})$, take n independent symmetric Markov processes $V_1, V_2, \ldots, V_n$ with values in R^d, each process starting according to some finite initial measure m. Let $p_t(x,y)$ be their common transition density function; $p_t(x,y) = p_t(y,x)$, $x, y \in R^d$, $t \geq 0$ and $\int_{R^d} p_t(x,y)dy = 1$, for each $x \in R^d$. The corresponding Green's function is given by

$$g(x,y) \equiv g^1(x,y), \quad g^\theta(x,y) = \int_0^\infty e^{-\theta t} p_t(x,y)dt, \tag{1}$$

If V is one of the processes in the sequence, then it can be thought of as describing the motion of a particle in the system. We now discuss the charge assigned to each of the n particles. Let the probability space be rich enough to support the sequences of i.i.d. random variables, $Y_{n,1}, Y_{n,2}, \ldots, Y_{n,n}$, which are independent of the V's and for which

$$P(Y_{n,i} > t) = \frac{1}{c_\alpha n} \int_t^\infty \frac{dx}{x^{1+\alpha}} = \frac{1}{\alpha c_\alpha n t^\alpha} \text{ for } t > (\alpha c_\alpha n)^{-1/\alpha}, \tag{2}$$

where $c_\alpha = \int_{R\setminus\{0\}}(1 - \cos x)dx/x^{1+\alpha}$. Let $\sigma_1, \sigma_2, \ldots, \sigma_i, \ldots$ be a sequence of i.i.d. Rademacher variables (on the same probability space), i.e., $P(\sigma_i = 1) = P(\sigma_i = -1) = 1/2$, which are independent of the V's as well as the Y's. Assign to each particle V_i a charge $\sigma_i Y_{n,i}$. We will call $\sigma_i Y_{n,i}$ the charge associated with the Markov particle V_i. The particle charges clearly belong to the domain of attraction of a stable law.

We now describe the evolution of the system in time. When particle i with charge $\sigma_i Y_{n,i}$ at time t passes through a point x in the space R^d, it leaves there a charge $\sigma_i Y_{n,i} e^{-t}$. Let $A \in \mathcal{B}(R^d)$ be a Borel set in the space R^d. We would like to find the amount of charge left in the set A after all particles have lost their charge and in the limit of increasing initial particle density, i.e., we are interested in finding the limiting distribution, as $n \to \infty$, of the normalized sum

$$\Phi_n(A) := \sum_{i=1}^{n} \int_0^\infty \sigma_i Y_{n,i} e^{-t} 1_A(V_i(t))dt.$$

More generally, define the bilinear form

$$\langle f, h\rangle \equiv \langle f, f\rangle_m := \int_{R^{3d}} m(da)dx\,dy\, f(x)g^2(a,x)g(x,y)f(y) \tag{3}$$

where g and g^2 were defined in (1). Let $S_d \equiv S_d(g)$ be the Sobolev space of C^∞ functions of finite norm, where the inner product in S_d is given by (3). For f in S_d we define the functional of the space of the paths of the Markov process V by

$$F_f(V) = \int_0^\infty e^{-t} f(V(t))dt \tag{4}$$

Then, we study the weak convergence, as $n \to \infty$, of the finite-dimensional distributions of the sum

$$\Phi_n(f) = \sum_{i=1}^{n} \sigma_i Y_{n,i} F_f(V_i) \tag{5}$$

Clearly, if $\langle |f|, |f|\rangle$ is finite for all functions f from the Schwartz space of C^∞ functions that decrease at infinity faster than any polynomial, we can take the Schwartz space as the index set of Φ_n. This is possible, for example, when the Markov process V is Brownian motion.

We now define the limiting field. A random variable X is said to have a symmetric α-stable (SαS) distribution if its characteristic function is given by

$$\mathbf{E}\exp\{itX\} = \exp\{-\sigma^\alpha |t|^\alpha\} \tag{6}$$

The SαS distribution is thus characterized by index $\alpha, 0 < \alpha \le 2$, and the scaling parameter σ. A stochastic process is SαS if all its finite-dimensional distributions are SαS, i.e., iff any linear combination of its components is an SαS random variable. On a separable finite measure space $(T, \mathcal{T}, \mu)$ let $X(\cdot)$ be a SαS Lévy random measure with control measure μ and Lévy measure $\nu(dx) = c_\alpha^{-1} dx/x^{-1-\alpha}$, where c^α is as in (2). This means that
(i) for every $A \in \mathcal{T}, X(A)$ is a SαS random variable with characteristic function

$$\begin{aligned}\mathbf{E}\exp\{itX(A)\} &= \exp\{-\mu(A)|t|^\alpha\},\\ &\equiv \exp\left\{-\int_A \int_{R\setminus\{0\}} (1-\cos tx)\nu(dx)\mu(ds)\right\}\end{aligned} \tag{7}$$

(ii) $X(\cdot)$ is independently scattered and σ-additive, i.e., for disjoint sets $A_1, \ldots, A_k, \ldots$ in $\mathcal{T}$, the random variables $X(A_1), \ldots, X(A_k)$ are independent and $X(\cup_{j=1}^\infty A_j) = \sum_{j=1}^\infty X(A_j)$.

We will denote by Xf the integral of f with respect to measure X and by $X^k f$ the multiple integral of order k with respect to measure X. The construction of these is given in Kallenberg and Szulga (1989).

Now we choose a specific separable finite measure space $(T, \mathcal{T}, \mu)$. Let $T = D([0,\infty))$ denote the path space of the R^d-valued Markov process V, let $\mathcal{B}(T) = \mathcal{T}$ denote the Borel σ-algebra of the cylinder sets in T, and let μ denote the measure induced by the process V. Thus, V, V_1, $V_2, \ldots$ are i.i.d. random variables uniformly distributed on $(T, \mathcal{T}, \mu)$. Take $X(\cdot)$ to be a SαS Lévy random measure on $(T, \mathcal{T}, \mu)$ with control measure μ.

A family of random variables $\{\Phi(f),\ f \in S_d\}$ is called a generalized random field if the following two conditions hold:
(a) Φ is a linear random functional, i.e., $\Phi(af + bh) = a\Phi(f) + b\Phi(h)$ a.s. for all f, $h \in S_d$, and all a, $b \in R$;
(b) Φ has a version with values in the dual space S_d'.
Recall from (4) the definition of the functional F_f on space T. We now consider a family of random variables

$$\{\Phi(f),\ f \in S_d\} \stackrel{D}{=} \{\int_T F_f(u) X(du),\ f \in S_d\} \tag{8}$$

Proposition 1. *Let $0 < \alpha < 2$. Then (8) defines the generalized stable random field on S_d such that for every $f \in S_d$ and $t \in R$,*

$$\begin{aligned} \mathbf{E}\exp\{it\Phi(f)\} &= \exp\left\{-|t|^\alpha \int_T |F_f(u)|^\alpha \mu(du)\right\}, \\ &\equiv \exp\left\{-|t|^\alpha \mathbf{E}\,|F_f(V)|^\alpha\right\} \end{aligned} \tag{9}$$

Theorem 2. *Let $0 < \alpha < 2$. As $n \to \infty$, the finite-dimensional distributions of $\{\Phi_n(f),\ f \in S_d\}$ converge weakly to those of the stable random field $\{\Phi(f),\ f \in S_d\}$.*

In order to construct a kth order integral with respect to Lévy random measure X, we have to consider systems of k Markov processes. For a function f_k of k variables, define the functional

$$F_{f_k}(V_1, \ldots, V_k) := \int_0^\infty \cdots \int_0^\infty e^{-t_1 - \ldots - t_k} f_k(V_1(t_1), \ldots, V_k(t_k)) dt_1 \ldots dt_k \tag{10}$$

Since

$$\mathbf{E}\left|F_{f_k}(V_{i_1}, \ldots, V_{i_k})\right|^2 = \langle f_k, f_k\rangle, k \geq 1, \tag{11}$$

where

$$\langle f_k, h_k\rangle := \int_{R^{3dk}} m(d\mathbf{a}) f_k(\mathbf{x}) g^2(\mathbf{a}, \mathbf{x}) g(\mathbf{x}, \mathbf{y}) h_k(\mathbf{y}) d\mathbf{x}\, d\mathbf{y} \tag{12}$$

$$g(\mathbf{x}, \mathbf{y}) = g(x_1, y_1) \ldots g(x_k, y_k),\ g^2(\mathbf{a}, \mathbf{x}) = g^2(a_1, x_1) \ldots g^2(a_k, x_k) \tag{13}$$

(see Feldman and Rachev, 1993), we restrict our parameter set to functions from the space

$$S_d^k \equiv S_d^k(g) := \left\{f_k : f_k = f_k(x_1, \ldots, x_k) \text{ symmetric in } C^\infty(R^{dk}) \text{ with } \langle |f_k|, |f_k|\rangle < \infty\right\} \tag{14}$$

We are interested in studying the limiting distribution, as $n \to \infty$, of the normalized sum

$$\Psi_n(f_k) := \sum_{i \leq i_1 < \ldots < i_k \leq n} \sigma_{i_1} \ldots \sigma_{i_k} Y_{ni_1} \ldots Y_{ni_k} F_{f_k}(V_{i_1}, \ldots, V_{i_k}) \tag{15}$$

Proposition 3. *Let $0 < \alpha < 2$. The field*

$$\{\Psi(f_k), f_k \in S_d^k\} \stackrel{D}{=} \{X^k F_{f_k}, f_k \in S_k^d\}$$
$$\equiv \left\{ \int_T \cdots \int_T F_{f_k}(u_1, \ldots, u_k) X(du_1) \ldots X(du_k),\ f_k \in S_k^d \right\} \quad (16)$$

is well defined and is a generalized random field on S_d^k.

Theorem 4. *Let $0 < \alpha < 2$. As $n \to \infty$, the finite-dimensional distributions of $\left\{\Psi_n(f_k),\ f_k \in S_d^k\right\}$ converge weakly to those of $\left\{\Psi(f_k),\ f_k \in S_d^k\right\}$.*

3. PROOFS

We will now present the proofs of the results stated in Section 2. We start with the following lemma.

Lemma 5. *For $0 < \alpha < 2$ and every $f \in S_d$,*

$$\int_T |F_f(u)|^\alpha \mu(du) \le (\langle f, f \rangle)^{\alpha/2} < \infty \quad (17)$$

Proof. Use Lyapunov's inequality (p. 191 of Shiryayev, 1984) and (11) to get

$$\int_T |F_f(u)|^\alpha \mu(du) \equiv \mathbf{E}\left|F_f(V)\right|^\alpha \le \left(\mathbf{E}\left|F_f(V)\right|^2\right)^{\alpha/2} = (\langle f, f \rangle)^{\alpha/2} < \infty$$

The proof is complete.

Proof of Proposition 1. The functional $F_f(V) = \int_0^\infty e^{-t} f(V(t))dt \equiv F(f, V)$ is jointly measureable on $S_d \times T$. Moreover, for each fixed $f \in S_d$, $F_f(V) \in L^\alpha(T)$ (by Lemma 5). Thus, (cp. Rosinski, 1986) the integral in (8) is well defined. Therefore Φ has the integral representation given by (8) and (cf., (3.2.2) Samorodnitsky and Taqqu, 1990), its distribution is specified by (9). It remains to be shown that Φ is a generalized field. Clearly, Φ is linear. By Corollary 4.2 of Walsh (1986), continuity in probability assures that the field has values in the dual space S'_d.

By Theorem 2, p. 254 of Shiryayev (1984), and the linearity of Φ, it is enough to show that

$$\pi_j(t) \to 0 \text{ in } S_d \text{ as } j \to \infty \Rightarrow E|\Phi(\pi_j)|^p \to 0, \quad p > 0. \quad (18)$$

By property 1.2.10, p. 16 of Samorodnitsky and Taqqu (1990) for $0 < p < \alpha$

$$\mathbf{E}|\Phi(\pi_k)|^p = (c(p, \alpha))^p \left\{ \int_T |F_{\pi_k}(u)|^\alpha \mu(du) \right\}^{p\backslash\alpha}$$

where $c(p, \alpha)$ is a constant that depends on α and p, and hence does not affect the convergence. Thus (18) follows immediately from Lemma 5 and the proof is complete.

Proof of Theorem 2. By Proposition 3.1 in Szulga (1992) with $k = 1$, it is clear that as $n \to \infty$, $\Phi_n(f)$ converges weakly to the stable integral XF_f which in turn is equal, in distribution, to $\{\Phi(f),\ f \in S_d\}$ (see Proposition 1). Use the Cramér-Wold device (p. 49 of Billingsley, 1968) and the linearity of Φ_n and Φ to complete the proof.

For $f_k \in S_d^k$, define the functional $\tilde{F}_{f_k}(V_1, \ldots, V_k)$ as:

$$\tilde{F}_{f_k}(V_1, \ldots, V_k) = \begin{cases} F_{f_k}(V_1, \ldots, V_k) & \text{if all } k \text{ arguments are distinct} \\ 0 & \text{otherwise} \end{cases} \quad (19)$$

Since the probability that any two or more of the paths of the Markov processes V_1, $V_2, \ldots, V_n$ are identical is zero, it follows that

$$\left\{\tilde{F}_{f_k}(V_1,\ldots,V_k),\ f_k \in S_d^k\right\} \stackrel{D}{=} \left\{F_{f_k}(V_1,\ldots,V_k),\ f_k \in S_d^k\right\} \tag{20}$$

Consider the set $\Omega_0 = \{\omega : V_1(\omega) \neq \ldots \neq V_k(\omega)\}$. Clearly, $P(\Omega_0) = 1$. Moreover, when restricted to the set Ω_0, the functionals $F_{f_k}(V_1,\ldots,V_k)$ and $\tilde{F}_{f_k}(V_1,\ldots,V_k)$ are identical. Therefore, for the next two results in this section, we will restrict our attention to Ω_0, and denote the functionals as $F_{f_k}(V_1,\ldots,V_k)$.

Proof of Proposition 3. The functional $F_{f_k}(V_1,\ldots,V_k)$ is jointly measurable on $S_d^k \times T^k$. Since the functions $f_k(u_1,\ldots,u_k) = f(u_1)\ldots f(u_k)$ are dense in S_d^k, it suffices to prove the Proposition for such f_k. Note that in this case

$$F_{f_k}(u_1,\ldots,u_k) = F_f(u_1)\ldots F_f(u_k) \tag{21}$$

Following Kallenberg and Szulga (1989), we set ζ to be a Poisson process on $R\backslash\{0\}\times T$, with intensity $\nu \times \mu$, so that ζ is constructed from the jumps of the process X. We denote its symmetrized version by $\tilde{\zeta}$. Then the following representation of the integral holds

$$X^k F_{f_k} = \tilde{\zeta}^k(LF_{f_k}) \text{ a.s.} \tag{22}$$

Here L is the operator defined on the space of functionals on T^k as

$$LF_{f_k} \equiv LF_{f_k}(u_1,\ldots,u_k;x_1,\ldots,x_k) = x_1\ldots x_k F_{f_k}(u_1,\ldots,u_k),$$

where $x_i \in R\setminus\{0\}$, $u_i \in T$, $i = 1,\ldots,k$ and the integral on the right of (22) exists if F_{f_k} belongs to the class $\mathcal{L}$, where

$$\mathcal{L} = \left\{F_{f_k} : \zeta^k(LF_{f_k})^2 < \infty \text{ a.s.}\right\}.$$

Let $\zeta_1,\ldots,\zeta_k$ be independent copies of ζ and note that by Theorem 3.4 in Kallenberg and Szulga (1989), the condition $\zeta^k(LF_{f_k})^2 < \infty$ a.s. is equivalent to $\zeta_1\ldots\zeta_k(LF_{f_k})^2 < \infty$ a.s. By the independence of $\zeta_1,\ldots,\zeta_k$, (21), and Lemma 2.2 in Kallenberg and Szulga (1989), this is equivalent to

$$\int_T\int_{R\backslash\{0\}}\left((x^2F_f^2)\wedge 1\right)\nu(dx)\mu(du) < \infty \text{ a.s.}$$

which in its turn follows if the following two conditions hold:

$$\text{(i)} \int_T\int_{R\backslash\{0\}}\left((x^2\wedge 1)F_f^2\right)\nu(dx)\mu(du) < \infty \text{ a.s.}$$

$$\text{(ii)} \int_T\int_{R\backslash\{0\}}\left((x^2\wedge 1)(F_f^2\vee 1)\right)\nu(dx)\mu(du) < \infty \text{ a.s.}$$

Since $\int_{R\backslash\{0\}}(x^2\wedge 1)\nu(dx) < \infty$ for Lévy measure ν on $R\setminus\{0\}$ and (11) holds, (i)-(ii) are satisfied. Thus, the multiple integral $X^kF_{f_k}$ exists. Linearity of Ψ in f_k follows from the linearity of F_{f_k}. To show that Ψ has a version in $(S_d^k)'$ follow the proof of Proposition 1, keeping in mind that (cf. (1.6) of Kallenberg and Szulga, 1989):

$$E(\tilde{\zeta}^k(LF_{f_k}))^2 = \sum_{m=1}^{k} k!\binom{k}{m}^2\left\{\nu^m\left(\nu^{k-m}(x_1\ldots x_k)\right)^2\right\}\left\{\mu^m\left(\mu^{k-m}\left(F_f(u_1)\ldots F_f(u_k)\right)\right)^2\right\}$$

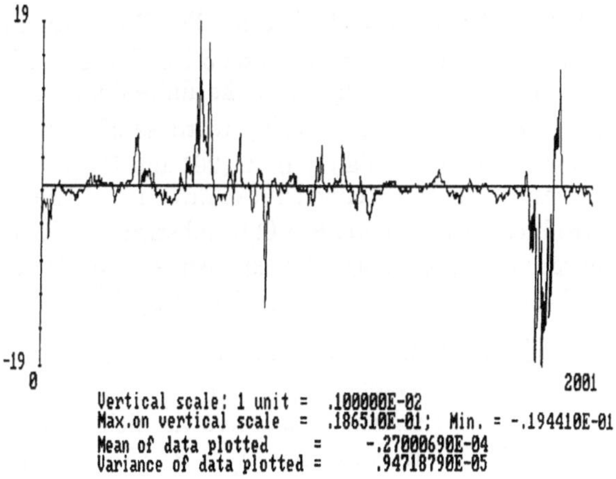

Fig. 6.1. Stable Process, $\alpha = 1.1$

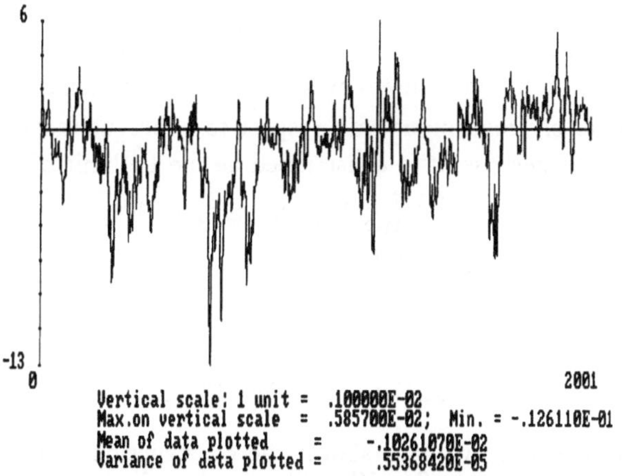

Fig. 6.2. Stable Process, $\alpha = 1.5$

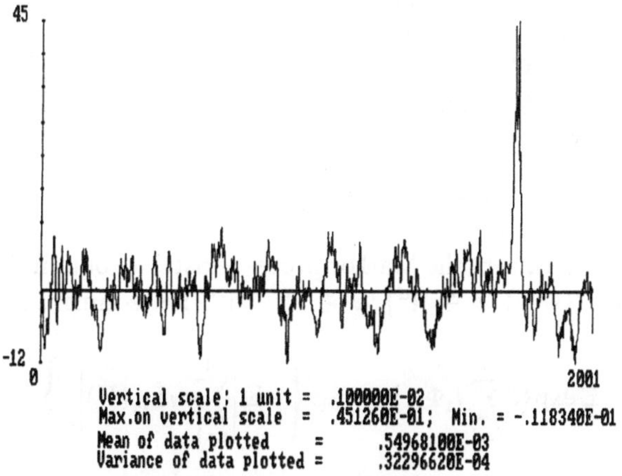

Fig. 6.3. Stable Process, $\alpha = 1.95$

where $\mu(F_f(u))^2 \equiv \int_T |F_f(u)|^2 \mu(du) = \langle f, f\rangle < \infty$. The Proposition is proved.

Proof of Theorem 4. It is clear from Proposition 3.1 in Szulga (1992), the Cramér-Wold device, and the linearity of $\Psi_n(f_k)$ that the finite-dimensional distributions of the field $\{\Psi_n(f_k),\ f_k \in S_d^k\}$, converge weakly to those of $\{X^k \tilde{F}_{f_k},\ f_k \in S_d^k\}$, as $n \to \infty$. The proof now follows immediately from (20) and Proposition 3.

Remark. The above results can be extended in two directions. The first one, which is quite trivial, is to introduce the parameter $\theta > 0$, to substitute the Green's function g^θ for g wherever it appeared, and to change the exponent e^{-t} to $e^{-\theta t}$ in the definitions (4) and (10). We will denote the resulting functionals by F_f^θ and the spaces of the test functions by $S_d^{\theta,k}$, $S_d^{\theta,1} \equiv S_d^\theta$. It is clear that all previous results hold under these changes; we will call the limiting families of fields $\{\Phi^\theta, \theta > 0\}$ and $\{\Psi^\theta, \theta > 0\}$.

The above results also hold for the case where the fields are indexed by measures. Let us briefly explain how to construct $F_{\gamma_k}^\theta \equiv F_{\gamma_k}^\theta(X_{i_1}, \ldots, X_{i_k})$ when γ_k is a symmetric measure on R^{dk}, $k \geq 1$, such that

$$\langle \gamma_k, \gamma_k \rangle_\theta = \int m(d\mathbf{a}) g^{2\theta}(\mathbf{a}, \mathbf{x}) g^\theta(\mathbf{x}, \mathbf{y}) \gamma_k(d\mathbf{x}) \gamma_k(d\mathbf{y}) < \infty. \tag{23}$$

We denote the class of such measures by $\mathcal{M}^{\theta,k} = \mathcal{M}^k(g^\theta)$, $\mathcal{M}^{\theta,1} \equiv \mathcal{M}^\theta$. Theorem 2.1 in Adler and Epstein (1987) guarantees existence of the functional $F_{\gamma_k}^\theta$ for each $\gamma_k \in \mathcal{M}^{\theta,k}$. If γ_k is absolutely continuous with respect to Lebesgue measure with density f_k then $F_{\gamma_k}^\theta = F_{f_k}^\theta$. Otherwise, it is constructed as the L^2 limit of path integrals of the form $F_{\gamma_k,\delta}^\theta := \int_{R_+^k} e^{-\theta(t_1+\ldots+t_k)} b_{\gamma_k,\delta}^\theta(V_1(t_1), \ldots, V_k(t_k)) dt_1 \ldots dt_k$. Here, $\gamma_\delta^\theta(d\mathbf{x}) = b_{\gamma_k,\delta}^\theta(\mathbf{x}) d\mathbf{x}$, and $b_{\gamma_k,\delta}^\theta(\mathbf{x}) = \int_{R^{kd}} e^{-\theta(\delta_1+\ldots+\delta_k)} p_{\delta_1}(x_1, y_1) \ldots p_{\delta_k}(x_k, y_k) \gamma_k(d\mathbf{y})$. Thus, γ_δ^θ is a smoothed version of γ_k.

4. POINT-INDEXED STABLE RANDOM FIELDS

In the previous section we have shown the construction of the stable random fields indexed by functions $f \in S_d$. We will now show how the same construction allows us to obtain stable fields on R^d. To do so, we would like to apply Theorem 2 to the case of measure $\gamma_x(dz) = \delta_x(z)dz$, where $x \in R^d$. Of course, this is possible only when $\gamma_x \in \mathcal{M}^1$, i.e., when (23) holds. This is true, for example, for Brownian motion on R^1. The corresponding functional $F_{\gamma_x}(V)$ is then the exponentially weighted local time $L_x(V)$ of the process V at point x.

Denote by Φ_x, $x \in R^d$, the stable random field obtained by applying the measure-variant of Theorem 2 to the sums of the functional $L_x(V)$. Then Φ_x, $x \in R^d$, has the integral representation

$$\Phi(x) \equiv \Phi_x = \int_T L_x(u) X(du).$$

So, whenever the local time of the Markov process exists, we obtain the point-indexed stable random field $\{\Phi(x),\ x \in R^d\}$, $x \in R^d$, with finite-dimensional distributions given by

$$\mathbf{E} \exp\{i \sum_{i=1}^n t_i \Phi_{x_i}\} = \exp\left\{ -\mathbf{E} \left| \sum_{i=1}^n t_i L_{x_i}(V) \right|^\alpha \right\}$$

Clearly, when a point-indexed field $\{\Phi(x),\ x \in R^d\}$ exists, we can always create a S_d indexed version of it by setting $\Phi(f) = \int_{R^d} \Phi(x) f(x) dx$.

In a similar manner, for those processes for which measure $\gamma_2(dx_1, dx_2) = \delta_x(x_1)\delta(x_1 - x_2)dx_1dx_2$ is of finite norm, one can use exponentially weighted intersection local times $L_x(V_1, V_2)$ to define the point-indexed random fields $\{\Psi(x),\ x \in R^d\}$ as limits of sums $\sum_{1 \le i_1 < i_2 \le n} \sigma_{i_1}\sigma_{i_2} Y_{n,i_1} Y_{n,i_2} L_x(V_{i_1}, V_{i_2})$. By Theorem 4

$$\Psi(x) \stackrel{D}{=} \int_T \int_T L_x(u_1, u_2) X(du_1) X(du_2).$$

5. RENORMALIZABLE STABLE FIELDS

The concept of renormalizability of families of generalized random fields was introduced in Adler and Epstein (1987) as an extension of the property of self-similarity. There the authors considered conditions for existence of this property for some Gaussian families and their multiple integrals. Here we modify their definition of renormalizability to include rescaling of an additional parameter and study this property for the families of stable fields that are obtained as limits of functionals of the paths of self-similar Markov processes.

In the following, we will assume that the initial measure of the Markov process is

$$m(dx) = \frac{1}{|B|} \mathbf{1}_B(x) dx \tag{24}$$

where $B \subset R^d$ is a fixed set of finite Lebesgue measure $|B|$.

Definition. *A process V is said to be self-similar with index β, if for any $\eta > 0$, V and the process*

$$\mathop{}^{\beta}_{\eta} V(t) := \eta^{-1} \tilde{V}(\eta^{\beta} t) \tag{25}$$

are identical in distribution, where $\tilde{V}$ is a process (on another probability space) which has the same transition probabilities as V but initial measure $\tilde{m}(dx) = \frac{1}{|\eta B|} \mathbf{1}_{\eta B}(x) dx$.

Note that the initial measure has to be changed and the set B has to be scaled in order to preserve the distribution of the starting points after scaling $\tilde{V}$ by the factor η. For any function $f(\mathbf{x})$ on R^{dk} and $\tau > 0$, define

$$\mathop{}^{\tau}_{\eta} f(\mathbf{x}) := \eta^{-\tau} f(\eta^{-1}\mathbf{x}). \tag{26}$$

Lemma 6. *Let V be a Markov process which is self-similar with index β, and let $q \in S_d^{\theta,k}$, $k \ge 1$. Then*

$$F_q^{\theta}(V_1, \ldots, V_k) \stackrel{D}{=} F^{\theta\eta^{-\beta}}_{\mathop{}^{\tau}_{\eta} q} \left(\tilde{V}_1, \ldots, \tilde{V}_k\right), \quad \tau = k\beta \tag{27}$$

Proof. Using the self-similarity property of V, and then making the change of variables $\eta^{\beta} t_i = s_i$, we get

$$\begin{aligned} F_q^{\theta}(V_1, \ldots, V_k) &= \int_{R_+^k} e^{-\theta(t_1 + \ldots + t_k)} q(V_1(t_1), \ldots, V_k(t_k)) dt_1 \ldots dt_k \\ &\stackrel{D}{=} \int_{R_+^k} e^{-\theta(t_1 + \ldots + t_k)} q(\eta^{-1}\tilde{V}_1(\eta^{\beta} t_1), \ldots, \eta^{-1}\tilde{V}_k(\eta^{\beta} t_k)) dt_1 \ldots dt_k \\ &= \int_{R_+^k} e^{-\theta\eta^{-\beta}(s_1 + \ldots + s_k)} q(\eta^{-1}\tilde{V}_1(s_1), \ldots, \eta^{-1}\tilde{V}_k(s_k)) \eta^{-k\beta} ds_1 \ldots ds_k \\ &\stackrel{D}{=} F^{\theta\eta^{-\beta}}_{\mathop{}^{\tau}_{\eta} q} \left(\tilde{V}_1, \ldots, \tilde{V}_k\right), \end{aligned} \tag{28}$$

with $\tau = k\beta$. This proves the lemma.

Let us denote by $X^m(.)$, the SαS Lévy random measure $X(.)$, on $(T, \mathcal{T}, \mu^m)$, with control measure μ^m. The index m indicates that the process V starts according to the initial measure m. Further, let $X^{m,k}f$ denote the multiple integral of order k, of the function f with respect to $X^m(.)$. When $k = 1$, we set $X^{m,1}f \equiv X^m f$. We define the following two families of stable fields:

$$\Phi^{\theta,m}(q) \stackrel{D}{=} \int_T F_q^\theta(u) X^m(du) \tag{29}$$

$$\Phi^{\theta\eta^{-\beta}}\left(\begin{smallmatrix}\tau \\ \eta\end{smallmatrix} q\right) \stackrel{D}{=} \int_T F^{\theta\eta^{-\beta}}_{\substack{\tau \\ \eta} q}(u) X^{\tilde{m}}(du) \tag{30}$$

Definition. *The family of random fields $\{\Phi^{\theta,m},\ \theta > 0\}$ is renormalizable with renormalizing parameters (τ, ρ) if, for every $\eta > 0$,*

$$\Phi^{\theta,m} \stackrel{D}{=} \underset{\eta}{\overset{\tau}{}}\Phi^{\theta\eta^{-\rho},\tilde{m}} \tag{31}$$

where

$$\underset{\eta}{\overset{\tau}{}}\Phi^{\theta\eta^{-\rho},\tilde{m}}(q) = \Phi^{\theta\eta^{-\rho},\tilde{m}}\left(\begin{smallmatrix}\tau \\ \eta\end{smallmatrix} q\right), \quad \underset{\eta}{\overset{\tau}{}}q(x) := \eta^{-\tau} q(\eta^{-1}x) \tag{32}$$

If the $\Phi^{\theta,m}$ are measure indexed, set

$$\underset{\eta}{\overset{\tau}{}}\Phi^{\theta\eta^{-\rho},m}(\gamma) = \Phi^{\theta\eta^{-\rho},m}\left(\begin{smallmatrix}\tau \\ \eta\end{smallmatrix} \gamma\right), \quad \underset{\eta}{\overset{\tau}{}}\gamma(A) := \eta^{-\tau+dk}\gamma(\eta^{-1}A), \quad \gamma \in \mathcal{M}^{\theta,k}. \tag{33}$$

Theorem 7. *Let $(V(t), m)$ be a self-similar Markov process with index β, and let g^θ, $\theta > 0$, be the corresponding Green function. The family of stable random fields $\{\Phi^{\theta,m},\ \theta > 0\}$ is renormalizable with parameters (β, β).*

Proof. Using (9) and Lemma 6 with $\tau = \beta$, one obtains:

$$\begin{aligned}
\mathbf{E}\exp\left\{it\Phi^{\theta\eta^{-\beta},\tilde{m}}\left(\begin{smallmatrix}\beta \\ \eta\end{smallmatrix} q\right)\right\} &= \exp\left\{-|t|^\alpha \mathbf{E}\left|F^{\theta\eta^{-\beta}}_{\substack{\beta \\ \eta} q}(\tilde{V})\right|^\alpha\right\} \\
&= \exp\left\{-|t|^\alpha \mathbf{E}\left|F_q^\theta(V)\right|^\alpha\right\} \\
&= \mathbf{E}\exp\left\{it\Phi^{\theta,m}(q)\right\}, \quad t \in R
\end{aligned}$$

and the proof is complete.

Our next result establishes the renormalizability of the family $\{\Psi^{\theta,m}(f_k), \theta > 0\}$ of multiple integrals of the fields $\{\Phi^{\theta,m},\ \theta > 0\}$.

Theorem 8. *Let $k > 1$. The family $\{\Psi^{\theta,m},\ \theta > 0\}$ is renormalizable with parameters $(k\beta, \beta)$.*

Proof. Let "$\Rightarrow$" denote weak convergence. By Theorem 4,

$$\sum_{1 \le i_1 < \ldots < i_k \le n} \sigma_{i_1} \ldots \sigma_{i_k} Y_{ni_1} \ldots Y_{ni_k} F^\theta_{q_k}(V_{i_1}, \ldots, V_{i_k}) \Rightarrow \Psi^{\theta,m}(q_k) \quad \text{as } n \to \infty,$$

$$\sum_{1 \le i_1 < \ldots < i_k \le n} \sigma_{i_1} \ldots \sigma_{i_k} Y_{ni_1} \ldots Y_{ni_k} F^{\theta\eta^{-\beta}}_{\substack{k\beta \\ \eta} q_k}(\tilde{V}_{i_1}, \ldots, \tilde{V}_{i_k}) \Rightarrow \Psi^{\theta\eta^{-\beta},\tilde{m}}(\begin{smallmatrix}k\beta \\ \eta\end{smallmatrix} q_k) \quad \text{as } n \to \infty.$$

By Proposition 3, $\Psi^{\theta,m}(q_k) \stackrel{D}{=} X^{m,k}F^\theta_{q_k}$, $\Psi^{\theta\eta^{-\beta},\tilde{m}}(\begin{smallmatrix}k\beta \\ \eta\end{smallmatrix} q_k) \stackrel{D}{=} X^{\tilde{m},k}F^{\theta\eta^{-\beta}}_{\substack{k\beta \\ \eta} q_k}$ and by Lemma 6, $F_q^\theta(V_1, \ldots, V_k) \stackrel{D}{=} F^{\theta\eta^{-\beta}}_{\substack{k\beta \\ \eta} q}(\tilde{V}_1, \ldots, \tilde{V}_k)$ Thus, we have.

$$\Psi^{\theta,m}_{q_k} \stackrel{D}{=} \Psi^{\theta\eta^{-\beta},\tilde{m}}_{\substack{k\beta \\ \eta} q_k}$$

and the theorem is proved.

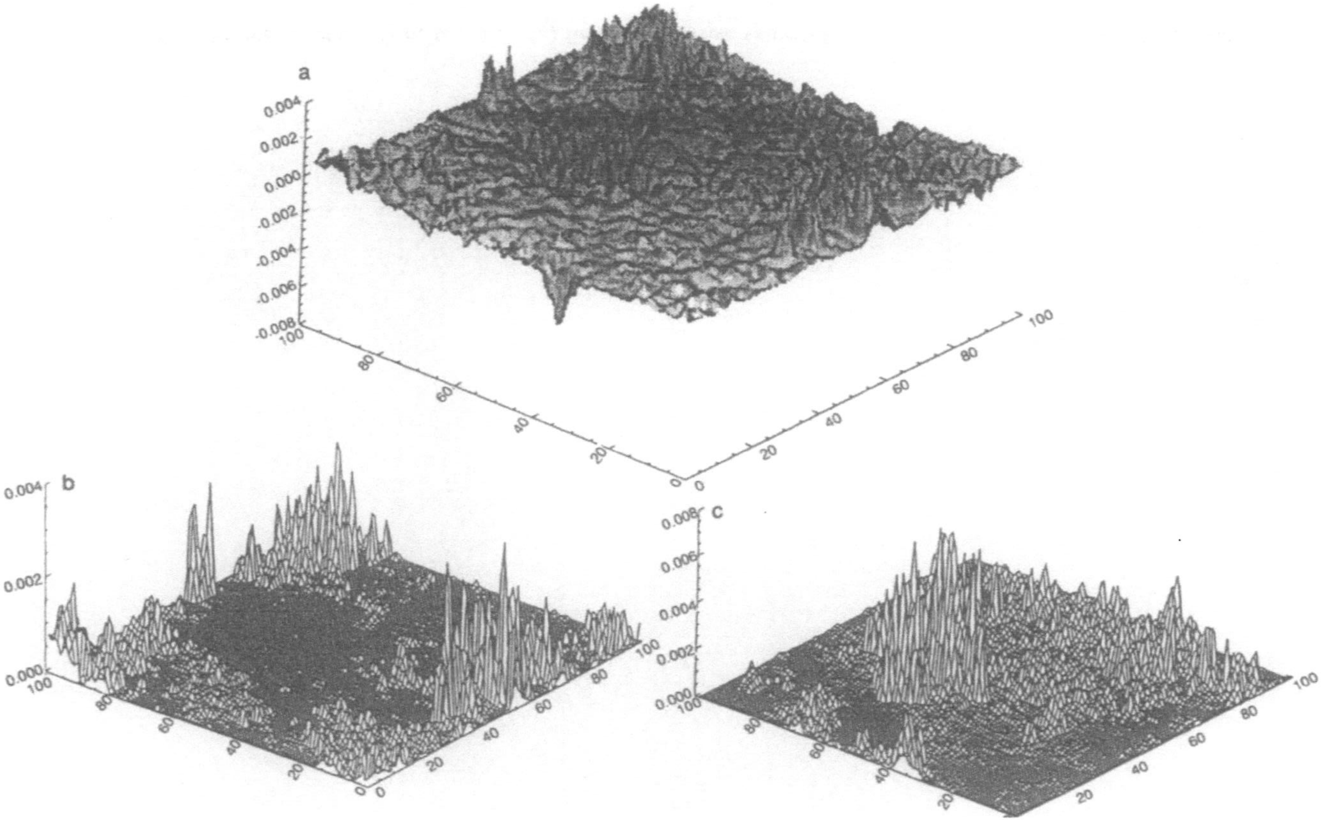

Fig. 6.4. Stable Random Field, $\alpha = 1.1$: (a) field, (b) positive values of the field, (c) negative values of the field

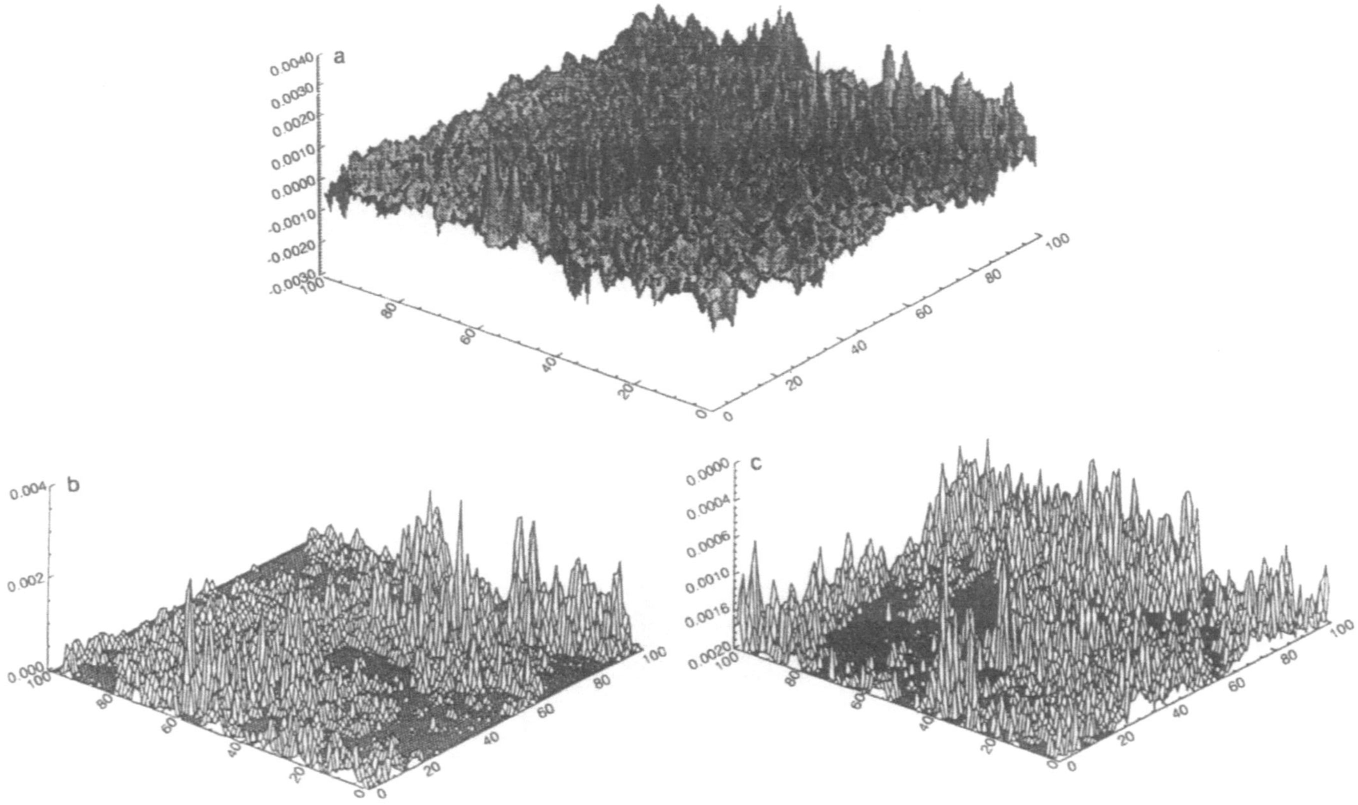

Fig. 6.5. Stable Random Field, $\alpha = 1.5$: (a) field, (b) positive values of the field, (c) negative values of the field

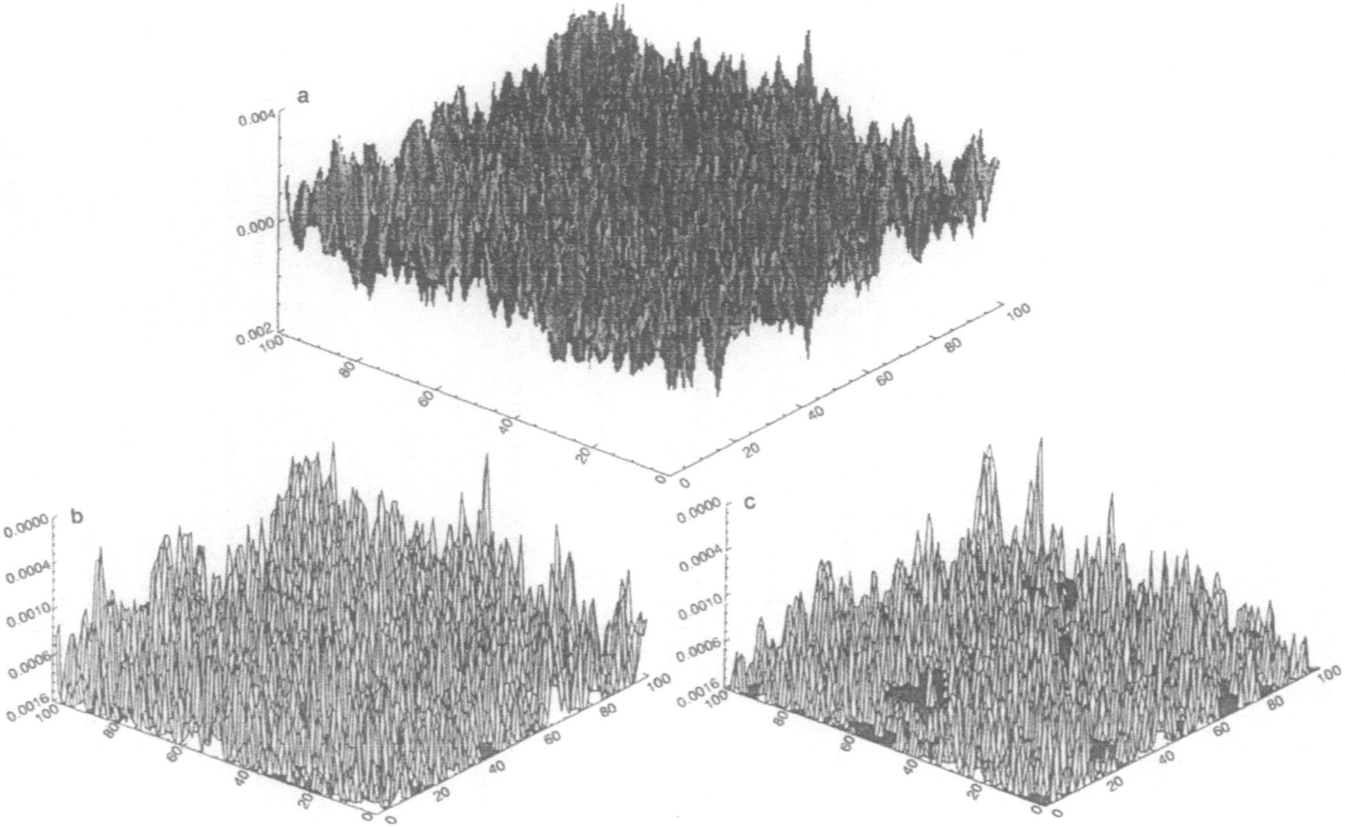

Fig. 6.6. Stable Random Field, $\alpha = 1.95$: (a) field, (b) positive values of the field, (c) negative values of the field

6. SIMULATION OF STABLE RANDOM FIELDS

Here we provide graphs of some R^d-indexed stable fields for $d = 1$ and 2. The simulations were based on results of Theorem 2 and discussion in Section 4, which allow one to build the fields Φ_x from the sums of functionals of the Markov processes V. We would like to explain briefly how these simulations were done.

We simulate symmetric random walks S_k on the integer lattice Z^d, $d = 1, 2$, i.e.,

$$S_0 = 0, \quad S_k = \xi_1 + \ldots + \xi_k, \; k = 1, 2, \ldots,$$

where, the ξ_i are i.i.d. Rademacher random variables. Their initial measure is $m(dx) = \mathbf{1}_B(x)dx$, where $B \subset R^d$ is a fixed set of finite Lebesgue measure $|B|$ to be specified later. Following a discussion in Dynkin (1988), use the functional of the random walk

$$F_{l,m}(S) = \frac{1}{m}\sum_{i=1}^{mT} e^{-i/m}\delta_l\left(\frac{1}{\sqrt{m}}S_i\right), \; l \in \frac{1}{\sqrt{m}}Z^d \tag{34}$$

to approximate the value of the exponentially weighted local time L_l of the Brownian motion (which exists for $d = 1$). This is a good choice for several reasons: As $m \to \infty$, the finite-dimensional distributions of the scaled random walk $\{S_{mt}/\sqrt{m}, t \in \{0, 1/m, 2/m, \ldots\}\}$ on the lattice $Z^d/\sqrt{m}$ converge to those of the Brownian motion on R^d. We would like to have a functional of $S/\sqrt{m}$ which approximates the functional of Brownian motion

$$F_f(V) = \int_0^\infty e^{-t} f(V(t))dt. \tag{35}$$

Since the value e^{-20} is less than 10^{-9}, for computational purposes we restrict the interval of integration in (35) to $[0, T]$, with $T = 20$. Dynkin (1988) gives the approximation as

$$F_{f,m}(S) = \frac{1}{m}\sum_{i=1}^{Tm} e^{-i/m} f(S_i/\sqrt{m}).$$

In order to obtain point-indexed fields, we replace the function f by the Dirac delta-function at l. Finally, we use sums

$$\Phi_n(l) := \sum_{i=1}^{n} \sigma_i Y_{n,i} F_{l,m}(S),$$

to approximate values of the stable field $\Phi(l)$.

For the case $d = 1$, we have chosen $m = 1000$, $n = 2000$, and $B = [-1000, 1000]$. We circularized the interval, i.e., when the random walk wanders off the left hand-side of $[-1000, 1000]$, it immediately reappears at the right hand-side and vice-versa. To fix the value of m we first obtained graphs of (34) for one fixed realization of the random walk (but different values of m) and concluded that the graph does not change much for $m > 1000$. To choose the value of n we fixed $\alpha = 1.95$ and $m = 1000$ and generated independently $n = 1500$ values of the stable process at a particular point. This histogram was very similar to the histograms of 1500 values from a SαS distribution with $\alpha = 1.95$ and $\sigma = 0.014678$ generated by the software package Splus. We also estimated the values of the parameters α and σ using a method based on McCullough (1986). The value of α was close to 1.95 and $\sigma = 0.014678$.

For $d = 2$ we took $m = 1500$, $n = 4000$ and $B = [0, 100) \times [0, 100)$. As for $d = 1$, we circularized the region.

To generate the stable random variates $Y_{n,i}$ that satisfy (2) we followed the algorithm in Bratley, Fox and Schrage (1987), which generates SαS random variates with scaling factor $\sigma = 1$, and then multiplied the values by $n^{-1/\alpha}$. The value of α is chosen before simulating the stable and Rademacher random variates.

The results of our simulations are presented in Figures 6.1-6.6. Figures 6.1-6.3 present SαS processes $\Phi(x)$, $x = l/m$, $l = -1000, -999, \ldots, 1000$ for three values of the parameter $\alpha = 1.1, 1.5$ and 1.95. The points $l = 0$ and 2001 on the graphs in the Figures correspond to the points $l = -1000$ and $x = 1000$, respectively. The graphs of the stable fields $\Phi(x)$, $x \in [0, 100) \times [0, 100)$ for $\alpha = 1.1$, 1.5 and 1.95 are given in Figures 6.4, 6.5, and 6.6. Each of these figures contains three graphs. The graph at the top is the graph of the stable random field. The graph at the bottom left is of the positive values of the field plotted separately; and the graph at the bottom right is of the negative values of the field. It is clear from the graphs that when α is close to 1, the field is fairly flat; as α increases the fields become more peaked.

Acknowledgements: The authors would like to thank Svetlozar Rachev for helpful discussions of stable distributions and processes. We would like to thank Makoto Maejima for the excellent lectures on stable processes that he gave at the University of California, Santa Barbara, and Dr. Jerzy Szulga for sharing the results of his research with us. We are extremely grateful to Dr. Phillip Feldman for letting us use his uniform random number generator package and also for his invaluable suggestions at various stages of the simulations. We would also like to thank Dr. Benny Cheng for providing us with his program which estimates the parameters of an α-stable distribution.

REFERENCES

Adler, R.J. (1989), Fluctuation theory for systems of signed and unsigned particles with interaction mechanisms based on interaction local times, *Advances in Applied Probability*, **21**, 331–356.

Adler, R.J. and Epstein, R. (1987), A central limit theorem for Markov paths and some properties of Gaussian random fields, *Stochastic Processes and their Applications*, **25**, 157–202.

Billingsley, P. (1968), *Convergence of Probability Measures*, John Wiley and Sons Inc., New York.

Blumenthal, R.M. and Getoor, R.K. (1964), Local times for Markov processes, *Z. Wahrscheinlichkeitstheorie*, **3**, 50–74.

Bratley, P., Fox. B.L. and Schrage, L.E. (1987), *A Guide to Simulation*, Springer-Verlag, New York.

Cambanis, S., Samorodnitsky, G. and Taqqu, M.S. (1990), *Stable Processes and Related Topics: A Selection of Papers from the Mathematical Sciences Institute Workshop*, Birkhaüser, Boston.

Dynkin, E.B. (1981), Additive functionals of several time-reversible Markov processes, *Journal of Functional Analysis*, **42**, 64–101.

Dynkin, E.B. (1988), Self-intersection gauge for random walks and for Brownian motion, *Annals of Probability*, **16**, 1–59.

Epstein, R. (1989), Some central limit theorems for functionals of the Brownian sheet, *Annals of Probability*, **17**, 538–558.

Feldman, R.E. and Rachev, S.T. (1993), U-statistics of random-size samples and limit theorems for systems of Markovian particles with non-Poisson initial distributions, to appear in *Annals of Probability*.

Halmos, P.R. (1950), *Measure Theory*, D. Van Nostrand Company, Inc., New York.

Janicki, A. and Weron, A. (1991), *Simulations and Ergodic Behavior of α-stable Stochastic Processes: A survey, Manuscript.*

Kallenberg, O. and Szulga, J. (1989), Multiple integration with respect to Poisson and Lévy Processes, *Probability Theory and Related Fields*, **83**, 101–134.
Krishnakumar, N. (1993), *Limit Theorems for Functionals of Markov Processes and Renormalizable Stable Fields*, Ph. D. Thesis, University of California, Santa Barbara.
Maejima, M. (1990), *Lectures on Self-Similar Stable Processes, Manuscript.*
Major, Peter (1981), Multiple Wiener-Ito integrals, *Lecture Notes in Mathematics*, **849**.
McCullough, J.H. (1986), Simple consistent estimators of stable distribution parameters, *Commun. Statist.-Simula.*, **14**, 1109–1136.
Revuz, D. and Yor, M. (1991), *Continuous Martingales and Brownian Motion*, Springer-Verlag, Berlin.
Rosinski, J. (1986), On stochastic integral representation of stable processes with sample paths in Banach spaces, *Journal of Multivariate Analysis*, **20**, 277–302.
Rosinski, J. and Woyczynski, W.A. (1984), On Itô-Stochastic integration with respect to p-stable motion: inner clock, integrability of sample paths, double and multiple integrals, *Annals of Probability*, **14**, 271–286.
Samorodnitsky, G. and Taqqu, M.S. (1990), *Stable Random Processes, Manuscript.*
Shiryayev, A.N. (1984), *Probability*, Springer-Verlag, New York.
Szulga, J. (1992), Limit distributions of U-statistics resampled by symmetric stable laws, *Probability Theory and Related Fields*, **94**, 83–90.
Walsh, J.B. (1986), An introduction to stochastic partial differential equations, in *Springer Lecture Notes in Math.*, **1180**, 265–439.
Weron, A. (1984), Stable processes and measures: A survey, in *Springer Lecture Notes in Math.*, **1080**, 306–364.

STABLE MODELS IN TESTABLE ASSET PRICING

Bertrand Gamrowski[1,*] and Svetlozar T. Rachev[2,**]

[1] École Polytechnique, Palaiseau CEDEX, France
[2] University of California, Santa Barbara, USA

ABSTRACT

The goal of this paper is to provide tests of Capital Asset Pricing Model (CAPM) and Arbitrage Pricing Theory (APT) in the case where asset prices follow symmetric Pareto-stable distributions. For the CAPM, we give a computable form of the "beta", that can be deduced from Fama's and Ross' works in this field, or from a direct proof. For the APT, we study an asymptotic stable version and we provide an original testing procedure. In both cases, our work is possible after we have established mathematical properties concerning the structure of spaces of symmetric Pareto-stable random variables.

1. INTRODUCTION

Capital Asset Pricing Model was introduced by Sharpe and Lintner [33,34] in 1964–65. It states that under reasonable market assumptions, the mean return of an asset, i, can be written

$$E(\tilde{R}_i) - \rho_0 = \beta_{im}(E(\tilde{R}_m) - \rho_0) \tag{1.1}$$

where ρ_0 is the return of the riskless asset, $\tilde{R}_m$ is the random return of the market portfolio (i.e., the portfolio of all marketed assets) and β_{im}—known as the "beta" of asset i—is $\frac{\mathrm{cov}(\tilde{R}_i;\tilde{R}_m)}{\mathrm{var}(\tilde{R}_m)}$.

CAPM was the first attempt to explain the asset returns behaviour (with one factor) and it has since known multiple theoretical developments. Merton [35] added a temporal dimension to CAPM by modeling asset returns by a diffusion process. Black, in his "zero-beta" version [36], widened assumptions on risk-free borrowing.

* This work was carried out while B. Gamrowski was visiting the Department of Statistics and Applied Probability, University of California, Santa Barbara. Grateful acknowledgement is made for hospitality.

** This work was supported in part by NSF grants DMS-89 02330 and DMS-91 03452. S.T. Rachev's work was also sponsored at the Laboratoire d'Econométrie de l'École Polytechnique by the French Ministère de la Recherche et de la Technologie with a grant for a "foreign researcher of international stature".

And Chamberlain [5,37] showed that the hypothesis of normality could be replaced with the weaker one of finite variance. But neither CAPM nor its extensions seemed really satisfactory when empirically tested (see e.g. [2, 3, 42]).

All above-referenced papers consider square integrability or– even more– normality of asset returns. Yet it has been very often questioned that prices should lie in L_2. The assumption of normality for returns may thus introduce inconsistencies in test. Pareto-stable (or α-stable) laws* are often proposed as better alternatives to normal laws for modeling returns (see e.g. [21]). Fama [13] tried to take this criticism into account and established CAPM in the case where asset returns are symmetric α-stable** and can be written

$$\tilde{R}_i = \rho_i + b_i\tilde{\delta} + \tilde{\epsilon}_i \tag{1.2}$$

where $\tilde{\delta}$ and $\tilde{\epsilon}_i$ are independent and symmetric α-stable. Fama showed that in this case, the coefficient "beta" in (1.1) is

$$\beta_{im} = \frac{1}{\sigma(\tilde{R}_m)} \frac{\partial\sigma(\tilde{R}_m)}{\partial(\lambda_{im}}, \tag{1.3}$$

where $\tilde{R}_m = \Sigma_i\lambda_{im}\tilde{R}_i$ ($\Sigma_i\lambda_{im} = 1$) models the return of the market portfolio and $\sigma(\cdot)$ is the scale parameter of the considered asset. Ross [26] also claimed that a CAPM-like formula would still stand with stable-distributed returns, should restriction (1.2) not be satisfied. But he did not give in this general case any expression of the beta similar to (1.3). The main drawback in the works of Fama and Ross is that it is in fact not possible to compute the beta, and consequently to test the theory in the stable case. Our first goal will be to give a computable expression of the beta in CAPM that will enable empirical tests, in the case where asset returns follow symmetric α-stable laws.

On the other hand, partly in response to CAPM empirical failures, Ross [25] suggested a multi-factor linear pricing model. Arbitrage Pricing Theory was born. APT implies that if the return $\tilde{R}_i$ of the i-th asset is assumed to be of the form $\rho_i + \beta_{i1}\tilde{\delta}_1 + \cdots + \beta_{ik}\tilde{\delta}_k + \tilde{\epsilon}_i$ (here the $\tilde{\delta}_j$'s are the *factors* and the $\tilde{\epsilon}_i$'s are the *idiosyncratic risks*), then, under usual assumptions, the mean return ρ_i of the i-th asset is approximately

$$\rho_0 + \beta_{i1}\gamma_1 + \cdots + \beta_{ik}\gamma_k \tag{1.4}$$

(here the γ_j's are *risk premia* linked to the factors). The idea contained in APT is that the asset mean return is not bound to its total variance, but only to the part of variance due to the market, since the idiosyncratic part can be diversified.

Two theories have evolved from Ross' work. The first is the so-called Asymptotic APT and was rigorously worked out by Huberman [15]. It considers a sequence of economies with a growing number of assets, but where the distance between the vector of real returns and the vector of explained returns (1.4) is bounded. The second is the Equilibrium APT (see [38, 39, 40]), where restrictions are set on the returns or on the utility functions of the agents implying that ρ_i is approximately (1.4).

In both theories, asset returns still lie in L_2. In 1984, Connor [7] introduced a general theory which encompassed an Equilibrium APT as well as the mutual fund separation theory (which was already some kind of extension of CAPM). In Connor's work, asset returns are still in L_2 but in 1988, Milne [20] further extended this theory, assuming that the asset returns belonged to just any normed vector space. Consequently, Milne's theory implies an Equilibrium α-stable APT. Nevertheless a stable version of APT has not been tested yet: tests of APT usually make the assumption of Gaussian returns. Our second goal will thus be to write an Asymptotic α-stable APT and a test of the α-stable APT.

* Appendix 1 provides all necessary definitions and properties of stable laws.

** In Fama's paper, $1 < \alpha < 2$. From now on, we will always assume that $1 < \alpha < 2$. $\alpha \leq 1$ gives different properties to stable laws and is never empirically met in modelling financial returns.

In section 2, we derive the CAPM beta from Fama's [13] and Ross' [27] works and we give a direct proof of the CAPM and one of an Asymptotic APT in the case of symmetric α-stable returns. And in section 3, we describe testing procedures for stable CAPM and APT. Appendix I sums up the necessary facts on α-stable laws. Appendix II contains the proofs of new properties of symmetric α-stable laws that we use in this paper.

2. STABLE LAWS IN PRICING MODELS

In paragraph A, we give a *computable* beta to Fama's and Ross' CAPMs. Paragraphs B and C contain respectively the proofs of CAPM and APT in the stable case.

A. Fama [13] was the first—by making assumption (1.1), as well as the usual assumptions of risk-averter consumers and the existence of a riskless-asset—to show a stable CAPM formula:

$$E(\tilde{R}_i) - \rho_0 = \frac{1}{\sigma(\tilde{R}_m)} \frac{\partial \sigma(\tilde{R}_m)}{\partial \lambda_{im}} (E(\tilde{R}_m) - \rho_0).$$

According to Appendix I, $\sigma(\tilde{R}_m) = (v_\alpha(\tilde{R}_m))^{1/\alpha}$, where $v_\alpha(.)$ is the variation of the considered α-stable random variable (r.v.). Hence,

$$\frac{1}{\sigma(\tilde{R}_m)} \frac{\partial \sigma(\tilde{R}_m)}{\partial \lambda_{im}} = \frac{1}{\alpha v_\alpha(\tilde{R}_m)} \frac{\partial v_\alpha(\tilde{R}_m)}{\partial \lambda_{im}}.$$

In addition, $v_\alpha(\tilde{m}) = \int_{S_n} |\Sigma_j \lambda_{jm} s_j|^\alpha \Gamma(ds)$ where Γ is the spectral measure of $\begin{pmatrix} \tilde{R}_1 \\ \vdots \\ \tilde{R}_n \end{pmatrix}$, which implies $\frac{\partial v_\alpha(\tilde{R}_m)}{\partial \lambda_{im}} = \alpha \int_{S_n} s_i (\Sigma_j \lambda_{jm} s_j)^{\langle \alpha - 1 \rangle} \Gamma(ds) = \alpha [\tilde{R}_i; \tilde{R}_m]_\alpha$. And thus, the coefficient in Fama's CAPM can be re-written $\frac{[\tilde{R}_i;\tilde{R}_m]_\alpha}{v_\alpha(\tilde{R}_m)}$. As in Cheng and Rachev [43], we can estimate this coefficient (of course with a traditional approximation of the market portfolio).

Now, we are going to extract the same expression of the beta from Ross' mutual fund separation theory [26] (but this time without restriction (1.1)). According to his theorem 2 (p. 267) and adding the assumption of the existence of a riskless asset, the return of the i-th asset can be written

$$\tilde{R}_i = \rho_i + b_i \tilde{\delta} + \tilde{\epsilon}_i, \tag{2.1}$$

where $\tilde{\delta} - E(\tilde{\delta})$ and the $\tilde{\epsilon}_i$'s are symmetric α-stable ($s\alpha s$) and satisfy $E(\tilde{\epsilon}_i|\tilde{\delta}) = 0$ and $\tilde{R}_m = \rho_0 + \tilde{\delta}$. We know, thanks to Kanter [44], that

$$E(\tilde{\epsilon}_i|\tilde{\delta}) = \frac{[\tilde{\epsilon}_i;\tilde{\delta}]_\alpha}{v_\alpha(\tilde{\delta})} (\tilde{\delta} - E(\tilde{\delta})).$$

Therefore, $E(\tilde{\epsilon}_i|\tilde{\delta}) = 0$ implies $[\tilde{\epsilon}_i; \tilde{\delta}]_\alpha = 0$. Combining the above results leads to $b_i = \frac{[\tilde{R}_i;\tilde{R}_m]_\alpha}{v_\alpha(\tilde{R}_m)}$. Ross [26] also shows that $E(\tilde{R}_i) - \rho_0 = b_i(E(\tilde{R}_m) - \rho_0)$ and his mutual fund separation theory implies that under the assumptions

(a) the preference of the agents has a Von-Neumann Morgenstern representation (with a monotone increasing concave function),
(b) there exists a riskless asset,
(c) the market portfolio has a non-zero variance,

we have the Sharpe-Lintner formula

$$E(\tilde{R}_i) - \rho_0 = \beta_{im}(E(\tilde{R}_m) - \rho_0)$$

with

$$\beta_{im} = \frac{[\tilde{R}_i;\tilde{R}_m]_\alpha}{v_\alpha(\tilde{R}_m)}.$$

Note that all *sas* vectors can be written of the form (2.1) with $E(\tilde{\epsilon}_i|\tilde{\delta}) = 0$. Fama's case is a restricted one since it was obtained under the more demanding hypothesis of independence of $\tilde{\delta}$ and of the $\tilde{\epsilon}_i$'s.

Since variation and covariation in *sas* spaces play the respective roles of variance and covariance in L_2, $\frac{[\tilde{R}_i;\tilde{R}_m]_\alpha}{v_\alpha(\tilde{R}_m)}$ is a not surprising generalization of the L_2-coefficient $\frac{\text{cov}(\tilde{R}_i;\tilde{R}_m)}{\text{var}(\tilde{R}_m)}$.

B. We shall follow the general idea of the proof of the L_2-CAPM given by Duffie ([9], p. 93) and due to Chamberlain [5] to prove our *sas* CAPM. This *sas* CAPM is equivalent to what we have deduced from Ross' mutual fund separation theory in Paragraph A, but its interest lies in the fact that it provides a proof using extensively properties of the covariation of stable vectors. Our proof is made possible by the properties that we set in Appendix 2.

We now introduce the definitions and a proposition that are essential for the theorem.

Definitions. For a given probability space $(\Omega;\mathcal{F};P)$ we define $\mathcal{E}$, the *exchange economy*† by $((S_\alpha;\succeq_i;\tilde{\omega}_i), i \in \mathcal{I})$, where S_α is the set of linear combinations of deterministic variables and all *sas* r.v.'s on $(\Omega;\mathcal{F};P)$. For each agent $i \in \mathcal{I}$:

- S_α is the choice space,
- $\succeq_i$ is the preference relation,
- $\tilde{\omega}_i$ is the initial endowment.

Let $A \subset S_\alpha$ be a set of assets for $\mathcal{E}$. We assume that A is such that the marketed space $M = \text{span } A$ is finite-dimensional. Let p be a linear pricing functional on M. We define the *α-stable* return $\tilde{R}_x$ by $\frac{\tilde{x}}{p(\tilde{x})}$.

A preference relation $\succeq$ will be called *α-strictly variation averse*‡ if

$$\forall \tilde{x}_1, \tilde{x}_2 \in S_\alpha / E(\tilde{x}_2) = 0 \text{ and } v_\alpha(\tilde{x}_1 + \tilde{x}_2) > v_\alpha(\tilde{x}_1),$$

then

$$\tilde{x}_1 \succ \tilde{x}_1 + \tilde{x}_2.$$

Similarly to the L_2-case, it is easy to demonstrate the following proposition.

Proposition.

$$\beta_{x\pi} = \frac{E(\tilde{R}_x)-\rho_0}{E(\tilde{R}_\pi)-\rho_0} \tag{2.2}$$

where:

- $\beta_{\xi_1\xi_2} = \frac{[\tilde{R}_{\xi_1};\tilde{R}_{\xi_2}]_\alpha}{v_\alpha(\tilde{R}_{\xi_1})}$,
- $\tilde{\pi}$ is the unique asset§ such that $p(\cdot) \equiv f_\alpha(\cdot;\tilde{\pi})$,
- $\rho_0 = \frac{1}{E(\tilde{\pi})}$.

Theorem. (Capital Asset Pricing Model for stable returns)

Let us consider an equilibrium for the exchange economy $(\mathcal{E};A)$ above defined. Suppose:

(a) the preference relation of each agent is α-strictly variation averse,
(b) the endowment $\tilde{\omega}_i$ of each agent i is in the marketed space,

† For the notion of exchange economy, we refer the reader to Duffie [9], p. 40.

‡ Fama [13] shows that any risk averse preference relation with Von-Neumann Morgenstern representation is α-strictly variation averse.

§ See Appendix II, corollary of lemma 3.

(c) the riskless asset 1_Ω has a non-zero market value,
(d) the market portfolio $\tilde{m} = \Sigma_i \tilde{\omega}_i$ has non-zero variation,
then, any asset $\tilde{x} \in M$ with non-zero market value satisfies

$$E(\tilde{R}_x) - \rho_0 = \beta_{xm}(E(\tilde{R}_m) - \rho_0). \tag{2.3}$$

Proof. We just have to show that the market portfolio is in span $\{1_\Omega; \tilde{\pi}\}$. This is sufficient to prove that $\beta_{x\pi}(E(\tilde{R}_\pi) - \rho_0) = \beta_{xm}(E(\tilde{R}_m) - \rho_0)$ and then, from (2.2), we can deduce (2.3).

Let $\tilde{x}$ be an equilibrium allocation and $\tilde{x}_0 = E(\tilde{x}) + \lambda_x \tilde{\pi}_0$ where $\tilde{\pi}_0 = \tilde{\pi} - E(\tilde{\pi})$ and $\lambda_x = \frac{[\tilde{x};\tilde{\pi}]_\alpha}{v_\alpha(\tilde{\pi})}$. We have $E(\tilde{x}) = E(\tilde{x}_0)$ and $p(\tilde{x}) = p(\tilde{x}_0)$. Thus, because of (a), $v_\alpha(\tilde{x}_0) = v_\alpha(\tilde{x})$. Moreover, $[\tilde{x} - \tilde{x}_0; \tilde{x}_0]_\alpha = 0$. Consequently, because of lemma 1 of Appendix II, $\tilde{x} = \tilde{x}_0$, which implies $\tilde{m} \in \text{span}\{1_\Omega; \tilde{\pi}_0\}$.

Q.E.D.

C. The proofs that we give now for the stable Asymptotic APT are close to the ones given by Huberman [15] in the L_2-case and do not present any additional difficulty. They merely use the new notion of variation instead of variance.

We consider a sequence of economies. In the n-th economy, there are n assets whose returns are modeled by r.v.'s $\tilde{R}_i^n$ $(i = 1, \cdots, n)$ on a probability space $(\Omega; \mathcal{F}; \mathcal{P})$. In our Paretian-stable setting we assume that $\tilde{R}_i^n$ is generated by a k-factor linear model of the form

$$\rho_i^n + \beta_{i1}^n \tilde{\delta}_1^n + \cdots + \beta_{ik}^n \tilde{\delta}_k^n + \tilde{\epsilon}_i^n \tag{2.4}$$

with

$$E(\tilde{\delta}_j^n) = 0. \tag{2.5}$$

$$\text{The } \tilde{\epsilon}_i' \text{s are } sas \text{ and independent.} \tag{2.6}$$

$$\forall i, v_\alpha(\tilde{\epsilon}_i) \leq V. \tag{2.7}$$

Using matrix notation, we re-write $\tilde{R}^n = \rho^n + \beta^n \tilde{\delta}^n + \tilde{\epsilon}^n$.

Definitions. θ^n is an arbitrage portfolio if ${}^t\theta^n e^n = 0$ with $e^n = \begin{pmatrix} 1 \\ \vdots \\ 1 \end{pmatrix}$. The return on a portfolio θ^n is ${}^t\theta^n \tilde{R}^n$. An *α-arbitrage* $(1 < \alpha \leq 2)$ is a subsequence (denoted by n for simplicity sake) of arbitrage portfolios whose returns satisfy $\lim_{n\to\infty} E({}^t\theta^n \tilde{R}^n)$ $= +\infty$ and $\lim_{n\to\infty} v_\alpha({}^t\theta^n \tilde{R}^n) = 0$. For $\alpha = 2$, this is the classical arbitrage notion.

Theorem. (Arbitrage Pricing Theory for stable returns 1)

Suppose that the returns on the risky investments satisfy conditions (2.4)–(2.7) and that there is no α-arbitrage, then

$$\exists A \in \Re \text{ and } \forall n, \quad \exists \rho_0^n, \gamma_1^n, \cdots, \gamma_k^n \in \Re / \sum_{i=1}^{n} |\rho_i^n - \rho_0^n - \sum_{j=1}^{k} \beta_{ij}^n \gamma_j^n|^\alpha \leq A.$$

Proof. Project ρ^n on the subspace spanned by $e^n = \begin{pmatrix} 1 \\ \vdots \\ 1 \end{pmatrix}$ and $\beta_j^n = \begin{pmatrix} \beta_{1j}^n \\ \vdots \\ \beta_{nj}^n \end{pmatrix}$ $(j = 1, \cdots, k)$.

$$\exists \rho_0^n \in \Re, \gamma^n \in \Re^k \text{ and } \theta^n \in \Re^n / \ \rho^n = \rho_0^n e^n + \beta^n \gamma^n + \theta^n$$

with ${}^t\theta^n e^n = 0$ and ${}^t\theta^n \beta^n = 0$. Then $\|\theta^n\|_\alpha^\alpha = \sum_i |\theta_i^n|^\alpha = \sum_i |\rho_i^n - \rho_0^n - {}^t\beta_i^n \gamma^n|^\alpha$. Let us suppose that the theorem is false. Consequently there exists a subsequence (still denoted by n) such that $\lim_{n\to\infty} \|\theta^n\|_\alpha = +\infty$. Let $p \in]-2; -1[$ and consider the portfolio $\zeta^n = \lambda_n \theta^n$ where $\lambda_n = ||\theta^n||_\alpha^p$. The return of ζ^n is ${}^t\zeta^n \tilde{R}^n = {}^t\zeta^n \theta^n + {}^t\zeta^n \tilde{\epsilon}^n = ||\theta^n||_\alpha^p ||\theta^n||_2^2 + ||\theta^n||_\alpha^p {}^t\theta^n \tilde{\epsilon}^n$.

$$E({}^t\zeta^n \tilde{R}^n) = ||\theta^n||_\alpha^p ||\theta^n||_2^2 \geq ||\theta^n||_\alpha^{p+2} \to +\infty.$$

Furthermore,

$$v_\alpha({}^t\zeta^n \tilde{R}^n) = [||\theta^n||_\alpha^p {}^t\theta^n \tilde{\epsilon}^n ; ||\theta^n||_\alpha^p {}^t\theta^n \tilde{\epsilon}^n]_\alpha,$$

$$v_\alpha({}^t\zeta^n \tilde{R}^n) = ||\tilde{\theta}||_\alpha^{p\alpha} \sum_i |\theta_i^n|^\alpha [\tilde{\epsilon}_i^n ; \tilde{\epsilon}_i^n]_\alpha,$$

$$v_\alpha({}^t\zeta^n \tilde{R}^n) \leq ||\theta^n||_\alpha^{(p+1)\alpha} V \to 0.$$

This contradicts the hypothesis of non-arbitrage.

Q.E.D.

We now examine a stationary model in which (2.4) is replaced with

$$\tilde{R}^n = \rho^n + \beta\tilde{\delta}^n + \tilde{\epsilon}^n. \tag{2.4'}$$

In the stationary theorem, we no longer consider a sequence of economies but one economy with an infinite number of assets. Yet the empirical interpretation remains the same.

Theorem. (Arbitrage Pricing Theory for stable returns 2)

Suppose that the returns on the risky investments satisfy the conditions (2.4′), (2.5)–(2.7), and that there is no α-arbitrage, then,

$$\exists \rho_0, \gamma_1, \cdots, \gamma_k \in \Re / \sum_{j=1}^{+\infty} \left| \rho_i - \rho_0 - \sum_{j=1}^{k} \beta_{ij}\gamma_j \right|^\alpha < +\infty.$$

Proof. The arguments are basically the same as in the traditional case (see Huberman [15]).

3. HOW TO TEST STABLE PRICING MODELS

A. For both models, the main interest is to see whether an $\alpha < 2$ lessens the empirical failures of tests run under the L_2-assumption.

We start with CAPM. Cheng and Rachev [43] provide a mean for computing the beta. It should thus be easy to reproduce the first test of CAPM (see e.g., [42, 3]). For example, Miller and Scholes [42] show that $\Delta_x = E(\tilde{R}_x) - \rho_0 - \beta_{xm}(E(\tilde{R}_m) - \rho_0)$ tends to be positive when β_{xm} is low, and negative when β_{xm} is high (whereas it should be nil according to CAPM). The question is whether stable betas would be more satisfactory. In a second step one should check if a stable version of Black's zero-beta CAPM [36] better resists to a test like Blume and Friend's [3].

APT's case is more complicated. We would like to see for example if Reinganum's size effect [23]¶ is weakened under stable assumptions. For this, we first have to compute the matrix $\beta(N \times k)$ that best fits into (2.4) for a given large number of assets N with assumptions (2.5)–(2.7). This requires a multivariate method. We cannot hope to mimic Ross' method [24]. Indeed, it uses Jöreskog's factor analysis maximum likelihood estimators [17] and a factor analysis-like stable method would require knowledge of the distribution of the empirical covariation matrix $\hat{V}_\alpha(\tilde{x})$, while

¶Reinganum [23] showed that in the case of small firms, the real return often happened to be bigger than the explained return whereas it tended to be the opposite in the case of large firms.

we have only asymptotic estimators of this distribution (see [43]) at our disposal. We provide in Paragraph C an original testing procedure for a stable APT, but first, we develop in Paragraph B two theoretical points that make this procedure possible.

B. Without any restriction for the model (2.4)–(2.7), we assume that rank $\beta = k$ and that $\{\tilde{\delta}_1; \cdots; \tilde{\delta}_k\}$ is a free family of r.v.'s. One should notice that our linear model (2.4) can take the equivalent form $\tilde{R} = \rho + (\beta P^{-1})(P\tilde{\delta}) + \tilde{\epsilon}$ where the exponent N is omitted for simplicity sake and P is any inversible $k * k$ matrix. Consequently, our goal is not really to determine a particular β but the set span β = span βP^{-1} (where span β designs the spanning of the columns β_j of β) that do not depend on the inversible matrix P.

The following properties show that we can compute the factor loadings if we know the spectral measure of $\tilde{R}^0 = \beta\tilde{\delta}$ and then that the observation of $\tilde{R}$ enables us to deduce the spectral measure of $\tilde{R}^0$. Our procedure will be built from these two facts.

Property 1.

$$\text{span } \beta = \text{span } \nabla r^0$$

where $r^0(\theta)$ is the risk of a portfolio θ on $\tilde{R}^0$ and span ∇r^0 = span $\{\nabla r^0(\theta), \theta \in \Re^N\}$.

Proof.

$$e^{-r^0(\theta)} = Ee^{i^t\theta\tilde{R}^0} = Ee^{i^t\theta\beta\tilde{\delta}} = Ee^{i^t({}^t\beta\theta)\tilde{\delta}}$$

and thus $r^0(\theta) = \int_{S_k} |^t\theta\beta s|^\alpha \Gamma_\delta(ds)$ where Γ_δ is the spectral measure of $\tilde{\delta}$.

$$\frac{\partial r^0}{\partial \theta_i} = \alpha \sum_j \beta_{ij} \int_{S_k} s_j({}^t\theta\beta s)^{\langle\alpha-1\rangle}\Gamma_\delta(ds),$$

$$\nabla r^0(\theta) = \alpha \sum_j \left[\tilde{\delta}_j; {}^t\theta\beta\tilde{\delta}\right]_\alpha \beta_j.$$

Thus, span $\nabla r^0 \subset$ span β.

To show the other inclusion, using lemma 2 of Appendix II, let us denote P an inversible matrix that transforms $\tilde{\delta}$ into $\tilde{\delta}' = P\tilde{\delta}$ such that $V_\alpha(\tilde{\delta}')$ is inversible. Since rank $\beta = k$, $\forall \ell \in \{1; \cdots; k\}$, $\exists \theta^\ell /\ {}^t\theta^\ell\beta\tilde{\delta} = \tilde{\delta}'_\ell$.

$$\nabla r^0(\theta^\ell) = \alpha \sum_j \left[\tilde{\delta}_j; \tilde{\delta}'_\ell\right]_\alpha \beta_j = \alpha\beta \begin{pmatrix} [\tilde{\delta}_1; \tilde{\delta}'_\ell]_\alpha \\ \vdots \\ [\tilde{\delta}_k; \tilde{\delta}'_\ell]_\alpha \end{pmatrix} = \alpha\beta P^{-1} P \begin{pmatrix} [\tilde{\delta}_1; \tilde{\delta}'_\ell]_\alpha \\ \vdots \\ [\tilde{\delta}_k; \tilde{\delta}'_\ell]_\alpha \end{pmatrix},$$

$$(\nabla r^0(\theta^1); \cdots; \nabla r^0(\theta^k)) = \alpha\beta P^{-1} V_\alpha(\tilde{\delta}'),$$

$$\beta = \frac{1}{\alpha}(\nabla r^0(\theta^1); \cdots; \nabla r^0(\theta^k)) V_\alpha(\tilde{\delta}')^{-1} P.$$

Thus, span $\beta \subset$ span ∇r^0.

Q.E.D.

Observing $\tilde{R}^0$ would enable us to compute ∇r^0. Then, we could carry a component analysis on the vectors $\nabla r(\theta^1), \cdots, \nabla r(\theta^\eta)$ where η is large and $\theta^1, \cdots \theta^\eta$ are arbitrarily chosen. We remind the reader that component analysis enables checking the hypothesis that a set of vectors belongs to a linear subspace. Component analysis would give us at the same time the appropriate number of factors in our model and the directions of the loading vectors.

Unfortunately, we do not observe $\tilde{R}^0$ but $\tilde{R} = \rho + \tilde{R}^0 + \tilde{\epsilon}$. Thus, we cannot estimate directly $\Gamma_0 = \Gamma_{\tilde{R}^0}$ and r^0 but only $\Gamma = \Gamma_{\tilde{R}}$ and $r(\theta) = v_\alpha({}^t\theta\tilde{R})$. Nonetheless, we known that $\Gamma = \Gamma_0 + \Gamma_\epsilon$ where Γ_ϵ is the spectral measure of $\tilde{\epsilon}$. Since $\tilde{\epsilon}$ is a vector of idiosyncratic risks, its spectral measure is composed of Dirac masses on the poles $(0;\cdots;0;1;0;\cdots;0)$ and $(0;\cdots;0;-1;0;\cdots;0)$. Thus, the idiosyncratic risks will appear in Γ only as Dirac masses on the poles. We can empirically remove these Dirac masses in Γ, but the problem now is to check whether Γ_0 also has a Dirac mass on one of the poles. Property 2 will solve this problem.

Property 2. Assuming (2.4)–(2.7), $\tilde{R}$ can be re-written

$$\tilde{R} = \rho + \beta'\tilde{\delta}' + \tilde{\epsilon}' \tag{2.4''}$$

where $\tilde{\delta}' = \begin{pmatrix} \tilde{\delta}'_1 \\ \vdots \\ \tilde{\delta}'_{k'} \end{pmatrix}$ $(k' \leq k)$, the $\tilde{\epsilon}_i$'s are idiosyncratic, span $\beta' \subset$ span β and $\tilde{R}^{0\prime} = \beta'\tilde{\delta}'$ has no Dirac distribution on the poles.

Remark. Passing from (2.4) to (2.4″) will in fact remove the idiosyncratic part of the β_i's, and at the same time, the Dirac masses on the poles in Γ_0. Applying APT theorem to (2.4) and (2.4″) is not contradictory but just says that the risk premia in the "idiosyncratic directions" of the factors are null.

Proof of Property 2. Let us suppose that Γ_0 has a Dirac mass $\lambda/2$ on $(1;0;\cdots;0)$. Then, $r^0(\theta) = \lambda|\theta_1|^\alpha + \int_{S_N} |^t\theta s|^\alpha \overline{\Gamma}_0(ds)$, where $\overline{\Gamma}_0$ is still a spectral measure, $r^0(\theta) \geq \lambda|\theta_1|^\alpha$. On the other hand, $r^0(\theta) = \int_{S_k} |^t\theta\beta s|^\alpha \Gamma_\delta(ds)$. Let $\theta^1, \cdots, \theta^{N-k}$ be the directions of (span $\beta)^\perp$. $r^0(\theta^i) = 0$ implies $\theta_1^i = 0$ for all $i = 1, \cdots, N-k$, and thus $e_1 = \begin{pmatrix} 1 \\ 0 \\ \vdots \\ 0 \end{pmatrix} \in$ span β. So, we can choose an inversible matrix P such that $\beta' = \beta P^{-1}$, $\tilde{\delta}' = P\tilde{\delta}$ and $\beta'_1 = e_1$. Let us consider $\tilde{R}^{0\prime} = \tilde{R}^0 - \tilde{\nu} e_1$ where $\tilde{\nu}$ is independent and $\tilde{\nu} \sim S_\alpha(\lambda^{1/\alpha})$. $\tilde{R}^{0\prime}$ no longer has a Dirac mass on the pole e_1. If we rename $\beta'_{11}\tilde{\delta}'_1 - \tilde{\nu}$ by $\tilde{\delta}'_1$, and if we define $\tilde{\epsilon}'_1 = \tilde{\epsilon}_1 + \tilde{\nu}$, $\tilde{\epsilon}'_i = \tilde{\epsilon}_i (i > 1)$, then we have $\tilde{R} = \rho + \beta'\tilde{\delta}' + \tilde{\epsilon}'$ where $\tilde{\epsilon}'$ is idiosyncratic, and the characteristic function of $\tilde{R}^{0\prime} = \beta'\tilde{\delta}'$ has no Dirac mass on the first pole.

Property 2 is obtained by repeating the same argument with each pole.

Q.E.D.

C. We are now able to describe a procedure of an APT test.

Step 1. Find the index of stability and the spectral measure of $\tilde{R}$ with time series of observations.

Step 2. Obtain the spectral measure of $\tilde{R}^0$ by "removing" the Dirac masses on the poles.

Step 3. Compute the gradient of r^0 for a large number of portfolios.

Step 4. Get span β by a component analysis on the set of the gradients.

Step 5. Estimate the risk premia by a regression of ρ on $\begin{pmatrix} 1 \\ \vdots \\ 1 \end{pmatrix}$ and the β_j's.

3. COMMENTS

One could argue that not the returns but the log-changes of the returns are empirically stated to be αs. But provided that the observed price changes are short-term (e.g., daily), the assumption of $s\alpha s$ returns is reasonable.

Symmetry is a restriction in the models we have just developed but in our opinion, the most restrictive and most regrettable assumption is that returns in CAPM and idiosyncratic risks have the same index of stability. As well as we know that the returns are not Gaussian, we know that they do not have either the same index of stability. But, until we have an inner product on $L_p (p < 2)$ or on the set of all stable distributions, with interesting properties, we cannot expect better. It is, however, not more restrictive than the hypothesis of square-integrability.

Computing work will be done in a forthcoming paper.

APPENDIX I: BASIC FACTS ON PARETO-STABLE LAWS

A random variable (r.v.) $\tilde{R}$ is stable if for all $a, b > 0$, there exist $c > 0$ and $d \in \Re$ such that $a\tilde{R}^1 + b\tilde{R}^2 \overset{d}{=} c\tilde{R} + d$ (where $\tilde{R}^1$ and $\tilde{R}^2$ are independent copies of $\tilde{R}$). This implies the existence of an $\alpha(0 < \alpha \leq 2)$—called the *index of stability*—such that $c = (a^\alpha + b^\alpha)^{\frac{1}{\alpha}}$. If $\alpha = 2$, $\tilde{R}$ is a Gaussian r.v.

$\tilde{R}$ is said to be *symmetric α-stable* (*sas*) if $\tilde{R} \overset{d}{=} -\tilde{R}$. The characteristic function of a *sas* r.v. is of the form $\phi_{\tilde{R}}(\theta) = Ee^{i\tilde{R}\theta} = e^{-\sigma^\alpha|\theta|^\alpha}$, where σ is the scale parameter. We will denote $\tilde{R} \sim S_\alpha(\sigma)$ or $\tilde{R} \sim sas$. All linear combinations of *sas* r.v.'s (with the same α) are *sas* r.v.'s. The set of all *sas* r.v.'s on a probability space has thus a structure of vector space.

A stable n-dimensional vector $\tilde{R}$ is defined in the same way by $\forall a, b > 0$, there exist $c > 0$ and $d \in \Re^n$ such that $a\tilde{R}^1 + b\tilde{R}^2 \overset{d}{=} c\tilde{R} + d$ (where $\tilde{R}^1$ and $\tilde{R}^2$ are independent copies of $\tilde{R}$). $\tilde{R}$ is *sas* if $\tilde{R} \overset{d}{=} -\tilde{R}$, if, and only if, all its components are *sas*, if, and only if, its characteristic function can be written

$$\phi_{\tilde{R}}(\theta) = Ee^{i^t\theta\tilde{R}} = \exp\left\{-\int_{S_n} |^t\theta s|^\alpha \Gamma(ds)\right\}$$

where Γ is a finite, symmetric (unique for $\tilde{R}$) measure on the n-dimensional unit sphere S_n. Γ is called the *spectral measure* of $\tilde{R}$.

Samorodnitsky and Taqqu [28] provide useful tools on spaces of *sas* r.v.'s when $1 < \alpha < 2$. They define an inner-product which they call *covariation*: if $\begin{pmatrix} \tilde{R}_1 \\ \tilde{R}_2 \end{pmatrix}$ is *sas* and Γ is its spectral measure, then the covariation is $[\tilde{R}_1; \tilde{R}_2]_\alpha = \int_{S_2} s_1 s_2{}^{\langle\alpha-1\rangle}\Gamma(ds)$, where the *signed power* $\xi^{\langle k\rangle}$ is defined for all reals ξ and k as $\xi^{\langle k\rangle} = |\xi|^k \text{sign}(\xi)$. One can notice that $\frac{d}{d\xi}|\xi|^\alpha = \alpha\xi^{\langle\alpha-1\rangle}$. The covariation is linear in its first term and satisfies the following properties in the second term:

- $\forall \lambda \in \Re$, $[\tilde{R}_1; \lambda\tilde{R}_2]_\alpha = \lambda^{\langle\alpha-1\rangle}[\tilde{R}_1; \tilde{R}_2]_\alpha$,
- if $\tilde{R}_2$ and $\tilde{R}_3$ are independent, then, $[\tilde{R}_1; \tilde{R}_2 + \tilde{R}_3]_\alpha = [\tilde{R}_1; \tilde{R}_2]_\alpha + [\tilde{R}_1; \tilde{R}_3]_\alpha$.

If $\tilde{R}_1$ and $\tilde{R}_2$ are independent, then, $[\tilde{R}_1; \tilde{R}_2]_\alpha = 0$ (but the opposite is usually false). In general $[\tilde{R}_1; \tilde{R}_2]_\alpha \neq [\tilde{R}_2; \tilde{R}_1]_\alpha$. Last, $\|\tilde{R}\|_\alpha = [\tilde{R}; \tilde{R}]_\alpha^{\frac{1}{\alpha}}$ is a norm that satisfies $|[\tilde{R}_1; \tilde{R}_2]_\alpha| \leq \|\tilde{R}_1\|_\alpha \|\tilde{R}_2\|_\alpha^{\alpha-1}$, and such that $\|\tilde{R}\|_\alpha$ is the scale parameter of $\tilde{R}$.

Since covariation is not linear in its second term, it does not endow a set of *sas* r.v.'s with a Hilbert space structure. When $1 < \alpha < 2$, $[\tilde{R}_1; \tilde{R}_2]_\alpha = 0$ is not a sufficient condition of independence, i.e., of orthogonality, as in the case where $\alpha = 2$. $[\tilde{R}_1; \tilde{R}_2]_\alpha = 0$ is only equivalent to the notion of *James-orthogonality* ($\tilde{R}_2 \perp_J \tilde{R}_1$) [43] defined by $\forall \lambda \in \Re$, $\|\lambda\tilde{R}_1 + \tilde{R}_2\|_\alpha \geq \|\tilde{R}_2\|_\alpha$. The success of L_2 in mathematical finance is partly bound to its Hilbertian structure. Samorodnitsky and Taqqu [28] however help us to translate a part of the L_2-theory in the *sas* case, even if covariation is not as powerful as the usual covariance.

We will have to consider finite-dimensional spaces of *sas* r.v.'s (called *sas spaces*). Let M be the set of linear combinations of deterministic variables and of variables belonging to a *sas* space E_α. $\tilde{R} \in M$ is still α-stable but no longer necessarily symmetric. $\tilde{R} - E(\tilde{R})$ is *sas*. For $\tilde{R}_1, \tilde{R}_2 \in M$, we will define the covariation as

$[\tilde{R}_1;\tilde{R}_2]_\alpha = [\tilde{R}_1 - E(\tilde{R}_1);\tilde{R}_2 - E(\tilde{R}_2)]_\alpha$. We also define $v_\alpha : M \to \Re, \tilde{R} \mapsto v_\alpha(\tilde{R}) = [\tilde{R};\tilde{R}]_\alpha$ and $f_\alpha : M^2 \to \Re, (\tilde{R}_1;\tilde{R}_2) \mapsto E(\tilde{R}_1)E(\tilde{R}_2) + [\tilde{R}_1;\tilde{R}_2]_\alpha$. $[\cdot;\cdot]_\alpha$, $v_\alpha(\cdot)$ and $f_\alpha(\cdot;\cdot)$ here play roles respectively similar to those of covariance, variance and scalar product in L_2.

$V_\alpha(\tilde{R})$ is the matrix with general term $[\tilde{R}_i;\tilde{R}_j]_\alpha$ and is called covariation matrix.

Cheng and Rachev [45] define the risk of a portfolio $\theta = \begin{pmatrix} \theta_1 \\ \vdots \\ \theta_N \end{pmatrix}$ of assets with returns $\tilde{R} = \begin{pmatrix} \tilde{R}_1 \\ \vdots \\ \tilde{R}_N \end{pmatrix}$ as $r(\theta) = v_\alpha({}^t\theta\tilde{R})$. This risk function proves useful in laying down properties of Appendix II and the APT test.

APPENDIX II: LEMMAS ON ORTHOGONALITY

The above properties resemble weakened Hilbertian properties. Whereas the definition of James orthogonality ensures that if

$$\tilde{R}_2 \perp_J \tilde{R}_1 - \tilde{R}_2 \text{ and } \tilde{R}_2 \neq \tilde{R}_1 \qquad (A.1)$$

then $||\tilde{R}_1|| \geq ||\tilde{R}_2||$, lemma 1 shows below that in the case of *sas* spaces, (A.1) implies $||\tilde{R}_1|| > ||\tilde{R}_2||$. Lemma 2 establishes the existence in any *sas* space of a basis $\tilde{e} = (\tilde{e}_1;\cdots;\tilde{e}_n)$ where $\tilde{e}_i \perp_J \tilde{e}_j$ if $j > i$. Such a basis could be called a James-orthogonal basis. The proof of lemma 2 comes from a process similar to Schmidt orthogonalization. Lemma 3 provides a link between any *sas* space and its dual space through covariation.

Claim. Let $\begin{pmatrix} \tilde{R}_1 \\ \tilde{R}_2 \end{pmatrix}$ be a *sas* vector and Γ its spectral measure. If

$$\int_{S_2} |s_1||s_2|^{\alpha-1}\Gamma(ds) = \left(\int_{S_2} |s_1|^p\Gamma(ds)\right)^{\frac{1}{p}} \left(\int_{S_2} |s_2|^{(\alpha-1)_q}\Gamma(ds)\right)^{\frac{1}{q}} \qquad (A.2)$$

(with p and q conjungate), then $\tilde{R}_1$ and $\tilde{R}_2$ are proportional r.v.'s.

Proof. (A.2) with "$\leq$" follows a Hölder's inequality. The equality is possible only if Γ is concentrate on a set where $|s_1|$ and $|s_2|^{(\alpha-1)}$ are proportional, i.e., a set $\{\bar{s};-\bar{s}\}$ containing two opposite points of the unit circle.

The characteristic function of $\begin{pmatrix} \tilde{R}_1 \\ \tilde{R}_2 \end{pmatrix}$ is $Ee^{i(\theta_1\tilde{R}_1+\theta_2\tilde{R}_2)} = \exp\{-k|\theta_1\bar{s}_1 + \theta_2\bar{s}_2|^\alpha\}$. Without loss of generality, we can pick $\bar{s}_1 \neq 0$. $Ee^{i(\theta_1\tilde{R}_1+\theta_2\tilde{R}_2)} = \exp\{-\sigma^\alpha|\theta_1 + \lambda\theta_2|^\alpha\}$ with $\lambda = \frac{\bar{s}_2}{\bar{s}_1}$ and $\sigma = k^{\frac{1}{\alpha}}|\bar{s}_1|$. Therefore,

$$\forall\theta_1,\theta_2, Ee^{i(\theta_1\tilde{R}_1+\theta_2\tilde{R}_2)} = Ee^{i(\theta_1+\lambda\theta_2)\tilde{R}} = Ee^{i(\theta_1\tilde{R}+\theta_2(\lambda\tilde{R}))},$$

which implies $\tilde{R}_2 = \lambda\tilde{R}_1$.

Q.E.D.

Lemma 1. If $\tilde{R}_1$ and $\tilde{R}_2$ are two non-zero *sas* r.v.'s, then, $[\tilde{R}_1 - \tilde{R}_2;\tilde{R}_2]_\alpha = 0$ implies

$$v_\alpha(\tilde{R}_1) = v_\alpha(\tilde{R}_2) \Rightarrow \tilde{R}_1 = \tilde{R}_2. \qquad (A.3)$$

Proof. Let us suppose that $[\tilde{R}_1 - \tilde{R}_2;\tilde{R}_2]_\alpha = 0$ and $v_\alpha(\tilde{R}_1) = v_\alpha(\tilde{R}_2)$.

$$[\tilde{R}_1;\tilde{R}_2]_\alpha = [\tilde{R}_1 - \tilde{R}_2;\tilde{R}_2]_\alpha + [\tilde{R}_2;\tilde{R}_2]_\alpha = v_\alpha(\tilde{R}_2). \qquad (A.4)$$

Combining (A.3) and (A.4), we get $[\tilde{R}_1;\tilde{R}_2]_\alpha = v_\alpha(\tilde{R}_1)^{\frac{1}{\alpha}} v_\alpha(\tilde{R}_2)^{1-\frac{1}{\alpha}}$, which implies (A.2) (with $p = \alpha$). Thus, $\exists\lambda/\tilde{R}_1 = \lambda\tilde{R}_2$. $v_\alpha(\tilde{R}_1) = |\lambda|^\alpha v_\alpha(\tilde{R}_2) = v_\alpha(\tilde{R}_2) \Rightarrow \lambda = 1$ ($\lambda = -1$ is made impossible by (A.4)). Thus, $\tilde{R}_1 = \tilde{R}_2$.

Q.E.D

Lemma 2. Any *sas* space E_α can be endowed with a basis $\tilde{e}$ such that $V_\alpha(\tilde{e})$ is inversible triangular.

Proof. Let $\tilde{R}$ be a basis of E_α.

Step 1. $\tilde{e}_1 = \tilde{R}_1$.

Step i $(i > 1)$.

Let $\tilde{f}(\lambda_1;\cdots;\lambda_{i-1}) = \tilde{f}(\lambda) = \tilde{R}_i - \sum_{j=1}^{i-1}\lambda_j\tilde{e}_j$.

$$v_\alpha(\tilde{f}(\lambda)) = \int_{S_i}\left|s_i - \sum_{j=1}^{i-1}\lambda_j s_j\right|^\alpha \Gamma_i(ds)$$

where Γ_i is the spectral measure of $\begin{pmatrix}\tilde{e}_1\\ \vdots\\ \tilde{e}_{i-1}\\ \tilde{R}_i\end{pmatrix}$.

$v_\alpha(\tilde{f}(\cdot))$ is differentiable and positive, and $\lim_{\|\lambda\|\to+\infty} v_\alpha(\tilde{f}(\lambda)) = +\infty$ (except in the case where the spectral measure is concentrate on a set of the form $\{s : \sum_{j=1}^{i-1}\bar{\lambda}_j s_j = 0\}$, which contradicts $\sum_{j=1}^{i-1}\bar{\lambda}_j\tilde{e}_j \neq 0$, and thus, that $\tilde{R}$ be a basis). Thus,

$$A_i = \arg\min_{\lambda\in\Re^{i-1}} v_\alpha(\tilde{f}(\lambda)) \neq \emptyset.$$

Let $\tilde{e}_i = \tilde{f}(\lambda^*)$ with $\lambda^* \in A_i$.

$$\frac{\partial}{\partial\lambda_k}v_\alpha(\tilde{f}(\lambda^*)) = 0 \Rightarrow \int_{S_i} s_k(s_i - \sum_j \lambda_j^* s_j)^{\langle\alpha-1\rangle}\Gamma_i(ds) = 0 \Rightarrow [\tilde{e}_k;\tilde{e}_i]_\alpha = 0, \forall k < i.$$

$V_\alpha(\tilde{e})$ is inversible triangular and $\tilde{e}$ is a basis.

Q.E.D.

Claim. If $\tilde{e}$ is a basis of a *sas* space, then $r(\theta) = v_\alpha({}^t\theta\tilde{e})$ is strictly convex.

Proof. $r(\theta) = \int_{S_k}|{}^t\theta s|^\alpha\Gamma_{\tilde{e}}(ds)$. Since $|\cdot|^\alpha$ is strictly convex, r is convex. We can have equality between $r(\lambda\theta^1 + (1-\lambda)\theta^2)$ and $\lambda r(\theta^1) + (1-\lambda)r(\theta^2)$ only if the spectral measure of $\Gamma_{\tilde{e}}$ is concentrate on a set such that ${}^t\theta^1_s = {}^t\theta^2 s$. This implies that $v_\alpha({}^t(\theta^1-\theta^2)\tilde{e}) = \int_{S_2}|{}^t(\theta^1-\theta^2)s|^\alpha\Gamma_{\tilde{e}}(ds) = 0$, and ${}^t(\theta^1-\theta^2)\tilde{e} = 0$, which contradicts the hypothesis that $\tilde{e}$ is a basis.

Q.E.D.

Lemma 3. Let E_α be a *sas* space and E_α^* its dual space (i.e., the set of all linear forms on E_α).

$$\forall p \in E_\alpha^*,\ \exists!\tilde{\pi}_0 \in E_\alpha/p(\cdot) \equiv [\cdot;\tilde{\pi}_0]_\alpha.$$

Proof. Let $\tilde{e}\begin{pmatrix}\tilde{e}_1\\ \vdots\\ \tilde{e}_k\end{pmatrix}$ be a basis of E_α and, $\forall\theta \in \Re^k, r(\theta) = r_{\tilde{e}}(\theta), q(\theta) = p({}^t\theta\tilde{e}), f(\theta) = \frac{1}{\alpha}r(\theta) - q(\theta)$. Since q is linear and r strictly convex and differentiable, f is strictly convex and differentiable. Let us consider $\begin{cases}\min f(\theta)\\ \theta\in\Re^k\end{cases}$. Since

$\lim_{\|\theta\|\to+\infty} f(\theta) = +\infty$ (because $\alpha > 1$ and $\tilde{e}$ is a basis), and f is strictly convex, $\exists!\theta^* \in \arg\min f$.

$$\forall i, \frac{\partial f}{\partial \theta_i}(\theta^*) = 0 = \frac{\partial}{\partial \theta_i}\left(\frac{1}{\alpha}\int_{S_k} |^t\theta^* s|^{\alpha}\Gamma_{\tilde{e}}(ds) - q(\theta^*)\right).$$

This implies

$$p(\tilde{e}_i) = \int_{S_k} s_i({}^t\theta^* s)^{\langle\alpha-1\rangle}\Gamma_{\tilde{e}}(ds) = [\tilde{e}_i; {}^t\theta^*\tilde{e}]_{\alpha}.$$

Let $\tilde{\pi}_0 = {}^t\theta^*\tilde{e}$. We have $p(\cdot) \equiv [\cdot; \tilde{\pi}_0]_{\alpha}$.

Q.E.D.

Corollary. Let M be the spanning of deterministic variables and of E_{α}, M its dual space.

$$\forall p \in M^*, \ \exists!\tilde{\pi} \in E_{\alpha}/\ p(\cdot) \equiv f_{\alpha}(\cdot; \tilde{\pi}).$$

REFERENCES

[1] A.R. ADMATI and P. PLEIDERER, Interpreting the factor risk premia in the arbitrage pricing theory, *J. Econ. Theory* **35** (1985), 191–195.

[2] J. AFFLECK-GRAVES and B. MAC DONALDS, Multivariate tests of asset pricing: the comparative power of alternative statistics, *J. Financial Quantitative Anal.* **25** (1990), 163–185.

[3] M. BLUME and I. FRIEND, A new look at the capital asset pricing model, *J. Finance* **28** (1973), 19–33.

[4] S.J. BROWN, The number of factors in security returns, *J. Finance* **44** (1989), 1247–1262.

[5] G. CHAMBERLAIN, Asset pricing in multiperiod securities markets, Unpublished, Department of Economics, University of Wisconsin, Madison.

[6] G. CHAMBERLAIN and M. ROTHSCHILD, Arbitrage, factor structure, and mean-variance analysis on large asset markets, *Econometrica* **51** (1983), 1281–1304.

[7] G. CONNOR, A unified beta pricing theory, *J. Econ. Theory* **34** (1984), 13–31.

[8] G. CONNOR and R.A. KORACZYK, Risk and return in an equilibrium A.P.T., *J. Financial Econom.* **21** (1988), 255–289.

[9] D. DUFFIE, "Security markets — Stochastic models", Academic Press, New York, 1988.

[10] P. DYBVIG and S.A. ROSS, Yes, the A.P.T. is testable, *J. Finance* **40** (1985), 1173–1188.

[11] E.F. FAMA, Portfolio analysis in a stable Paretian market, *Management Sci.*

[12] E.F. FAMA, Risk, return and equilibrium. Some clarifying comments, *J. Finance* **23** (1988), 29–40.

[13] E.F. FAMA, Risk, return and equilibrium, *J. Polit. Econ.* **78** (1970), 30–55.

[14] M.R. GIBBONS, S.A. ROSS and J. SHANKEN, A test of the efficiency of a given portfolio, *Econometrica* **57** (1989), 1121–1152.

[15] G. HUBERMAN, A simple approach to arbitrage pricing theory, *J. Econ. Theory* **28** (1982), 183–191.

[16] J. JOBSON and R. KORKIE, A performance interpretation of multivariate tests of asset set intersection spanning, and mean-variance efficiency, *J. Financial Quantitative Anal.* **24** (1989), 185–204.

[17] K.G. JORESKOG, Some contributions to maximum likelihood factor analysis, *Psychometrika* **32** (1967), 443–482.

[18] H.B. KAZEMI, A multiperiod asset pricing model with unobservable market portfolio: a note. *J. Finance* **43** (1988), 1015–1023.

[19] B.B. MANDELBROT, "The fractal geometry of nature". W.M. Freeman & Co., New York, 1977.
[20] F. MILNE, Arbitrage and diversification in a general equilibrium asset economy, *Econometrica* **56** (1988), 815–840.
[21] S. MITNIK and S.T. RACHEV, Modeling asset returns with alternative stable distributions and reply to comments on "Modeling asset returns with alternative stable distributions", *To appear in Econometric Reviews, Dec. 1993.*
[22] L.T. NIELSEN, Existence of equilibrium in CAPM, *J. Econ. Theory* **52** (1990), 223–231.
[23] M. REINGANUM, The Arbitrage Pricing Theory: is it testable?, *J. Finance* **37** (1982), 1129–1140.
[24] R. ROLL and S.A. ROSS, An empirical investigation of the Arbitrage Pricing Theory, *J. Finance* **35** (1980), 1073–1103.
[25] S.A. ROSS, The arbitrage theory of capital asset pricing, *J. Econ. Theory* **13** (1976), 341–360.
[26] S.A. ROSS, Mutual fund separation in financial theory — the separating distributions, *J. Econ. Theory* **17** (1978), 254–286.
[27] G. SAMORODNITSKY and M. TAQQU, Conditional moment and linear regression for stable processes, *Stochastic processes and the applications* **39** (1991), 183–199.
[28] G. SAMORODNITSKY and M. TAQQU, "Stable random processes", *To appear.*
[29] M.I. SCHNELLER, The arbitrage pricing theories: a synthesis and critical review, *Research in finance* **8** (1990), 1–21.
[30] W.F. SHARPE, Capital asset prices with and without negative holdings, *J. Finance* **46** (1991), 489–509.
[31] C. TRZCINKA, On the number of factors in the arbitrage pricing model, *J. Finance* **41** (1986), 347–368.
[32] K.C. WEI, An asset pricing theory unifying CAPM and APT, *J. Finance* **43** (1988), 881–892.
[33] J. LINTNER, The valuation of risk assets, and the selection of risky investments in stock portfolios and capital budgets, *Review of Economics and Statistics,* **47** (1965), 13–37.
[34] W. SHARPE, Capital asset prices: a theory of market equilibrium under conditions of risk, *J. Finance* **19** (1964), 425–442.
[35] R. MERTON, An intertemporal capital asset pricing model, *Econometrica* **41** (1973), 867–886.
[36] F. BLACK, Capital market equilibrium with restricted borrowing, *J. Business* **45** (1972), 444–454.
[37] G. CHAMBERLAIN, A characterization of the distributions that imply mean variance utility functions, *J. Econ. Theory* **29** (1983), 185–201.
[38] N.F. CHEN and J. INGERSOLL, Exact pricing in linear factor models with finitely many assets: a note, *J. Finance* **38** (1983), 483–496.
[39] P. DYBVIG, An explicit bound on individual assets' deviations from APT pricing in a finite economy, *J. Financial Econom.* **12** (1983), 483–496.
[40] M. GRINBLATT and S. TITMAN, Factor pricing in a finite economy, *J. Financial Econom.* **12** (1983), 497–507.
[41] R.C. JAMES, Inner products in normed linear spaces, *bulletin of A.M.S.* (1947), 559–566.
[42] F. BLACK, M. JENSEN and M. SCHOLES, The capital asset pricing model: some empirical tests, "Studies in the theory of capital markets", Ed. Jensen, New York, 1972.
[43] B.N. CHENG and S.T. RACHEV, Multivariate stable securities in financial markets, Technical Report, Dept. of Statistics and Appl.Probability, Univ. of California, Santa Barbara, CA 93106-3110, submitted to *Mathematical Finance.*
[44] M. KANTER, Linear sample spaces and stable processes, *J. Funct. Anal.* **9** (1972), 441–459.

A NOTE ON THE DEATH PROCESS WITH A RANDOM PARAMETER

J.M. Gani and G.R. Haynatzki

Department of Statistics & Applied Probability
University of California, Santa Barbara, CA 93106, U.S.A.

ABSTRACT

This note considers a pure death process with random death rate, subject to a 2-state Markov chain. The Kolmogorov forward equations lead to a formal solution in terms of Laplace transforms. The problem is a first step towards solving the carrier borne epidemic where the carriers are subject to this type of death process.

1. INTRODUCTION

Puri (1975, 1976) has studied stochastic processes under the influence of other stochastic processes. While his methods are powerful, only a few detailed examples of such processes are known. In this note, we shall consider what may possibly be one of the simplest such processes, namely a pure death process under the influence of a 2-state continuous time Markov chain.

The purpose of this work is to demonstrate the use of bivariate Markov chain methods in a practical situation. We can regard our case as modelling the death process of a cohort of individuals, whose death rates are governed by the random food supply. For similar processes, the reader is referred to Neyman, Park and Scott (1956).

Suppose that the death rate, Y(t), of a pure death process is a Markov chain in continuous time such that Y(t) can take the values μ_1 or μ_2 ($\mu_2 > \mu_1 > 0$) which reflect the abundance or scarcity of the food supply.

Let the infinitesimal transition probabilities for this process be

$$\begin{aligned} P\{Y(t+\delta t) = \mu_2 \mid Y(t) = \mu_1\} &= \lambda_1\delta t + o(\delta t), \\ P\{Y(t+\delta t) = \mu_1 \mid Y(t) = \mu_1\} &= 1 - \lambda_1\delta t - o(\delta t), \\ P\{Y(t+\delta t) = \mu_1 \mid Y(t) = \mu_2\} &= \lambda_2\delta t + o(\delta t), \\ P\{Y(t+\delta t) = \mu_2 \mid Y(t) = \mu_2\} &= 1 - \lambda_2\delta t - o(\delta t), \end{aligned} \tag{1.1}$$

where $\lambda_1, \lambda_2 > 0$ are usually distinct. The probability of more than one event in $(t, t+\delta t)$ is $o(\delta t)$. Then, following the usual treatment of this two-state Markov chain, it is known that starting with $Y(0) = \mu_1$, the state probabilities are given by

$$\begin{aligned} p_1(t) = p_{11}(t) = P\{Y(t) = \mu_1 | Y(0) = \mu_1\} &= \frac{\lambda_1}{\lambda_1+\lambda_2} e^{-(\lambda_1+\lambda_2)t} + \frac{\lambda_2}{\lambda_1+\lambda_2}, \\ p_2(t) = p_{12}(t) = P\{Y(t) = \mu_2 | Y(0) = \mu_1\} &= \frac{\lambda_1}{\lambda_1+\lambda_2}\left(1 - e^{-(\lambda_1+\lambda_2)t}\right), \end{aligned} \tag{1.2}$$

with Laplace transforms $\hat{p}_i(s) = \int_0^\infty e^{-st} p_i(t)dt$, $\mathrm{Re}(s) \ge 0$, $i=1,2$, of the form

$$\hat{p}_1(s) = \frac{\lambda_1}{\lambda_1+\lambda_2} \cdot \frac{1}{s+\lambda_1+\lambda_2} + \frac{\lambda_2}{\lambda_1+\lambda_2} \cdot \frac{1}{s},$$
$$\hat{p}_2(s) = \frac{\lambda_1}{\lambda_1+\lambda_2} \cdot \frac{1}{s} - \frac{1}{\lambda_1+\lambda_2} \cdot \frac{1}{s+\lambda_1+\lambda_2}. \tag{1.3}$$

We shall now study the pure death process, whose death parameter $Y(t)$ follows this Markov chain, using bivariate Markov chain methods. These may prove useful in more complex processes of the birth-death type.

2. THE DEATH PROCESS WITH DEATH RATE Y(t)

Let us consider the process $\{Y(t), X(t); t \ge 0\}$ with initial conditions $Y(0) = \mu_1$, $X(0) = N$, subject to the following transition probabilities:

$$P\{Y(t+\delta t) = \mu_1, X(t+\delta t) = j-1 | Y(t) = \mu_1, X(t) = j\} = \mu_1 j \delta t + o(\delta t),$$
$$P\{Y(t+\delta t) = \mu_2, X(t+\delta t) = j-1 | Y(t) = \mu_2, X(t) = j\} = \mu_2 j \delta t + o(\delta t),$$
$$P\{Y(t+\delta t) = \mu_2 | Y(t) = \mu_1\} = \lambda_1 \delta t + o(\delta t),$$
$$P\{Y(t+\delta t) = \mu_1 | Y(t) = \mu_2\} = \lambda_2 \delta t + o(\delta t),$$

which are adequate to define our process. Probabilities of more than one event will be of order $o(\delta t)$. Figure 1 indicates a possible realization of the process.

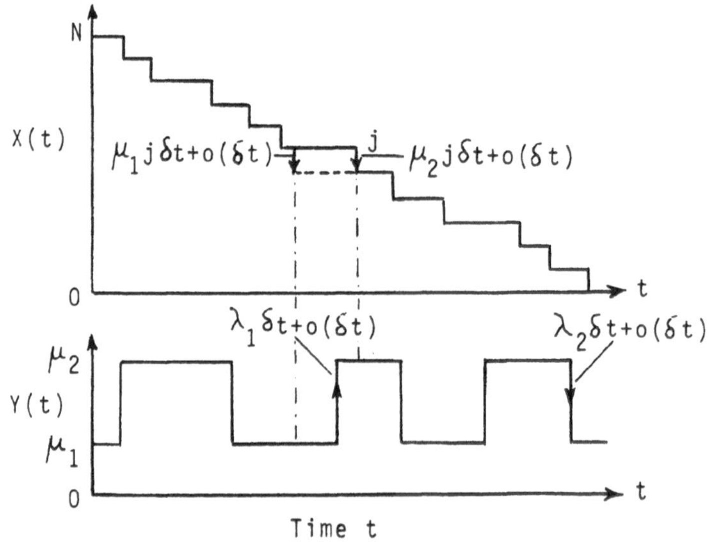

Figure 1. The bivariate Markov chain $\{Y(t), X(t); t \ge 0\}$.

We now derive the Kolmogorov forward equations for this process. We denote the state probabilities by

$$q_{1j}(t) = P\{Y(t) = \mu_1, X(t) = j | Y(0) = \mu_1, X(0) = N\},$$
$$q_{2j}(t) = P\{Y(t) = \mu_2, X(t) = j | Y(0) = \mu_1, X(0) = N\},$$

where $q_{ij}(t)$ indicates that at time t, $Y(t) = \mu_i$ (i=1,2), and $X(t) = j$ (j=0,1,...,N).

Using the standard methods, we derive the Kolmogorov forward equations

$$
\begin{aligned}
q'_{1j} &= -(\lambda_1 + j\mu_1)q_{1j} + (j+1)\mu_1 q_{1,j+1} + \lambda_2 q_{2j} \\
q'_{2j} &= -(\lambda_2 + j\mu_2)q_{2j} + (j+1)\mu_2 q_{2,j+1} + \lambda_1 q_{1j}
\end{aligned}
\tag{2.1}
$$

where j=0,1,...,N with $q_{1,N+1} = q_{2,N+1} = 0$ and q'_{ij} is the derivative of $q_{ij}(t)$ with respect to time t.

Transforming these equations to equations for the Laplace transforms $\hat{q}_{ij}(s) = \int_0^\infty q_{ij}(t)e^{-st}dt$, (Re $s \geq 0$), we can readily see from (2.1) that for j=0,1,...,N-1,

$$
\begin{aligned}
(s + \lambda_1 + j\mu_1)\hat{q}_{1j} &= (j+1)\mu_1 \hat{q}_{1,j+1} + \lambda_2 \hat{q}_{2j}, \\
(s + \lambda_2 + j\mu_2)\hat{q}_{2j} &= (j+2)\mu_2 \hat{q}_{2,j+1} + \lambda_1 \hat{q}_{1j},
\end{aligned}
\tag{2.2}
$$

while for j = N, since $q_{1N}(0) = 1$,

$$
\begin{aligned}
(s + \lambda_1 + N\mu_1)\hat{q}_{1N} - 1 &= \lambda_2 \hat{q}_{2N}, \\
(s + \lambda_2 + N\mu_2)\hat{q}_{2N} &= \lambda_1 \hat{q}_{1N}.
\end{aligned}
\tag{2.3}
$$

We may regard (2.2) and (2.3) as a set of linear equations in the $\hat{q}_{ij}$ which can be solved starting with $\hat{q}_{1N}$, $\hat{q}_{2N}$. The latter yield the solutions

$$
\begin{aligned}
\hat{q}_{1N} &= \frac{s + \lambda_2 + N\mu_2}{(s + \lambda_1 + N\mu_1)(s + \lambda_2 + N\mu_2) - \lambda_1\lambda_2}, \\
\hat{q}_{2N} &= \frac{\lambda_1}{(s + \lambda_1 + N\mu_1)(s + \lambda_2 + N\mu_2) - \lambda_1\lambda_2}.
\end{aligned}
\tag{2.4}
$$

Now we proceed to invert the Laplace transforms (2.4) and consider the general form of (2.2)-(2.3) in order to give an idea of the structure of the relevant probabilities $\{q_{ij}(t)\}$.

3. INVERTING TRANSFORMS; STRUCTURE OF THE PROCESS

It is clear that the Laplace transforms (2.4) can be readily broken up into simpler components as follows. Let us write

$$
\begin{aligned}
(s + \lambda_1 + N\mu_1)(s + \lambda_2 + N\mu_2) - \lambda_1\lambda_2 &= s^2 + (\lambda_1 + \lambda_2 + N\mu_1 + N\mu_2)s \\
&+ (N^2\mu_1\mu_2 + N\lambda_1\mu_2 + N\lambda_2\mu_1) = (s + \zeta_{1N})(s + \zeta_{2N})
\end{aligned}
\tag{3.1}
$$

with the distinct negative roots: $-\zeta_{1N}$,$-\zeta_{2N}$ given by

$$
\zeta_{1N,2N} = +\frac{1}{2}(\lambda_1 + \lambda_2 + N\mu_1 + N\mu_2) \pm \frac{1}{2}\sqrt{(\lambda_1 + \lambda_2 + N\mu_1 + N\mu_2)^2 - 4(N^2\mu_1\mu_2 + N\lambda_1\mu_2 + N\lambda_2\mu_1)}\ .
\tag{3.2}
$$

The values $\zeta_{1N} > \zeta_{2N}$ will both be positive real numbers since

$$
D_N = (\lambda_1 + \lambda_2 + N\mu_1 + N\mu_2)^2 - 4(N^2\mu_1\mu_2 + N\lambda_1\mu_2 + N\lambda_2\mu_1) \geq 0
$$

holds. We can see this easily if we write D_N as

$$(\lambda_1+\lambda_2)^2+(N\mu_1+N\mu_2)^2+2(\lambda_1+\lambda_2)(N\mu_1+N\mu_2)-4N^2\mu_1\mu_2-4N\lambda_1\mu_2-4N\lambda_2\mu_1$$

$$=(\lambda_1+\lambda_2)^2+(N\mu_1+N\mu_2)^2-4N^2\mu_1\mu_2+2N\lambda_1\mu_1-2N\lambda_1\mu_2+2N\lambda_2\mu_2-2N\lambda_2\mu_1$$

$$=\{(\lambda_1-\lambda_2)+(N\mu_1-N\mu_2)\}^2+4\lambda_1\lambda_2 ,$$

which is clearly positive. Then ζ_{1N} is clearly positive, and it is obvious that the positive square root of D_N will be smaller than $(\lambda_1+\lambda_2+N\mu_1+N\mu_2)$ so that $\zeta_{2N}>0$.

Let us now find $q_{1N}(t)$ and $q_{2N}(t)$. The Laplace transforms can, after some calculations, be written as

$$\hat{q}_{1N}(s)=\frac{1}{\zeta_{1N}-\zeta_{2N}}\left\{\frac{\zeta_{1N}-\lambda_2-N\mu_2}{s+\zeta_{1N}}-\frac{\zeta_{2N}-\lambda_2-N\mu_2}{s+\zeta_{2N}}\right\} ,$$

$$\hat{q}_{2N}(s)=\frac{\lambda_1}{\zeta_{1N}-\zeta_{2N}}\left\{-\frac{1}{s+\zeta_{1N}}+\frac{1}{s+\zeta_{2N}}\right\} , \tag{3.3}$$

and their inverses are

$$q_{1N}(t)=\frac{(\zeta_{1N}-\lambda_2-N\mu_2)e^{-\zeta_{1N}t}-(\zeta_{2N}-\lambda_2-N\mu_2)e^{-\zeta_{2N}t}}{\zeta_{1N}-\zeta_{2N}} ,$$

$$q_{2N}(t)=\lambda_1\frac{e^{-\zeta_{2N}t}-e^{-\zeta_{1N}t}}{\zeta_{1N}-\zeta_{2N}} . \tag{3.4}$$

We may for convenience write (3.3) as

$$\hat{q}_{1N}(s)=\frac{A_{2N}}{s+\zeta_{1N}}-\frac{B_{2N}}{s+\zeta_{2N}} ,$$

$$\hat{q}_{2N}(s)=C_N\left(-\frac{1}{s+\zeta_{1N}}+\frac{1}{s+\zeta_{2N}}\right) , \tag{3.5}$$

where these constants are fully defined by (3.3).

Let us now define

$$\hat{Q}_j=\begin{bmatrix}\hat{q}_{1j}\\ \hat{q}_{2j}\end{bmatrix},\quad \Lambda=\begin{bmatrix}\lambda_1 & -\lambda_2\\ -\lambda_1 & \lambda_2\end{bmatrix},\quad M=\begin{bmatrix}\mu_1 & 0\\ 0 & \mu_2\end{bmatrix},\quad I=\begin{bmatrix}1 & 0\\ 0 & 1\end{bmatrix};$$

we can then see that (2.2) and (2.3) can be expressed as

$$(sI+\Lambda+jM)\hat{Q}_j=(j+1)M\hat{Q}_{j+1} , \qquad j=0,1,\dots,N-1,$$

$$(sI+\Lambda+NM)\hat{Q}_N=E , \quad E'=[1,0].$$

The entire system can be represented as follows

$$\begin{bmatrix} sI+\Lambda+NM & & & & \\ -NM & sI+\Lambda+(N-1)M & & & \\ & & \ddots & & \\ & & -2M & sI+\Lambda+M & 0 \\ & & & -M & sI+\Lambda \end{bmatrix} \begin{bmatrix} \hat{Q}_N \\ \hat{Q}_{N-1} \\ \vdots \\ \hat{Q}_1 \\ \hat{Q}_0 \end{bmatrix} = \begin{bmatrix} E \\ 0 \\ \vdots \\ 0 \\ 0 \end{bmatrix} . \quad (3.6)$$

With $\hat{Q}_N$ known from (2.4) or (3.3) we see that

$$(sI+\Lambda+(N-1)M)\hat{Q}_{N-1} = NM\hat{Q}_N$$

whence

$$\hat{Q}_{N-1} = (M^{-1}(sI+\Lambda)+(N-1)I)^{-1}N\hat{Q}_N \; .$$

It follows therefore, that

$$\hat{Q}_j = \frac{N!}{j!}\{\prod_{k=j}^{N-1}(M^{-1}(sI+\Lambda)+kI)^{-1}\}(sI+\Lambda+NM)^{-1}E \quad (3.7)$$

$$= \frac{N!}{j!}\{(M^{-1}(sI+\Lambda)+jI)^{-1}(M^{-1}(sI+\Lambda)+(j+1)I)^{-1}\ldots$$

$$(M^{-1}(sI+\Lambda)+(N-1)I)^{-1}\}(sI+\Lambda+NM)^{-1}E \; .$$

Consider the matrix

$$(M^{-1}(sI+\Lambda)+kI)^{-1} = \begin{bmatrix} \dfrac{s+\lambda_1+\mu_1 k}{\mu_1} & \dfrac{-\lambda_2}{\mu_1} \\ \dfrac{-\lambda_1}{\mu_2} & \dfrac{s+\lambda_2+\mu_2 k}{\mu_2} \end{bmatrix}^{-1}$$

$$= D_k^{-1} \begin{bmatrix} \dfrac{s+\lambda_2+\mu_2 k}{\mu_2} & \dfrac{\lambda_2}{\mu_1} \\ \dfrac{\lambda_1}{\mu_2} & \dfrac{s+\lambda_1+\mu_1 k}{\mu_1} \end{bmatrix}, \quad (3.8)$$

where

$$D_k = \frac{(s+\lambda_1+\mu_1 k)(s+\lambda_2+\mu_2 k)}{\mu_1 \mu_2} \; .$$

This is similar in structure to the D_N already considered for $\hat{Q}_N$ in (3.3).

Let us now set $(s+\lambda_1+\mu_1 k)(s+\lambda_2+\mu_2 k)-\lambda_1\lambda_2=(s+\zeta_{1k})(s+\zeta_{2k})$, with $\zeta_{1k}>\zeta_{2k}>0$ exactly as before. Then $(M^{-1}(sI+\Lambda)+kI)^{-1}$ can be written as

$$(M^{-1}(sI+\Lambda)+kI)^{-1}=\begin{bmatrix} \dfrac{(s+\lambda_2+\mu_2 k)\mu_1}{(s+\zeta_{1k})(s+\zeta_{2k})} & \dfrac{\lambda_2\mu_2}{(s+\zeta_{1k})(s+\zeta_{2k})} \\ \dfrac{\lambda_1\mu_1}{(s+\zeta_{1k})(s+\zeta_{2k})} & \dfrac{(s+\lambda_1+\mu_1 k)\mu_2}{(s+\zeta_{1k})(s+\zeta_{2k})} \end{bmatrix} \tag{3.9}$$

$$=\begin{bmatrix} \mu_1\left(\dfrac{A_{2k}}{s+\zeta_{1k}}-\dfrac{B_{2k}}{s+\zeta_{2k}}\right) & \dfrac{\lambda_2\mu_2}{\lambda_1}C_k\left(-\dfrac{1}{s+\zeta_{1k}}+\dfrac{1}{s+\zeta_{2k}}\right) \\ \mu_1 C_k\left(-\dfrac{1}{s+\zeta_{1k}}+\dfrac{1}{s+\zeta_{2k}}\right) & \mu_2\left(\dfrac{A_{1k}}{s+\zeta_{1k}}-\dfrac{B_{1k}}{s+\zeta_{2k}}\right) \end{bmatrix},$$

where

$$A_{2k}=\frac{\zeta_{1k}-\lambda_2-k\mu_2}{\zeta_{1k}-\zeta_{2k}}\ ,\quad B_{2k}=\frac{\zeta_{2k}-\lambda_2-k\mu_2}{\zeta_{1k}-\zeta_{2k}}\ ,\ \text{and}$$

$$A_{1k}=\frac{\zeta_{1k}-\lambda_1-k\mu_1}{\zeta_{1k}-\zeta_{2k}}\ ,\quad B_{1k}=\frac{\zeta_{2k}-\lambda_1-k\mu_1}{\zeta_{1k}-\zeta_{2k}}\ ,\quad C_k=\frac{\lambda_1}{\zeta_{1k}-\zeta_{2k}}\ .$$

Note that for k=0, $\zeta_{10}=\lambda_1+\lambda_2, \zeta_{20}=0$, and our matrix above becomes

$$\begin{bmatrix} \dfrac{\mu_1\lambda_1}{\lambda_1+\lambda_2}\cdot\dfrac{1}{s+\lambda_1+\lambda_2}+\dfrac{\mu_1\lambda_2}{s(\lambda_1+\lambda_2)} & \dfrac{\lambda_2\mu_2}{\lambda_1+\lambda_2}\left(-\dfrac{1}{s+\lambda_1+\lambda_2}+\dfrac{1}{s}\right) \\ \dfrac{\mu_1\lambda_1}{\lambda_1+\lambda_2}\left(-\dfrac{1}{s+\lambda_1+\lambda_2}+\dfrac{1}{s}\right) & \dfrac{\mu_2\lambda_2}{\lambda_1+\lambda_2}\cdot\dfrac{1}{s+\lambda_1+\lambda_2}+\dfrac{\mu_2\lambda_1}{(\lambda_1+\lambda_2)s} \end{bmatrix}. \tag{3.10}$$

It follows that

$$\hat{Q}_j=\frac{N!}{j!}\prod_{k=j}^{N-1}\begin{bmatrix} \mu_1\left(\dfrac{A_{2k}}{s+\zeta_{1k}}-\dfrac{B_{2k}}{s+\zeta_{2k}}\right) & \dfrac{\lambda_2\mu_2}{\mu_1}C_k\left(-\dfrac{1}{s+\zeta_{1k}}+\dfrac{1}{s+\zeta_{2k}}\right) \\ \mu_1 C_k\left(-\dfrac{1}{s+\zeta_{1k}}+\dfrac{1}{s+\zeta_{2k}}\right) & \mu_2\left(\dfrac{A_{1k}}{s+\zeta_{1k}}-\dfrac{B_{1k}}{s+\zeta_{2k}}\right) \end{bmatrix}\begin{bmatrix} \dfrac{A_{2N}}{s+\zeta_{1N}}-\dfrac{B_{2N}}{s+\zeta_{2n}} \\ C_N\left(-\dfrac{1}{s+\zeta_{1N}}+\dfrac{1}{s+\zeta_{2N}}\right) \end{bmatrix},$$

with every $A_{2k}, B_{2k}, A_{1k}, B_{1k}, C_k$ well defined.

Products of the type

$$\frac{1}{s+\zeta_{1k}}\cdot\frac{1}{s+\zeta_{1,k+1}},\frac{1}{s+\zeta_{2k}}\cdot\frac{1}{s+\zeta_{1,k+1}},\frac{1}{s+\zeta_{1k}}\cdot\frac{1}{s+\zeta_{2,k+1}},\frac{1}{s+\zeta_{2k}}\cdot\frac{1}{s+\zeta_{2,k+1}}$$

will occur in each matrix multiplication, but each of these can be readily reduced to a form similar to

$$\frac{1}{s+\zeta_{1k}} \cdot \frac{1}{s+\zeta_{1,k+1}} = \frac{1}{\zeta_{1,k+1}-\zeta_{1k}}\left(\frac{1}{s+\zeta_{1k}} - \frac{1}{s+\zeta_{1,k+1}}\right)$$

so that at each step we deal only with sums of exponentials. The evaluation may prove tedious, but in principle, we now have the probability distribution of the process $\{Y(t), X(t); t \geq 0\}$. We illustrate the procedure in an example later.

We may be interested in the distribution of the time T until all individuals in the population become extinct. This will be given by

$$f(t)dt = P\{t \leq T \leq t+dt\} = [q_{11}(t)\mu_1 + q_{21}(t)\mu_2]dt \quad ,$$

with Laplace transform

$$\int_0^\infty f(t)e^{-st}dt = \hat{q}_{11}\mu_1 + \hat{q}_{21}\mu_2 \quad .$$

4. CONCLUDING REMARKS

We demonstrate by a simple example the methods outlined in Section 3; these can be implemented by a computer algorithm. Let us consider a population of size $N = 2$ at time $t = 0$, with parameters $\lambda_1 = \lambda_2 = 2$, and $\mu_1 = 1$, $\mu_2 = 4$. Then from

$$(s+\lambda_1+2\mu_1)(s+\lambda_2+2\mu_2) - \lambda_1\lambda_2$$
$$= (s+4)(s+10) - 4 = (s+7+\sqrt{13})(s+7-\sqrt{13})$$

we obtain $\zeta_{12} = 7 + \sqrt{13}$ and $\zeta_{22} = 7 - \sqrt{13}$. Hence after evaluating the relevant coefficients in (3.3), we find

$$\hat{q}_{12}(s) = \frac{1}{2\sqrt{13}}\left(\frac{\sqrt{13}-3}{s+7+\sqrt{13}} + \frac{\sqrt{13}+3}{s+7-\sqrt{13}}\right),$$
$$\hat{q}_{22}(s) = \frac{1}{\sqrt{13}}\left(-\frac{1}{s+7+\sqrt{13}} + \frac{1}{s+7-\sqrt{13}}\right). \tag{4.1}$$

Again, from

$$(s+\lambda_1+\mu_1)(s+\lambda_2+\mu_2) - \lambda_1\lambda_2$$
$$= (s+3)(s+6) - 4 = (s+7)(s+2)$$

we obtain $\zeta_{11} = 7$, $\zeta_{21} = 2$. Hence

$$\begin{bmatrix}\hat{q}_{11}(s)\\ \hat{q}_{21}(s)\end{bmatrix} = 2\begin{bmatrix}\frac{1}{5}\left(\frac{1}{s+7}+\frac{4}{s+2}\right) & \frac{8}{5}\left(-\frac{1}{s+7}+\frac{1}{s+2}\right)\\ \frac{2}{5}\left(-\frac{1}{s+7}+\frac{1}{s+2}\right) & \frac{4}{5}\left(\frac{4}{s+7}+\frac{1}{s+2}\right)\end{bmatrix}\begin{bmatrix}\hat{q}_{12}(s)\\ \hat{q}_{22}(s)\end{bmatrix},$$

so that

$$\hat{q}_{11}(s)=\frac{1}{5\sqrt{13}}\left(\frac{1}{s+7}+\frac{4}{s+2}\right)\left(\frac{\sqrt{13}-3}{s+7+\sqrt{13}}+\frac{\sqrt{13}+3}{s+7-\sqrt{13}}\right)$$

$$+\frac{16}{5\sqrt{13}}\left(-\frac{1}{s+7}+\frac{1}{s+2}\right)\left(-\frac{1}{s+7+\sqrt{13}}+\frac{1}{s+7-\sqrt{13}}\right)$$

$$=\frac{8}{5}\left(\frac{1}{s+2}\right)+\frac{2}{5}\left(\frac{1}{s+7}\right)+\frac{2(1-\sqrt{13})}{13+5\sqrt{13}}\left(\frac{1}{s+7+\sqrt{13}}\right)+\frac{2(1+\sqrt{13})}{13-5\sqrt{13}}\left(\frac{1}{s+7-\sqrt{13}}\right),$$

$$\hat{q}_{21}(s)=\frac{2}{5\sqrt{13}}\left(-\frac{1}{s+7}+\frac{1}{s+2}\right)\left(\frac{\sqrt{13}-3}{s+7+\sqrt{13}}+\frac{\sqrt{13}+3}{s+7-\sqrt{13}}\right)$$

$$+\frac{8}{5\sqrt{13}}\left(\frac{4}{s+7}+\frac{1}{s+2}\right)\left(-\frac{1}{s+7+\sqrt{13}}+\frac{1}{s+7-\sqrt{13}}\right)$$

$$=-\frac{4}{5}\left(\frac{1}{s+7}\right)+\frac{4}{5}\left(\frac{1}{s+2}\right)+\frac{2}{\sqrt{13}}\left(\frac{1}{s+7+\sqrt{13}}\right)-\frac{2}{\sqrt{13}}\left(\frac{1}{s+7-\sqrt{13}}\right). \tag{4.2}$$

Finally from

$$(s+\lambda_1)(s+\lambda_2)-\lambda_1\lambda_2=(s+2)(s+2)-4=(s+4)s$$

we obtain $\zeta_{10}=4$, $\zeta_{20}=0$. Hence

$$\begin{bmatrix}\hat{q}_{10}(s)\\ \hat{q}_{20}(s)\end{bmatrix}=\begin{bmatrix}\frac{1}{2}\left(\frac{1}{s+4}+\frac{1}{s}\right) & 2\left(-\frac{1}{s+4}+\frac{1}{s}\right)\\ \frac{1}{2}\left(-\frac{1}{s+4}+\frac{1}{s}\right) & 2\left(\frac{1}{s+4}+\frac{1}{s}\right)\end{bmatrix}\begin{bmatrix}\hat{q}_{11}\\ \hat{q}_{21}\end{bmatrix},$$

so that

$$\hat{q}_{10}(s)=\frac{1}{2}\left(\frac{1}{s+4}+\frac{1}{s}\right)\left[\frac{8}{5}\left(\frac{1}{s+2}\right)+\frac{2}{5}\left(\frac{1}{s+7}\right)+\frac{2(1-\sqrt{13})}{13+5\sqrt{13}}\left(\frac{1}{s+7+\sqrt{13}}\right)\right.$$

$$\left.+\frac{2(1+\sqrt{13})}{13-5\sqrt{13}}\left(\frac{1}{s+7-\sqrt{13}}\right)\right]$$

$$+2\left(-\frac{1}{s+4}+\frac{1}{s}\right)\left[-\frac{4}{5}\left(\frac{1}{s+7}\right)+\frac{4}{5}\left(\frac{1}{s+2}\right)+\frac{2}{\sqrt{13}}\left(\frac{1}{s+7+\sqrt{13}}\right)\right.$$

$$\left.-\frac{2}{\sqrt{13}}\left(\frac{1}{s+7-\sqrt{13}}\right)\right]$$

$$=0.5\left(\frac{1}{s}\right)+0.5\left(\frac{1}{s+4}\right)-1.6\left(\frac{1}{s+2}\right)-0.4\left(\frac{1}{s+7}\right)$$

$$+0.0840\left(\frac{1}{s+7+\sqrt{13}}\right)+0.9160\left(\frac{1}{s+7-\sqrt{13}}\right) \tag{4.3}$$

Here we have calculated the values of the coefficients numerically to avoid surds of increasing size. This is what one would do in any practical situation. The value of the Laplace transform $\hat{q}_{20}(s)$ can be derived from

$$\hat{q}_{12}(s)+\hat{q}_{22}(s)+\hat{q}_{11}(s)+\hat{q}_{21}(s)+\hat{q}_{10}+\hat{q}_{20}(s)=\frac{1}{s},$$

so we shall not write it out.

The methods outlined can be extended in principle to the case where the death rate $\{Y(t);\ t\geq 0\}$ can take any of k values $\mu_1,\ldots,\mu_k$ and forms a Markov chain where, for example, the relevant infinitesimal probabilities are

$$P\{Y(t+\delta t)=\mu_{i+1}|Y(t)=\mu_i\}=\lambda_{i,i+1}\delta t+o(\delta t),\quad 1\leq i\leq k-1\ ,$$

$$P\{Y(t+\delta t)=\mu_{i-1}|Y(t)=\mu_i\}=\lambda_{i,i-1}\delta t+o(\delta t),\quad 2\leq i\leq k\ ,$$

with $Y(0)=\mu_1$. Instead of $\hat{Q}_j$ being 2-rowed, as in (3.6),it now becomes k-rowed, and the equations for the Laplace transforms of

$$q_{ij}(t)=P\{Y(t)=\mu_i,X(t)=j|Y(0)=\mu_1,X(0)=N\}\ ,$$

when j = N, and i = 1,2,...,k, are

$$s\hat{q}_{iN}-\delta_{i1}=-\left(\lambda_{i,i+1}+\lambda_{i,i-1}+N\mu_i\right)\hat{q}_{iN}+\lambda_{i+1,i}\hat{q}_{i+1,N}+\lambda_{i-1,i}\hat{q}_{i-1,N}\ ,$$

with δ_{i1} being the Kronecker delta, and the $\lambda_{i,i+1}$, $\lambda_{i,i-1}$ are zero if i+1 or i-1 are outside the range (1,...,k).

Similarly, for j = 1,2,...,N-1, and i = 1,2,...,k

$$s\hat{q}_{ij}=-\left(\lambda_{i,i+1}+\lambda_{i,i-1}+j\mu_i\right)\hat{q}_{ij}+\lambda_{i+1,i}\hat{q}_{i+1,j}+\lambda_{i-1,i}\hat{q}_{i-1,j}+(j+1)\mu_i\hat{q}_{i,j+1}\ ,$$

with the same restriction on the rates $\lambda_{i,i+1}$, $\lambda_{i,i-1}$ as before.

A matrix equation similar to (3.6) for the $\hat{Q}_j$ can be constructed, and the $\hat{q}_{ij}$ formally evaluated.

Thus the concept of the bivariate Markov chain has served us well in resolving this relatively simple problem, as it has also for Weiss' (1965) epidemic with carriers (for details see Bailey, 1975). While recognizing that the problem presented is not too difficult, it is only the first step to a more serious research project. This is the analysis of the stochastic carrier borne epidemic model for which the carrier death process is precisely that described above. The bivariate Markov chain model, which this note describes, gives hope that one can analyze the carrier borne epidemic influenced by such a carrier death process, through a trivariate Markov chain. This is the subject of our next investigation.

REFERENCES

Puri, P.S. (1975), A linear birth and death process under the influence of another process, *J. Appl. Prob.*, 12, 1-17.

Puri, P.S. (1976), A stochastic process under the influence of another arising in the theory of epidemics, *Stoch. Proc. Applicns.*, 2, 235-256.

Bailey, N.T.J. (1975), *The Mathematical Theory of Infectious Diseases*, Griffin, London.

Neyman, J., Park, T. and Scott, E.L. (1956), Struggle for existence. The Tribolium model: biological and statistical aspects, *Proc. Third Berkeley Symp. Math. Stat. Prob.*, 4, 41-79.

Weiss, G.H. (1965), On the spread of epidemics by carriers, *Biometrics*, 21, 481-490.

STOCHASTIC APPROXIMATION UNDER DEPENDENT NOISES, DETECTING SIGNALS AND ADAPTIVE CONTROL

Oleg N. Granichin

Mathematics and Mechanics
Department
St.Petersburg University
St.Petersburg, 198904, Russia

1. INTRODUCTION

Let $f_\alpha(\theta)$ be a family of unknown functions from a set $\Theta' \subset \mathbb{R}^N$ into $\mathbb{R}^1$, $\alpha \in A \subset \mathbb{R}^M$ be a random parameter with the distribution $\mathbb{P}_\alpha$ and with the mean value $\bar{\alpha}$. The function $f(\theta)=f_{\bar{\alpha}}(\theta)$ is assumed to have unique minimum in Θ' at an internal point θ_*.

We consider the following problem: to estimate

$$\theta_* = \underset{\Theta'}{argmin}\, f(\theta)$$

from given measurements (or observations)

$$y_n = f_{\alpha_n}(x_n) + \xi_n, n=0,1,2,\ldots, \qquad (1.1)$$

where n-dimensional measurement points $\{x_n\}$ (design of an experiment) are generated by a random source, $\{\xi_n\}$ are unmeasurable noises (measurements errors), $\{\alpha_n\}$ are random unmeasurable variables with the identical distribution $\mathbb{P}_\alpha$.

The distinction of the present statement from the stochastic approximation in classical framework is, at first, in the fact that measurement points $\{x_n\}$ do not coincide exactly with estimates generated by the algorithm. Secondly, in the literature, the problem of estimating the

Approximation, Probability, and Related Fields, Edited by
G. Anastassiou and S.T. Rachev, Plenum Press, New York 1994

minimum point of the fixed function $f(\theta)$ is mainly treated whereas, in the present paper, a set of functions $f_\alpha(\theta)$ is analyzed.

For the consistency analysis of the estimate $\{\theta_n\}$, the matrix $\sum_{i=1}^{n}\Phi_i\Phi_i^T$ made of the stochastic regressors for (1.1) is of great importance. It has been stated in earlier papers of Ljung(1977), Moore(1978), Solo(1979), Chen and Guo(1985)) that the persistent excitation condition (which sounds that the ratio of the maximum to the minimum of eigenvalues of $\sum_{i=1}^{n}\Phi_i\Phi_i^T$ is bounded or, maybe, its growth is bounded in some or other way) is the key to be sure in the strong consistency. If there is almost nothing to say about statistical properties of measurement errors $\{\xi_n\}$ or if they are given by a definite function (systematic errors) then we cannot obtain consistent estimates directly from the least-square estimation method or from the stochastic-gradient algorithm. In Granichin(1989, 1992a), it has been proposed to include the probing noise $\{\zeta_n\}$ in the measurement channel, namely, points of measurements are chosen from previous estimates with additional random measurable noises $n^{-1/2}\zeta_n$ decreasing with increasing discrete time. Here, a new scheme of measurements $Y_n=\zeta_n(f(\theta_{n-1}+n^{-1/2}\zeta_n)+\xi_n)$, $n=0,1,2,\ldots$ was suggested instead of the standard one. Variables Y_n are to be measured in a direct way or can be calculated from real measurements. Under very general proposals, the persistent excitation condition holds true and we may use in this problem the stochastic-gradient algorithms following to Robbins and Monro(1951) or to Kiefer and Wolfowitz(1952). Such a stochastic approximation algorithm with additive probing noises has recently been investigated by Polyak and Tsybakov(1990) for independent measurement noises and fixed function $f(\theta)$ and by Granichin(1989, 1992a, 1992b) for some special cases with dependent measurement noises. Polyak and Tsybakov(1990) have shown that this algorithm had an optimum minimax rate of convergence in wide variety of algorithms.

In all the previous papers of the author, the basic condition provided the strong consistency of estimates was in independence of both the probing noises ζ_n and the measurement errors ξ_n. In the present paper, a more general situation is considered. The approach described above remains. But, in addition, the probing noises ζ_n and the measurement errors ξ_n are allowed to be dependent quantities with a correlation which tends to zero with respect to the time.

It should be emphasized that the problem statement including a family of regression functions has just been studied by Granichin(1992b). The reason for this step lies in practical needs. For instance, there is a problem of testing a target by a series of pulses and it is known that the target properties have a statistical behavior (radioactive decay, for instance). So the resulting signal at the receiver is characterized not only by the fixed parameters of a target (chemical composition and so on), pulses parameters and measurement noises but also by some probability characteristics of the target. The plan of the paper is the following. Section 2 gives a motivation of the general problem statement. Section 3 deals with a general stochastic approximation algorithm with probing noises. The probability one convergence and the mean square convergence are proved by the condition of pertaining to zero a correlation between probing and measurement noises.

Sections 4-7 illuminate practical capabilities of the method proposed. In section 4, the problem of signal detection with dependent noises is considered. The measurement channel is described by a linear model with additive noises. The algorithm of the stochastic approximation with additive probing noises is used for finding the estimates of the unknown signal parameters.

Problems of adaptive control and identification of unknown parameters of a dynamic system are considered in sections 5 and 6. The dynamic system operates in a discrete time and it is described by a linear difference equation with additive bounded arbitrary noises (the ARMAX model is considered) . According to Granichin and Fomin (1986) as well as to Chen and Guo(1986), the identification technique relies on noncorrelatedness probing signals in the feedback loop. In Chen and Guo(1986), the least-square identification algorithm was studied and its rate of convergence was analyzed. In this paper, the stochastic approximation algorithm with additive probing noises is considered. It will be shown in what follows, that its convergence rate is better than for the least-square algorithm. The worth of the identification method which is proved under very general conditions, consists in a conceptual application to the solution of adaptive control problems where current estimates of unknown system parameters are employed in constructing control functions. The intensity of probing signal decreases with time right up to zero what leads to the consistent estimates of unknown parameters and to the design of a limited extremely optimum control with a minimax loss function. Thus, asymptotically, the adaptive control algorithm achieves the same performance as it could be achieved if the system parameters were known.

2. PROBLEM STATEMENT

Let $f_{\alpha}(\theta):\Theta'\longrightarrow\mathbb{R}^1$, $\Theta'\subset\mathbb{R}^N$ be a family of nonlinear functions, $x_1,x_2,\ldots$ be measurement points, $n=0,1,2,\ldots$ be time series, and

$$y_n=f_{\alpha_n}(x_n)+\xi_n,$$

be results of measurements of function $f_{\alpha_n}(\cdot)$ at the space points x_n with random unobservable noises ξ_n, $\{\alpha_n\}$ be random unobservable quantities with the identical distribution $\mathbb{P}_{\alpha}$ and the mean value $\bar{\alpha}$, $\alpha_n\in\mathcal{A}\subset\mathbb{R}^M$, $\mathcal{A}$ be a set.

Now, we formulate the main assumptions.

1.For any $\alpha\in\mathcal{A}$, the function $f_{\alpha}(\theta)\in C^{\ell}$ is the ℓ times differentiable function with the ℓ-th derivative satisfying on Θ' Hoelder's conditions with the constant h, $0<h\leq 1$, so that

$$|f_{\alpha}(z)-\sum_{|m|\leq\ell}\frac{1}{m!}D^{m}f_{\alpha}(\theta)(z-\theta)^{m}|\leq L|z-\theta|^{\beta},$$

where $\beta=\ell+h$, $m=(m_1,\dots,m_N)\in\mathbb{N}^N$ is a multiplex index, $m_i\geq 0$, $|m|=m_1+\dots+m_N$, $m!=m_1!\dots m_N!$, $z=(z_1,\dots,z_N)\in\mathbb{R}^N$, $z^m=z_1^{m_1}\dots z_N^{m_N}$, $D^m=\partial^{|m|}/\partial z_1^{m_1}\dots\partial z_N^{m_N}$.

2.The function $f(\theta)=f_{\bar{\alpha}}(\theta)$ $(\bar{\alpha}=\mathbb{E}\alpha)$ has the unique minimum in Θ' at the internal point θ_* and, for arbitrary $\theta\in\Theta'$,

$$\int(\theta-\theta_*,\nabla f_\alpha(\theta))d\mathbb{P}_\alpha\geq\delta|\theta-\theta_*|^2.$$

3.For any point $\theta\in\Theta'$ and for $\alpha\in\mathcal{A}$,

$$|\nabla f_\alpha(\theta)-\nabla f_\alpha(\theta_*)|<A|\theta-\theta_*|.$$

4.For arbitrary $\alpha\in\mathcal{A}$ $|f_\alpha(\theta_*)|\leq B$, $|\nabla f_\alpha(\theta_*)|\leq C$.
Here, $|x|$ denotes the Euclidean norm, $(\cdot,\cdot)$ is the inner product, and A,B,C,L,δ $(A>h)$ are positive constants.

The minimum point θ_* of the function $f(\theta)$ should be estimated from given measurements y_n, $n=0,1,2,\dots$.

3. GENERAL ALGORITHM

Let $\{\zeta_n\}$, $n=0,1,2,\dots$ be random, independent (on each other) and identically distributed (i.i.d.) k-dimensional values with the distribution $\mathbb{P}_\zeta$ on $\mathbb{R}^K$.

The following recursive algorithm for building up $\{x_n\}$ and $\{\theta_n\}$ is proposed:

$$\begin{aligned}&\theta_{-1}=b, && x_n=\theta_{n-1}+\phi_n(\zeta_n),\\ &\theta_n=P_{\Theta_n}(\theta_{n-1}-y_n\psi_n(\zeta_n)), && y_n=f_{\alpha_n}(x_n)+\xi_n.\end{aligned}\tag{3.1}$$

Here, Θ_n is the sequence of convex closed bounded sets, which contain the point θ_* starting with a fixed $n>>1$, P_{Θ_n} is the projection operator on Θ_n, θ_n is a current approximation to θ_*, $\phi_n(z),\psi_n(z):\mathbb{R}^K\longrightarrow\mathbb{R}^N$, $n=0,1,2,\dots$ are sequences of vector-functions satisfying

$$\int\Big(\sum_{|m|\leq\ell}\frac{1}{m!}D^m f_\alpha(\theta)\phi_n(z)^m\Big)\psi_n(z)d\mathbb{P}_\zeta(z)=C_D n^{-1}\nabla f_\alpha(\theta)\tag{3.2}$$

for any $\theta\in\Theta'$ and

$$\int \psi_n(z)d\mathbb{P}_\zeta(z)=0, \quad \int (\psi_n(z),\psi_n(z))d\mathbb{P}_\zeta(z)\le C_\psi n^{-2+1/\beta},$$

$$\int (\phi_n(z),\phi_n(z))^\beta d\mathbb{P}_\zeta(z)\le C_\phi^\beta n^{-1}, \quad |\phi_n(z)|^2\le C_\phi, \tag{3.3}$$

where C_D, C_ϕ, C_ψ are constants. At any discrete time n, we may measure or calculate y_n and $\psi_n(\zeta_n)$. If we know the convex closed bounded set Θ: $\theta_*\in\Theta$, then we can put $\Theta_n=\Theta$ for any n.

Theorem 1. Let conditions 1-4 from section 2 hold true for a family of functions $f_\alpha(\cdot)$; $\beta\ge2$. Equations (3.2) and (3.3) are justified and $\delta C_D>(\beta-1)/\beta$, $diam(\Theta_n)=d_n=o(n^{(\beta-1)/(2\beta)})$ as $n\longrightarrow\infty$. For any fixed n, ζ_n and $\xi_0,\ldots,\xi_{n-1},\alpha_0,\ldots,\alpha_n$ are independent; the correlation between $\psi_n(\zeta_n)$ and ξ_n is decreasing to zero as $n\longrightarrow\infty$ as follows

$$|\mathbb{E}\xi_n\psi_n(\zeta_n)|\le C_\chi n^{-1-(\beta-1)/(2\beta)}, \quad \mathbb{E}(\xi_n\psi_n(\zeta_n))^2\le\grave{\sigma}^2 n^{-2+1/\beta}. \tag{3.4}$$

Then, for the estimates generated by algorithm (3.1), one has

$$\mathbb{E}\{|\theta_n-\theta_*|^2\}n^{(\beta-1)/\beta} < \infty. \tag{3.5}$$

Moreover, if $diam(\Theta_n)=d_n=o(n^{(\beta-1)/(4\beta)})$ as $n\longrightarrow\infty$ then $\theta_n\longrightarrow\theta_*$ as $n\longrightarrow\infty$ with probability one.

For instance, one can note that conditions (3.2)-(3.3) for the distribution $\mathbb{P}_\zeta$ and for the functions $\phi_n(\cdot),\psi_n(\cdot)$ are correct if $\{\zeta_n\}$ are random independent uniformly distributed on $[-1/2,1/2]^N$ values and functions $\phi_n(\cdot)$ and $\psi_n(\cdot)$ are represented as (Polyak and Tsybakov(1990))

$$\phi_n(z)=\phi n^{-1/(2\beta)}z, \quad \psi_n(z)=\psi n^{-1+1/(2\beta)}K(z), \tag{3.6}$$

where $\phi=\sqrt{C_\phi}$, $\psi=\sqrt{C_\psi}=C_D$; the vector-function $K(z)$ is determined by orthogonal Legendre's polynomials $p_j(\cdot)$, $j=0,1,\ldots,\ell$ (Katkovnik(1985), Polyak and Tsybakov(1990)) :

$$K_i(z)=K_0(z_i)\prod_{m\neq i}\tilde{K}(z_m) \quad , \; i=1,\dots, N \quad , \; z=(z_1,\dots,z_N)\subset\mathbb{R}^N \; ,$$

$$K_0(u)=\sum_{j=0}^{\ell} a_j p_j(u) \; , \qquad a_j=p_j'(0)/\int_{-1/2}^{1/2} p_j^2(u)du \; , \qquad (3.7)$$

$$\tilde{K}(u)=\sum_{j=0}^{\ell-1} b_j p_j(u) \; , \qquad b_j=p_j(0)/\int_{-1/2}^{1/2} p_j^2(u)du \; .$$

For initial values $\ell=1,2$ (i.e. $2\leq\beta\leq3$) we have

$$K_0(u)=12u, \; \tilde{K}(u)=1, \; |u|\leq 1/2.$$

For next values $\ell=3,4$ (i.e. $3<\beta\leq5$) we have

$$K_0(u)=5u(15-84u^2), \; \tilde{K}(u)=9/4-15u^2, \; |u|\leq1/2.$$

Further, for $\ell=5,6$ (i.e. $5<\beta\leq7$), we have

$$K_0(u)=5u(15.1875-87.5u^2+2079u^4),$$

$$\tilde{K}(u)=3.515625-65.625u^2+236.25u^4, \; |u|\leq1/2.$$

It should be mentioned that there are two reasons to use more general definition of the functions $\phi_n(\cdot)$ and $\psi_n(\cdot)$ than (3.6). The former is in that it may be useful in some applications do not have a fixed dependence of $\phi_n(\cdot)$ and $\psi_n(\cdot)$ on n which is present in (3.6). The latter consists in understanding that the problem statement in itself may lead to a specific type of the distribution $\mathbb{P}_\zeta$. So, we cannot choose this one following to Polyak and Tsybakov(1990) as uniform distribution on a cube. Besides, the realization of the algorithm requires a way to (measure) calculate $\psi_n(\zeta_n)$ at every n-th step.

The proof of theorem 1 is given in appendix B.

4. DETECTING THE SIGNAL WITH DEPENDENT NOISE

Now, we can study a problem of detection of a signal which is measured coupled with dependent noises. Let the signal $\{\zeta_n\}$ be a sequence of random observable independent values uniformly distributed on the range $[-1/2,1/2]$. The mean value of the random sequence $\{\alpha_n\}$ should be found from given measurements

$$y_n=\alpha_n\zeta_n+v_n \ , \ n=0,1,2,..., \tag{4.1}$$

Here, $\{v_n\}$ are measured noises, $\{\alpha_n\}$ are random independent values identically distributed on the segment $\mathcal{A}=[\alpha^{(1)},\alpha^{(2)}]$ with the distribution $\mathbb{P}_\alpha$ and the mean value $\bar{\alpha}$. The signal occurs for any $\bar{\alpha}\neq0$. At $\bar{\alpha}=0$, there is no signal.

One starts with the problem of estimating $\bar{\alpha}$ from given measurements y_n, $n=0,1,2,...$.

Lemma 1. Let $\beta\geq2$, $\beta\in\mathbb{N}$, $\{\zeta_n\}$ be a sequence of random observable independent values uniformly distributed on the range $[-1/2,1/2]$; $\{\alpha_n\}$ are random independent identically distributed on segment $\mathcal{A}=[\alpha^{(1)},\alpha^{(2)}]$ values with the distribution $\mathbb{P}_\alpha$ and the mean value $\bar{\alpha}$. For any $n=0,1,2,...$, ζ_n and $v_0,...,v_{n-1}$ are independent quantities. ζ_n and α_n are also independent. The correlation between $K(\zeta_n)$ and v_n is decreasing to zero as $n\longrightarrow\infty$ as follows

$$|\mathbb{E}\{v_nK(\zeta_n)\}|\leq C_\chi n^{-(\beta-1)/(2\beta)}, \tag{4.2}$$

and mean square growth of v_n is lower than $n^{1/\beta}$

$$\mathbb{E}\{v_n^2\}\leq C_v n^{1/\beta}. \tag{4.3}$$

Then, the algorithm

$$\begin{aligned} &\theta_{-1}=(\alpha^{(1)}+\alpha^{(2)})/2, \\ &\theta_n=P_{\mathcal{A}}\left(\theta_{n-1}-n^{-1+\frac{1}{2\beta}}K(\zeta_n)\left(\frac{1}{2}(\theta_{n-1}+n^{-\frac{1}{2\beta}}\zeta_n)^2-y_n\right)\right)\end{aligned} \tag{4.4}$$

provides the probability one convergence and the mean square convergence of the sequence $\{\theta_n\}$ to $\bar{\alpha}$. The mean square convergence rate of the algorithm is $O(n^{-(\beta-1)/\beta})$. Here, $P_{\mathcal{A}}$ is a projection operator on $\mathcal{A}$, the scalar function $K(\cdot)$ is determined by (3.6) and (3.7), i.e.

$$K(z)=\sum_{j=0}^{\beta-1}a_jp_j(z), \ z\in\mathbb{R}^1, \ a_j=p_j'(0)/\int p_j^2(z)dz, \tag{4.5}$$

$p_j(\cdot)$, $j=0,...,\beta$ are the orthogonal Legendre's polynomials of order j, C_X, C_V, a_j, $j=0,\dots,\beta$ are constants.

The proof of this lemma follows from theorem 1. Let $f_\alpha(\theta)=\frac{1}{2}(\theta-\alpha)^2$ $:f_\alpha(\theta):[\alpha^{(1)},\alpha^{(2)}]\longrightarrow\mathbb{R}^1$ be a family of quadratic functions. For this family, conditions 1-4 are valid at any $\beta\geq 2$ and $\delta=1$. The function $f(\theta)$ is $f(\theta)=\frac{1}{2}(\theta-\bar{\alpha})^2$. Consider the sequence of points in $\mathbb{R}^N$ $\{x_n\}$: $x_n=\theta_{n-1}+n^{-1/(2\beta)}\zeta_n$, $n=0,1,2,\dots$ and the sequence of observable values $\{Y_n\}$: $Y_n=\frac{1}{2}x_n^2-n^{-1/(2\beta)}y_n$. Functions $\phi_n(z)=n^{-1/(2\beta)}z$ and $\psi_n(z)=n^{-1+1/(2\beta)}K(z)$ satisfy (3.2) and (3.3) with $C_D=C_\phi=C_\psi=1$. For measurements Y_n, $n=1,2,\dots$, we obtain

$$Y_n=\frac{1}{2}(\theta_{n-1}+n^{-1/(2\beta)}\zeta_n)^2-n^{-1/(2\beta)}\zeta_n\alpha_n-n^{-1/(2\beta)}v_n=$$

$$=\frac{1}{2}((\theta_{n-1}+n^{-1/(2\beta)}\zeta_n)-\alpha_n)^2+\xi_n=f_{\alpha_n}(x_n)+\xi_n,\ n=1,2,\dots,$$

where $\xi_n=-\frac{1}{2}\alpha_n^2+\theta_{n-1}\alpha_n-v_n n^{-1/(2\beta)}$. For any $n=1,2,\dots$, ζ_n and $\xi_0,\dots,\xi_{n-1}$ are independent quantities since ζ_n and $v_0,\dots,v_{n-1},\alpha_n$ are independent as well. Condition (3.4) follows from (4.2) and (4.3). We have

$$|\mathbb{E}\xi_n\psi_n(\zeta_n)|=n^{-1+1/(2\beta)}|\mathbb{E}(-\frac{1}{2}\alpha_n^2+\theta_{n-1}\alpha_n-v_n n^{-1/(2\beta)})K(\zeta_n)|=$$

$$=n^{-1+1/(2\beta)}|\mathbb{E}(-\frac{1}{2}\alpha_n^2+\theta_{n-1}\alpha_n)||\mathbb{E}K(\zeta_n)|+n^{-1}|\mathbb{E}v_nK(\zeta_n)|\leq$$

$$\leq C_X n^{-1-(\beta-1)/(2\beta)}$$

and

$$\mathbb{E}(\xi_n\psi_n(\zeta_n))^2\leq n^{-2+1/\beta}(3/2\max(|\alpha^{(1)}|^2,|\alpha^{(2)}|^2)+C_V)\leq n^{-2+1/\beta}\sigma^2$$

with a constant $\sigma^2=3/2\max(|\alpha^{(1)}|^2,|\alpha^{(2)}|^2)+C_V$. From the estimation algorithm (4.4), we have

$$\theta_n=P_{\mathcal{A}}(\theta_{n-1}-n^{-1+1/(2\beta)}K(\zeta_n)Y_n),$$

so that the type of algorithm (4.4) is similar to (3.1). The point $\bar{\alpha}$ is the minimum point of the function $f(\theta)=f_{\bar{\alpha}}(\theta)$. Hence, Lemma 1 is proved.

From the practical point of view, this statement corresponds to testing a body with statistical behavior by a series of pulses. One may consider a transient relativistic electronic beam incident upon a target. The target reaction displayed in stimulated emission contains standard instantaneous part which is experienced for any chemical composition of the body and a delayed part indicating a presence of a specific chemical element.

Let us assume that the pulse length is Δt_p, the injection process is switched on at $t=t_0$, the repetition frequency is $1/\Delta t_h$. A number of electrons contained by n-th pulse is $N^e+\zeta_n$; N^e is a mean value of electrons in pulses. A detector (receiver) is switched on at moment $t_0+\Delta t_d$ where Δt_d is a delayed time defined by the distance transducer-receiver and by the mean pulse energy. In what follows, the receiver also operates in a pulsed regime with the same working frequency $1/\Delta t_h$. Its working time is Δt_s whereas the ratio $\Delta t_p/\Delta t_s$ may have arbitrary order of magnitude which is defined by technical details. The value Δt_h is taken to be large enough in order to separate the responses from different pulses. A signal response on record is connected with a number of delayed particles incident upon the receiver N_n^d and being due to the n-th electronic pulse with $N^e+\zeta_n$ electrons. Furthermore, at the n-th step, the detector points to a number of side particles N_n^{ext} because of solar noise or other particle fluxes including jamming which distort a primary signal structure. The total number of particles incident upon the receiver during the n-th step is $y_n=N_n^d+N_n^{ext}$.

Note that we often don't know statistical properties of N_n^{ext} and we cannot therefore use in this problem usual estimation methods. Under appropriate conditions, one can show that a number of delayed particles measured at the n-th step is $N_n^d=\alpha_n(N^e+\zeta_n)$. The values α_n $(0\leq\alpha_n\leq\alpha_{max})$ are determined by efficiency of particle registration, which clearly depends on the detector type, on the volume angle with the top on the target and, at last, on the angle between the direction toward the detector and the direction of outer normal to the body surface. This angle depends on the instantaneous body position. The subscript n indicates this fact and, besides, the values α_n also depend on statistical properties of the body. If the body has no specific element then $\alpha_n=0$. Thus, we have

$$y_n=\alpha_n(N^e+\zeta_n)+N_n^{ext}=\alpha_n\zeta_n+v_n.$$

Here, v_n denotes $\alpha_n N^e+N_n^{ext}$. So we obtain exactly the same problem statement as equation (4.1) outlined above. Moreover, correlation between random values ζ_n and v_n is equal to zero in wide range of cases.

Under the assumptions made, one may consider it well grounded to state the conditions of lemma 1 to be valid. Therefore, in order to answer the question of whether the response contains the delayed particle emission or not the algorithm (4.4) is applicable if one chooses the interval as $A=[0,\alpha_{max}]$ and the initial approximation as $\theta_0=\alpha_{max}/2$.

In the example considered, the deviation ζ_n of the electronic beam power from its mean value N^e plays a role of a probing noise. The exact measurement of quantity ζ_n probably can be made in the experiment described without difficulty. Therefore, all the variables in the right-hand side of equation (4.4) may be calculated or measured based on the real time observation data and on the previous values of the parameter estimates. If the sequence $\{\theta_n\}$ converges to zero that means that the target response to the primary signal (pulsed electronic beam) does not include delayed emission. If not, then a specific element is present inside the target.

5. IDENTIFICATION OF DYNAMICS SYSTEM UNKNOWN PARAMETERS

Consider the problem of parameters identification for a scalar dynamics system. The system is described by the scalar linear difference equation with additive bounded noises

$$y_s+a_*^{(1)}y_{s-1}+\ldots+a_*^{(p)}y_{s-p}=b_*^{(1)}u_{s-1}+\ldots+b_*^{(q)}u_{s-q}+v_s\ , \quad (5.1)$$

for $s=1,2,\ldots$. Here, $\{y_s\}$ are output observations, $\{u_s\}$ are input actions, $\{v_s\}$ are bounded unobservable noises: $|v_s|\le C_v$.

Let

$$\tau_*=(-a_*^{(1)},\ldots,-a_*^{(p)},b_*^{(1)},\ldots,b_*^{(q)})^T$$

be the $(p+q)$-dimensional vector of unknown parameters, T be a determined set containing τ_*, $\tau_*\in T$. The vector τ_* should be estimated from measurements of outputs y_s and inputs u_s for $s=0,1,2,\ldots$.

The solution of the problem suggested by Granichin and Fomin(1986) requires the introduction of a vector function $\theta: T\longrightarrow\Theta\subset\mathbb{R}^{p+q}$: $\theta(\tau)=\mathfrak{A}^{-1}(\tau)\mathfrak{B}(\tau)$, where $\mathfrak{A}(\tau)$ is the $(p+q)\times(p+q)$ matrix, $\mathfrak{B}(\tau)$ is the $(p+q)\times 1$ vector

$$\mathfrak{A}(\tau)=\begin{pmatrix} 1, & 0, & \ldots\ldots\ldots\ldots & 0, & 0\\ a^{(1)}, & 1, & 0\ldots\ldots\ldots\ldots & 0, & 0\\ \ldots & \ldots & \ldots\ldots\ldots\ldots & \ldots & \ldots\\ 0, & 0, & \ldots a^{(p)}, & \ldots a^{(1)}, & 1 \end{pmatrix},\quad \mathfrak{B}(\tau)=\begin{pmatrix} b^{(1)}\\ \vdots\\ b^{(q)}\\ \vdots\\ 0\end{pmatrix}$$

Let the image of the set T $\Theta=\theta(T)$ be a parallelepiped, i.e. $\Theta=\Theta^{(1)}\times\ldots\times\Theta^{(p+q)}$, where $\Theta^{(i)}=[\theta^{(i)}_{min},\theta^{(i)}_{max}]$, $i=1,\ldots,p+q$ are closed segments in $\mathbb{R}^1$. We denote by P_i the projection operator on $\Theta^{(i)}$. If, for any $\tau\in T$, the corresponding polynomials

$$a(\lambda,\tau)=1+a^{(1)}\lambda+\ldots+a^{(p)}\lambda^p,\quad b(\lambda,\tau)=b^{(1)}\lambda+\ldots+b^{(q)}\lambda^q$$

have no common roots then there exists the continuous inverse function $\tau:\Theta\longrightarrow T$, $\tau(\cdot)=\theta^{-1}(\cdot)$ (see Granichin and Fomin(1986)). In what follows it will be assumed that $\theta_*=\theta(\tau_*)\in\Theta$.

In stochastic adaptive-control system, a performance index of the limited type or long-run average type is widely used at the moment. In such a case, external perturbations added to the (out)input of the system and decreasing with time do not vary the performance index. In order the sequence of the system stochastic regressors to satisfy the persistent excitation condition we introduce a random subsidence. If its covariance matrix tends to zero then one may expect that both chances are available: optimality of the control and a consistency of estimates. Strictly speaking, this approach called "the attenuating excitation technique" (Chen and Guo(1986), Granichin and Fomin(1986)) requires the following.

Let $\beta\ge 2$; $\{\zeta_n\}$ be a random independent uniformly distributed on the segment $[-1/2,1/2]$ values. $\{\zeta_n\}$ and $\{v_n\}$ are taken to be independent. The attenuating excitation technique consists in choosing the control strategy in the form

$$u_s=\bar{u}_s+w_s \quad , \tag{5.2}$$

instead of $u_s=\bar{u}_s$. Here, the sequence $\{\bar{u}_s\}$ is formed by a closed-loop strategy as follows

$$\bar{u}_s=\bar{U}_s(\bar{u}_0,\dots,\bar{u}_{s-1},y_0,\dots,y_s),\ s=1,2,\dots,\ \bar{u}_0=U_0(y_0), \tag{5.3}$$

$\bar{U}^\infty=\{\bar{U}_0(\cdot),\bar{U}_1(\cdot),\dots\}$ is a sequence of functions, i.e. the control itself. w_s is the random item

$$\begin{aligned} w_{n(p+q)} &= n^{-1/(2\beta)}R_n\zeta_n,\ n=0,1,2,\dots, \\ w_{n(p+q)+i} &= 0 \qquad\quad ,\ i=1,\dots,p+q-1, \end{aligned} \tag{5.4}$$

where the observable sequence $\{R_n\}$ satisfies the following conditions:

$$\begin{aligned} &\mathbb{E}\{|\bar{u}_{n(p+q)+i}|\mathcal{F}_n\}\le C_uR_n, \quad i=1,\dots,p+q-1, \\ &1+\sum_{k=0}^{p-1}|y_{n(p+q)-k}|+\sum_{k=1}^{q-1}|u_{n(p+q)-k}| \le C_RR_n \end{aligned} \tag{5.5}$$

with constants C_u,C_R. Here, $\mathcal{F}_n$ is the σ-algebra generated by $\{u_0,\dots,u_{n(p+q)-1},y_0,\dots,y_{n(p+q)},\zeta_1,\dots,\zeta_n\}$, $\mathbb{E}\{\cdot|\mathcal{F}_n\}$ is the conditional expectation.

The additive random item in equation (5.2) will ensure the inclusion of probing noises into the estimation algorithm. It is easily seen from equation (5.4) that this item is added not at every step but only one time per $p+q$ steps.

Theorem 2. Let $\beta\geq 2$, $\{\zeta_n\}$ be a random independent uniformly distributed on the segment $[-1/2,1/2]$ values. $\{\zeta_n\}$ and $\{v_n\}$ are independent. Control strategy (5.2) for the dynamics system (5.1) which is given by $u_s=\bar{u}_s+w_s$ with $\{u_s\}$, $\{w_s\}$ from (5.3),(5.4) satisfies conditions (5.5). A set T is given containing the vector of the system unknown parameters τ_*; $\tau_*\in T$. A set Θ is obtained from the set T by using the function $\theta(\cdot)$, $\Theta=\theta(T)= =\Theta^{(1)}\times\ldots\times\Theta^{(p+q)}$ is a parallelepiped in space $\mathbb{R}^{p+q}$. For any $\tau\in T$, the polynomials $a(\lambda,\tau)$, $b(\lambda,\tau)$ have no common roots.

Then, for any $\beta\geq 2$, the identification algorithm

$$\theta^{(i)}_{-1}=(\theta^{(i)}_{max}-\theta^{(i)}_{min})/2,\ i=1,\ldots,p+q,$$

$$\theta^{(i)}_n=P_i\left(\theta^{(i)}_{n-1}-\frac{K(\zeta_n)}{n^{1-\frac{1}{2\beta}}}\left(\frac{\zeta_n^2}{2n^{\frac{1}{\beta}}}-\frac{y_{n(p+q)+i}-\sum_{j=1}^{i}\theta^{(j)}_{n-1}u_{n(p+q)-j+i}}{R_n}\right)\right),\qquad(5.6)$$

$$\tau_{n(p+q)+i-1}=\tau(\theta_{n-1}),\ \theta_n=(\theta^{(1)}_n,\ldots,\theta^{(p+q)}_n),\ n=0,1,2,\ldots.$$

provides the almost sure and the mean square convergence $\{\tau_s\}$ to τ_* as $n\longrightarrow\infty$. Besides, the rate of convergence of the estimates $\{\theta_n\}$ to vector $\theta_*=\theta(\tau_*)$ is $O(n^{-(\beta-1)/\beta})$. Here, the function $K(z)$ was determined previously by equation (4.5).

The proof of theorem 2 is given in appendix B.

6. ADAPTIVE OPTIMAL CONTROL

If the sequence $\{R_n\}$ is bounded then the random item in equation (5.2) tends to zero as $n\longrightarrow\infty$. That gives us a possibility to construct an optimally limited control strategy.

We now apply the results obtained in section 5 for system (5.1) with arbitrary bounded noises v_s, $s=1,2,\ldots$, $|v_s|\leq C_v$, to the adaptive-control problem with the loss function $\vartheta=\overline{\lim}_{s\longrightarrow\infty}|y_s|$. The feedback control actions u_s are assumed to be measurable with respect to the σ-algebra generated by $\{y_0,\ldots,y_s,u_0,\ldots,u_s,\zeta_1,\ldots,\zeta_{[s/(p+q)]}\}$, where $[\cdot]$ is the entire function. The control problem we treat is an adaptive one because u_s is not permitted to be an explicit function of the coefficients of τ_*, but that only depends on these quantities through the observations.

Assume that we know the vector τ_*. Barabanov and Granichin(1984) and Granichin(1990) have considered the problem of finding the minimax optimally limited stabilizing linear regulator for the scalar dynamic system which is described by equation (5.1) with arbitrary bounded measurement noises v_s, $s=1,2,\ldots$. The optimum regulator has been determined by the linear equation

$$u_s+c_{\tau_*}^{(1)}u_{s-1}+\ldots+c_{\tau_*}^{(g)}u_{s-g}=d_{\tau_*}^{(0)}y_s+\ldots+d_{\tau_*}^{(h)}y_{s-h},\ s=1,2,\ldots. \quad (6.1)$$

Denote by $c(\lambda,\tau_*)=1+c_{\tau_*}^{(1)}+\ldots+c_{\tau_*}^{(g)}$ and by $d(\lambda,\tau_*)=d_{\tau_*}^{(0)}+\ldots+d_{\tau_*}^{(h)}$ corresponding polynomials. Let Ω be a set of all the linear stabilizing regulators for system (5.1) satisfying

$$\sup_{s=1,2,\ldots} |y_s|+|u_s|<\infty. \quad (6.2)$$

In the papers cited, the control objective was to design a linear feedback control in order to make $\{u_s\}$ and $\{y_s\}$ bounded and to satisfy the inequality

$$\overline{\lim_{s\to\infty}}|y_s| \le \vartheta(\tau_*)=\inf_{\mathcal{U}^\infty\in\Omega}\ \sup_{\substack{|v_s|\le C_v\\ s=1,2,\ldots}}\ \overline{\lim_{s\to\infty}}|y_s|. \quad (6.3)$$

It has been shown by Barabanov and Granichin(1984) how to construct the minimax optimum linear regulator in the continuous way with respect to the system parameters if the polynomial $b(\lambda,\tau_*)$ has no roots on the unit circle $S_1=\{\lambda\in\mathbb{C}, |\lambda|=1\}$.

Revert to the problem of adaptive optimum control whose object is to propose an algorithm of the choice of control actions u_s. The measurements $y_0,\ldots,y_s$, the previous control actions $u_0,\ldots,u_{s-1}$ and the estimates $\tau_1,\ldots,\tau_s$ of the vector of unknown parameters τ_* obtained for all the steps including the step under consideration are given. We also have values $\zeta_0,\ldots,\zeta_{[s/(p+q)]}$. The requirements are in the uniform boundedness of y_s and u_s (inequality (6.2)) and in carrying out the control objective (6.3).

It must be emphasized again that the control actions u_s, $s=0,1,2,..$ cannot directly depend on the vector of unknown parameters τ_* for system (5.1). By an indirect way, these depend on τ_* through the observations of the system behavior. They are affected by earlier control actions.

In order to solve the problem, the identification technique is often used (Goodwin , Ramandge and Caines(1981), Lai and Wei(1986), Chen and Guo(1986), Granichin and Fomin(1986) and others). At every s-th step, estimates τ_s of the vector τ_* are calculated. Then we substitute these into the algorithm of the optimum control for a imaginable system whose parameters coincide with the current estimate τ_s. As a result, we find from here the control actions u_s.

In passing to the details, the estimates τ_s are to be found from algorithm (5.6) described in section 5 based on attenuation excitation technique. Here, the control actions are defined by equations (5.2) and (5.4). To choose the constants R_n in equation (5.4), we enlist the following reasons. It has been already stated in this section that the minimax optimum regulator continuously depends on τ. If the set T is compact and, for any $\tau \in T$, the polynomial $b(\lambda,\tau)$ has no roots on the unit circle S_1 then there exist constants G,H that $\forall \tau \in T$ $deg(c(\lambda,\tau)) \le G$, $deg(d(\lambda,\tau)) \le H$. Here, $deg(\cdot)$ denotes the polynomial degree. Thus, the constants R_n are proposed to be

$$R_n = 1 + \sum_{k=0}^{\max(G,p)-1} |y_{n(p+q)-k}| + \sum_{k=1}^{H-1} |\bar{u}_{n(p+q)-k}| + \sum_{k=1}^{q-1} |u_{n(p+q)-k}| .$$

Substituting this result into equation (5.4) gives the values of random item w_s in (5.2).

In order to find another part of the control action $\bar{u}_s$ (the control itself) we use an algorithm of building up the control action for the system described by the same equation as (5.1) but with the vector of parameters τ_s. That formally coincides with (6.1). The only distinction may be in that the algorithm (6.1) written now for the vector τ_s can be characterized by other coefficients and by the different number of previous observations and control actions.

Following by Granichin and Fomin(1986), one can show that the control strategy described satisfies the conditions from the equation (5.5). If the rest of conditions of theorem 2 is preserved this means that the estimates sequence $\{\tau_s\}$ converges to τ_* with probability one. Then, it may be shown (see also Granichin and Fomin(1986)) that the control objective (6.2),(6.3) is reached. So, the following theorem is proved.

Theorem 3. Let $\beta \geq 2$, $\{\zeta_n\}$ be a random independent uniformly distributed on the segment $[-1/2, 1/2]$ values. $\{\zeta_n\}$ and $\{v_n\}$ are independent. A set T is given containing the vector of the system unknown parameters τ_*; $\tau_* \in T$. A set Θ is obtained from the set T by using the function $\theta(\cdot)$; $\Theta = \theta(T) = \Theta^{(1)} \times \ldots \times \Theta^{(p+q)}$ is a parallelepiped in space $\mathbb{R}^{p+q}$. For any $\tau \in T$, the polynomials $a(\lambda, \tau)$, $b(\lambda, \tau)$ have no common roots, polynomial $b(\lambda, \tau)$ has no roots on unit circle S_1. For the dynamics system (5.1), the control strategy (5.2) is given by $u_s = \bar{u}_s + w_s$, where

$$w_{n(p+q)} = \frac{1 + \sum_{k=0}^{\max(\underline{G}, p)-1} |y_{n(p+q)-k}| + \sum_{k=1}^{H-1} |\bar{u}_{n(p+q)-k}| + \sum_{k=1}^{q-1} |u_{n(p+q)-k}|}{n^{2\beta}} \zeta_n$$

$$w_{n(p+q)+i} = 0 \qquad , \; i = 1, \ldots, p+q-1, \; n = 0, 1, 2, \ldots,$$

and

$$\bar{u}_s + c^{(1)}_{\tau_s} \bar{u}_{s-1} + \ldots + c^{(g)}_{\tau_s} \bar{u}_{s-g} = d^{(0)}_{\tau_s} y_s + \ldots + d^{(h)}_{\tau_s} y_{s-h}, \; s = 0, 1, \ldots$$

The current estimates $\{\tau_s\}$ are defined from the identification algorithm (5.6), polynomials $c(\lambda, \tau_s)$, $d(\lambda, \tau_s)$ correspond to the minimax optimum regulator of system (5.1) with the vector of parameters τ_s.

Then, first, the sequence of estimates $\{\tau_s\}$ converges to τ_* in the mean square sense and with the probability one. Second, the control objective (6.2),(6.3) is achieved.

7. CONCLUSIONS

A new stochastic approximation algorithm has been proposed. The distinction from the ordinary stochastic approximation is in that the points of measurements are chosen with additive random observable noises - so called probing noises. The probability one convergence and the mean square convergence are proved under assumption of decreasing to zero the correlation between probing and measured noises. This method is applied to the problem of the detection of the signal measured coupled with dependent noises as well as for the problem of dynamic system unknown parameters identification. The identification technique relies on noncorrelatedness probing signals in the feedback loop. The worth of this parametric identification method has been proved under very general conditions, what makes it possible to apply the method to the solution of adaptive control problems where current estimates of unknown system parameters are made use in the generation of control functions.

8. ACKNOWLEDGEMENTS

I wish to thank Prof. V.N. Fomin for helpful discussion and Prof. B.T. Polyak for his comments on an earlier draft of the manuscript.

Appendix A

This lemma will be used in the proof of theorem 1.

Lemma 2. If the sequence $\{e_n\}$ of nonnegative values satisfies the inequality

$$e_{n+1} \le (1-c/n+o(n^{-1}))e_n + d/n^{p+1},\ d>0,\ p>0,\ c>p$$

then

$$\overline{\lim_{n\to\infty}}\ n^p e_n \le d(c-p)^{-1}.$$

Proof of this lemma immediately follows from Polyak(1987) (see ch.2, Lemma 4).

Appendix B

Proof of the theorem 1. For sufficiently large n, $\theta_*\in\Theta_n$. Then, using the projection property we obtain

$$|\theta_n-\theta_*|^2\le|\theta_{n-1}-\theta_*-y_n\psi_n(\zeta_n)|^2.$$

Applying the conditional expectation at given $\theta_0,\ldots,\theta_{n-1}$ we have from this inequality

$$\mathbb{E}\{|\theta_n-\theta_*|^2|\theta_0,\ldots,\theta_{n-1}\} \le |\theta_{n-1}-\theta_*|^2-$$

$$-2(\theta_{n-1}-\theta_*,\mathbb{E}\{y_n\psi_n(\zeta_n)|\theta_0,\ldots,\theta_{n-1}\}) + \tag{B.1}$$

$$+\mathbb{E}\{|y_n|^2|\psi_n(\zeta_n)|^2|\theta_0,\ldots,\theta_{n-1}\} .$$

Since conditions 3,4 for the family $f_\alpha(\cdot)$ hold we obtain

$$f_\alpha^2(\theta_{n-1}+\phi_n(z)) \le 2B^2+4C^2(|\theta_{n-1}-\theta_*|^2+|\phi_n(z)|^2)+$$

$$+8A^2(|\theta_{n-1}-\theta_*|^4+|\phi_n(z)|^4)$$

uniformly in $\alpha\in\mathcal{A}$.

For the last item in (B.1), one has

$$\mathbb{E}\{|y_n|^2|\psi_n(\zeta_n)|^2|\theta_0,\ldots,\theta_{n-1}\}\le$$

$$\le 2\Bigg(\int\!\!\int f_\alpha^2(\theta_{n-1}+\phi_n(z))|\psi_n(z)|^2 d\mathbb{P}_\zeta(z)d\mathbb{P}_\alpha+$$

$$+\mathbb{E}\{(\xi_n\psi_n(\zeta_n))^2|\theta_0,\ldots,\theta_{n-1}\}\Bigg)\le \tag{B.2}$$

$$\le-(8C^2+16A^2d_n^2)n^{-2+1/\beta}|\theta_{n-1}-\theta_*|^2+$$

$$+2\mathbb{E}\{(\xi_n\psi_n(\zeta_n))^2|\theta_0,\ldots,\theta_{n-1}\}+(4B^2+8C^2C_\phi+16A^2C_\phi^2)C_\psi n^{-2+1/\beta}.$$

Consider the second item in the right-hand side of equation (B.1). Since $\xi_0,\ldots,\xi_{n-1},\alpha_0,\ldots,\alpha_n$ and ζ_n are independent quantities, then, according to conditions (3.2)-(3.3) for $\phi_n(\cdot),\psi_n(\cdot),\mathbb{P}_\zeta$ and property 1 for the family of functions $f_\alpha(\cdot)$, we get

$$-2(\theta_{n-1}-\theta_*,\mathbb{E}\{y_n\psi_n(\zeta_n)|\theta_0,\ldots,\theta_{n-1}\}) =$$

$$=-2(\theta_{n-1}-\theta_*,\iint(f_\alpha(\theta_{n-1}+\phi_n(z))\ \psi_n(z)d\mathbb{P}_\zeta(z)d\mathbb{P}_\alpha\ ) -$$

$$-2(\theta_{n-1}-\theta_*,\mathbb{E}\{\xi_n\psi_n(\zeta_n)|\theta_0,\ldots,\theta_{n-1}\})\le$$

$$\le -2c_D\int(\theta_{n-1}-\theta_*,\nabla f_\alpha(\theta_{n-1}))n^{-1}d\mathbb{P}_\alpha +2\Big(LC\sqrt{C_\phi^\beta C_\psi}\ n^{-1+1/(2\beta)}+$$

$$+n^{1/2}\mathbb{E}\{\xi_n\psi_n(\zeta_n)|\theta_0,\ldots,\theta_{n-1}\}\Big)n^{-1/2}|\theta_{n-1}-\theta_*|\ .$$

But

$$En^{-1/2}|\theta_{n-1}-\theta_*|\ \le\ E^2\varepsilon^{-1}/4+\varepsilon n^{-1}|\theta_{n-1}-\theta_*|^2$$

for any $E,\varepsilon>0$. According to property 2 for the family of functions $f_\alpha(\cdot)$, we find

$$-2(\theta_{n-1}-\theta_*,\mathbb{E}\{y_n\psi_n(\zeta_n)|\theta_0,\ldots,\theta_{n-1}\})\le-(2\delta C_D-\varepsilon)n^{-1}|\theta_{n-1}-\theta_*|^2+$$

$$+\ \varepsilon^{-1}\Big(n(\mathbb{E}\{\xi_n\psi_n(\zeta_n)|\theta_0,\ldots,\theta_{n-1}\})^2+L^2C^2C_\phi^\beta C_\psi n^{-2+1/\beta}\Big). \tag{B.3}$$

Let $\varepsilon=\delta C_D$. It follows from (B.1)-(B.3) that

$$\mathbb{E}\{|\theta_n-\theta_*|^2|\theta_0,\ldots,\theta_{n-1}\}\ \le$$

$$\le|\theta_{n-1}-\theta_*|^2(1-\delta C_D n^{-1}+(8C^2+16A^2d_n^2)n^{-1-(\beta-1)/\beta})+\gamma_n, \tag{B.4}$$

where

$$\gamma_n=\Big((4B^2+8C^2C_\phi+16A^2C_\phi^2)C_\psi+L^2C^2\delta^{-1}C_D^{-1}C_\phi^\beta C_\psi\Big)n^{-2+1/\beta}+$$

$$+\delta^{-1}C_D^{-1}n(\mathbb{E}\{\xi_n\psi_n(\zeta_n)|\theta_0,\ldots,\theta_{n-1}\})^2+\mathbb{E}\{(\xi_n\psi_n(\zeta_n))^2|\theta_0,\ldots,\theta_{n-1}\}.$$

Condition (3.4) leads to the inequality $\mathbb{E}\gamma_n \le C_\gamma n^{-2+1/\beta}$ with the constant C_γ given by

$$C_\gamma=(4B^2+8C^2C_\phi+2\sigma^2+16A^2C_\phi^2)C_\psi+\delta^{-1}C_D^{-1}(L^2C^2C_\phi^\beta C_\psi+C_\chi^2).$$

Now, the application of unconditional expectation to both sides of inequality (B.4) yields

$$\mathbb{E}\{|\theta_n-\theta_*|^2\} \le \mathbb{E}\{|\theta_{n-1}-\theta_*|^2\}(1-\delta C_D n^{-1}+o(n^{-1}))+C_\gamma n^{-2+1/\beta}.$$

Since $\delta C_D>(\beta-1)/\beta$, from lemma 2 on numerical sequences (see appendix A), it follows that

$$\mathbb{E}\{|\theta_n-\theta_*|^2\}\longrightarrow 0 \text{ a.s. as } n\longrightarrow\infty \text{ and } \mathbb{E}\{|\theta_n-\theta_*|^2\}n^{(\beta-1)/\beta}<\infty.$$

Consider the random sequence $\{\nu_n\}$

$$\nu_n=\nu_n(\theta_0,\ldots,\theta_{n-1})=-\sum_{k=1}^{n-1}(\gamma_k+(8C^2+16A^2d_k^2)|\theta_k-\theta_*|^2)(k+1)^{-1+(\beta-1)/\beta}.$$

That is a supermartingal since $\mathbb{E}\{\nu_n|\nu_1,\ldots,\nu_{n-1}\}\le\nu_{n-1}\le 0$. If $d_n=o(n^{(\beta-1)/(4\beta)})$ then $\mathbb{E}|\nu_n|<\infty$ and, according to the classical Doob's theorem (Doob(1953), theorem 4.1s, p.324) $\nu_n\longrightarrow\nu_*$ as $n\longrightarrow\infty$ with probability one, where ν_* is a constant. For the sequence $\{\eta_n\}$: $\eta_n=|\theta_n-\theta_*|^2-\nu_n$ we also have from equation (B.4)

$$\mathbb{E}\{\eta_n|\eta_1,\ldots,\eta_{n-1}\}\le|\theta_{n-1}-\theta_*|^2(1-\delta C_D n^{-1}+(8C^2+16A^2d_n^2)n^{-1-(\beta-1)/\beta})-$$

$$-\nu_n\le|\theta_{n-1}-\theta_*|^2-\nu_{n-1}=\eta_{n-1}.$$

Since $\mathbb{E}(\eta_n-|\eta_n|)>-\infty$ then it follows from Doob's theorem (Doob(1953)) that there is η_*: $\eta_n\longrightarrow\eta_*$ as $n\longrightarrow\infty$ with probability one. Since $\mathbb{E}\{|\theta_n-\theta_*|^2\}\longrightarrow 0$ as $n\longrightarrow\infty$ we obtain $|\theta_n-\theta_*|\longrightarrow 0$ as $n\longrightarrow\infty$ with probability one.

Appendix C

Proof of theorem 2. Consider equation (5.1) at the

discrete time "moment" $s=n(q+p)+i$, $i=1,2,\ldots,p+q$, $n=0,1,\ldots$. Eliminating the variables $y_{s-1},\ldots,y_{s-i+1}$ from the left-hand side of equation (5.1), we find according to definition of the function $\theta(\cdot)$

$$y_s=\sum_{j=1}^{i}\theta_*^{(j)}u_{s-j+1}+\bar{v}_s^{(i)}. \tag{C.1}$$

Here, the values of $\bar{v}_s^{(i)}$ are determined by $v_s,\ldots,v_{s-i+1}$ and by a linear function of $y_{s-i},\ldots,y_{s-i-p+1}$, $u_{s-i-1},\ldots,u_{s-i-q+1}$. Since $|v_s|\le C_V$ and since the second condition of (5.5) holds the relation $|\bar{v}_s^{(i)}|\le C_{\bar{V}}R_n$ is valid with the constant $C_{\bar{V}}$ depending on C_V,C_R. We consider $p+q$ sequences of observable values $\{Y_n^{(i)}\}$, $i=1,2,\ldots,p+q$

$$Y_n^{(i)}=n^{1/(2\beta)}\left(\frac{1}{2}(\theta_{n-1}^{(i)})^2+\left(y_{n(p+q)+i}-\sum_{j=1}^{i}\theta_{n-1}^{(j)}\bar{u}_{n(p+q)-j+i}\right)/R_n\right).$$

Equations (C.1),(5.4) give

$$Y_n^{(i)}=n^{\frac{1}{2\beta}}\left(\frac{\sum_{j=1}^{i}(\theta_*^{(j)}-\theta_{n-1}^{(j)})\bar{u}_{n(p+q)-j+i}+\theta_*^{(j)}w_{n(p+q)}+\bar{v}_{n(p+q)+i}^{(i)}}{R_n}+\right.$$

$$\left.+\frac{1}{2}(\theta_{n-1}^{(i)})^2\right)=\theta_*^{(i)}\zeta_n+v_n^{(i)},\quad i=1,\ldots,p+q,$$

where

$$v_n^{(i)}=n^{1/(2\beta)}\left(\sum_{j=1}^{i-1}(\theta_*^{(j)}-\theta_{n-1}^{(j)})\bar{u}_{n(p+q)-j+i}\right)/R_n+$$
$$+n^{1/(2\beta)}\left(\left((\theta_{n-1}^{(i)}-\theta_*^{(i)})\bar{u}_{n(p+q)}+\bar{v}_{n(p+q)+i}^{(i)}\right)/R_n+\frac{1}{2}(\theta_{n-1}^{(j)})^2\right). \tag{C.2}$$

Observable sequences $\{Y_n^{(i)}\}$, $i=1,2,\ldots,p+q$ satisfy equation (4.1) with $\alpha_n\equiv\theta_*^{(i)}$ and $v_n=v_n^{(i)}$. For any $i=1,2,\ldots,p+q$, the sequence $\{\theta_n^{(i)}\}$ generated by algorithm (5.6) is equivalent to the sequence of estimates generated by equation (4.4)

$$\theta_n^{(i)}=P_i\left(\theta_{n-1}^{(i)}-n^{-1}K(\zeta_n)\left(\frac{1}{2}(n^{1/(4\beta)}\theta_{n-1}+n^{-1/(4\beta)}\zeta_n)^2-Y_n^{(i)}\right)\right)=$$

$$=P_i\left(\theta_{n-1}^{(i)}-n^{-1+1/(2\beta)}K(\zeta_n)\left(\frac{1}{2}(\theta_{n-1}^{(i)})^2+n^{-1/(2\beta)}\theta_{n-1}^{(i)}\zeta_n+n^{-1/\beta}\zeta_n^2-\right.\right.$$

$$\left.\left.-\frac{\sum_{j=1}^{i}(\theta_*^{(j)}-\theta_{n-1}^{(j)})\bar{u}_{n(p+q)-j+i}+\theta_*^{(j)}w_{n(p+q)}+\bar{v}^{(i)}_{n(p+q)+i}}{R_n}-\frac{(\theta_{n-1}^{(i)})^2}{2}-\theta_{n-1}^{(i)}\zeta_n\right)\right)=$$

$$=P_i\left(\theta_{n-1}^{(i)}-\frac{K(\zeta_n)}{n^{1-\frac{1}{2\beta}}}\left(\frac{\zeta_n^2}{2n^{\frac{1}{\beta}}}-\frac{y_{n(p+q)+i}-\sum_{j=1}^{i}\theta_{n-1}^{(j)}u_{n(p+q)-j+i}}{R_n}\right)\right).$$

For sequences $\{v_n^{(i)}\}$, $i=1,2,\dots,p+q$, we have

$$\mathbb{E}|v_n^{(i)}|\le n^{1/(2\beta)}\left(C_{\bar{v}}+\frac{1}{2}\max\{|\theta_{min}^{(i)}|^2,|\theta_{max}^{(i)}|^2\}+C_u\left(\sum_{j=1}^{i}|\theta_{max}^{(i)}-\theta_{min}^{(i)}|\right)\right).$$

Therefore, conditions (4.3) of lemma 1 hold true for any $i=1,2,\dots,p+q$.

The method of mathematical induction is used in the proof.

Let $i=1$. The first item in the right-hand side of equation (C.2) is equal to zero. Sequences $\{v_n^{(1)}\}$ and $\{\zeta_n\}$ are independent. All the conditions of Lemma 1 hold for the observable sequence $\{Y_n^{(1)}\}$. Hence, we obtain $\theta_n^{(1)}\longrightarrow\theta_*^{(1)}$ as $n\longrightarrow\infty$ with probability one and $\mathbb{E}\{|\theta_n^{(1)}-\theta_*^{(1)}|^2\}<C^{(1)}n^{-(\beta-1)/\beta}$ with a constant $C^{(1)}$.

Assume that there is a number i that we have $\theta_n^{(j)}\longrightarrow\theta_*^{(j)}$ as $n\longrightarrow\infty$ with probability one and $\mathbb{E}\{|\theta_n^{(j)}-\theta_*^{(j)}|^2\}<C^{(j)}n^{-(\beta-1)/\beta}$ with constants $C^{(j)}$, $j=1,2,\dots,i-1$. From (5.5) and (C.2), we obtain

$$|\mathbb{E}\{v_n^{(i)}K(\zeta_n)\}|\le n^{1/(2\beta)}\left|\mathbb{E}\left\{\left(\sum_{j=1}^{i-1}(\theta_{n-1}^{(j)}-\theta_*^{(j)})\bar{u}_{n(p+q)-j+i}\right)/R_n\right\}\right|\le$$

$$\le C_u\left(\sum_{j=1}^{i-1}|\theta^{(1)}_{max}-\theta^{(1)}_{min}|\right)n^{-(\beta-1)/(2\beta)},$$

i.e. condition (4.2) for the sequences $\{v^{(1)}_n\}$ hold true. All the conditions of Lemma 1 are valid for the observable sequence $\{Y^{(1)}_n\}$ too. Hence, we obtain $\theta^{(1)}_n \longrightarrow \theta^{(1)}_*$ as $n\longrightarrow\infty$ with probability one and $\mathbb{E}\{|\theta^{(1)}_n-\theta^{(1)}_*|^2\}<C^{(1)}n^{-(\beta-1)/\beta}$ with the constant $C^{(1)}$. Thus, we have $\theta_n\longrightarrow\theta_*$ as $n\longrightarrow\infty$ with probability one. The mean square convergence rate is $O(n^{-(\beta-1)/\beta})$. The function $\tau(\cdot)$ is the continuous one. Hence, the sequence $\{\tau_s\}:\tau_s=\tau(\theta_{[s/(p+q)]-1})$ converges to τ_* with the probability one. Theorem 2 has proved.

REFERENCES

Barabanov A.Ye.,Granichin O.N.,1984, Optimal controller for a linear plant with bounded noise. *Automat. Remote Control*, 45, No.5, part 1, 578-584.

Doob J.L.,1953, Stochastic Processes. New York. Tonh Wiley & Sons. 654.

Chen H.F., Guo L.,1985, Adaptive control with recursive identification for stochastic linear systems. *Advances in Control and Dynamic Systems*, 24.

Chen H.F., Guo L.,1986, Convergence rate of least-squares stostochastic systems. *Int. Journal of Control*, 44, No 5, 1459-1477.

Goodwin G.C., Ramandge P.J., Caines P.E.,1981, Discrete time stochastic adaptive control. *SIAM J. Contr. Optimiz.*, 19, 829-853.

Granichin O.N., Fomin V.N.,1986, Adaptive control using test signals in the feedback channel. *Automat. Remote Control*, 47, No.2, part 2, 238-248.

Granichin O.N.,1989, A stochastic recursive procedure with dependent noises in the observation that uses sample perturbations in the Input. *Vestnik Leningrad Univ. Math.*, v.22, No.1, 27-31.

Granichin O.N.,1990, Construction of a suboptimal controller of a linear object with limited noise. *Automat. Remote Control*, 51, No.2, part 1, 184-187.

Granichin O.N.,1992a, Stochastic approximation with sample

perturbations in the input. *Automat. Remote Control*,53, No. 2, 100-110.

Granichin O.N.,1992b, Unknown function minimum point estimation under dependent noise. *Problems Inform. Transmission* ,28, No. 2, 90-99.

Lai T.L., Wei C.-Z.,1986, Extended least squares and their applications to adaptive control and prediction in linear systems. *IEEE Trans. on Automatic Control*, AC-31, No. 10, 899-907.

Ljung L.,1977, Analysis of recursive stochastic algorithms. *IEEE Trans. Auto. Control*, AC-22, 551-575.

Katkovnik V.,Ya. Linear Estimates and Stochastic Optimization Problems. Moscow. 1984.

Kiefer J., Wolfowitz J.,1952, Statistical estimation on the maximum of a regression function. *Ann. Math. Statist.* v.23, 462-466.

Moore J.B.,1978, On strong consistency of least squares identification algorithms. *Automatica*. 14, 505-509.

Polyak B.T.,1987, Introduction to Optimization. Optim. Software, New York.

Polyak B.T., Tsybakov A.B.,1990, Optimal orders of accuracy for search algorithms of stochastic optimization. *Problems Inform. Transmission*, v.26, No.2, 126-133.

Robbins H., Monro S.,1951, A stochastic approximation method. *Ann. Math. Statist.* v.22, 400-407.

Solo V.,1979, The Convergence of AML. *IEEE Trans.on Automatic Control*, AC-24, 958-962.

UNIFORM NORM OF POLYNOMIALS CONSTRAINED WITH MANY ZEROS IN TWO DISJOINT INTERVALS

Matthew He

Department of Mathematics, Nova University
Ft. Lauderdale, FL 33314, U.S.A. E-mail: hem@polaris.nova.edu

ABSTRACT. For nonnegative integers (s_1, s_2, m), consider the set of all real or complex polynomials of the form

$$P_n(x) = \prod_{i=1}^{s_1}\left(x - x_i^{(s_1)}\right)\prod_{j=1}^{s_2}\left(x - x_j^{(s_2)}\right)\sum_{k=0}^{m} a_k x^k,$$

where $s_1 \geq \theta_1(s_1 + s_2 + m) = \theta_1 n > 0$ and $s_2 \geq \theta_2(s_1 + s_2 + m) = \theta_2 n > 0$, $\theta_1 > 0$, $\theta_2 > 0$ and having s_1, s_2 zeros constrained so that $x_i^{(s_1)} \in [d, 1]$, $i = 1, 2, ..., s_1$, and $x_j^{(s_2)} \in [-1, c]$, $j = 1, 2, ..., s_2$, respectively. For all $P_n(x)'s$ bounded on $[-1, 1]$, we found the smallest interval $[a, b] \subseteq [-1, 1]$ such that

$$||P_n||_{[-1,1]} = ||P_n||_{[a,b]}.$$

1. INTRODUCTION

Lorentz [4] studied polynomials on $[0, 1]$ that vanish at zero with high order. That is, he considered polynomials of the form

$$P_n(x) = x^s \sum_{k=0}^{m} a_k x^k,$$

where $s \geq \theta(m + s)$, $\theta > 0$. He proved that if $P_n's$ are bounded on $[0, 1]$, then

$$||P_n(x)||_{[0,1]} = ||P_n(x)||_{[\theta^2,1]}.$$

The set $[\theta^2, 1]$ is called minimal essential set or the support of the extremal measure [8]. Polynomials of the form

$$P_n(x) = (x+1)^{s_1}(x-1)^{s_2}\sum_{k=0}^{m} a_k x^k,$$

with $s_1 \geq \theta_1(s_1 + s_2 + m)$ zeros at $x = -1$ and $s_2 \geq \theta_2(s_1 + s_2 + m)$ zeros at $x = 1$ were extensively studied in [1, 2, 5, 6, 7]. It was proved in [5] that if $P_n(x)'s$ are bounded on $[-1, 1]$, then

$$||P_n(x)||_{[-1,1]} = ||P_n(x)||_{[a,b]},$$

1991 *Mathematics Subject Classification.* 30C15, 41A10.

Key words and phrases. Uniform norm, incomplete polynomials, minimal set.

This paper is in final form and no version of it will be submitted for publication elsewhere.

where

$$a := a(\theta_1, \theta_2) = \mu\nu - \sqrt{(1-\mu^2)(1-\nu^2)}, \tag{1.1}$$

$$b := b(\theta_1, \theta_2) = \mu\nu + \sqrt{(1-\mu^2)(1-\nu^2)}, \tag{1.2}$$

with $\mu = \theta_1 + \theta_2, \nu = \theta_2 - \theta_1$. Lachance and Saff [2] investigated the polynomials of the following form

$$\prod_{i=1}^{s}(x - x_i)\sum_{k=0}^{m} \alpha_k x^k, \tag{1.3}$$

where $s \geq \theta(m+s) > 0$, with s zeros in a single interval $[-1, c]$ and proved that if $|P_n(x)| \leq M$ for $[-1, 1]$, then

$$\| P_n(x) \|_{[-1,1]} = \| P_n(x) \|_{[2\theta^2-1,1]} .$$

In this note, we generalize the form (1.3) to the polynomials of the form

$$\prod_{i=1}^{s_1}\left(x - x_i^{(s_1)}\right)\prod_{j=1}^{s_2}\left(x - x_j^{(s_2)}\right)\sum_{k=0}^{m} \alpha_k x^k, \tag{1.4}$$

where $s_1 \geq \theta_1(s_1 + s_2 + m) > 0$ and $s_2 \geq \theta_2(s_1 + s_2 + m) > 0$, $\theta_1 > 0$, $\theta_2 > 0$ and having s_1, s_2 zeros constrained to lie in two disjoint intervals $[d, 1]$ and $[-1, c]$, respectively. We find the minimal essential set for the uniform norm of $P_n(x)$.

In §2 we introduce some needed notation and state our main result. The proofs of some useful lemmas and the main result will be given in §3.

2. Main Result

To state our main result, we need the following notation. Let

$$w = \phi(z) = \frac{\sqrt{z-a} + \sqrt{z-b}}{\sqrt{z-a} - \sqrt{z-b}},$$

for some suitable branch of the square root function, where $-1 < a < b < 1$ as in (1.1) and (2.6). The function $\phi(z)$ maps the exterior of $[a, b]$ onto the exterior of $| w | \leq 1$. Define as in [3]

$$G(z; \theta_1, \theta_2) := \begin{cases} |\phi(z)| \left|\frac{\phi(z)-\phi(1)}{\phi(1)\phi(z)-1}\right|^{\theta_1} \left|\frac{\phi(z)-\phi(-1)}{\phi(-1)\phi(z)-1}\right|^{\theta_2}, & \text{if } z \in \overline{C} \setminus [a, b], \\ 1, & \text{otherwise.} \end{cases}$$

Fixing $\theta_1 > 0$, $\theta_2 > 0$ and $\theta_1 + \theta_2 < 1$ and considering $G(x; \theta_1, \theta_2)$ as a function of a real variable x, it was proved in [3] that $G(x; \theta_1, \theta_2)$ is strictly increasing on $(-1, a) \cup (1, \infty)$ and strictly decreasing on $(b, 1) \cup (-\infty, -1)$.

Consequently, there exist two unique real numbers $\rho := \rho(\theta_1, \theta_2) > 1$ and $\sigma := \sigma(\theta_1, \theta_2) > 1$ satisfying

$$G(-\rho; \theta_1, \theta_2) = 1 = G(\sigma; \theta_1, \theta_2). \tag{2.5}$$

We now state our main result.

Theorem 2.0.1. *Let a and b be as in* (1.1) *and* (2.6). *Let*

$$P_n(x) = \prod_{i=1}^{s_1}\left(x - x_i^{(s_1)}\right)\prod_{j=1}^{s_2}\left(x - x_j^{(s_2)}\right)\sum_{k=0}^{m}\alpha_k x^k,$$

where $s_1 \geq \theta_1(s_1+s_2+m) = \theta_1 n$ *and* $s_2 \geq \theta_1(s_1+s_2+m) = \theta_2 n$,

$$\frac{\rho-\sigma+2}{\rho+\sigma} =: \lambda_1 < d \leq x_i^{(s_1)} \leq 1,$$

where $i = 1, 2, \ldots, s_1$,

$$-1 \leq x_j^{(s_2)} \leq c < \lambda_2 := \frac{\rho-\sigma-2}{\rho+\sigma}$$

where $j = 1, 2, \ldots, s_2$, *and* ρ, σ *are positive constants defined in* (2.5).

If $| P_n(x) | \leq M$ *for* $a \leq x \leq b$, *then*

$$| P_n(x) | < M, \tag{2.6}$$

for $x \in (\frac{1+\rho}{2}c + \frac{1-\rho}{2}d, a) \cup (b, \frac{1-\sigma}{2}c + \frac{1+\sigma}{2}d)$.

In particular, (2.6) *implies*

$$\| P_n \|_{[-1,1]} = \| P_n \|_{[a,b]} \,. \tag{2.7}$$

Moreover, this result is best possible in the following senses:

(i) *the inequality* (2.6) *does not in general hold on any open interval properly containing* $(\frac{1+\rho}{2}c + \frac{1-\rho}{2}d, a)$ *or* $(b, \frac{1-\sigma}{2}c + \frac{1+\sigma}{2}d)$;
(ii) *if* $c > \lambda_2$ *or* $d < \lambda_1$, *then* (2.7) *is no longer true.*

An immediate consequence of the Theorem 2.0.1 is the following well known result.

Corollary 2.0.2. *If* $x_i^{(s_1)} = 1$, *for* $i = 1, 2, \ldots, s_1$; $x_j^{(s_2)} = -1$, *for* $j = 1, 2, \ldots, s_2$ *in Theorem 2.0.1, then* $P_n(x) = (x-1)^{s_1}(x+1)^{s_2}\sum_{k=0}^{m}\alpha_k x^k$. *If* $| P_n(x) | \leq M$ *for* $a \leq x \leq b$, *then*

$$|P_n(x)| < M, \ \forall x \in (-\rho, a) \cup (b, \sigma). \tag{2.8}$$

In particular, (2.8) *implies that*

$$\| P_n(x) \|_{[-1,1]} = \| P_n(x) \|_{[a,b]} \,. \tag{2.9}$$

Furthermore, this result is best possible in the sense that the inequality (2.9) *does not in general hold on any open interval properly containing* $(-\rho, a) \cup (b, \sigma)$.

3. Lemmas and Proofs

In order to prove our main result we need the following lemmas.

Lemma 3.0.3. *Let $G(t;\theta_1,\theta_2)$ be defined as in §2, and consider the function*

$$g(t) := G(t;\theta_1,\theta_2)/|1-t|^{\theta_1}|1+t|^{\theta_2}$$

as a function of t for each fixed θ_1, θ_2. Then $g(t)$ has two removable discontinuous points at $t=-1$, $t=1$ and is strictly decreasing on $(-\infty,a)$ and strictly increasing on (b,∞).

Proof. It follows from Theorem 3.4 in [5] that $G(t)$ has the following potential representation

$$G(t;\theta_1,\theta_2) = \frac{1}{\Delta}\left\{|1-t|^{\theta_1}|1+t|^{\theta_2}\exp\left((1-\theta_1-\theta_2)\int_a^b log|x-t|d\mu^*(x)\right)\right\},$$

where $\Delta := \frac{1}{2}\sqrt{(1+\mu)^{1+\mu}(1-\mu)^{1-\mu}(1+\nu)^{1+\nu}(1-\nu)^{1-\nu}}$, μ and ν are defined as in (1.1) and (1.2), and

$$(1-\theta_1-\theta_2)\mu^*(x) = \frac{\sqrt{(x-a)(b-x)}}{\pi(1-x^2)}dx.$$

Hence for $x\in[a,b]$,

$$g(t) = \frac{1}{\Delta}\exp\left((1-\theta_1-\theta_2)\int_a^b log|x-t|d\mu^*(x)\right).$$

Since the potential $\int_a^b log|x-t|d\mu^*(x)$ is continuous for all $t\in C$, we see that $g(t)$ is continuous at $t=-1$, and $t=1$. Also $g(t)>0$ for $t\in(-\infty,a)\cup(b,\infty)$.

Next, we prove the monotone property of $g(t)$. For $t\in(-\infty,a)$,

$$g'(t) = -g(t)(1-\theta_1-\theta_2)\int_a^b \frac{1}{x-t}d\mu^*(x) < 0.$$

For $t\in(b,\infty)$,

$$g'(t) = g(t)(1-\theta_1-\theta_2)\int_a^b \frac{1}{t-x}d\mu^*(x) > 0.$$

Thus $g(t)$ is decreasing on $(-\infty,a)$ and increasing on (b,∞). This completes the proof of Lemma 3.0.3. □

The next lemma gives a crude estimate for polynomials of the form (1.4).

Lemma 3.0.4. *Let $P_n(x)$ be a polynomial of degree at most n of the form* (1.4). *Suppose that $b<d\le x_i^{(s_1)}\le 1$, $i=1,2,...,s_1$, $-1\le x_j^{(s_2)}\le c<a$, $j=1,2,...,s_2$. Then we have*

$$\begin{aligned}|P_n(z)| \;\le\;& M\left\{G\left(\frac{(z-1)+(z-c)}{1-c};\theta_1,\theta_2\right)\right\}^n\\ &\times\prod_{i=1}^{s_1}\left|\frac{(b-1)(z-x_i^{(s_1)})}{(z-1)(b-x_i^{(s_1)})}\right|\times\prod_{j=1}^{s_2}\left|\frac{(b-c)(z-x_j^{(s_2)})}{(z-c)(b-x_j^{(s_2)})}\right|\end{aligned}$$

for $z \neq c$ and $z \neq 1$ in C, and

$$|P_n(z)| \leq M\left\{G\left(\frac{(z+1)+(z-d)}{1+d};\theta_1,\theta_2\right)\right\}^n \times \prod_{i=1}^{s_1}\left|\frac{(a-d)(z-x_i^{(s_1)})}{(z-d)(a-x_i^{(s_1)})}\right| \times \prod_{j=1}^{s_2}\left|\frac{(a+1)(z-x_j^{(s_2)})}{(z+1)(a-x_j^{(s_2)})}\right|$$

for $z \neq d$ and $z \neq -1$ in C.

Proof. Let $P_n(x) = \prod_{i=1}^{s_1}\left(x-x_i^{(s_1)}\right)\prod_{j=1}^{s_2}\left(x-x_j^{(s_2)}\right)\sum_{k=0}^{m}\alpha_k x^k$. Define

$$\Phi(x) := \prod_{i=1}^{s_1}\frac{(x-1)}{(x-x_i^{(s_1)})}\prod_{j=1}^{s_2}\frac{(x-c)}{(x-x_j^{(s_2)})},$$

$$\Psi(x) := \prod_{i=1}^{s_1}\frac{(x-d)}{(x-x_i^{(s_1)})}\prod_{j=1}^{s_2}\frac{(x+1)}{(x-x_j^{(s_2)})}.$$

Then

$$\Phi'(x) = \Phi(x)\left[\sum_{i=1}^{s_1}\frac{1-x_i^{(s_1)}}{(x-1)(x-x_i^{(s_1)})} + \sum_{j=1}^{s_2}\frac{c-x_j^{(s_2)}}{(x-c)(x-x_j^{(s_2)})}\right],$$

and

$$\Psi'(x) = \Psi(x)\left[\sum_{i=1}^{s_1}\frac{d-x_i^{(s_1)}}{(x-d)(x-x_i^{(s_1)})} + \sum_{j=1}^{s_2}\frac{-(1+x_j^{(s_2)})}{(x+1)(x-x_j^{(s_2)})}\right].$$

We observe that

$$\Phi(x) > 0,\ \Phi'(x) > 0,\ \forall x \in [a,b],$$

and

$$\Psi(x) > 0,\ \Psi'(x) < 0,\ \forall x \in [a,b].$$

Thus for $x \in [a,b]$, we have

$$0 < \Phi(x) \leq \Phi(b)$$

and

$$0 < \Psi(x) \leq \Psi(a).$$

Set

$$P(x) = \Phi(x)P_n(x) = (x-1)^{s_1}(x-c)^{s_2}\sum_{k=0}^{m}\alpha_k x^k,$$

and

$$Q(x) = \Psi(x)P_n(x) = (x-d)^{s_1}(x+1)^{s_2}\sum_{k=0}^{m}\alpha_k x^k.$$

Considering simple linear transformations and applying Theorem 2.5 in [3], we obtain

$$|P(z)| \leq M\Phi(b)\left\{G\left(\frac{(z-1)+(z-c)}{1-c};\theta_1,\theta_2\right)\right\}^n,$$

$$|Q(z)| \le M\Psi(a)\left\{G\left(\frac{(z+1)+(z-d)}{1+d};\theta_1,\theta_2\right)\right\}^n$$

for all $z \in C$. Thus, Lemma 3.0.4 follows from the definitions of $P(x)$ and $Q(x)$. □

Lemma 3.0.5. *Let $P_n(x)$ be a polynomial as in Lemma 3.0.4, and $\lambda_1 < d \le x_i^{(s_1)} \le 1$, $i = 1,2,\ldots,s_1$, $-1 \le x_j^{(s_2)} \le c < \lambda_2$, $j = 1,2,\ldots,s_2$, where $s_1 := \theta_1(m+\theta_1+\theta_2) > 0$, $s_2 := \theta_2(m+s_1+s_2) > 0$, and λ_1 and λ_2 are defined as in Theorem 2.0.1 in §2.*

If $|P_n(x)| \le M$ for $a \le x \le b$, then

$$|P_n(x)| \le M\eta^n, \ \alpha \le x \le \beta,$$

where $0 < \eta < 1$ and $[\alpha,\beta] \subset (\frac{1+\rho}{2}c + \frac{1-\rho}{2}d, a) \cup (b, \frac{1-\sigma}{2}c + \frac{1+\sigma}{2}d)$.

In particular, if $c = -1$ and $d = 1$, then

$$|P_n(x)| \le M\eta^n, \ \forall \alpha \le x \le \beta,$$

where $0 < \eta < 1$, $[\alpha,\beta] \subset (-\rho, a) \cup (b, \sigma)$.

Proof. It suffices to consider the case when

$$[\alpha,\beta] \subset \left(\frac{1+\rho}{2}c + \frac{1-\rho}{2}d, a\right).$$

We need to consider only the case that $[-1,c] \subset [\alpha,\beta]$, otherwise, we let $\alpha' := min\{-1,\alpha\}$ and $\beta' := max\{c,\beta\}$. Hence,

$$[-1,c] \subset [\alpha',\beta'] \subset \left(\frac{1+\rho}{2}c + \frac{1-\rho}{2}d, a\right).$$

We divide interval $[\alpha,\beta]$ into two subintervals $[\alpha,c)$ and $[c,\beta]$.

For $c \le x \le a$, $i = 1,2,\ldots,s_1$, and $j = 1,2,\ldots,s_2$, we have

$$\left|\frac{(a-d)(x-x_i^{(s_1)})}{(x-d)(x-x_i^{(s_1)})}\right| \le 1$$

and

$$\left|\frac{(a+1)(x-x_j^{(s_2)})}{(x+1)(a-x_j^{(s_2)})}\right| \le 1.$$

These can be seen by considering functions

$$f(x) = \frac{(a-d)(x-x_i^{(s_1)})}{(a-x_i^{(s_1)})(x-d)}$$

and

$$g(x) = \frac{(a+1)(x-x_j^{(s_2)})}{(x+1)(a-x_j^{(s_2)})}$$

for $c \le x \le a$.

Since $G(x;\theta_1,\theta_2)$ is increasing on $[-1,a]$, using Lemma 3.0.4, we get

$$|P_n(x)| \le M\left\{G\left(\frac{2\beta+1-d}{1+d};\theta_1,\theta_2\right)\right\}^n$$

for $c \le x \le \beta$.

For $\alpha \le x < c$, $i = 1, 2, ..., s_1$, $j = 1, 2, ..., s_2$, we have

$$\left|\frac{(b-c)(x-x_j^{(s_2)})}{(x-c)(b-x_j^{(s_2)})}\right| \le \left|\frac{c-\alpha}{x-c}\right|$$

and

$$\left|\frac{(b-1)(x-x_i^{(s_1)})}{(x-1)(b-x_i^{(s_1)})}\right| \le \left|\frac{\alpha-1}{x-1}\right|.$$

Applying Lemma 3.0.4, we get

$$\begin{aligned} |P_n(x)| &\le M\left\{G\left(\frac{(x-1)+(x-c)}{1-c};\theta_1,\theta_2\right)\right\}^n \left|\frac{\alpha-1}{x-1}\right|^{s_1} \left|\frac{c-\alpha}{x-c}\right|^{s_2} \\ &= M\left\{G\left(\frac{(x-1)+(x-c)}{1-c};\theta_1,\theta_2\right)\right\}^n \left|\frac{\alpha-1}{x-1}\right|^{n\theta_1} \left|\frac{c-\alpha}{x-c}\right|^{n\theta_2} \\ &= M\left\{\frac{G(\frac{(x-1)+(x-c)}{1-c};\theta_1,\theta_2)}{|x-1|^{\theta_1}|x-c|^{\theta_2}}\right\}^n |\alpha-1|^{n\theta_1}|c-\alpha|^{n\theta_2}. \end{aligned}$$

Since $g(t)$ is decreasing on $(-\infty, a)$ from Lemma 3.0.3, we have, when $x = \alpha$,

$$|P_n(x)| \le M\left\{G\left(\frac{2\alpha-1-c}{1-c};\theta_1,\theta_2\right)\right\}^n.$$

It follows that

$$| P_n(x) | \le M\eta_1^n, \ \forall \alpha \le x \le \beta,$$

where $\eta_1 := max\left\{G(\frac{2\beta+1-d}{1+d};\theta_1,\theta_2), G(\frac{2\alpha-1-c}{1-c};\theta_1,\theta_2)\right\}$.

Similarly, we can prove that when $[\alpha,\beta] \subset (b, \frac{1-\sigma}{2}c + \frac{1+\sigma}{2}d)$,

$$| P_n(x) | \le M\eta_2^n, \ \forall \alpha \le x \le \beta,$$

where $\eta_2 = max\left\{G(\frac{2\alpha-1-c}{1-c};\theta_1,\theta_2), G(\frac{2\beta+1-d}{1+d};\theta_1,\theta_2)\right\}$. This completes the proof of Lemma 3.0.5. □

We now use the above lemmas to prove our main theorem.

Proof. For $-1 \le c < \lambda_2$ and $\lambda_1 < d \le 1$, (2.6) follows from Lemma 3.0.5. We remark that (2.6) is also true when $d = \lambda_1$ and $c = \lambda_2$, for $c = \lambda_2$, $d = \lambda_1$ imply that

$$\frac{1+\rho}{2}c + \frac{1-\rho}{2}d = -1,$$
$$\frac{1-\sigma}{2}c + \frac{1+\sigma}{2}d = 1.$$

For $x \in (-1, a) \cup (b, 1)$, (2.6) follows from Theorem 2.5 in [3].

We now verify the sharpness statements.

(i) We will show that $\frac{1-\sigma}{2}c + \frac{1+\sigma}{2}d$ cannot be decreased and $\frac{1+\rho}{2}c + \frac{1-\rho}{2}d$ cannot be increased in general.

Let $J_k(z) \in \prod \left(s_1^{(k)}, s_2^{(k)}, m\right)$, where

$$\prod \left(s_1^{(k)}, s_2^{(k)}, m\right) = \left\{ J_k(z) = (x-1)^{s_1^{(k)}} (x+1)^{s_2^{(k)}} \sum_{i=0}^{m} \alpha_{i,k} z^i \right\},$$

$s_1^{(k)} \geq \theta_1(s_1^{(k)} + s_2^{(k)} + m)$, $s_2^{(k)} \geq \theta_2(s_1^{(k)} + s_2^{(k)} + m)$, be the Chebyshev polynomial as in [3] with

$$\|J_k(z)\|_{[-1,1]} = 1 \tag{3.10}$$

and

$$|J_k(z)|^{1/n_k} \to G(z; \theta_1, \theta_2), \ k \to \infty, \tag{3.11}$$

where $n_k = s_1^{(k)} + s_2^{(k)} + m$, $z \in C \setminus [a, b]$ and $J_k(1) = J_k(-1) = 0$. It follows from [3] that

$$\|J_k^*(x)\|_{[c,d]} = \left\| J_k\left(\frac{(x-c)+(x-d)}{d-c}\right) \right\|_{[-1,1]} \leq 1$$

and that $J_k^*(x)$ has a zero of order $s_1^{(k)}$ at $x = d$ and a zero of order $s_2^{(k)}$ at $x = c$ respectively. Furthermore, for $x < \frac{1-\sigma}{2}c + \frac{1+\sigma}{2}d$ or $x > \frac{1+\rho}{2}c + \frac{1-\rho}{2}d$ we have

$$|J_k^*(x)|^{1/n_k} = |J_k(t)|^{1/n_k},$$

where $t = [(x-c)+(x-d)]/(d-c) > \sigma$ or $t = [(x-c)+(x-d)]/(d-c) < -\rho$. It follows from (3.10) and (3.11) that, for such x, we have

$$\lim_{k\to\infty} |J_k^*(x)|^{1/n_k} = \lim_{k\to\infty} |J_k(t)|^{1/n_k} = G(t; \theta_1, \theta_2) > 1,$$

which completes the proof of assertion (i).

(ii) If $b \leq d < \lambda_1$ or $\lambda_2 < c \leq a$, then in considering the same sequence of

$$J_k^*(x) = J_k\left(\frac{(x-c)+(x-d)}{d-c}\right)$$

as that in the proof of statement (i), we have

$$\|J_k^*\|_{[-1,1]}^{1/n_k} \geq |J_k^*(-1)|^{1/n_k} = \left| J_k\left(\frac{-2-(d+c)}{d-c}\right) \right|^{1/n_k} = |J_k(-\rho-\epsilon)|^{1/n_k}$$

for some $\epsilon > 0$, and

$$\|J_k^*\|_{[-1,1]}^{1/n_k} \geq |J_k^*(1)|^{1/n_k} = \left| J_k\left(\frac{2-(d+c)}{d-c}\right) \right|^{1/n_k} = |J_k(\sigma+\delta)|^{1/n_k}$$

for some $\delta > 0$. Then

$$\liminf_{k\to\infty} \|J_k^*\|_{[-1,1]} \geq G(-\rho-\epsilon; \theta_1, \theta_2) > 1,$$

and

$$\liminf_{k\to\infty} \|J_k^*\|_{[-1,1]} \geq G(\sigma+\delta; \theta_1, \theta_2) > 1.$$

This contradicts $\|J_k^*\|_{[-1,1]} \leq 1$. □

REFERENCES

1. M. He and X. Li, *Uniform convergence of polynomials associated with varying Jacobi weights,* Rocky Mountain Journal of Mathematics, 21(1991), 281–300.
2. M. A. Lachance and E. B. Saff, *Bounds for algebraic polynomials with zeros in an interval,* Canadian Math. Soc. Conf. Proc., Vol.3 (1983), 227–237.
3. M. A. Lachance, E, B, Saff and R. S. Varga, *Bounds for incomplete polynomials vanishing at both endpoints of an interval,* Constructive approaches to mathematics models, C. V. Coffiman and G. J. Fix, Eds. Academic Press, New York 1979, 421–437.
4. G. G. Lorentz, *Approximation by incomplete polynomials, Pade and rational approximation: Theory and applications,* E. B. Saff and R. S. Varga, Eds. Academic Press, New York 1977, 289–302.
5. E.B. Saff, J.L. Ullman and R.S. Varga, *Incomplete polynomials: an electrostatics approach,* in *Approximation Theory III* E.W. Cheney, Ed. Academic Press, New York 1980, 769–782.
6. E.B. Saff and R.S. Varga, *The Sharpness of Lorentz's Theorem on incomplete polynomials,* Trans. Amer. Math. Soc., 249(1978), 163–186.
7. E.B. Saff and R.S. Varga, *Uniform approximation by incomplete polynomials,* Internat. J. Math. Soc., 1(1978), 407–420.
8. V. Totik, *Approximation by polynomials,* 1993 (manuscript).

ADMISSIBLE LOCATION OF SAMPLE POINTS OF INTERPOLATION BY BIVARIATE C^1 QUADRATIC SPLINES

Tian-Xiao He

Department of Mathematics
Illinois Wesleyan University
Bloomington, IL 61702-2900

ABSTRACT

In this paper, we will discuss the admissible location of sample points of C^1 quadratic spline interpolation in $S_2^1(\Delta_{MN}^{(2)})$. General and simple poisedness conditions have been obtained by using smooth and conformality conditions expressed in terms of Bézier coefficients, the techniques of univariate spline interpolation, and theory of multivariate polynomial interpolations. First, we construct poised sets on an arbitrary rectangular cell $R_{ij} = [x_i, x_{i+1}] \otimes [y_j, y_{j+1}]$, that is, to find the location of sample points which admits unique Lagrange interpolation in $S_2^1(\Delta_{11}^{(2)}, R_{ij})$. Then, by means of the so-called flows from a source cell R_{i_0,j_0} in which the poised set of sample points has been obtained, we can arrange the location of remaining sample points in other cells R_{ij} along the flows.

1. INTRODUCTION

Let $V \subset C(D)$, $D \subset R^s$, $s \geq 1$, be a vector space with dimension n, and $\mathbf{X}_n = \mathbf{x}^1, \cdots, \mathbf{x}^n$ be a set of n distinct points in D. The sample point set $\mathbf{X}_n$ is said to be poised with respect to V, or equivalently, the problem of interpolation from V at the sample point set $\mathbf{X}_n$ is said to be unisolvent, if for any given set of data $\{y_1, \cdots, y_n\} \subset \mathbf{R}$, there exists one and only one $f \in V$ such that $f(\mathbf{x}^i) = y_i$, $i = 1, \cdots, n$. For instance, if $V = \pi_d^s$, the collection of all polynomials in s variables of degree d, its dimension is given by the binomial coefficient

$$n = \binom{d+s}{s}.$$

Although any sample point set $\mathbf{X}_n$ in $\mathbf{R}$ is always poised with respect to π_d^1, the multivariate interpolation problem from π_d^s, $s > 1$ is not always unisolvent for an arbitrary sample point set $\mathbf{X}_n$ in $\mathbf{R}^2$ since the Haar condition corresponding to the univariate functions cannot be generalized to the multivariate setting. The necessary and sufficient conditions for the interpolation problem from π_d^2 to be unisolvent are shown in Chung and Yao [3], Liang [5], and Zhou, Chang, and the author [8]. An algebraic (polynomial) curve of degree d is a curve with expression of series $\Sigma_{i=1}^{n} c_i p_i(x,y)$, $n = \binom{d+s}{s}$, here $\{p_i(x,y) : i = 1, \cdots, n\}$ is a basis of π_d^2. By using the notion of algebraic curves, a necessary and sufficient condition of poised interpolation knots with respect to π_d^2 was given by [8] in the following form.

Theorem A. A set of sample points $\mathbf{X}_n = \{(x_i, y_i), i = 1, \cdots, n\}$, $n = \binom{d+2}{2}$, admits unique Lagrange interpolation in π_d^2 if and only if $\mathbf{X}_n$ does not lie on any algebraic polynomial curve of degree d.

According to Theorem A, one can develop many methods to give the location of sample points in $\mathbf{R}^2$ that guarantees poisedness of the interpolation knots with respect to π_d^2. For instance, on d+1 arbitrary distinct straight lines, which are labeled as q_i, $i = 1, \cdots, d+1$, we take i sample points on the ith straight line q_i, $i = 1, \cdots, d+1$, such that all $\binom{d+2}{2}$ points are distinct. Then it is easily checked that the sample point set $\mathbf{X}_n$ is indeed a poised set for π_d^2 by using well-known Bezout Theorem in Algebraic Geometry.

If we consider a univariate spline space V, then the poisedness of the interpolation knots is completely characterized by the Schoenberg and Whitney condition. But the Schoenberg and Whitney condition does not directly generalize to the multivariate spline spaces. An example on the interpolation in the space of bivariate linear splines on a triangulation was first studied by Chui, Wang, and the author [2]. Although the results in [2] also seem to indicate that a poisedness condition which is both necessary and sufficient is probably very difficult to obtain, several necessary conditions and some sufficient conditions which can be easily used in applications were given.

As to the poisedness condition of sample points of interpolation in higher order spline spaces, such as a C^1 quadratic spline space, say, we have two obvious obstacles. First, the dimension of the spline space may be unstable with respect to the geometry of the grid partition (cf. Morgon and Scott's example in [6]), and second, locally supported splines may not exist in general. Hence we have to be satisfied with a study of the poisedness problem for the spline spaces under some special triangulations, such as the type-1 triangulation $\Delta_{MN}^{(1)}$ and the type-2 triangulation $\Delta_{MN}^{(2)}$ of $R = [a,b] \otimes [c,d]$, which are constructed from the rectangular grid of R,

$$R_{ij} = [x_i, x_{i+1}] \otimes [y_j, y_{j+1}], i = 0, 1, \cdots, M-1; j = 0, 1, \cdots, N-1$$

by drawing in all positive diagonals and all two diagonals in each R_{ij}, respectively. Denote the C^1 quadratic bivariate spline space over R under the triangulation $\Delta_{MN}^{(1)}$ and $\Delta_{MN}^{(2)}$ by $S_2^1(\Delta_{MN}^{(1)}) = S_2^1(\Delta_{MN}^{(1)}, R)$ and $S_2^1(\Delta_{MN}^{(2)}) = S_2^1(\Delta_{MN}^{(2)}, R)$, respectively. The poisedness problem of the interpolation knots with respect to $S_2^1(\Delta_{MN}^{(2)})$ was first studied by Chui and the author [1], and later by Nürnberger and Riessinger [7]. Also Dæham and Lyche [4] considered the corresponding problem in $S_2^1(\Delta_{MN}^{(1)})$. It is still very difficult to get a condition which is both sufficient and necessary for the poisedness of C^1 quadratic spline interpolation knots with respect to $S_2^1(\Delta_{MN}^{(2)})$. Although two necessary conditions on the poisedness problem were given in [1], sufficient conditions were only given a little consideration.

In this paper, we discuss the admissible location of sample points of the interpolation in $S_2^1(\Delta_{MN}^{(2)})$. General and simple sufficient conditions for the poisedness have been obtained by using smooth and conformality conditions of bivariate splines expressed in terms of Bézier coefficients, the techniques of univariate spline interpolation, and theory of multivariate polynomial interpolation. First, we construct poised sets on an arbitrary cell $R_{ij} = [x_i, x_{i+1}] \otimes [y_j, y_{j+1}]$, that is, to find the location of sample points which admit unique Lagrange interpolation in $S_2^1(\Delta_{11}^{(2)}, R_{ij})$. Then, by means of the so-called flows from a source cell R_{i_0,j_0} in which the poised set of sample points has been obtained, we can arrange the location of remaining sample points in the other cells R_{ij} along the flows. Our problem about the location of sample points and corresponding necessary poisedness conditions are presented in section 2. Section 3 gives general sufficient conditions for the poised knots of interpolation from $S_2^1(\Delta_{11}^{(2)}, R_{ij})$. The main results are included in Theorem 1 and 2 that are derived from several Lemmas. In section 4, we discuss the construction of the poised set of interpolation in $S_2^1(\Delta_{MN}^{(2)}, R)$ by using the notion "flow" and give the procedures to locate the sample points.

2. THE LOCATION PROBLEM OF INTERPOLATION KNOTS IN $S_2^1(\Delta_{MN}^{(2)}, R_{ij})$

We now discuss locations of sample points for the interpolation problem from $S_2^1(\Delta_{MN}^{(2)})$ such that they are poised. A very modest situation will be discussed in the following. After putting sample points at the (M+1)(N+1) vertices

$$P_{ij} = (x_i, y_j), \qquad 0 \le i \le M,\, 0 \le j \le N,$$

where are we allowed to place the remaining

$$dim\, S_2^1(\Delta_{MN}^{(2)}) - (M+1)(N+1) = M+N+2$$

sample points?

From Chui and author [1], there are two necessary conditions that arise from the theory of bivariate polynomial interpolation (cf. Theorem A).

(N.1) At most six sample points are allowed in the closure of any triangular cell, and if six points are placed there, then they must not lie on any algebraic curve of degree two.

(N.2) At most three sample points are allowed to be collinear in the closure of any triangular cell.

The sufficient poisedness conditions are more complicated. In order to find them, we need the following smooth conditions and conformality conditions of splines in $S_2^1(\Delta_{11}^{(2)}, R_{ij})$.

There are two types of vertices in the triangulation $\Delta_{MN}^{(2)}$ as shown in Fig. 1 and Fig. 2, respectively, where the interior vertices in Fig. 1 and Fig. 2 are $((x_i + x_{i+1})/2, (y_j + y_{j+1})/2)$ and (x_i, y_j), respectively. We label the Bézier coefficients of the C^1 piecewise quadratic polynomial $s \in S_2^1(\Delta_{MN}^{(2)}, R)$ on all the triangular cells in both figures. Obviously, the Bézier coefficients of adjacent polynomials at the same points on the common edges must agree for continuity. For the Fig. 1, four more smooth conditions shown as follows are needed for C^1 continuity:

$$\text{(1)} \qquad \begin{cases} a_6 + b_6 = 2a_4, \\ b_6 + c_6 = 2b_5, \\ c_6 + d_6 = 2c_4, \\ d_6 + a_6 = 2d_5. \end{cases}$$

From the above smooth conditions (1) we have

$$a_6 + b_6 + c_6 + d_6 = 2a_4 + 2c_4.$$

Since $a_4 + c_4 = 2a_1$ from the C^1 continuity, a C^1 conformality condition for the interior vertex shown in Fig. 1 is obtained as follows.

(2) $$a_6 + b_6 + c_6 + d_6 = 4a_1.$$

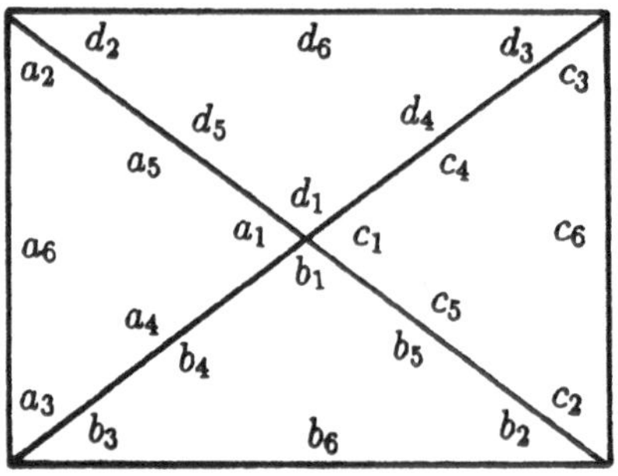

Figure. 1. Bézier coefficients of s centered at the vertex $(\frac{1}{2}(x_i + x_{i+1}), \frac{1}{2}(y_j + y_{j+1}))$

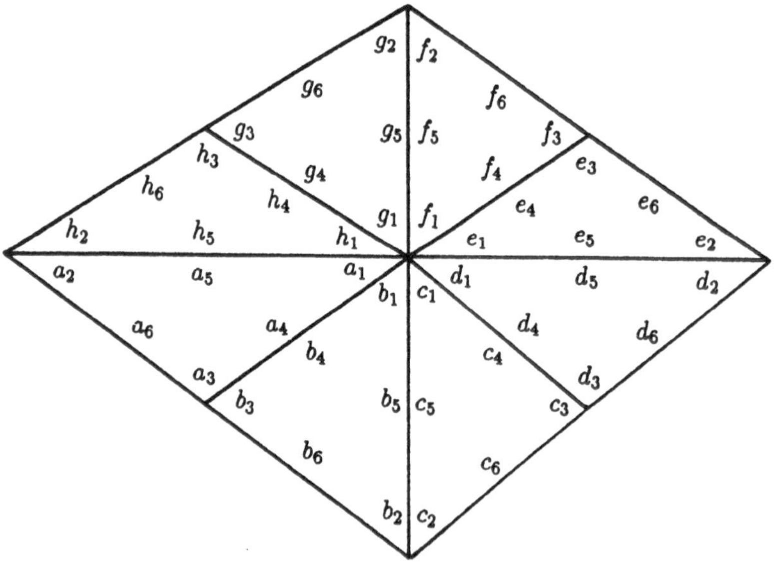

Figure. 2. Bézier coefficients of s centered at the vertex (x_i, y_j)

Similarly, for the Fig. 2, we have the following 10 smooth conditions for C^1 continuity.

(3) $$\begin{cases} 2B_j h_6 + 2B'_j a_6 - a_2 = a_5, \\ 2B_j e_6 + 2B'_j d_6 - d_2 = d_5, \\ 2A_i f_6 + 2A'_i g_6 - f_2 = f_5, \\ 2A'_i b_6 + 2A_i c_6 - b_2 = c_5, \end{cases}$$

(4) $$\begin{cases} f_5 + a_5 = 2h_4, \\ f_5 + d_5 = 2e_4, \\ d_5 + c_5 = 2c_4, \\ a_5 + c_5 = 2a_4, \end{cases}$$

(5) $$\begin{cases} B_j f_5 + B'_j b_5 = a_1, \\ A_i e_5 + A'_i a_5 = a_1, \end{cases}$$

where $A_i = (x_i - x_{i-1})/(x_{i+1} - x_{i-1})$, $B_j = (y_j - y_{j-1})/(y_{j+1} - y_{j-1})$, $A'_i = 1 - A_i$, $B'_j = 1 - B_j$. Also two conformlity conditions can be obtained from (3)-(5):

(6) $$\begin{cases} 2A_iB_je_6 + 2A_iB'_jd_6 + 2A'_iB_jh_6 + 2A'_iB'_ja_6 - A_id_2 - A'_ia_2 = a_1, \\ 2A_iB_jf_6 + 2A'_iB_jg_6 + 2A'_iB'_jb_6 + 2A_iB'_jc_6 - B_jf_2 - B'_jb_2 = a_1. \end{cases}$$

3. SUFFICIENT POISEDNESS CONDITIONS

To formulate sufficient conditions of the poisedness, it is enough to show that the zero function is the only function that interpolates the zero data. Let us first consider the location of poised sets of sample points of interpolation in $S_2^1(\Delta_{11}^{(2)}, R_{ij})$, $R_{ij} = [x_i, x_{i+1}] \otimes [y_j, y_{j+1}] \subset R$. As shown before, four sample points are indicated at four corners, and we wish to place

$$dim\, S_2^1(\Delta_{11}^{(2)}, R_{ij}) - 4 = 8 - 4 = 4$$

more sample points on R_{ij} such that these eight sample points are poised with respect to $S_2^1(\Delta_{11}^{(2)}, R_{ij})$. For this purpose, we may refer to Fig. 3, and apply the conditions (1) and (2) to determine the poisedness of the interpolation knots on R_{ij}.

Since the zero data has been given at four corner points, from conditions (1) and (2), the Bézier coefficients of the corresponding interpolation spline in $S_2^1(\Delta_{11}^{(2)}, R_{ij})$ can be shown in terms of parameters f, g, h, and k as in Fig. 3. In the following we will give some admissible positions for the remaining four sample points that guarantee poisedness; that is, to make all parameters f, g, h, and k be zero.

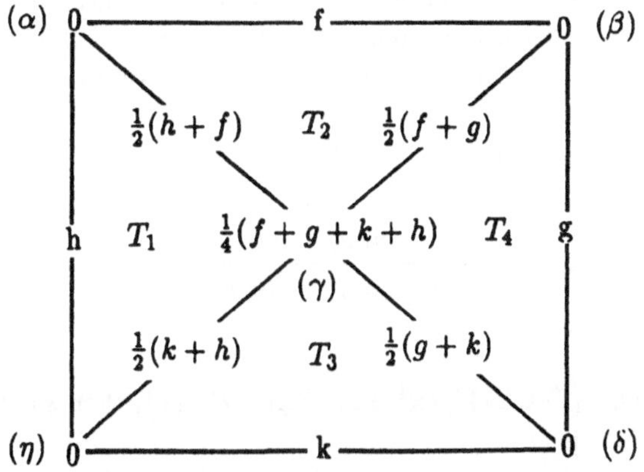

Figure. 3. Bézier coefficients of interpolation spline in $S_2^1(\Delta_{11}^{(2)}, R_{ij})$

In Fig. 3, we denote the triangles $\langle P_{i,j}, P_{i+1,j}, \frac{1}{2}(P_{i+1,j}+P_{i,j+1})\rangle$, $\langle P_{i+1,j}, P_{i+1,j+1}, \frac{1}{2}(P_{i+1,j}+P_{i,j+1})\rangle$, $\langle P_{i+1,j+1}, P_{i,j+1}, \frac{1}{2}(P_{i+1,j}+P_{i,j+1})\rangle$, and $\langle P_{i,j+1}, P_{ij}, \frac{1}{2}(P_{i+1,j}+P_{i,j+1})\rangle$ by T_i, i=1,2,3, and 4, respectively, and the barycentric coordinates of $\mathbf{x}=(x,y)$ relative to the triangle T_1, T_2, T_3, and T_4 by (u_1,v_1,w_1), (u_2,v_2,w_2), (u_3,v_3,w_3), and (u_4,v_4,w_4), respectively.

The proofs of the following lemmas will illustrate our general method to construct poised sets of interpolation in $S_2^1(\Delta_{11}^{(2)}, R_{ij})$.

Lemma 1. Let $s \in S_2^1(\Delta_{11}^{(2)}, R_{ij})$ vanish on the boundary of R_{ij}. Then $s \equiv 0$.

Proof. Since $s \in S_2^1(\Delta_{11}^{(2)}, R_{ij})$ vanishes on the boundary of R_{ij}, then $f=g=h=k=0$. Thus, from Fig. 3, we have $s \equiv 0$.

Lemma 2. Let $s \in S_2^1(\Delta_{11}^{(2)}, R_{ij})$ vanish on any three edges, and s has another zero point on R_{ij}. Then $s \equiv 0$.

Proof. W.L.O.G., we assume that $s|_{x=x_i}=0$, $s|_{x=x_{i+1}}=0$, $s|_{y=y_{j+1}}=0$, $\mathbf{x}\in T_1$. Then $f=g=h=0$ and $\mathbf{x}=(u_1,v_1,w_1)$, $u_1,v_1,w_1 \geq 0$ $u_1+v_1+w_1=1$. Thus

$$s(\mathbf{x}) = \frac{1}{4}k(w_1)^2 + 2ku_1v_1 + ku_1w_1 + kv_1w_1 = 0.$$

Because of $u_1,v_1,w_1 \geq 0$ and $u_1+v_1+w_1=1$, $\frac{1}{4}(w_1)^2+2u_1v_1+u_1w_1+v_1w_1 \neq 0$. Then we have $k=0$ and hence, $s\equiv 0$.

Lemma 3. Denote $E_{i+1,j}=\{(x,y): x=x_{i+1},\ y_j \leq y \leq y_{j+1}\}$ and $\hat{E}_{i,j+1}=\{(x,y): y=y_{j+1},\ x_i \leq x \leq x_{i+1}\}$. Let $s \in S_2^1(\Delta_{11}^{(2)}, R_{ij})$, $s|_{E_{i+1,j}}\equiv 0$, $s|_{\hat{E}_{i,j+1}}\equiv 0$, and $s|_{R_{ij}} \not\equiv 0$. If s has a zero point $\mathbf{x}^1=(u_1,v_1,w_1)\in T_1$, then there exists a zero line $L_{1,\mathbf{x}^1}$ of s with respect to point $\mathbf{x}^1$, which passes from $P_{i+1,j+1}$ to some point $\mathbf{x}^2=(u_2,v_2,w_2)$ that lies on the line from $P_{i,j+1}$ to $P_{i+1,j}$ and satisfies

$$u_2 = 1-w_2, \qquad v_2 = 0,$$

and

$$w_2 = \frac{4u_1w_1+(w_1)^2}{4u_1w_1+(w_1)^2+2u_1v_1+v_1w_1}.$$

In addition, if $s\neq 0$ on $\{T_1\cup T_4\}\setminus\{\mathbf{x}^1, P_{i,j}, P_{i+1,j}, P_{i,j+1}\}$, then $s \neq 0$ on $\{T_2\cup T_3\}\setminus\{L_{1,\mathbf{x}^1}\cup E_{i+1,j}\cup \hat{E}_{i,j+1}\}$.

Similarly, if s has a zero point $\mathbf{x}^4=(u_4,v_4,w_4)\in intT_4$, then there exists a zero line $L_{2,\mathbf{x}^4}$ of s with respect to point $\mathbf{x}^4$, which passes from $P_{i+1,j+1}$ to a point $\mathbf{x}^3=(u_3,v_3,w_3)$ that lies on the line from $P_{i,j+1}$ to $P_{i+1,j}$ and satisfies

$$u_3 = 0, \qquad v_3 = 1-w_3,$$

and

$$w_3 = \frac{4v_2w_2+(w_2)^2}{4v_2w_2+(w_2)^2+2u_2v_2+u_2w_2}.$$

In addition, if $s\neq 0$ on $\{T_1\cup T_4\}\setminus\{\mathbf{x}^4, P_{i,j}, P_{i+1,j}, P_{i,j+1}\}$, then $s \neq 0$ on $\{T_2\cup T_3\}\setminus\{L_{2,\mathbf{x}^4}\cup E_{i+1,j}\cup \hat{E}_{i,j+1}\}$.

Proof. Since the proof of the first part and the second part of the Lemma are similar, we only need to prove the first part.

From $s(\mathbf{x}^1) = 0$, $\mathbf{x}^1 = (u_1, v_1, w_1)$, by using the Bézier coefficients of s shown in Fig. 3, we have

$$s(\mathbf{x}^1) = 2ku_1v_1 + kv_1w_1 + (h+k)u_1w_1 + \frac{1}{4}(h+k)(w_1)^2 = 0. \tag{7}$$

If there exists another zero point $\mathbf{x}$ of s on $T_2 \setminus E_{i+1,j}$, then $s(\mathbf{x}) = 0$. Denote the line from $P_{i+1,j+1}$ to $\mathbf{x}$ by L. Since $\frac{\partial s}{\partial x}(P_{i+1,j+1}) = \frac{\partial s}{\partial y}(P_{i+1,j+1}) = 0$, then $\frac{\partial s}{\partial L}(P_{i+1,j+1}) = 0$. Note that $s(P_{i+1,j+1}) = 0$, $s(\mathbf{x}) = 0$. Thus we have $s|_L \equiv 0$. Denote the intersection point of the extension of L and the line from $P_{i,j+1}$ to $P_{i+1,j}$ by $\mathbf{x}^2 = (u_2, v_2, w_2)$. Then

$$ku_2w_2 + \frac{1}{4}(h+k)(w_2)^2 = 0. \tag{8}$$

From (7) and (8), we obtain the system

$$\begin{bmatrix} 4u_1w_1 + (w_1)^2 & 8u_1v_1 + 4v_1w_1 \\ (w_2)^2 & 4u_2w_2 \end{bmatrix} \begin{bmatrix} h \\ k \end{bmatrix} = 0. \tag{9}$$

Since $s \not\equiv 0$, then $(h,k) \neq 0$. Note $u_2 = 1 - w_2$. Thus the determinant of the coefficient matrix of (9) is zero; that is,

$$w_2 = \frac{4u_1w_1 + (w_1)^2}{4u_1w_1 + (w_1)^2 + 2u_1v_1 + v_1w_1}. \tag{10}$$

Note that $4u_1w_1 + (w_1)^2 + 2u_1v_1 + v_1w_1 > 4u_1w_1 + (w_1)^2 \geq 0$, since $u_1, v_1, w_1 \geq 0$ and $u_1 + v_1 + w_1 = 1$. Thus, $0 \leq w_2 \leq 1$, and $\mathbf{x}^2 = (u_2, v_2, w_2) = (1 - w_2, 0, w_2)$ indeed lies on the line from $P_{i,j+1}$ to $P_{i+1,j}$. Hence, the line $L_{1,\mathbf{x}^1}$ from $P_{i+1,j+1}$ to $\mathbf{x}^2 = (1 - w_2, 0, w_2)$, where w_2 is shown in (10), is a zero line of s. Obviously, there does not exist any other zero point of s on $T_2 \setminus \{E_{i+1,j} \cup L_{1,\mathbf{x}^1}\}$. Otherwise, the line from $P_{i+1,j+1}$ to the extra zero point of s must pass through $\mathbf{x}^2 = (1 - w_2, 0, w_2)$, where w_2 is given as in (10), because of $s \not\equiv 0$. Thus, this extra zero point of s lies on the line $L_{1,\mathbf{x}^1}$. This contradiction shows that there do not exist other zero points on $T_2 \setminus \{E_{i+1,j} \cup L_{1,\mathbf{x}^1}\}$.

Next, if there exists a zero point $\mathbf{x}$ of s on $T_3 \setminus \hat{E}_{i,j+1}$, then we will deduce a contradiction. Assume the intersection of the line though $P_{i+1,j+1}$ and $\mathbf{x}$ and the line from $P_{i,j+1}$ to $P_{i+1,j}$ is $\mathbf{x}^3 = (u_3, v_3, w_3)$, then we have $u_3 = 0$ and $v_3 = 1 - w_3$. Similarly, the line from $P_{i+1,j+1}$ to $\mathbf{x}^3$ must be a zero line of s and we denote it by $\hat{L}_{1,\mathbf{x}^1}$. Thus from $s|_{\mathbf{x}^1} = 0$ and $s|_{\mathbf{x}^3}$, we have a linear system. Note that $s \not\equiv 0$. The determinant of the coefficient matrix of this linear system is

$$det \begin{bmatrix} 4u_1w_1 + (w_1)^2 & 8u_1v_1 + 4v_1w_1 \\ 4 - 3w_3 & 4w_3 - 4 \end{bmatrix} = 0,$$

or equivalently,

$$w_3 = \frac{4u_1w_1 + (w_1)^2 + 8u_1v_1 + 4v_1w_1}{4u_1w_1 + (w_1)^2 + 6u_1v_1 + 3v_1w_1}.$$

Since $u_1, v_1, w_1 \geq 0$ and $u_1 + v_1 + w_1 = 1$, then $4u_1w_1 + (w_1)^2 + 8u_1v_1 + 4v_1w_1 > 4u - 1w_1 + (w_1)^2 + 6u_1v_1 + 3v_1w_1 > 0$. Thus $w_3 > 1$. But this is impossible because

$\mathbf{x}^3 \in T_3 \setminus \hat{E}_{i,j+1}$. This contradiction shows that there does not exist any zero point of s on $T_3 \setminus \hat{E}_{i,j+1}$, if $s \not\equiv 0$, $s|_{E_{i+1,j}} = 0$, $s|_{\hat{E}_{i,j+1}} = 0$, and s has a zero point $\mathbf{x}^1$ in T_1. Hence, if $s \neq 0$ on $\{T_1 \cup T_4\} \setminus \{\mathbf{x}^1, P_{ij}, P_{i+1,j}, P_{i,j+1}\}$, then the set of zero points of s is only $L_{1,\mathbf{x}^1} \cup E_{i+1,j} \cup \hat{E}_{i,j+1}$. The proof has been completed.

Lemma 4. Let $s \in S_2^1(\Delta_{11}^{(2)}, R_{ij})$ vanish on the two edges $E_{i+1,j}$ and $\hat{E}_{i,j+1}$ of R_{ij}, where $E_{i+1,j}$ and $\hat{E}_{i,j+1}$ are defined in Lemma 3. Assume that s has two more zero points $\mathbf{x}^1$ and $\mathbf{x}^2$ on $\{T_2 \cup T_3\} \setminus \{E_{i+1,j} \cup \hat{E}_{i,j+1}\}$, which are noncollinear with $P_{i+1,j+1}$. Then $s \equiv 0$.

Proof. It is sufficient to prove the case where $\mathbf{x}^1$ and $\mathbf{x}^2 \in T_2 \setminus \hat{E}_{i,j+1}$. The other cases can be proved similarly. Denote $\mathbf{x}^1 = (u_2, v_2, w_2)$ and $\mathbf{x}^2 = (\hat{u}_2, \hat{v}_2, \hat{w}_2)$. Since $s|_{E_{i+1,j}} = s|_{\hat{E}_{i,j+1}} = 0$, then $f = g = 0$. From the assumption $s(\mathbf{x}^1) = s(\mathbf{x}^2) = 0$, we obtain the system

$$\begin{bmatrix} \frac{1}{4}(w_2)^2 & \frac{1}{4}(w_2)^2 + u_2 w_2 \\ \frac{1}{4}(\hat{w}_2)^2 & \frac{1}{4}(\hat{w}_2)^2 + \hat{u}_2 \hat{w}_2 \end{bmatrix} \begin{bmatrix} h \\ k \end{bmatrix} = 0 \tag{11}$$

Since $\mathbf{x}^1$ and $\mathbf{x}^2$ are noncollinear with $P_{i+1,j+1}$, the determinant of the coefficient matrix of system (11) is

$$\begin{aligned} &\left[\frac{1}{4}(w_2)^2(\hat{u}_2\hat{w}_2) - \frac{1}{4}(\hat{w}_2)^2(u_2 w_2)\right] \\ =&\frac{1}{4} w_2 \hat{w}_2 \left[\hat{u}_2 w_2 - \hat{w}_2 u_2\right] \neq 0. \end{aligned}$$

Thus $h = k = 0$, and $s \equiv 0$.

Lemma 5. Let $s \in S_2^1(\Delta_{11}^{(2)}, R_{ij})$ vanish on the edge $\hat{E}_{i,j+1}$ of R_{ij}, where $\hat{E}_{i,j+1}$ is defined in Lemma 3. Assume that s has three more zero points $\mathbf{x}^1$, $\mathbf{x}^2$, and $\mathbf{x}^3$ on $T_3 \setminus \hat{E}_{i,j+1}$ which are noncollinear. Then $s \equiv 0$.

Proof. Since $s|_{\hat{E}_{i,j+1}} = 0$, then $f = 0$. Let $\mathbf{x}^1 = (u_3, v_3, w_3)$, $\mathbf{x}^2 = (\hat{u}_3, \hat{v}_3, \hat{w}_3)$, and $\mathbf{x}^3 = (\bar{u}_3, \bar{v}_3, \bar{w}_3)$. Since $s(\mathbf{x}^i) = 0$, $i = 1, 2, 3$, we have a linear system with unknown values g, h, and k. It is easy to see that the determinant of the corresponding coefficient matrix is

$$\frac{w_3 \hat{w}_3 \bar{w}_3}{4} \begin{vmatrix} u_3 & \hat{u}_3 & \bar{u}_3 \\ v_3 & \hat{v}_3 & \bar{v}_3 \\ w_3 & \hat{w}_3 & \bar{w}_3 \end{vmatrix}.$$

Obviously, this value is not zero, because $\mathbf{x}^i$, $i = 1, 2, 3$ are not noncollinear. Thus $g = h = k = 0$, and $s \equiv 0$.

From Lemmas 1 - 5, we can present some admissible positions of the sample points $\mathbf{X}_8 = \{\mathbf{x}^1, \cdots, \mathbf{x}^8\}$ that guarantee poisedness. Here, we always assume that $\mathbf{x}^1, \cdots, \mathbf{x}^4$ lie at the four vertices of R_{ij}.

Theorem 6. In each situation of the following list, $\mathbf{X}_8 = \{\mathbf{x}^1, \cdots, \mathbf{x}^8\}$ is a poised set of interpolation in $S_2^1(\Delta_{11}^{(2)}, R_{ij})$. The notations $E_{i+1,j}$, $\hat{E}_{i,j+1}$, $L_{1,\cdot}$, and $L_{2,\cdot}$ are defined as the same as those in Lemma 3.

(S.1) Place four points $\mathbf{x}^5$, $\mathbf{x}^6$, $\mathbf{x}^7$, and $\mathbf{x}^8$ on four different edges of R_{ij}.

(S.2) Place three points $\mathbf{x}^5$, $\mathbf{x}^6$, and $\mathbf{x}^7$ on three different edges of R_{ij}, and place $\mathbf{x}^8$ anywhere in R_{ij} so that (N.2) is satisfied.

(S.3) Place two points $\mathbf{x}^5$ and $\mathbf{x}^6$ on the two different edges of R_{ij} : $E_{i+1,j}$ and $\hat{E}_{i,j+1}$. Also place $\mathbf{x}^7$ in T_1, $\mathbf{x}^8$ anywhere in $\{T_2 \cup T_3\} \setminus \{L_{1,\mathbf{x}^7} \cup E_{i+1,j} \cup \hat{E}_{i,j+1}\}$; or place $\mathbf{x}^7$ in T_4, $\mathbf{x}^8$ anywhere in $\{T_2 \cup T_3\} \setminus \{L_{2,\mathbf{x}^7} \cup E_{i+1,j} \cup \hat{E}_{i,j+1}\}$.

(S.4) Place two points $\mathbf{x}^5$ and $\mathbf{x}^6$ on the two different edges $E_{i+1,j}$ and $\hat{E}_{i,j+1}$ of R_{ij}, and place $\mathbf{x}^7$ and $\mathbf{x}^8$ anywhere on $\{T_2 \cup T_3\} \setminus \{E_{i+1,j} \cup \hat{E}_{i,j+1}\}$, so long as they are noncollinear with $P_{i+1,j+1}$.

(S.5) Place point $\mathbf{x}^5$ on the edge $\hat{E}_{i,j+1}$ of R_{ij}, and place $\mathbf{x}^6$, $\mathbf{x}^7$, and $\mathbf{x}^8$ anywhere on $T_3 \setminus \hat{E}_{i,j+1}$ provided they are noncollinear.

Obviously, we have several symmetric forms of each of the situations (S.3), (S.4), and (S.5). For instance, in (S.3), $E_{i+1,j}$ and $\hat{E}_{i,j+1}$ can be changed into any other two adjacent edges of R_{ij}, and the admissible positions of $\mathbf{x}^5, \cdots, \mathbf{x}^8$ should be changed correspondingly by using symmetry.

The sufficient conditions (S.2) and (S.4) shown in Theorem 6 have been expressed in Chui and the author's joint paper [1].

Denote $T_3 \cup T_4 \subset R_{ij}$ by T_{ij}^1 and $T_1 \cup T_2 \subset R_{ij}$ by T_{ij}^2. Obviously, the dimension of $S_2^1(\Delta_{11}^{(2)}, T_{ij}^k)$, $k = 1$, or 2, is 7, where $\Delta_{MN}^{(2)}$ is still the type-2 triangulation on the R_{ij} that can be considered as an extension of T_{ij}^1 or T_{ij}^2. From Theorem 6, we can present some admissible position of sample points $\mathbf{X}_7 = \{\mathbf{x}^1, \cdots, \mathbf{x}^7\}$ that guarantee poisedness of interpolation knots. Here, for convenience, we only describe the results on T_{ij}^1. By symmetry, the corresponding results on T_{ij}^2 can be expressed easily.

Theorem 7. In each situation of the following list, $\mathbf{X}_7 = \{\mathbf{x}^1, \cdots, \mathbf{x}^7\}$ is a poised set of interpolation in $S_2^1(\Delta_{11}^{(2)}, T_{ij}^1)$, where $\mathbf{x}^1$, $\mathbf{x}^2$, and $\mathbf{x}^3$ are always placed at the three vertices $P_{i,j}$, $P_{i,j+1}$, and $P_{i+1,j+1}$ of T_{ij}^1. The notations $E_{i+1,j}$ and $\hat{E}_{i,j+1}$ are defined in Lemma 3.

(S.6) Place two points $\mathbf{x}^4$ and $\mathbf{x}^5$ on the two different edges E_{ij} and $\hat{E}_{i,j+1}$ of T_{ij}^1, and place $\mathbf{x}^6$ and $\mathbf{x}^7$ anywhere in $T_{ij}^1 \setminus \{E_{ij} \cup \hat{E}_{i,j+1}\}$ so long as they are noncollinear with $P_{i,j+1}$.

(S.7) Place point $\mathbf{x}^4$ on the edge E_{ij} of T_{ij}^1, and place $\mathbf{x}^5$, $\mathbf{x}^6$, and $\mathbf{x}^7$ anywhere on $T_\ell \setminus E_{ij}$ for $\ell = 3$ or 4 provided that they are noncollinear.

Similar to the proof of the sufficient conditions (S.4) and (S.5), we can prove that (S.6) and (S.7) give poised sets by using the C^1 smooth conditions on T_{ij}^1.

When a poised set $\mathbf{X}_7$ is obtained for $T_{ij}^1 \subset R_{ij}$, then $\mathbf{X}_8 = \mathbf{X}_7 \cup \{P_{i+1,j}\}$ is a poised set for interpolation in $S_2^1(\Delta_{11}^{(2)}, R_{ij})$. Similarly, if a poised set $\mathbf{X}_7$ is obtained for $T_{ij}^2 \subset R_{ij}$, then $\mathbf{X}_8 = \mathbf{X}_7 \cup \{P_{i,j+1}\}$ is also a poised set for interpolation in $S_2^1(\Delta_{11}^{(2)}, R_{ij})$.

Of course, there are many other admissible positions for poised set, but in general, they are much more complicated than those shown in Theorems 6 and 7.

4. FLOWS AND POISED INTERPOLATION KNOTS

We now return to the situation of constructing a poised set with respect to $S_2^1(\Delta_{MN}^{(2)}) = S_2^1(\Delta_{MN}^{(2)}, R)$, where the first (M+1)(N+1) sample points have been placed at (M+1)(N+1) vertices P_{ij}, $i = 0, 1, \cdots, M$, $j = 0, 1, \cdots, N$. To state the admissible locations of the other M+N+2 sample points, we find it convenient to use the notion of source and flow. Fix an arbitrary rectangle R_{i_0,j_0} (or T_{i_0,j_0}^k, $k = 1$ or 2), $0 \le i_0 \le M-1$, $0 \le j_0 \le N-1$, that is called the source. Then the notion "flow" with respect to the partition $\Delta_{MN}^{(2)}$: $R = \Sigma_{i=0}^{M-1}\Sigma_{j=0}^{N-1} R_{ij}$, $M, N > 1$, is described as follows.

Definition 8. A flow F with respect to the partition $\Delta_{MN}^{(2)}$ on R from a source R_{i_0,j_0} is a union of finite sets F_k, $k = 1, \cdots, m$, $m \in \mathbf{Z}_+$:

$$F = \bigcup_{k=1}^{m} F_k,$$

where the set F_k, $k = 1, \cdots, m$, are called branches consisting of finite cells of triangles $F_{k\ell}$, $\ell = 1, \cdots, n_k$, of $\Delta_{MN}^{(2)}$ of R; i.e.,

$$F_k = \{F_{k\ell},\ \ell = 1, \cdots, n_k\},$$

where
(i) $F_{11} \cap R_{i_0,j_0} \neq \phi$;
(ii) $F_{k\ell} \cap F_{k,\ell+1} \neq \phi$ for all $\ell = 1, \cdots, n_k - 1$;
(iii) for any positive integer k, $2 \leq k \leq m$, there exist cells $F_{k'\ell}$, $k' < k$, $1 \leq \ell \leq n_{k'}$, such that $F_{k1} \cap F_{k'\ell} \neq \phi$;
and
(iv) for any s and t, $0 \leq s \leq M-1$, $0 \leq t \leq N-1$, there exist F_{k_s} and F_{k_t}, $1 \leq k_s$, $k_t \leq m$, satisfying

$$F_{k_s} \bigcap \{\bigcup_{j=0}^{N-1} R_{sj}\} \neq \phi$$

and

$$F_{k_t} \bigcap \{\bigcup_{i=0}^{M-1} R_{it}\} \neq \phi,$$

respectively.

Obviously, we have many choices for $F_{k\ell}$. Since (M+1)(N+1) sample points have been placed at the vertices P_{ij}, $0 \leq i \leq M$, $0 \leq j \leq N$, the best two choices are $F_{k\ell} = R_{i_{k\ell}j_{k\ell}}$, for some $(i_{k\ell}j_{k\ell}) \in I = \{(i,j) : 0 \leq i \leq M-1, 0 \leq j \leq N-1\}$, and $F_{k\ell} = T^k_{i_{k\ell}j_{k\ell}}$, for some $(i_{k\ell}j_{k\ell}) \in I$, $k = 1$ or 2.

Thus, as the first choice, we have the branches

$$F_k^1 = \{R_{i_{k1}j_{k1}}, \cdots, R_{i_{kn_k}j_{kn_k}}\}$$

and the flow

$$F^1 = \bigcup_{k=1}^{m_1} F_k^1.$$

As the second choice, we have the branches

$$F_k^2 = \{T^{\ell_{k1}}_{i_{k1}j_{k1}}, \cdots, T^{\ell_{kn_k}}_{i_{kn_k}j_{kn_k}}\}$$

and the flow

$$F^2 = \bigcup_{k=1}^{m_2} F_k^2,$$

where $\ell_{kj} = 1$ or 2, $1 \leq j \leq n_k$, $1 \leq k \leq m_2$.

We may place other M+N+2 sample points in $\Delta_{MN}^{(2)}$ along the flow F^1 or F^2 by using (N.1), (N.2), (S.1)-(S.7), and the smooth conditions (3)-(5).

If the restrictions $s|_{E_{i+1,j}}$ or $s|_{\hat{E}_{i,j+1}}$ of s on the edges $E_{i+1,j}$ or $\hat{E}_{i,j+1}$ of R_{ij} are obtained, then we say that $E_{i+1,j}$ or $\hat{E}_{i,j+1}$ is a determined edge of R_{ij} with respect to s, or, for convenience, we say that $E_{i+1,j}$ or $\hat{E}_{i,j+1}$ is already determined. Of course, if $E_{i+1,j}$ is a determined edge of R_{ij} with respect to s, then it is also a determined edge of $R_{i+1,j}$ with respect to s; if $\hat{E}_{i,j+1}$ is a determined edge of R_{ij} with respect to s, then it is also a determined edge of $R_{i,j+1}$ with respect to s. Since $s|_{E_{i+1,j}}$ and $s|_{\hat{E}_{i,j+1}}$ are univariate quadratic polynomials, three values of s at three different sample points on $E_{i+1,j}$ or $\hat{E}_{i,j+1}$ will determine $s|_{E_{i+1,j}}$ or $s|_{\hat{E}_{i,j+1}}$, respectively. Furthermore, for any fixed i_0, $0 \le i_0 \le M$, if N+1 sample points of s have been placed at the vertices $P_{i_0,j}$, $0 \le j \le N$, and s is determined on an edge E_{i_0+1,j_0} of R_{i_0,j_0} or R_{i_0+1,j_0} for certain j_0, $0 \le j_0 \le N-1$, then by the C^1 smooth conditions (5), s is determined on the edge $E_{i_0+1,j}$ of $R_{i_0,j}$ or $R_{i_0+1,j}$, for all j, $0 \le j \le N-1$. Similarly, for any fixed j_0, $0 \le j_0 \le N$, if M+1 sample points of s have been placed at the vertices P_{i,j_0}, $0 \le i \le M$, and $\hat{E}_{i_0,j_0+1}$ is a determined edge of R_{i_0,j_0} or R_{i_0,j_0+1} for a i_0, $0 \le i_0 \le M-1$, then all $\hat{E}_{i,j_0+1}$, $0 \le i \le M-1$, are determined with respect to s.

From the source R_{i_0,j_0}, we can place the sample points of interpolation in $\Delta_{MN}^{(2)}$ along the flow F^1 by using the following rules.

Procedure 9.

(i) Place one sample point at each of the vertices P_{ij} of $\Delta_{MN}^{(2)}$ on R.

(ii) Choose a source rectangle R_{i_0,j_0}, where (i_0,j_0) is chosen from the set $I = \{(i,j) : 0 \le i \le M-1, 0 \le j \le N-1\}$ arbitrarily, and place four additional sample points in R_{i_0,j_0} so that the conditions (N.1), (N.2), and one of the conditions (S.1)-(S.5) are satisfied.

(iii) Place sample points on the flow starting from $R_{i_{11}j_{11}}$. We distinguish two cases:

(a) If $R_{i_{11}j_{11}} \cap R_{i_0,j_0} = E_{i_0,j_0}$, E_{i_0+1,j_0}, $\hat{E}_{i_0,j_0}$, or $\hat{E}_{i_0,j_0+1}$, then E_{i_0,j_0}, E_{i_0+1,j_0}, $\hat{E}_{i_0,j_0}$, or $\hat{E}_{i_0,j_0+1}$ is determined. Hence, place one additional sample point anywhere in $R_{i_{11}j_{11}}$ with the exception of the following three edges of $R_{i_{11}j_{11}}$: the determined edge and other two edges that are adjacent to it.

(b) If $R_{i_{11}j_{11}} \cap R_{i_0,j_0} = P_{i_0j_0}$, P_{i_0+1,j_0}, P_{i_0,j_0+1} or P_{i_0+1,j_0+1}, then place two additional sample points anywhere in $R_{i_{11}j_{11}}$ which satisfy (S.3) or (S.4) or their symmetric forms.

Then we construct sample points along the first branch from $R_{i_{11}j_{11}}$, through $R_{i_{12}j_{12}}, \cdots$, to $R_{i_{1n_1}j_{1n_1}}$. Next, we construct sample points along the second branch from $R_{i_{21}j_{21}}$, through $R_{i_{22}j_{22}}, \cdots$, to $R_{i_{2n_2}j_{2n_2}}$ and so on. In general, if we come to the rectangle $R_{i_{k\ell}j_{k\ell}}$ along the flow F^1, then we can arrange the sample points according to the following three cases (c), (d), and (e). For convenience, we need some notations. Denote $\hat{I} = \{(s,t) : 1 \le t \le n_s, 1 \le s \le m\}$. For any fixed $k, \ell \in \hat{I}$, we denote

$$K_{k\ell} = \{(s,t) : (s,t) \in \hat{I};\ either\ s < k,\ or\ s = k\ and\ t < \ell\}.$$

Let

$$X_{k\ell} = \{(x_{i_{st}}, y) : (s,t) \in K_{k\ell},\ c \le y \le d\}$$
$$Y_{k\ell} = \{(x, y_{j_{st}}) : (s,t) \in K_{k\ell},\ a \le x \le b\}.$$

From the definition of the flow, there must exist a pair $(\bar{k},\bar{\ell}) \in K_{k\ell}$ such that $R_{i_{\bar{k}\bar{\ell}}j_{\bar{k}\bar{\ell}}} \cap R_{i_{k\ell}j_{k\ell}} = \bar{E}_{k\ell} \ne \phi$. If $meas\, \bar{E}_{k\ell} \ne 0$, then denote two edges of $R_{i_{k\ell}j_{k\ell}}$ that are adjacent to $\bar{E}_{k\ell}$ by $\tilde{E}_{k\ell}$ and $\tilde{E}'_{k\ell}$, respectively, and the 4th edge of $R_{i_{k\ell}j_{k\ell}}$ by $\bar{E}'_{k\ell}$. If $meas\,\bar{E}_{k\ell} = 0$, then denote two edges of $R_{i_{k\ell}j_{k\ell}}$ which intersect at $\bar{E}_{k\ell}$ by $\bar{E}^1_{k\ell}$ and $\tilde{E}^2_{k\ell}$, and the two opposite edges of $\bar{E}^1_{k\ell}$ and $\tilde{E}^2_{k\ell}$ by $\bar{E}^{1\prime}_{k\ell}$ and $\tilde{E}^{2\prime}_{k\ell}$, respectively.

(c) If either of the following cases holds, then no more sample points are placed in $R_{i_{k\ell}j_{k\ell}}$:

(c.1) $meas\,\bar{E}_{k\ell} \neq 0$, and $\bar{E}'_{k\ell} \subset X_{k\ell} \cup Y_{k\ell}$.

(c.2) $meas\,\bar{E}_{k\ell} = 0$, and $\bar{E}^{1\prime}_{k\ell}$ and $\tilde{E}^{2\prime}_{k\ell} \subset X_{k\ell} \cup Y_{k\ell}$.

(d) If either of the following cases holds, then one additional sample point is placed anywhere in $R_{i_{k\ell}j_{k\ell}}$ with the exception of three determined edges:

(d.1) $meas\,\bar{E}_{k\ell} \neq 0$, and $\bar{E}'_{k\ell} \not\subset X_{k\ell} \cup Y_{k\ell}$, where the three determined edges are $\bar{E}_{k\ell}$, $\tilde{E}_{k\ell}$, and $\tilde{E}'_{k\ell}$.

(d.2) $meas\,\bar{E}_{k\ell} = 0$, and either $\bar{E}^{1\prime}_{k\ell}$ or $\tilde{E}^{2\prime}_{k\ell} \subset X_{k\ell} \cup Y_{k\ell}$, where the three determined edges are $\bar{E}^1_{k\ell}$, $\tilde{E}^2_{k\ell}$, and $\bar{E}^{1\prime}_{k\ell}$ or $\tilde{E}^{2\prime}_{k\ell}$.

(e) If $meas\,\bar{E}_{k\ell} = 0$, and both $\bar{E}^{1\prime}_{k\ell}$ and $\tilde{E}^{2\prime}_{k\ell} \not\subset X_{k\ell} \cup Y_{k\ell}$, then place two additional sample points anywhere in $R_{i_{k\ell}j_{k\ell}}$ with two determined edges $\bar{E}^1_{k\ell}$ and $\tilde{E}^2_{k\ell}$, provided that the sample points satisfy condition (S.3), (S.4), or their symmetric forms.

Obviously, in step (i) of the Procedure 9, (M+1)(N+1) sample points are arranged; in step (ii), 4 additional sample points are placed in the source R_{i_0,j_0}; in step (iii), the number of additional sample points placed along the flow F^1 is exactly equal to the number of horizontal and vertical grid lines minus 4, the number of edges of R_{i_0,j_0}; i.e., it is equal to $(M+1)+(N+1)-4 = M+N-2$. Thus the total number of sample points are

$$(M+1)(N+1) + 4 + (M+N-2) = MN + M + N + 3 = dim\, S^1_2(\Delta^{(2)}_{MN}, R).$$

On the other hand, when the zero data is given at all sample points which are constructed by Procedure 9, we have $s|_{R_{i_0,j_0}} \equiv 0$ and $s|_{F^1} \equiv 0$. Thus $s = 0$ on each edge of any R_{ij} in $\Delta^{(2)}_{MN}$, because the flow F^1 passes through all horizontal and vertical grid lines of $\Delta^{(2)}_{MN}$. Hence, $s \equiv 0$ on R and we have the following theorem.

Theorem 10. Procedure 9 gives rise to a poised set X_n of $n = MN + M + N + 3$ sample points of the interpolation from $S^1_2(\Delta^{(2)}_{MN}, R)$.

From R_{i_0,j_0}, along flow F^2, we can place sample points in $\Delta^{(2)}_{MN}$ according to the following rules.

Procedure 11.

(i) Place one sample point at each of the vertices P_{ij} of $\Delta^{(2)}_{MN}$ on R.

(ii) Choose a source rectangle R_{i_0,j_0}, where (i_0, j_0) is chosen from $I = \{(i,j) : 0 \le i \le M-1, 0 \le j \le N-1\}$ arbitrarily, and place four additional sample points in R_{i_0,j_0} so that conditions (N.1), (N.2), and one of the conditions (S.6)-(S.7) is satisfied.

(iii) Place sample points on the flow F^2 starting from $T^{\ell_{11}}_{i_{11}j_{11}}$. We distinguish three cases:

(a) If $T^{\ell_{11}}_{i_{11}j_{11}} \cap R_{i_0,j_0} = E_{i_0,j_0}$, E_{i_0+1,j_0}, $\bar{E}_{i_0,j_0}$, or $\bar{E}_{i_0,j_0+1}$, then E_{i_0,j_0}, E_{i_0+1,j_0}, $\bar{E}_{i_0,j_0}$, or $\bar{E}_{i_0,j_0+1}$ is determined. Hence place one additional sample point anywhere in $R_{i_{11}j_{11}}$ with the exception of the following three edges of $R_{i_{11}j_{11}}$: the determined edge and other two edges that are adjacent to it.

(b) If $T^{\ell_{11}}_{i_{11}j_{11}} \cap R_{i_0,j_0} = P_{i_0,j_0+1}$ or P_{i_0+1,j_0}, then place two additional sample points anywhere in $T^{\ell_{11}}_{i_{11}j_{11}}$ provided that they satisfy condition (S.6) or its symmetric forms.

(c) If $T^{\ell_{11}}_{i_{11}j_{11}} \cap R_{i_0,j_0} = P_{i_0,j_0}$ or P_{i_0+1,j_0+1}, then place three additional sample points anywhere in $T^{\ell_{11}}_{i_{11}j_{11}}$ provided that they satisfy the condition (S.7) or its symmetric forms.

Similar to the Procedure 9, we construct sample points along the first branch from $T^{\ell_{11}}_{i_{11}j_{11}}$, through $T^{\ell_{12}}_{i_{12}j_{12}}, \cdots$, to $T^{\ell_{1n_1}}_{i_{1n_1}j_{1n_1}}$. Next, we construct sample points along the second branch from $T^{\ell_{21}}_{i_{21}j_{21}}$, through $T^{\ell_{22}}_{i_{22}j_{22}}, \cdots$, to $T^{\ell_{2n_2}}_{i_{2n_2}j_{2n_2}}$ and so on. In general, if we come to the $T^{\ell_{k\ell}}_{i_{k\ell}j_{k\ell}}$ along the flow F^2, then we can construct the additional sample points in $T^{\ell_{k\ell}}_{i_{k\ell}j_{k\ell}}$ according to the following three cases. The notations $\bar{I}$, $K_{k\ell}$, $X_{k\ell}$, $Y_{k\ell}$ are defined in Procedure 9. From the definition of a flow, there must exist a pair $(\bar{k},\bar{\ell}) \in K_{k\ell}$ such that $T^{\ell_{\bar{k}\bar{\ell}}}_{i_{\bar{k}\bar{\ell}}j_{\bar{k}\bar{\ell}}} \cap T^{\ell_{k\ell}}_{i_{k\ell}j_{k\ell}} = \bar{E}_{k\ell} \neq \phi$. If $(\bar{k}\bar{\ell}) \neq (k,\ell)$ and $meas\, \bar{E}_{k\ell} \neq 0$, then denote by $\tilde{E}_{k\ell}$ the edge of $T^{\ell_{k\ell}}_{i_{k\ell}j_{k\ell}}$ which is perpendicular to $\bar{E}_{k\ell}$, $\bar{E}'_{k\ell}$ and $\tilde{E}'_{k\ell}$ the two edges of $R_{i_{k\ell}j_{k\ell}}$ that are opposite to the edges $\bar{E}_{k\ell}$ and $\tilde{E}_{k\ell}$ respectively. If $(\bar{k}\bar{\ell}) \neq (k,\ell)$ and $meas\, \bar{E}_{k\ell} = 0$, then denote by $\bar{E}^1_{k\ell}$ and $\tilde{E}^2_{k\ell}$ the two edges of $T^{\ell_{k\ell}}_{i_{k\ell}j_{k\ell}}$ that intersect at $\bar{E}_{k\ell}$. And the two opposite edges of $\bar{E}^1_{k\ell}$ and $\tilde{E}^2_{k\ell}$ are denoted by $\bar{E}^{1\prime}_{k\ell}$ and $\tilde{E}^{2\prime}_{k\ell}$, respectively.

(d) If one of the following three cases holds, then no more sample points are placed in $T^{\ell_{k\ell}}_{i_{k\ell}j_{k\ell}}$:

(d.1) $(\bar{k},\bar{\ell}) = (k,\ell)$.

(d.2) $(\bar{k},\bar{\ell}) \neq (k,\ell)$, $meas\, \bar{E}_{k\ell} \neq 0$, and $\bar{E}'_{k\ell} \subset X_{k\ell} \cup Y_{k\ell}$.

(d.3) $(\bar{k},\bar{\ell}) \neq (k,\ell)$, $meas\, \bar{E}_{k\ell} = 0$, and $\bar{E}^{1\prime}_{k\ell}$ and $\tilde{E}^{2\prime}_{k\ell} \subset X_{k\ell} \cup Y_{k\ell}$.

(e) If $(\bar{k},\bar{\ell}) \neq (k,\ell)$, $meas\, \bar{E}_{k\ell} \neq 0$, and $\bar{E}'_{k\ell} \not\subset X_{k\ell} \cup Y_{k\ell}$, then place one additional sample point anywhere in $T^{\ell_{k\ell}}_{i_{k\ell}j_{k\ell}}$ with the exception of three determined edges: $\bar{E}_{k\ell}$, $\tilde{E}_{k\ell}$, and $\tilde{E}'_{k\ell}$.

(f) If $(\bar{k},\bar{\ell}) \neq (k,\ell)$, $meas\, \bar{E}_{k\ell} = 0$, and $\bar{E}^{1\prime}_{k\ell}$ and $\tilde{E}^{2\prime}_{k\ell} \not\subset X_{k\ell} \cup Y_{k\ell}$, then we distinguish two cases:

(f.1) If $\bar{E}_{k\ell} = P_{i_{k\ell}+1,j_{k\ell}}$ or $P_{i_{k\ell},j_{k\ell}+1}$, then place two additional sample points anywhere in $T^{\ell_{k\ell}}_{i_{k\ell}j_{k\ell}}$ provided (S.6) or its symmetric forms is satisfied.

(f.2) If $\bar{E}_{k\ell} \neq P_{i_{k\ell}+1,j_{k\ell}}$ or $P_{i_{k\ell},j_{k\ell}+1}$, then place one additional sample point anywhere in $T^{\ell_{k\ell}}_{i_{k\ell}j_{k\ell}}$ and place another additional sample point anywhere in the component of $T^{\ell_{k\ell}}_{i_{k\ell}j_{k\ell}}$ with respect to $R_{i_{k\ell}j_{k\ell}}$, which two additional sample points satisfy the conditions (S.3), (S.4), or their symmetric forms.

Similarly, we can prove that Procedure 11 gives a poised set of sample points. Thus we have the following.

Theorem 12. Procedure 11 gives rise to a poised set X_n of $n = MN + M + N + 3$ sample points of the interpolation from $S^1_2(\Delta^{(2)}_{MN}, R)$.

Procedure 11 is designed for some polygonal regions.

REFERENCES

[1] C.K. Chui and T.X. He, On location of sample points in C^1 quadratic bivariate spline interpolation, in "Numerical Methods of Approximation Theory", Vol. 8, (L. Collatz, G. Meinardus, and G. Nürnberger, Eds.), Birkhäuser, Basel, 1987. pp.30-43.

[2] C.K. Chui, T.X. He, and R.H. Wang, Interpolation by bivariate linear splines, in Alfred Haar Memorial Conference, (J. Szabados and K. Tandori, Eds.), North-Holland, Amsterdam, 1986. pp.247-255.

[3] K.C. Chung and T.H. Yao, On lattices admitting unique Lagrange interpolations, *SIAM J. Numer. Anal.*, **14** (1977), 735-741.

[4] M. Dæhlen and T. Lyche, Bivariate interpolation with quadratic box splines, *Math. Comp.*, **51** (1988), 219-230.

[5] S.Z. Liang, On the poised nodes of multivariate interpolations, *J. Jilin University (Natural Sciences)*, 1 (1979), 27-32.
[6] J. Morgan and R. Scott, The dimension of the space of C^1 piecewise polynomials, unpublished manuscript, 1975.
[7] G. Nürnberger and T.H. Riessinger, Lagrange and Hermite interpolation by bivariate splines, Fakultät für Mathematik und Informatik, Universität Mannheim, 1990.
[8] Y. S. Zhou, Y.T. Chang, and T.X. He, On multivariate interpolations, *Engineering Mathematics*, 1 (1984), 12-16.

ON A BERNSTEIN-TYPE OPERATOR OF BLEIMANN, BUTZER AND HAHN III

C. Jayasri[1] and Y. Sitaraman[2]

[1]Department of Mathematics
University of Kerala
Kariavattom 695581, India

[2]Kentucky Wesleyan College
Owensboro, KY 42301, USA

1. INTRODUCTION

The operators

$$L_n(f;x) = \sum_{k=0}^{n} P_{n,k}(x) f\left(\frac{k}{n-k+1}\right) \ (x \geq 0, n \in N)$$

with $P_{n,k}(x) = \binom{n}{k} x^k(1+x)^{-n}$ were introduced by Bleimann, Butzer and Hahn [3]. These operators are defined on $C[0,\infty)$, the space of continuous functions on the unbounded interval $[0,\infty)$ and have the property that they converge to f uniformly on any finite interval $[a,b]$ contained in $[0,\infty)$ provided f is bounded and continuous on $[0,\infty)$. Using probabilistic arguments Khan [10] simplified and sharpened some of the results given in [3]. In [11] Kahn further showed that $L_n(f;x) \geq L_{n+1}(f;x) \geq \ldots \geq f(x)$ if f is convex; $L_n(f;x)$ is itself convex if f is a non-increasing convex function and $L_n(f;x) \in Lip^\alpha(A)$ $(0 \leq \alpha \leq 1)$ if and only if $f \in Lip^\alpha(A)$ $(0 \leq \alpha \leq 1)$. The authors obtained in [8] the largest subclass of $C[0,\infty)$ on which (L_n) defines a pointwise approximation process. The behaviour of the rational functions $L_n(f;z)$ for complex values of z outside $[0,\infty)$ has also been determined in [8]. Totik [13] solved the saturation and so called non-optimal approximation problem for (L_n). A local saturation theorem for the operators (L_n) making use of the parabolic technique of De Vore [5] was obtained by the authors in [9]. A new Bernstein-type operator associated with the Polya distribution which includes as particular case the operator (L_n) is introduced by Adell et al. in [2] and the approximation properties of this operator concerning rate of convergence, preservation of Lipschitz constant and monotonic convergence under convexity are given. For a study of Bernstein-type operators, including (L_n), using (i) limit theorems of probability theory and (ii) stochastic process one may refer to [4] and [1], respectively. The aim of this note is to prove a local direct theorem and a local inverse theorem for (L_n) when f belongs to the subclass $\mathcal{F} = \left\{C[0,\infty)|e^{Ax}\right\}$ of $C[0,\infty)$ where $\mathcal{F} = \{f \in C[0,\infty)$: for each $A > 0,\ f(x) = O(1)e^{Ax}$ on $[0,\infty)\}$.

2. A LOCAL DIRECT THEOREM FOR $(\mathbf{L_n})$

Definition 2.1 Suppose $0 < \gamma \leq 2$. Let I be a subinterval of $\mathcal{R}$ and let $f \in C(I)$. We say that f belongs to $Lip^*\gamma$ on I if $\omega_2(\delta; f; I) = O(1)\delta^\gamma (0 < \delta \leq 1)$ where $\omega_2(\delta; f; I) = \sup_{x,x+2h \in I} |\Delta_h^2(f;x)|$, $0 < h \leq d$, and $\Delta_h^2(f;x) = f(x+2h) - 2f(x+h) + f(x)$.

Lemma 2.2 ([12], p. 121) Suppose $f \in C[a,b]$ and that $f(a) = f(b) = 0$. Then f has an extension f^* onto $(-\infty, \infty)$ which is periodic of period $2(b-a)$ such that $\omega_2(\delta; f; (-\infty,\infty)) \leq 5\omega_2(\delta; f; [a,b]),\ (0 < \delta \leq \frac{b-a}{2})$.

Lemma 2.3 ([9], Lemma 2.4) If an element $f \in \mathcal{F}$ vanishes on an open interval (a,b) contained in $(0,\infty)$, then $n^2 L_n(f;x) = O(1)$ $(n \in N)$ uniformly on compact subintervals of (a,b).

Theorem 2.4 Let $A \geq 0$. There exists some constant $M > 0$ such that for all positive integers $n \geq x(1+x)^2$ and for f in $\{C[0,\infty)|x^A\}$

$$\frac{L_n(f;x) - f(x)}{(1+x)^A} \leq M\left\{\|f\|_A \frac{x}{n} + \omega^*\left(\frac{\phi(x)}{n^{1/2}}; f; A\right)\right\}$$

where $\|f\|_A = \sup_{x \geq 0} \frac{|f(x)|}{(1+x)^A}$, $\phi(x) = x^{1/2}(1+x)$ and $\omega^*(\delta; f; A) = \sup_{\substack{0<h\leq\delta \\ 0\leq x<\infty}} \frac{|\Delta_h^2(f;x)|}{(1+x)^A}$.

The proof of this theorem is similar to that of Theorem 3.5 of [7]. The corollary given below follows easily from the theorem.

Corollary 2.5 Suppose f belongs to $\left\{C[0,\infty)|x^A\right\}$ for some $A \geq 0$. Assume that for some γ such that $0 < \gamma \leq 1$

$$\frac{\Delta_h^2(f;x)}{(1+x)^A} = O(1)h^{2\gamma}\ (0 < h \leq 1,\ 0 \leq x < \infty). \text{ Then}$$

$$\frac{L_n(f;x) - f(x)}{(1+x)^A} = O(1)\left(\frac{x}{n}(1+x)^2\right)^\gamma\ (n \in N,\ x \geq 0).$$

Theorem 2.6 Suppose $f \in \mathcal{F}$ belongs also to $Lip^*\gamma$ on a compact subinterval $[a,b]$ of $[0,\infty)$. Then for $0 < \gamma \leq 2$, $n^{\gamma/2}\{L_n(f;x) - f(x)\} = O(1)$, $n \in N$ uniformly on compact subintervals of (a,b).

Proof: Let $I(x)$ be the unique linear function coinciding with f at a and b. For $a \leq x \leq b$, define $f_1(x) = f(x) - I(x)$. Then $f_1(x)$ belongs to $C[a,b]$ and $f_1(a) = f_1(b) = 0$. By Lemma 2.2 f_1 has a periodic extension f_1^* onto $(-\infty,\infty)$ satisfying the lemma. Since $f \in Lip^*\gamma$ on $[a,b]$ and I is linear, $f_1 = f - I$ also belongs to $Lip^*\gamma$ on $[a,b]$ and consequently $\omega_2(\delta; f_1^*; (-\infty,\infty)) = O(1)\delta^\gamma (0 < \delta \leq 1)$. In particular

$$\Delta_h^2(f_1^*;x) = O(1)h^\gamma (x \geq 0,\ 0 < h \leq 1). \tag{2.1}$$

Notice that f_1^* is a continuous bounded function on $(-\infty,\infty)$ and f_1^* restricted to $[0,\infty)$ belongs to the space $C_B[0,\infty)$ of continuous bounded functions on $[0,\infty)$. By (2.1) and Corollary 2.5 with $A = 0$, we then have

$$L_n(f_1^*;x) - f_1^*(x) = O(1)\left(\frac{x}{n}(1+x)^2\right)^{\frac{\gamma}{2}},\ (x > 0, n \in N). \tag{2.2}$$

Since $f_1^*(x) = f_1(x)$ on $[a,b]$, $f_1^* - f_1$ vanishes on the open interval (a,b). By Lemma 2.3, $L_n(f_1^* - f_1; x) = O(1)n^{-2}$, $(n \in N,\ a < a_1 \leq x \leq b_1 < b)$ and this along with (2.2) leads us to conclude that

$$L_n(f_1;x) - f_1(x) = O(1)n^{\frac{-\gamma}{2}},\ (n \in N,\ a_1 \leq x \leq b_1). \tag{2.3}$$

Notice further that if $I(x) = a'x + b'$, then

$$L_n(I;x) - I(x) = O(1)n^{-2} \ (n \in N, \ a \leq x \leq b). \tag{2.4}$$

Finally from (2.3) and (2.4), for all $n \in N$, we have $n^{\frac{\gamma}{2}}(L_n(f;x) - f(x)) = O(1)$ uniformly on $[a_1, b_1]$, where $[a_1, b_1]$ is an arbitrary compact subinterval of (a, b).

3. A LOCAL INVERSE THEOREM FOR ($\mathbf{L_n}$)

We first obtain some auxiliary results needed in the sequel.

Lemma 3.1 ([7], Lemma 3.1) The operators (L_n) have the following properties: (i) $L_n(1;x) = 1$; (ii) $L_n(t;x) = x - x\left(\frac{x}{1+x}\right)^n$; (iii) $L_n\left((t-x)^2(1+t)^A\right) = O(1)\frac{x(1+x)^{2+A}}{n}$; (iv) $L_n\left((1+t)^A : x\right) = O(1)(1+x)^A$. (In estimates (iii) and (iv), $A \geq 0$ is a fixed real number.)

Lemma 3.2 Let $0 < a < a_1 < b_1 < b < \infty$. Then there exists a function g having the following properties: g, g', g'' all belong to $C_B[0,\infty)$, $\operatorname{supp} g \subset [a,b]$ and $g(x) = 1$ on $[a_1, b_1]$.

Proof: Let $h(x)$ be the unique polynomial of degree less than or equal to 5 satisfying the following conditions: $h(0) = h'(0) = h''(0) = 0$, $h(1) = 1$; $h'(1) = 0$, $h''(1) = 0$. Define g on $[0,\infty)$ as follows: $g(x) = 0$ for $0 \leq x \leq a$; $g(x) = h\left(\frac{x-a}{a_1-a}\right)$ for $a < x \leq a_1$; $g(x) = 1$ for $a_1 < x \leq b_1$; $g(x) = h\left(\frac{b-x}{b-b_1}\right)$ for $b_1 < x \leq b$; and $g(x) = 0$ for $b < x < \infty$. This g has all the properties stated in this lemma.

Lemma 3.3 Let $0 < a < a_1 < b_1 < b < \infty$. Let $g \in C[0,\infty)$ with $\operatorname{supp} g \subset [a_1, b_1]$. Suppose for some γ, $0 < \gamma < 2$, $\|L_n g - g\|_{[a,b]} = O(1)n^{\frac{-\gamma}{2}}(n \in N)$. Then g belongs to $Lip^*\gamma$ on $[0,\infty]$.

Proof: From the assumption we conclude that

$$L_n(g:x) - g(x) = O(1)\left\{\frac{x(1+x)^2}{n}\right\}^{\frac{\gamma}{2}} \quad (n \in N, \ a \leq x \leq b). \tag{3.1}$$

If $x \in [0,\infty)$ and $x \notin [a,b]$, then

$$\begin{aligned} |L_n(g;x) - g(x)| &= |L_n(g;x)|, \text{ since } \operatorname{supp} g \subset [a_1, b_1] \\ &\leq M\frac{\|g\|}{\delta^2}\,\frac{x(1+x)^2}{n} \text{ by (iii) of Lemma (3.1)} \end{aligned}$$

where $\|g\| = \max_{0 \leq t < \infty}|g(t)|$, $\delta = \min\{a_1 - a, b - b_1\}$, $M > 0$ is some constant and $n \in N$.

Consequently

$$L_n(g;x) - g(x) = O(1)\left\{\frac{x(1+x)^2}{n}\right\}^{\frac{\gamma}{2}}, \ n \in N, \ x \in [0,\infty) - [a,b], \ n \geq x(1+x)^2. \tag{3.2}$$

Combining estimates (3.1) and (3.2), we get

$$L_n(g;x) - g(x) = O(1)\left\{\frac{x(1+x)^2}{n}\right\}^{\frac{\gamma}{2}}, \ (x \in [0,\infty), \ n \in N, \ n \geq x(1+x)^2). \tag{3.3}$$

Now it is possible to conclude that $\Delta_h^2(g;x) = O(1)h^\gamma(0 < h \leq 1, \ 0 < x < \infty)$ or g belongs to $Lip^*\gamma$ on $[0,\infty)$.

Theorem 3.4 Let $0 < \gamma < 2$. Suppose f in $\mathcal{F}$ satisfies the estimate $n^{\frac{\gamma}{2}}(L_n(f;x) - f(x)) = O(1)$ $(n \in N)$ uniformly on $[a,b]$ where $[a,b] \subset (0,\infty)$. Then f belongs to $Lip^*\gamma$ on $[a_1,b_1]$ whenever $a < a_1 < b_1 < b$.

Our approach to the proof of this theorem is similar to the one employed by Ditzian and May [6].

Proof: Case (i) $0 < \gamma \le 1$. Define f_1 on $[0,\infty)$ by $f_1(x) = f(x)$ for $0 \le x \le b+1$ and $f_1(x) = f(b+1)$ for $x > b+1$, then $f_1 \in C_B[0,\infty)$ and $f_1 - f$ vanishes on an open interval containing $[a,b]$. Hence by Lemma 2.3 $L_n(f_1 - f; x) = O(1)n^{-2}$ $(a \le x \le b,\ n \in N)$. This together with the hypothesis of the theorem and the fact that $f_1 = f$ on $[a,b]$ leads to

$$L_n(f_1;x) - f_1(x) = O(1)n^{\frac{-\gamma}{2}}\ (a \le x \le b,\ n \in N). \tag{3.4}$$

Let $a < a_2 < a_1 < b_1 < b_2 < b$. Then by Lemma 3.2 there exists a function g such that $g,\ g',\ g''$ all belong to $C_B[0,\infty)$, $\operatorname{supp} g \subset [a_2,b_2]$ and $g(x) = 1$ on $[a_1,b_1]$. Consider the function $f_1 g$. We have $f_1 g = f_1 = f$ on $[a_1,b_1]$ and

$$|L_n(f_1 g;x) - (f_1 g)(x)| \le \|g'\|\ \|f_1\|\ L_n(|t-x|;x) + \|g\|\ |L_n(f_1;x) - f_1(x)|, \tag{3.5}$$

Also $L_n(|t-x|;x) = O(1)\left\{\dfrac{x(1+x)^2}{n}\right\}^{1/2}$ by (iii) of Lemma 3.1

$$= O(1)n^{\frac{-\gamma}{2}}(a \le x \le b),\ n \in N) \tag{3.6}$$

since in this case $0 < \gamma \le 1$. Estimates (3.4), (3.5) and (3.6) together yield $\|L_n(f_1 g;x) - (f_1 g)(x)\|_{[a,b]} = O(1)n^{\frac{-\gamma}{2}}$ $(n \in N)$. Further $\operatorname{supp} f_1 g \subset \operatorname{supp} g \subset [a_2,b_2] \subset (a,b)$. Hence by Lemma 3.3 $f_1 g$ belongs to $Lip^*\gamma$ on $[0,\infty)$. Since $f_1 g = f_1 = f$ on $[a_1,b_1]$, f belongs to $Lip^*\gamma$ on $[a_1,b_1]$.

Proof: Case (ii) $1 < \gamma < 2$. As in case (i), replace f by a function f_1 such that $f_1 \in C_B[0,\infty)$ and $f = f_1$ on an open interval containing $[a,b]$. Then as in case (i) we have

$$L_n(f_1;x) - f_1(x) = O(1)n^{\frac{-\gamma}{2}}(n \in N,\ x \in [a,b]). \tag{3.7}$$

Since $\gamma > 1$, we conclude from the hypothesis that $\|L_n f - f\|_{[a,b]} = O(1)n^{-1/2}$ $(n \in N)$. Then by case (i) f belongs to Lip^*1 on compact subintervals of (a,b) and so for each α such that $0 < \alpha < 1$, f belongs to $Lip^*\alpha$ on compact subintervals of (a,b). Thus f and also f_1 belongs to $Lip\alpha$ on compact subintervals of (a,b) for each α where $0 < \alpha < 1$. Since in this case $0 < (\gamma - 1) < 1$, f_1 belongs to $Lip(\gamma-1)$ on compact subintervals of (a,b). Thus $f_1(t) - f_1(x) = O(1)\ |t-x|^{\gamma-1}$ $(t, x \in [a_4,b_4])$ where $a < a_4 < a_3 < a_2 < b_2 < b_3 < b_4 < b$. Let g be a function on $[0,\infty)$ such that $g,\ g',\ g'' \in C_B[0,\infty)$, $\operatorname{supp} g \subset [a_2,b_2]$ and $g(x) = 1$ on $[a,b]$. Then $|g(t) - g(x)| \le \|g'\|\ |t-x|$ $(t,x \in [0,\infty))$. Hence

$$(f_1(t) - f_1(x))\,(g(t) - g(x)) = O(1)|t-x|^{\gamma}(t,x \in [a_4,b_4]). \tag{3.8}$$

If $\delta = \min\{a_3 - a_4, b_4 - b_3\}$, then for all t in $[0,\infty)$ outside the interval $[a_4,b_4]$ and for all x in $[a_3,b_3]$, $|t-x| \ge \delta$ and

$$|(f_1(t) - f_1(x))\,(g(t) - g(x))| \le 4\|f_1\|\ \|g\| = O(1)\frac{(t-x)^2}{\delta^2}. \tag{3.9}$$

From (3.8) and (3.9) by (iii) of Lemma 3.1 we have

$$L_n\,(f_1(t) - f_1(x))\,(g(t) - g(x);x) = O(1)n^{\frac{-\gamma}{2}} \tag{3.10}$$

for all x in $[a_3,b_3]$ and $n \in N$.

Now

$$L_n(f_1g;x) - (f_1g)(x) = L_n\left((f_1(t) - f_1(x))\left(g(t) - g(x)\right);x\right) \\ + f_1(x)\left(L_n(g;x) - g(x)\right) + g(x)\left(L_n(f_1;x) - f_1(x)\right) \quad (3.11)$$

Since $g'' \in C_B[0,\infty)$, g belongs to Lip^*2 on $[0,\infty)$ and by Theorem 2.6

$$L_n(g;x) - g(x) = O(1)n^{-1} \ (a_3 \le x \le b_3,\ n \in N). \quad (3.12)$$

We now conclude from (3.7), (3.10), (3.11) and (3.12) that

$$\|L_n(f_1g) - (f_1g)\|_{[a_3,b_3]} = O(1)n^{\frac{-\gamma}{2}} \ (n \in N).$$

Further $f_1g \in C_B[0,\infty)$ and $\operatorname{supp} f_1g \subset \operatorname{supp} g \subset [a_2,b_2] \subset (a_3,b_3)$. Hence by Lemma 3.3, f_1g belongs to $Lip^*\gamma$ on $[0,\infty)$ and since $f_1g = f_1 = f$ on $[a_1,b_1]$, f belongs to $Lip^*\gamma$ on $[a_1,b_1]$.

REFERENCES

1. J.A. Adell and J. de la Cal, Using stochastic processes for studying Bernstein-type operators, Universidad de Zaragoza, Spain, preprint.
2. J.A. Adell, J. de la Cal, and M. San Miguel, Inverse Beta and generalized Bleimann-Butzer-Hahn operator, *J. Approx. Theory*, to appear.
3. G. Bleimann, P.L. Butzer, and L. Hahn, A Bernstein-type operator approximating continuous functions on the semi axis, *Indag. Math.* 42:255–262 (1980).
4. J. de la Cal and F. Luquin, A note on limiting properties of some Bernstein-type operators, *J. Approx. Theory* 69:322–329 (1992).
5. R.A. De Vore, "The approximation of continuous functions by positive linear operators," Lecture notes in Math., 293, Springer, Berlin (1972).
6. Z. Ditzian and C.P. May, L_p saturation and inverse theorems for certain Bernstein polynomials, *Indiana Univ. Math. J.* 25:733–751 (1976).
7. C. Jayasri and Y. Sitaraman, Direct and Inverse theorems for certain Bernstein-type operators, *Indian J. Pure Appl. Math.* 16:1495–1511 (1985).
8. C. Jayasri and Y. Sitaraman, On a Bernstein-type operator of Bleimann-Butzer and Hahn, *J. Comput. Appl. Math.*, to appear.
9. C. Jayasri and Y. Sitaraman, On a Bernstein-type operator of Bleimann-Butzer and Hahn II, *J. Analysis* 1:125–137 (1993).
10. R.A. Khan, A note on a Bernstein-type operator of Bleimann, Butzer and Hahn, *J. Approx. Theory* 53:295–303 (1988).
11. R.A. Khan, Some properties of a Bernstein-type operator of Bleimann, Butzer and Hahn, "Progress in Approximation Theory," Eds. P. Nevai and A. Pinkus, Academic Press, New York (1991).
12. A.F. Timan, "Theory of approximation of functions of a real variable," MacMillan, New York (1963).
13. V. Totik, Uniform approximation by Bernstein-type operators, *Indag. Math.* 46:87–93 (1984).

A PROBLEM OF ANALYTIC MONOTONE EXTENSION IN THREE DIMENSIONS

Shelby Kilmer, Xingping Sun and Xiang Ming Yu

Department of Mathematics
Southwest Missouri State University
Springfield, MO 65804

ABSTRACT

Let Ω denote the simplicial complex $[0,1]^d$ in $\mathbb{R}^d$. For a nonnegative integer p with $p \le d$, let $\Omega^{(p)}$ denote the p–skeleton of Ω. Let $d = 3$ and suppose that f is a monotone function on $\Omega^{(2)}$. We give construction schemes which extend f as a monotone function on Ω, while maintaining smoothness. Our results are related to those of Dahmen, DeVore and Micchelli for the case $d = 2$ and $p = 1$ and those of Kilmer, Sun and Yu for the case $d = 3$ and $p = 1$.

0. INTRODUCTION

Let $\mathbb{R}^d$ denote the d–dimensional Euclidean space, and $\mathbb{R}^d_+$ the set of all points in $\mathbb{R}^d$ having nonnegative components. For $\Omega \subset \mathbb{R}^d$, a function f is said to be monotone (nondecreasing) on Ω if for any pair x and y in Ω with $x - y \in \mathbb{R}^d_+$, we have $f(x) \ge f(y)$. The monotonicity is said to be strict if $f(x) > f(y)$, when $x \ne y$.

Motivated by the modeling of charge distribution for semiconductor design, Dahmen, DeVore and Micchelli [1] studied the following problem: Given a monotone function f on the boundary of Ω, denoted $\partial\Omega$, find a monotone F on Ω such that $F(x) = f(x)$ whenever $x \in \partial\Omega$. The function F is called a monotone extension of f. They proved the following theorem:

Theorem 0.1. Let Ω be a bounded subset of $\mathbb{R}^d$ having nonempty interior. There is no bounded linear operator $L:C(\partial\Omega) \to C(\Omega)$ such that Lf is a monotone extension whenever f is monotone.

1991 Mathematics subject classification: Primary 41A05, Secondary 41A63

In [1], several nonlinear methods of constructing monotone extensions were given. The case $\Omega = [0,1]^2$ received particular attention.

In practice it is important that F keep the smoothness of the boundary functions. This means the following: If the four boundary functions $f(x,0)$, $f(x,1)$, $f(0,y)$ and $f(1,y)$ have a certain degree of smoothness as univariate functions, then we require that F have the same degree of smoothness as a bivariate function. For example if $f(x,0)$, $f(x,1)$, $f(0,y)$ and $f(1,y)$ are analytic on their domains, then F must be analytic on $[0,1]^2$. We shall refer to such a function F as an *analytic monotone extension* (AME) of the given boundary functions.

To construct AMEs, Dahmen, DeVore and Micchelli [1] used some variations of the "blending" method. Inspired by their approach, Sun and Yu [3] gave several simple constructions of AMEs and discussed a general method of generating AMEs. In [2] details of this general method were presented.

Let Ω denote the simplicial complex $[0,1]^d$ in $\mathbb{R}^d$. For a nonnegative integer p with $p \leq d$, let $\Omega^{(p)}$ denote the p–skeleton of Ω. Consider $d > 2$ and $p < d$. The following question naturally arises: If f is a monotone function defined on $\Omega^{(p)}$, how can f be extended to a monotone function on Ω with the same degree of smoothness? A satisfactory answer to this question for the case $d = 3$ and $p = 1$ is also given in [2]. In this paper we give some results for the case $d = 3$ and $p = 2$.

1. PRELIMINARIES AND NOTATION

Let Ω denote $[0,1]^3$ throughout. ∂ will mean boundary, Π_P will represent the projection from $\mathbb{R}^3$ onto the plane P and Π_ξ will represent the projection from $\mathbb{R}^3$ onto the ξ–axis. For example $\Pi_{z=1}$ is the projection onto the plane $z = 1$.

Definitions.

i) Given $f: \partial\Omega \to [0,1]$, define the function $f^*: \partial\Omega \to \mathbb{R}$ by

$$f^* = f + f\circ\Pi_x + f\circ\Pi_y + f\circ\Pi_z - f\circ\Pi_{x=0} - f\circ\Pi_{y=0} - f\circ\Pi_{z=0}$$

ii) We say that g satisfies property (A) if the second mixed partials are continuous and satisfy that

$$g_{xy}(x,y,1) \geq \alpha_1\, g_y(1,y,1)\, g_x(x,1,1),$$

$$g_{xz}(x,1,z) \geq \alpha_2\, g_z(1,1,z)\, g_x(x,1,1)$$

and
$$g_{yz}(1,y,z) \geq \alpha_3\, g_z(1,1,z)\, g_y(1,y,1),$$

where α_1, α_2 and α_3 are constants such that $\alpha_1 + \alpha_2 + \alpha_3 = 2$.

iii) We say that f satisfies property (B) if f^* satisfies property (A).

2. SOME SPECIAL CASES

Theorem 2.1. Let $g\colon \partial\Omega \to [0,1]$ be onto, continuous and monotone. If $g(x,y,z) = 0$ whenever x, y or $z = 0$, and satisfies property (A), then g has an AME.

Proof. Define $G\colon \Omega \to \mathbb{R}$ by

$$G(x,y,z) = g\circ\Pi_{x=1} g\circ\Pi_{y=1}\circ\Pi_{z=1} + g\circ\Pi_{y=1} g\circ\Pi_{x=1}\circ\Pi_{z=1} + g\circ\Pi_{z=1} g\circ\Pi_{x=1}\circ\Pi_{y=1}$$
$$- 2\, g\circ\Pi_{x=1}\circ\Pi_{z=1}\, g\circ\Pi_{y=1}\circ\Pi_{z=1} g\circ\Pi_{x=1}\circ\Pi_{y=1}.$$

One checks that G extends g; for example,

$$G(1,y,z) = g(1,y,z)\, g(1,1,1) + g(1,1,z)\, g(1,y,1) + g(1,y,1)\, g(1,1,z)$$
$$- 2\, g(1,y,1)\, g(1,1,1)\, g(1,1,z)$$
$$= g(1,y,z)$$

Since projections and polynomials are analytic, the smoothness of G depends only on the smoothness of g, hence G is an analytic extension.

In order to see that G is monotone we will first find G_x.

$$G_x(x,y,z) = g(1,y,z)\, g_x(x,1,1) + g_x(x,1,z)\, g(1,y,1) + g_x(x,y,1)\, g(1,1,z)$$
$$- 2\, g_x(x,1,1)\, g(1,y,1)\, g(1,1,z)$$
$$= g_x(x,1,1)\left[g(1,y,z) - \alpha_1\, g(1,y,1)\, g(1,1,z)\right] \tag{2.1}$$
$$+ g(1,y,1)\left[g_x(x,1,z) - \alpha_2\, g_x(x,1,1)\, g(1,1,z)\right]$$
$$+ g(1,1,z)\left[g_x(x,y,1) - \alpha_3\, g_x(x,1,1)\, g(1,y,1)\right]$$

To see $G_x \geq 0$, consider φ defined by

$$\varphi(y,z) = g(1,y,z) - \alpha_1\, g(1,1,z)\, g(1,y,1).$$

Because g satisfies property (A), $\varphi_{yz} \geq 0$. Since $\varphi(y,0) = 0$, $\varphi_y(y,z) \geq 0$ for each $z \in [0,1]$ and we see φ_y is increasing. Now since $\varphi(0,z) = 0$, $\varphi(y,z) \geq 0$ for each $y,z \in [0,1]$ and we have

$$g(1,y,z) - \alpha_1\, g(1,y,1)\, g(1,1,z) \geq 0 \tag{2.2}$$

Now for a given x consider ψ defined by

$$\psi(z) = g_x(x,1,z) - \alpha_2\, g(1,1,z)\, g_x(x,1,1).$$

Since g satisfies property (A), $\psi'(z) \geq 0$. Since $\psi(0) = 0$, we see that $\psi(z) \geq 0$ for each $z \in [0,1]$. We conclude

$$g_x(x,1,z) - \alpha_2\, g_x(x,1,1)\, g(1,1,z) \geq 0. \tag{2.3}$$

Similarly

$$g_x(x,y,1) - \alpha_3\, g_x(x,1,1)\, g(1,y,1) \geq 0. \tag{2.4}$$

Combining (2.1), (2.2), (2.3) and (2.4) we obtain $G_x \geq 0$. Since G is symmetric in the three variables, we conclude that G is monotone and the proof is complete. □

Theorem 2.2. If $f: \partial\Omega \to [0,1]$ is onto, continuous and monotone, satisfies property (B) and

$$\begin{aligned} f_{xy}(x,y,1) &> f_{xy}(x,y,0) \geq 0 \\ f_{xz}(x,1,z) &> f_{xz}(x,0,z) \geq 0 \\ f_{yz}(1,y,z) &> f_{yz}(0,y,z) \geq 0, \end{aligned} \tag{2.5}$$

then f has an AME.

Proof. Using (2.5) and the Mean Value Theorem one can show that

$$f^*(1,1,1) > 0.$$

Let $g = f^*/f^*(1,1,1)$. Then g is continuous and $g(x,y,z) = 0$ when x, y or $z = 0$. Furthermore, when (x,y,z) lies on the planes $z = 1$, or $y = 1$,

$$g_x(x,y,z) = [f_x(x,y,z) + f_x(x,0,0) - f_x(x,y,0) - f_x(x,0,z)]/f^*(1,1,1).$$

On the plane $y = 1$, by the Mean Value Theorem, there exists $\xi \in (0,z)$ such that

$$\begin{aligned} g_x(x,1,z) &= [(f_x(x,1,z) - f_x(x,0,z)) - (f_x(x,1,0) - f_x(x,0,0))]/f^*(1,1,1) \\ &= z\,[f_{xz}(x,1,\xi) - f_{xz}(x,0,\xi)]/f^*(1,1,1) \\ &\geq 0 \quad \text{by (2.5).} \end{aligned}$$

The definition of g and the inequalities in (2.5) are symmetric with respect to the three variables, so similar reasoning yields the following inequalities:

$$\begin{aligned} g_x &\geq 0 \ \text{ on } y = 1 \text{ and } z = 1 \\ g_y &\geq 0 \ \text{ on } x = 1 \text{ and } z = 1 \\ g_z &\geq 0 \ \text{ on } x = 1 \text{ and } y = 1 \end{aligned} \tag{2.6}$$

Since g is zero on $y = 0$, $x = 0$, and $z = 0$, (2.6) implies the monotonicity of g. By

Theorem 2.1, we obtain G, an AME of g.

We define $F:\Omega \to \mathbb{R}$ by

$$F = f^*(1,1,1)\,G - f\circ\Pi_x - f\circ\Pi_y - f\circ\Pi_z + f\circ\Pi_{x=0} + f\circ\Pi_{y=0} + f\circ\Pi_{z=0}.$$

Clearly F is an analytic extension of f.

Moreover by the Mean Value Theorem there exists $\xi \in (0,y)$ such that

$$\begin{aligned}\frac{\partial}{\partial x}(f\circ\Pi_{z=0} - f\circ\Pi_x)(x,y,z) &= f_x(x,y,0) - f_x(x,0,0)\\ &= y\, f_{xy}(x,\xi,0) \geq 0, \text{ by (2.5)}\end{aligned}$$

In addition

$$\frac{\partial}{\partial x}(f\circ\Pi_{y=0}(x,y,z) - f\circ\Pi_z(x,y,z)) = f_x(x,0,z) \geq 0$$

and

$$\frac{\partial}{\partial x}(f\circ\Pi_{x=0}(x,y,z) - f\circ\Pi_y(x,y,z)) = 0.$$

Thus $F_x \geq 0$. Similar arguments show $F_y \geq 0$ and $F_z \geq 0$, and the proof is complete. □

Theorem 2.3. Let $f: \partial\Omega \to [0,1]$ be onto, continuous and monotone, with positive partials and continuous second partials. If $f\circ\Pi_{\xi=0} < f\circ\Pi_{\xi=1}$ for $\xi = x$, y and z, then there exists an onto strictly increasing function $\Phi: [0,1] \to [0,1]$ satisfying

$$(\Phi\circ f)_{xy}(x,y,1) > (\Phi\circ f)_{xy}(x,y,0) > 0$$

$$(\Phi\circ f)_{xz}(x,1,z) > (\Phi\circ f)_{xz}(x,0,z) > 0$$

$$(\Phi\circ f)_{yz}(1,y,z) > (\Phi\circ f)_{yz}(0,y,z) > 0.$$

Furthermore if there exists an AME for $\Phi\circ f$, then there exists one for f.

Proof. For $n > 1$, define Φ_n by

$$\Phi_n(t) = \frac{(t+1)^n - 1}{2^n - 1}, \quad \text{for } 0 \leq t \leq 1.$$

Each Φ_n is continuous with $\Phi_n(0) = 0$ and $\Phi_n(1) = 1$, so Φ_n is onto [0,1]. One Checks that $\Phi_n' > 0$, in order to see that Φ is monotone.

Because of continuity and compactness, there exists positive real numbers m, N, M and ε satisfying the following inequalities for all x, y, and z:

$$m \leq f_x(x,y,z) \leq M$$

$$m \leq f_y(x,y,z) \leq M$$

$$m \leq f_z(x,y,z) \leq M$$

$$-N \leq f_{yz}(x,y,z) \leq N$$

$$f(1,y,z) - f(0,y,z) \geq \varepsilon$$

From the last of these inequalities we see

$$\frac{f(1,y,z) + 1}{f(0,y,z) + 1} \geq 1 + \frac{\varepsilon}{f(0,y,z) + 1} \geq 1 + \frac{\varepsilon}{2}.$$

Since

$$(\Phi_n \circ f)_{yz} = \Phi'' f_y f_z + \Phi' f_{yz}$$
$$= \frac{n(n-1)(f+1)^{n-2}}{2^n - 1} f_y f_z + \frac{n(f+1)^{n-1}}{2^n - 1} f_{yz},$$

we have for all y and z,

$$\frac{(\Phi_n \circ f)_{yz}(1,y,z)}{(\Phi_n \circ f)_{yz}(0,y,z)} = \frac{(n-1)(f(1,y,z)+1)^{n-2} f_z(1,y,z) f_y(1,y,z) + (f(1,y,z)+1)^{n-1} f_{yz}(1,y,z)}{(n-1)(f(0,y,z)+1)^{n-2} f_z(0,y,z) f_y(0,y,z) + (f(0,y,z)+1)^{n-1} f_{yz}(0,y,z)}$$

$$= \left[\frac{f(1,y,z)+1}{f(0,y,z)+1}\right]^{n-2} \frac{(n-1) f_z(1,y,z) f_y(1,y,z) + (f(1,y,z)+1) f_{yz}(1,y,z)}{(n-1) f_z(0,y,z) f_y(0,y,z) + (f(0,y,z)+1) f_{yz}(0,y,z)}$$

$$\geq [1 + \tfrac{\varepsilon}{2}]^{n-2} \frac{(n-1)m^2 - 2N}{(n-1)M^2 + 2N} \to \infty, \tag{2.7}$$

as $n \to \infty$.

Now, choose n so that

$$\frac{(\Phi_n \circ f)_{yz}(1,y,z)}{(\Phi_n \circ f)_{yz}(0,y,z)} > 1 \text{ for all y and z}$$

and define $\Phi = \Phi_n$. It follows that

$$(\Phi \circ f)_{yz}(1,y,z) > (\Phi \circ f)_{yz}(0,y,z) \tag{2.8}$$

It follows from (2.7) and (2.8) that

$$(\Phi \circ f)_{yz}(0,y,z) > 0.$$

Similar arguments show that

$$(\Phi \circ f)_{xy}(x,y,1) > (\Phi \circ f)_{xy}(x,y,0) > 0$$

and

$$(\Phi \circ f)_{xz}(x,1,z) > (\Phi \circ f)_{xz}(x,0,z) > 0.$$

Finally, suppose G extends $\Phi \circ f$. Since Φ is strictly increasing, Φ^{-1} exists and $\Phi^{-1} \circ G$ extends f. □

3. A NEAR AME

Theorem 3.1. If $f\colon \partial\Omega \to [0,1]$ is onto, continuous and monotone with positive partials and continuous second partials and if $f \circ \Pi_{\xi=1} > f \circ \Pi_{\xi=0}$, for $\xi = x$, y or z, then f has a near AME, i.e., for any $\eta > 0$, there exists an analytic extension of f that is monotone on $[\eta, 1]^3$.

Proof. Let $\eta > 0$. In light of Theorem 2.3, we may assume without loss of generality the following:

$$\begin{aligned} &f_{xy}(x,y,1) > f_{xy}(x,y,0) > 0 \\ &f_{xz}(x,1,z) > f_{xz}(x,0,z) > 0 \\ &f_{yz}(1,y,z) > f_{yz}(0,y,z) > 0. \end{aligned} \tag{3.1}$$

Let $g = f^*/f^*(1,1,1)$. We see, from the proof of Theorem 2.2, that the function g satisfies all the conditions in Theorem 2.1, except possibly property (A). We will construct an onto strictly increasing function $\Psi_\eta: [0,1] \to [0,1]$ such that $\Psi_\eta \circ g$ has an analytic monotone extension to $[\eta, 1]^3$. It follows that g, and hence f will have an analytic monotone extension to $[\eta, 1]^3$.

For simplicity, we will construct Ψ_η so that the desired analytic extension of $\Psi_\eta \circ g$ has a positive partial with respect to x on $[\eta,1]^3$. It should then be evident how to construct Ψ_η so the partials with respect to y and z are positive as well. In order to do this, we first need the following technical claim.

Claim 1. There exists $\eta_0 \in (0,1)$ such that

$$\begin{aligned} &g_x(x,y,1) \geq \tfrac{5}{6}\, g_x(x,1,1), \\ &g_x(x,1,z) \geq \tfrac{5}{6}\, g_x(x,1,1) \\ \text{and}\quad &g(1,y,z) \geq \tfrac{1}{2} \end{aligned} \tag{3.2}$$

hold for all y and $z \in [1-\eta_0, 1]$.

Proof of Claim 1: Because of continuity and compactness, there is a $\delta > 0$ such that

$$\begin{aligned} &f_{xy}(x,y,1) - f_{xy}(x,y,0) \geq \delta \\ &f_{xz}(x,1,z) - f_{xz}(0,y,z) \geq \delta \\ &f_{yz}(1,y,z) - f_{yz}(0,y,z) \geq \delta. \end{aligned} \tag{3.3}$$

From the definition of g and the Mean–Value Theorem, we have

$$\begin{aligned} g_x(x,y,1) &= ([f_x(x,y,1) - f_x(x,y,0)] - [f_x(x,0,1) - f_x(x,0,0)])/f^*(1,1,1) \\ &= [f_{xy}(x,\xi,1) - f_{xy}(x,\xi,0)] \cdot y/f^*(1,1,1) \\ &\geq \delta y/f^*(1,1,1), \end{aligned} \tag{3.4}$$

for $0 \leq x,\ y \leq 1,\ 0 \leq \xi \leq y$.

Consider h defined by

$$h(x,y) = g_x(x,y,1) - \frac{5}{6} g_x(x,1,1)$$

We have

$$h(x,1) = \frac{1}{6} g_x(x,1,1) \geq \frac{1}{6} \delta/f^*(1,1,1) > 0,$$

for $0 \leq x \leq 1$, hence, from continuity and compactness, there is a $\eta_1 > 0$ such that if $0 \leq x \leq 1$ and $1 - \eta_1 \leq y \leq 1$, then $h(x,y) \geq 0$. It follows that

$$g_x(x,y,1) \geq \frac{5}{6} g_x(x,1,1).$$

Similarly we can find $\eta_2 > 0$ such that if $0 \leq x \leq 1$ and $1 - \eta_2 \leq z \leq 1$, then

$$g_x(x,1,z) \geq \frac{5}{6} g_x(x,1,1).$$

Since $g(1,1,1) = 1$ we also have $\eta_3 > 0$ such that if y and $z \in [1 - \eta_3, 1]$,

$$g(1,y,z) \geq \frac{1}{2}.$$

Taking $\eta_0 = \min\{\eta_1, \eta_2, \eta_3\}$ completes the proof of Claim 1.

We now continue with our construction of Ψ_η.

Claim 2. There exists an onto strictly increasing function $\Psi_\eta:[0,1] \to [0,1]$ such that the function $\bar{g}$ defined by $\bar{g} = \Psi_\eta \circ g$ satisfying the following:

$$\bar{g}_x(x,y,1) \geq 2\, \bar{g}_x(x,1,1)\, \bar{g}(1,y,1), \tag{3.5}$$

for $0 \leq y \leq 1 - \eta$ and $\eta \leq x \leq 1$; and

$$\bar{g}_x(x,1,z) \geq 2\bar{g}_x(x,1,1)\, \bar{g}(1,1,z), \tag{3.6}$$

for $0 \leq z \leq 1 - \eta$ and $\eta \leq x \leq 1$

Proof of Claim 2: Because of symmetry, we only prove (3.5).

Let

$$M = \max_{0 \leq x, y \leq 1} [g_x(x,1,1)\, g_y(1,y,1)],$$

$$\xi_1 = \min_{\substack{0 \leq y \leq 1-\eta \\ x \geq \eta}} [g(x,1,1) - g(x,y,1)],$$

$$\xi_2 = \min_{\substack{0 \leq z \leq 1-\eta \\ x \geq \eta}} [g(x,1,1) - g(x,1,z)]$$

and

$$\xi := \min(\xi_1, \xi_2).$$

By the Mean Value Theorem and the inequalities similar to (3.4) corresponding to g_y and g_z one can show $\xi > 0$. Define

$$\Psi_\eta(t) = \frac{1 - (1+t)^{-n}}{1 - 2^{-n}} \quad \text{for } 0 \leq t \leq 1,$$

where n is an integer that depends on g and η and will be chosen later. Then $\Psi_\eta:[0,1]\to[0,1]$ is an onto strictly increasing function. Notice that

$$\begin{aligned}\bar{g}_x(x,y,1) &= \Psi'_\eta(g(x,y,1))\, g_x(x,y,1) \\ &= \frac{n(1+g(x,y,1))^{-n-1}}{1-2^{-n}}\, g_x(x,y,1)\end{aligned}$$

and

$$\bar{g}_x(x,1,1) = \frac{n(1+g(x,1,1))^{-n-1}}{1-2^{-n}}\, g_x(x,1,1).$$

Then, from the definitions of ξ and M, inequality (3.4) and the fact that

$$0 \le g(x,1,1) \le 1,$$

we have for $0 \le y \le 1-\eta$ and $\eta \le x \le 1$,

$$\begin{aligned}\frac{\bar{g}_x(x,y,1)}{\bar{g}_x(x,1,1)} &= \left[\frac{1+g(x,y,1)}{1+g(x,1,1)}\right]^{-n-1} \frac{g_x(x,y,1)}{g_x(x,1,1)} \\ &\ge \frac{\delta y}{Mf^*(1,1,1)} \left[\frac{1+g(x,1,1)-\xi}{1+g(x,1,1)}\right]^{-n-1} \\ &\ge \frac{\delta y}{Mf^*(1,1,1)} (1-\tfrac{\xi}{2})^{-n-1}. \qquad (3.7)\end{aligned}$$

On the other hand, by the Mean Value Theorem there exists a nonnegative $\zeta \le y$ such that

$$\begin{aligned}\bar{g}(1,y,1) &= \bar{g}(1,y,1) - \bar{g}(1,0,1)\, \cdot \\ &= \bar{g}_y(1,\zeta,1)\cdot y \\ &= \Psi'_\eta(g(1,\zeta,1))\, g_y(1,\zeta,1)\, y \\ &= \frac{n(1+g(1,\zeta,1))^{-n-1}}{1-2^{-n}}\, g_y(1,\zeta,1) y \\ &\le \frac{n}{1-2^{-n}}\, My \le 2nMy, \qquad (3.8)\end{aligned}$$

Combining (3.7) and (3.8) yields for $0 \le y \le 1-\eta$ and $\eta \le x \le 1$,

$$\frac{\bar{g}_x(x,y,1)}{\bar{g}_x(x,1,1)} \ge \left[(1-\tfrac{\xi}{2})^{-n-1} \frac{\delta}{2n\ M^2 f^*(1,1,1)}\right] \bar{g}(1,y,1). \qquad (3.9)$$

Since $(1-\frac{\xi}{2})^{-n-1} \frac{\delta}{2nM^2} \to \infty$ as $n \to \infty$, we can choose n so that

$$(1-\tfrac{\xi}{2})^{-n-1} \frac{\delta}{2n\ M^2 f^*(1,1,1)} > 2.$$

Because we shall have an inequality similar to (3.9) for $\bar{g}_x(x,1,z)$, we can choose Ψ_η

satisfying (3.5) and (3.6). This completes the proof of Claim 2.

Let G be as in the proof of Theorem 2.1 with $\bar{g}$ in the place of g. Without loss of generality, we asume $\eta < \eta_0$. To show that $G_x \geq 0$ on $[\eta,1]^3$, we consider several cases.

Case (i): $0 \leq y \leq 1-\eta$ and $x \geq \eta$.

From (3.5), we have

$$G_x = \bar{g}(1,1,z)[\bar{g}_x(x,y,1) - 2\bar{g}_x(x,1,1)\bar{g}(1,y,1)]$$
$$+ \bar{g}(1,y,1)\bar{g}_x(x,1,z) + \bar{g}_x(x,1,1)\bar{g}(1,y,z)] \geq 0.$$

Case (ii): $0 \leq z \leq 1-\eta$ and $x \geq \eta$.

From (3.6), we have

$$G_x = \bar{g}(1,y,1)[\bar{g}_x(x,1,z) - 2\bar{g}_x(x,1,1)\bar{g}(1,1,z)]$$
$$+ \bar{g}(1,1,z)\bar{g}_x(x,y,1) + \bar{g}_x(x,1,1)\bar{g}(1,y,z) \geq 0$$

Case (iii): $1-\eta \leq y,\ z \leq 1$ and $0 \leq x \leq 1$.

First we show that if $g(1,y,z) \geq \frac{1}{2}$, then $\bar{g}(1,y,z) \geq \frac{1}{3}$. Indeed, by definition, we have

$$\bar{g}(1,y,z) = \frac{1-(1+g(1,y,z))^{-n}}{1-2^{-n}} \geq \frac{1-(1+\frac{1}{2})^{-n}}{1-2^{-n}}$$
$$= \frac{1-(\frac{2}{3})^{n}}{1-2^{-n}} \geq \frac{1-\frac{2}{3}}{1} = \frac{1}{3} \tag{3.10}$$

Since $0 < \eta < \eta_0$, from Claim 1, we have

$$\bar{g}_x(x,y,1) = \Psi'_\eta(g(x,y,1))\, g_x(x,y,1)$$
$$= \frac{n(1+g(x,y,1))^{-n-1}}{1-2^{-n}}\, g_x(x,y,1)$$
$$\geq \frac{5}{6}\,\frac{n(1+g(x,1,1))^{-n-1}}{1-2^{-n}}\, g_x(x,1,1)$$
$$= \frac{5}{6}\Psi'_\eta(g(x,1,1)) \cdot g_x(x,1,1)$$
$$= \frac{5}{6}\bar{g}_x(x,1,1) \geq \frac{5}{6}\bar{g}_x(x,1,1)\,\bar{g}(1,y,1). \tag{3.11}$$

Similarly we have

$$\bar{g}_x(x,1,z) \geq \frac{5}{6}\bar{g}_x(x,1,1)\,\bar{g}(1,1,z) \tag{3.12}$$

From (3.10), (3.11) and (3.12), we have

$$\begin{aligned} G_x &= \bar{g}_x(x,1,1)\bar{g}(1,y,z) + \bar{g}_x(x,1,z)\,\bar{g}(1,y,1) \\ &\quad + \bar{g}_x(x,y,1)\,\bar{g}(1,1,z) - 2\bar{g}_x(x,1,1)\,\bar{g}(1,y,1)\,\bar{g}(1,1,z) \\ &\geq \tfrac{1}{3}\bar{g}_x(x,1,1)\,\bar{g}(1,y,1)\,\bar{g}(1,1,z) + \tfrac{5}{6}\bar{g}_x(x,1,1)\,\bar{g}(1,1,z)\,\bar{g}(1,y,1) \\ &\quad + \tfrac{5}{6}\bar{g}_x(x,1,1)\,\bar{g}(1,y,1)\,\bar{g}(1,1,z) - 2\bar{g}_x(x,1,1)\,\bar{g}(1,y,1)\,\bar{g}(1,1,z) = 0 \end{aligned}$$

Thus $G_x \geq 0$ if $x \geq \eta$. This completes the proof. □

REFERENCES

1. W. Dahmen, R. DeVore and C. Micchelli, On monotone extension of boundary data, to appear in Numer. Math.

2. S. Kilmer, X. Sun and X. Yu, Analytic monotone extension in two and three dimensions, preprint.

3. X. Sun and X. Yu, Analytic monotone extension of boundary data, to appear in J. Approx. Theory.

ON THE JOINT ESTIMATION OF STABLE LAW PARAMETERS

L.B. Klebanov,[1] J.A. Melamed,[2] and S.T. Rachev[3]

[1]St. Petersburg University for Architecture and Civil Engineering, Russia
[2]University of Southern California, USA
[3]University of California, Santa Barbara, USA

ABSTRACT

Computationally simple estimators for the vector of parameters of a stable law are constructed based on the modified method of scoring [3]. The estimators are consistent, asymptotically normal, and asymptotically efficient in the class of estimators with partial information on the distribution.

1. INTRODUCTION

Recently, a number of papers have appeared that deal with the estimation of parameters of stable distributions, (see [4,5] and the references therein). Applications aside, this problem is of interest because the density functions of stable laws do not have a simple analytic form in general. Therefore, such traditional estimators as Bayesian or the maximum likelihood cannot be easily obtained (see [4]).

In this paper, we propose estimators for the vector $\underset{\sim}{\theta}= (\alpha, \beta, \gamma, \lambda)$ of parameters of a stable law with a distribution function (d.f.) $F(x;\underset{\sim}{\theta})$, whose characteristic function (ch.f.) (see [6, Introduction, (M)]) is given by

$$\varphi(t;\underset{\sim}{\theta}) = \begin{cases} \exp\{\lambda[it\gamma - |t|^\alpha + it(|t|^{\alpha-1} - 1)\beta\tan(\frac{\pi}{2}\alpha)]\} & \text{if } \alpha \neq 1, \\ \exp\{\lambda(it\gamma - |t| - i\frac{2}{\pi}\beta t\ln|t|)\} & \text{if } \alpha = 1. \end{cases} \tag{1}$$

$\varphi(t;\underset{\sim}{\theta})$ is continuous w.r.t. $\underset{\sim}{\theta}$. For the construction, we utilize the modified method of scoring (MMS) (see [3]). The constructed estimators are computationally simple and exploit the available information on $F(x;\underset{\sim}{\theta})$ to the utmost. They are asymptotically efficient in the class of estimators based on the same information. The estimators are consistent and asymptotically normal.

2. NOTATIONS AND RESULTS

Let $x_1, \ldots, x_n$ be a random sample of size n from a population with the d.f. $F(x;\underset{\sim}{\theta})$ and ch.f. given by (1). φ depends on an unknown vector parameter, $\underset{\sim}{\theta}=$

Approximation, Probability, and Related Fields, Edited by
G. Anastassiou and S.T. Rachev, Plenum Press, New York 1994

$(\theta_1, \theta_2, \theta_3, \theta_4) = (\alpha, \beta, \gamma, \lambda)$ where $0 < \alpha \le 2$, $-1 \le \beta \le 1$, $-A < \gamma < A$, $0 < \lambda < B$ for some $A > 0$ and $B > 0$.

Consider the problem of estimating $\underset{\sim}{\theta}$ on the basis of the given random sample. Let t_j, $j = 1, \ldots, k$, be k positive numbers. Let L be a linear space spanned by elements $\underset{\sim}{1}$, $\xi = \left((\cos(t_j x_1), \sin(t_j x_1))', j = 1, 2, \ldots, k\right)$. Let $\mathcal{A} = \mathcal{A}(\underset{\sim}{\theta}) = \|a_{pq}^{(r\ell)}\|$ be the matrix with elements,

$$a_{pq}^{(11)} = \tfrac{1}{2}\left[Re\varphi(t_p - t_q;\underset{\sim}{\theta}) + Re\varphi(t_p + t_q;\underset{\sim}{\theta})\right] - Re\varphi(t_p;\underset{\sim}{\theta})Re\varphi(t_q;\underset{\sim}{\theta}),$$

$$a_{pq}^{(12)} = \tfrac{1}{2}\left[Im\varphi(t_p + t_q;\underset{\sim}{\theta}) + Im\varphi(t_p - t_q;\underset{\sim}{\theta})\right] - Im\varphi(t_q;\underset{\sim}{\theta})Re\varphi(t_p;\underset{\sim}{\theta}),$$

$$a_{pq}^{(22)} = \tfrac{1}{2}\left[Re\varphi(t_p - t_q;\underset{\sim}{\theta}) - Re\varphi(t_p + t_q;\underset{\sim}{\theta})\right] - Im\varphi(t_p;\underset{\sim}{\theta})Im\varphi(t_q;\underset{\sim}{\theta}).$$

Let

$$E_{\underset{\sim}{\theta}}\zeta' = \left(E_{\underset{\sim}{\theta}} \cos(t_1 x_1), E_{\underset{\sim}{\theta}} \sin(t_1 x_1), \ldots, E_{\underset{\sim}{\theta}} \cos(t_k x_1), E_{\underset{\sim}{\theta}} \sin(t_k x_1)\right)$$
$$= \left(Re(\varphi(t_1;\underset{\sim}{\theta})), Im(\varphi(t_1;\underset{\sim}{\theta})), \ldots, Re(\varphi(t_k;\underset{\sim}{\theta})), Im(\varphi(t_k,\underset{\sim}{\theta}))\right),$$

where expectation is w.r.t. unknown parameter $\underset{\sim}{\theta}$.

$$\Lambda^{(a,b)}(\underset{\sim}{\theta}) = \begin{pmatrix} 0 & 0 & \frac{\partial}{\partial \underset{\sim}{\theta}_a} E_{\underset{\sim}{\theta}}\zeta' \\ 0 & 1 & O' \\ \frac{\partial}{\partial \underset{\sim}{\theta}_\zeta} E_{\underset{\sim}{\theta}}\zeta & O & \mathcal{A} \end{pmatrix}, \quad a, b = 1, 2, 3, 4$$

where O is the null column vector of appropriate dimension and $\Lambda_{0,j}^{(a)}(\underset{\sim}{\theta})$ is a cofactor of the element in the first line and (j+1)st column of the matrix $\Lambda^{(aa)}(\underset{\sim}{\theta})$.

The matrix $I(\underset{\sim}{\theta}; C) = \|I_{ab}(\underset{\sim}{\theta})\|$ with elements given by

$$I_{ab}(\underset{\sim}{\theta}) = -\frac{\det \Lambda^{(ab)}(\underset{\sim}{\theta})}{\det \mathcal{A}}, \quad a, b = 1, 2, 3, 4$$

is the matrix of Fisher information on $\underset{\sim}{\theta}$ contained in the linear space C (see [2]). It is clear that $\det \mathcal{A} > 0$ when $F(x;\underset{\sim}{\theta})$ has more than k points of growth. This follows from the fact that $\det \mathcal{A}$ is Gram's determinant of the linearly independent system $\underset{\sim}{1}, \zeta' - E_{\underset{\sim}{\theta}}\zeta'$.

Consider the empirical ch.f., $\varphi_n(t) = \frac{1}{n}\sum_{j=1}^{n} e^{itx_j}$ and let

$$\xi' = (\xi_j, j = 1, \ldots, 2k), \text{ where } \xi_j = \begin{cases} Re\left(\varphi_n(t_j)\right) & j = 1, \ldots, k \\ Im\left(\varphi_n(t_{j-k})\right) & j = k+1, \ldots, 2k. \end{cases}$$

We introduce the vector score function $\mathcal{J}_n(\underset{\sim}{\theta})$ (see [2]) with the components

$$J_n^{(a)}(\underset{\sim}{\theta}) = -n \sum_{j=1}^{2k} \frac{\Lambda_{0,j+1}^{(a)}(\underset{\sim}{\theta})}{\det \mathcal{A}(\underset{\sim}{\theta})}(\xi_j - E_{\underset{\sim}{\theta}}\xi_j), \quad a = 1, 2, 3, 4.$$

Let $\underset{\sim}{\theta}^* = (\theta_1^*, \theta_2^*, \theta_3^*, \theta_4^*)$ be a $\sqrt{n}$-consistent estimator of $\underset{\sim}{\theta}$. The process of constructing such an estimator is described below. Consider the corresponding MMS estimator of $\underset{\sim}{\theta}$ given by

$$\hat{\underset{\sim}{\theta}} = \underset{\sim}{\theta}^* - \left(\frac{1}{n}\frac{\partial \mathcal{J}_n(\underset{\sim}{\theta})}{\partial \underset{\sim}{\theta}}\bigg|_{\underset{\sim}{\theta}=\underset{\sim}{\theta}^*}\right)^{-1}\left(\frac{1}{n}\mathcal{J}_n(\underset{\sim}{\theta}^*)\right), \tag{2}$$

where $\left(\left.\frac{\partial \mathcal{J}_n(\underset{\sim}{\theta})}{\partial \underset{\sim}{\theta}}\right|_{\underset{\sim}{\theta}=\underset{\sim}{\theta}^*}\right)^{-1}$ is the matrix reciprocal of $\left\|\frac{\partial J_n^{(a)}(\underset{\sim}{\theta})}{\partial \theta_b}\right\|$, $a, b = 1, 2, 3, 4$.

Summarizing and applying [3] we have the following limiting result.

Proposition. *Let the d.f.* $F(x; \underset{\sim}{\theta})$ *have more than k points of growth. Let* $t_1, \ldots, t_k$ *be positive numbers such that* (i) *the functions* $\varphi(t_j; \underset{\sim}{\theta})$, $j = 1, \ldots, k$, $\varphi(t_p + t_q; \underset{\sim}{\theta})$, $p \neq q$, $p, q = 1, \ldots, k$ *are continuous w.r.t.* $\underset{\sim}{\theta}$ *and* $\varphi(t_j; \underset{\sim}{\theta})$, $j = 1, \ldots, k$, *have continuous w.r.t.* $\underset{\sim}{\theta}$ *partial derivatives of the first and second order;* (ii) $\det I(\underset{\sim}{\theta}; C) \neq 0$.

Then $\hat{\underset{\sim}{\theta}} \xrightarrow{P} \underset{\sim}{\theta}$ *as* $n \to \infty$, *and the random vector* $\sqrt{n}\left(\hat{\underset{\sim}{\theta}} - \underset{\sim}{\theta}\right)$ *is asymptotically normal* $\mathcal{N}(O, I^{-1}(\underset{\sim}{\theta}; C))$.

The ch.f. (1) has continuous partial derivatives w.r.t. $\underset{\sim}{\theta}$ up to the second order:

$$\frac{\partial \varphi(t; \underset{\sim}{\theta})}{\partial \alpha} = \begin{cases} \exp\{\lambda[it\gamma - |t|^\alpha + it(|t|^{\alpha-1} - 1)\beta \tan(\frac{\pi}{2} - \alpha)]\} \cdot \{-\lambda |t|^\alpha \ell n|t| \\ +i\lambda\beta t\{|t|^{\alpha-1}(\ell n|t| \cdot \tan(\frac{\pi}{2}\alpha) + \frac{\pi}{2}\cos^{-2}(\frac{\pi}{2}\alpha)) - \frac{\pi}{2}\cos^{-2}(\frac{\pi}{2}\alpha)]\}, & \text{if } \alpha \neq 1, \\ \exp\{\lambda(it\gamma - |t| - i\frac{2}{\pi}\beta t \ell n|t|)\} \cdot [-\lambda \ell n|t|(|t| + i\beta t \cdot \frac{1}{\pi}\ell n|t|)], & \text{if } \alpha = 1; \end{cases}$$

$$\frac{\partial^2 \varphi(t; \underset{\sim}{\theta})}{\partial \alpha^2} = \begin{cases} \exp\{\lambda[it\gamma - |t|^\alpha + it(|t|^{\alpha-1} - 1)\beta \tan(\frac{\pi}{2}\alpha)]\} \cdot [\{-\lambda|t|^\alpha \ell n|t| \\ +i\lambda\beta t[|t|^{\alpha-1}(\ell n|t| \cdot \tan(\frac{\pi}{2}\alpha) + \frac{\pi}{2}\cos^{-2}(\frac{\pi}{2}\alpha)) \\ -\frac{\pi}{2}\cos^{-2}(\frac{\pi}{2}\alpha)]\}^2 - \lambda|t|^\alpha \ell n^2|t| + i\lambda\beta t\{|t|^{\alpha-1}(\ell n^2|t| \cdot \tan(\frac{\pi}{2}\alpha) \\ +2\pi \ell n|t|/(1 + \cos\pi\alpha) + 2\pi^2 \sin\pi\alpha/(3 + 4\cos\pi\alpha + \cos 2\pi\alpha)) \\ -2\pi^2 \sin\pi\alpha/(3 + 4\cos\pi\alpha + \cos 2\pi\alpha)\}], & \text{if } \alpha \neq 1, \\ \exp\{\lambda(it\gamma - |t| - i\frac{2}{\pi}\beta t \ell n|t|)\} \cdot [\lambda^2 \ell n^2|t|(|t| + i\beta t \cdot \frac{1}{\pi}\ell n|t|)^2 \\ -\lambda|t|\ell n^2|t| + i\lambda\beta t \cdot \frac{2}{3\pi}(\pi^2 \ell n|t| - 2\ell n^3|t|)], & \text{if } \alpha = 1. \end{cases}$$

Let us now construct an auxiliary $\sqrt{n}$-consistent estimator of $\underset{\sim}{\theta}$. We first choose a $t_0 > 0$ such that $|\lambda\gamma t_0 + \lambda t_0(t_0^{\alpha-1} - 1)\beta \tan(\frac{\pi}{2}\alpha)| < \frac{\pi}{2}$ for all $0 < \alpha \leq 2$, $-1 \leq \beta \leq 1$, $-A < \gamma < A$, $0 < \lambda < B$.

Let $0 < \varepsilon < A$. Then since

$$\lim_{\alpha \to 1} \left|\left(t^{\alpha-1} - 1\right) \tan\left(\frac{\pi}{2}\alpha\right)\right| = \frac{2}{\pi}|\ell nt| \qquad \text{for } t > 0,$$

there exists a $\delta > 0$ such that for all $0 < \alpha \leq 2$ and $|\alpha - 1| < \delta$ we have

$$\left|\left(t^{\alpha-1} - 1\right) \tan\frac{\pi}{2}\alpha\right| < \frac{2}{\pi}|\ell nt| + \varepsilon.$$

One can verify now that

$$t_0^{(1)} = \max\left\{t : 0 < t < 1,\ 2ABt + Bt|\ln t| < \frac{\pi}{2}\right\}$$

satisfies the inequality

$$\left|\lambda\gamma t + \lambda t\left(t^{\alpha-1} - 1\right)\beta \tan\left(\frac{\pi}{2}\alpha\right)\right| < \frac{\pi}{2} \tag{3}$$

on the set where $\Theta_1 = \{\alpha \in (1 - \delta, 1 + \delta),\ \beta \in [-1, 1],\ \gamma \in (-A, A),\ \lambda \in (0, B)\}$.

Similarly, we choose $t_0^{(2)}$ such that,

$$0 < t_0^{(2)} < \min\left\{\left(\frac{\pi}{4B\tan\left(\frac{\pi}{2}(1 - \delta)\right)}\right)^{\frac{1}{1-\delta}}, \frac{\pi}{4AB}\right\}.$$

$t_0^{(2)}$ satisfies (3) on the set $\Theta_2 = \{\alpha \in (0, 1-\delta), \beta \in [-1,1], \gamma \in (-A, A), \lambda \in (0, B)\}$.
We now choose $t_0^{(3)} > 0$, such that

$$0 < t_0^{(3)} < \frac{\pi}{2\left[AB + B\tan\left(\frac{\pi}{2}(1-\delta)\right)\right]}$$

and $t_0^{(3)}$ satisfies (3) on the set $\Theta_3 = \{\alpha \in (1+\delta, 2), \beta \in [-1,1], \gamma \in (-A, A), \lambda \in (0, B)\}$.

Define $t_0 = \min\left(t_0^{(1)}, t_0^{(2)}, t_0^{(3)}\right)$. Let $\tau_1, \tau_2 \in (0, t_0)$. Using the method of [1], we construct $\underset{\sim}{\theta}^*$ as,

$$\alpha^* = \ln\left(\frac{\ln(|\varphi_n(\tau_1)|)}{\ln(|\varphi_n(\tau_2)|)}\right) / \ln\left(\frac{\tau_1}{\tau_2}\right)$$

$$\lambda^* = \exp\left\{[\ln\tau_1 \cdot \ln(-\ln|\varphi_n(\tau_2)|) - \ln\tau_2 \cdot \ln(-\ln|\varphi_n(\tau_1)|)] / \ln\frac{\tau_1}{\tau_2}\right\}$$

$$\beta^* = \frac{\frac{1}{\tau_1}\arctan\left(Im\left(\varphi_n(\tau_1)\right)/Re\left(\varphi_n(\tau_1)\right)\right) - \frac{1}{\tau_2}\arctan\left(Im\left(\varphi_n(\tau_2)\right)/Re\left(\varphi_n(\tau_2)\right)\right)}{\left(\tau_1^{\alpha^*-1} - \tau_2^{\alpha^*-1}\right)\lambda^*\tan\left(\frac{\pi}{2}\alpha^*\right)} \qquad (4)$$

$$\gamma^* = \frac{\left(\tau_1^{\alpha^*-1} - 1\right)\frac{1}{\tau_2}\arctan\left(Im\left(\varphi_n(\tau_2)\right)/Re\left(\varphi_n(\tau_2)\right)\right) - \left(\tau_2^{\alpha^*-1} - 1\right)\arctan\frac{Im\left(\varphi_n(\tau_1)\right)}{Re\left(\varphi_n(\tau_1)\right)}}{\left(\tau_1^{\alpha^*-1} - \tau_2^{\alpha^*-1}\right)\lambda^*}$$

It can be seen that the random vector $\sqrt{n}(\underset{\sim}{\theta}^* - \underset{\sim}{\theta})$ is asymptotically normal. Hence, $\underset{\sim}{\theta}^*$ is the $\sqrt{n}$-consistent estimator of $\underset{\sim}{\theta}$. Thus, the following assertion holds.

Theorem. *Let $t_1, \ldots, t_k$ be positive numbers such that* $\det I(\underset{\sim}{\theta}: \mathcal{L}) \neq 0$. *Then the MMS estimator $\hat{\underset{\sim}{\theta}}$ defined by relations (2), (4) possesses the following properties:*

(i) $\hat{\underset{\sim}{\theta}} \xrightarrow{P} \underset{\sim}{\theta}$ *as* $n \to \infty$;

(ii) the random vector $\sqrt{n}(\hat{\underset{\sim}{\theta}} - \underset{\sim}{\theta})$ *is asymptotically normal* $\mathcal{N}(O, I^{-1}(\underset{\sim}{\theta}; \mathcal{L}))$.

Remark 1. Since the covariance matrix of the limit law for the vector $\sqrt{n}(\hat{\underset{\sim}{\theta}} - \underset{\sim}{\theta})$ coincides with the informational bound in the Rao-Cramer inequality for estimators belonging to a linear space, then $\hat{\underset{\sim}{\theta}}$ is asymptotically efficient in the class of $\underset{\sim}{\theta}$ parameter estimators belonging to $\mathcal{L}$.

Remark 2. The ch.f. of stable law is often represented in forms other than the one given by (1). We list two of them below:

A. $$\varphi_A(t; \underset{\sim}{\theta}_A) = \exp\{\lambda_A(it\gamma_A - |t|^{\alpha_A} + itw_A(t; \alpha_A, \beta_A)\},$$

where

$$w_A(t; \alpha_A, \beta_A) = \begin{cases} |t|^{\alpha_A - 1}\beta_A \tan\left(\frac{\pi}{2}\alpha_A\right), & \text{if } \alpha_A \neq 1, \\ -\beta_A \frac{2}{\pi}\ln|t|, & \text{if } \alpha_A = 1. \end{cases}$$

B. $$\varphi_B(t; \underset{\sim}{\theta}_B) = \exp\{\lambda_B(it\gamma_B - |t|^{\alpha_B} \cdot w_B(t; \alpha_B, \beta_B))\},$$

where

$$w_B(t; \alpha_B, \beta_B) = \begin{cases} \exp\left(-i\frac{\pi}{2}\beta_B K(\alpha_B) sign t\right), & \text{if } \alpha_B \neq 1, \\ \frac{\pi}{2} + i\beta_B \ln|t| \cdot sign t, & \text{if } \alpha_B = 1, \end{cases}$$

$$K(\alpha_B) = \alpha_B - 1 + sign(1 - \alpha_B),$$

$\underset{\sim}{\theta}_u = (\alpha_u, \beta_u, \gamma_u, \lambda_u)$, $0 < \alpha_u \le 2$, $-1 \le \beta_u \le 1$, $-\infty < \gamma_u < \infty$, $0 < \lambda_u < \infty$, $u = A, B$.

By making use of the relations between the parameters in various forms, ([6], Introduction) based on the constructed MMS estimator $\hat{\underset{\sim}{\theta}}$ we write down the estimators for $\underset{\sim}{\theta}_A$ and $\underset{\sim}{\theta}_B$ (under the condition that the range of variation of these parametric vectors is identical to the one we have):

$$\hat{\underset{\sim}{\theta}}_A = \left(\hat{\alpha}_A, \hat{\beta}_A, \hat{\gamma}_A, \hat{\lambda}_A\right) : \hat{\alpha}_A = \hat{\theta}_1, \hat{\beta}_A = \hat{\theta}_2,$$
$$\hat{\gamma}_A = \hat{\theta}_3 - \hat{\theta}_2 + \tan\left(\frac{\pi}{2}\hat{\theta}_1\right), \hat{\lambda}_A = \hat{\theta}_4;$$
$$\hat{\underset{\sim}{\theta}}_B = \left(\hat{\alpha}_B, \hat{\beta}_B, \hat{\gamma}_B, \hat{\lambda}_B\right) : \hat{\alpha}_B = \hat{\theta}_1, \hat{\beta}_B = 2\arctan\left(\hat{\theta}_2 \tan\left(\frac{\pi}{2}\hat{\theta}_1\right)\right) / \left(\pi K\left(\hat{\theta}_1\right)\right),$$
$$\hat{\gamma}_B = \left(\hat{\theta}_3 - \hat{\theta}_2 \tan\left(\frac{\pi}{2}\hat{\theta}_1\right)\right) \cos\left[\arctan\left(\hat{\theta}_2 + \tan\left(\frac{\pi}{2}\hat{\theta}_1\right)\right)\right],$$
$$\hat{\lambda}_B = \hat{\theta}_4 / \cos\left[\arctan\left(\hat{\theta}_2 \tan\left(\frac{\pi}{2}\hat{\theta}_1\right)\right)\right].$$

Hence, it is clear that when the problem of estimating $\underset{\sim}{\theta}$ for the forms (A) or (B) is considered, one must ensure that consistent estimators can be constructed only if it is known that $\alpha = 1$ or $\alpha \neq 1$.

REFERENCES

[1] Fielitz, B.D. and Roselle, J.P., 1981, Method-of-moments estimators for stable distribution parameters, *Appl. Math. and Comput.* 8, 4, 303–320.

[2] Kagan, A.M., 1976, Fisher information contained in a finite dimensional linear space and the correct version of the method of moments (in Russian), *Problemy Peredaci Informacii*, 12, 2, 20–42.

[3] Klebanov, L.B. and Melamed, J.A., 1984, On stable estimation of parameters by the modified method of scoring, *Proc. Third Prague Symp. on Asymptotic Statistics*, 347–354.

[4] Mittnik, S. and Rachev, S.T., 1993, Modeling asset returns with alternative stable distributions, to appear in *Econometric Reviews*.

[5] Mittnik, S. and Rachev, S.T., 1993, Reply to comments on "Modeling asset returns with alternative stable distributions" and some extensions, to appear in *Econometric Reviews*.

[6] Zolotarev, V.M., 1986, One-dimensional Stable Distributions, Vol. 65 of Translation of Mathematical Monographs, American Mathematical Society, R.I.

REPRESENTATION AND PROPERTIES OF GEOMETRIC STABLE LAWS

Tomasz J. Kozubowski[1]

Department of Mathematics
University of Tennessee at Chattanooga
Chattanooga, TN 37403

1 INTRODUCTION

A random summation scheme, where the number of terms is geometrically distributed, is called a geometric summation scheme (geometric compound, geometric convolution) (Klebanow et al., 1984). Geometric convolutions naturally arise in many applied probability problems. In particular, they appear in queueing theory and reliability in connection to "regenerating processes with rare events" (Gertsbakh, 1984; Jacobs, 1986). Some recent results suggest that geometric compounds could provide useful models in economics (Kozubowski and Rachev, 1992).

The paper is organized as follows. After giving preliminary definitions in Section 2, we give the representation of densities of geometric stable laws in Section 3. The properties of geometric stable laws are presented in Section 4. Finally, we discuss the problem of unimodality in Section 5.

2 PRELIMINARY DEFINITIONS

We start with the definition of geometric stable random vectors given in Mitnik and Rachev (1991).

Definition 2.1 *A random variable Y with distribution function G is said to be geometric stable with respect to the summation scheme (in short, (geo+)-stable), if there exists a sequence of iid random variables $X_1, X_2, \ldots$, a geometric random variable $T(p)$ independent of all X_i, and constants $a=a(p)>0$ and $b=b(p) \in R$ such that*

$$a(p) \sum_{i=1}^{T(p)} (X_i + b(p)) \xrightarrow{d} Y, \quad as \quad p \to 0. \tag{1}$$

[1]Research partially supported by grant from CECA, University of Tennessee at Chattanooga.

Definition 2.2 *If $b(p)=0$ in* (1), *Y is called strictly geometric stable.*

Definition 2.3 (Domains of attraction) *The distribution H belongs to the domain of attraction of the (geo+)-stable distribution G, ($H \in D(G)$), if* (1) *holds with $X_i \sim H$, $Y \sim G$. ("$\sim$" here means "is distributed as").*

Geometric stable distributions are best characterized by their characteristic functions. We quote the following proposition (Mitnik and Rachev, 1991).

Proposition 2.1 *A nondegenerate distribution function is (geo+)-stable if and only if its ch.f. ψ has the form*

$$\psi(t) = (1 - ln\phi(t))^{-1}, \quad t \in R, \tag{2}$$

where $\phi(t)$ is the ch.f. of some α-stable distribution.

There are several representations of characteristic function of α-stable laws. Using the representation of ch.f. of stable laws given in Hall (1981) leads to the following form of the ch.f. ψ of (geo+)-stable distribution

$$\psi(t) = (1 + \lambda|t|^{\alpha} w_A(t,\alpha,\beta) - i\mu t)^{-1}, \qquad (A)$$

where

$$w_A(t,\alpha,\beta) = \begin{cases} 1 - i\beta\, tan(\pi\alpha/2) sgn(t) & \text{if } \alpha \neq 1 \\ 1 + i\beta(2/\pi) ln|t| sgn(t) & \text{if } \alpha = 1, \end{cases}$$

and

$$0 < \alpha \leq 2, \qquad -1 \leq \beta \leq 1, \qquad -\infty < \mu < \infty, \qquad \lambda \geq 0. \tag{3}$$

Often λ is written as $\lambda = \sigma^{\alpha}$. The reason is that for $\alpha \neq 1$, stable random variables form a location-scale family, where σ is a scale parameter and μ is a location parameter. An alternative representation of stable laws (Devroye, 1986) leads to the following form of the ch.f. ψ of (geo+)-stable distributions

$$\psi(t) = (1 + \lambda|t|^{\alpha} w_B(t,\alpha,\beta) - i\mu t)^{-1}, \qquad (B)$$

where

$$w_B(t,\alpha,\beta) = \begin{cases} exp(-i\beta(\pi/2)K(\alpha) sgn(t)) & \text{if } \alpha \neq 1 \\ 1 + i\beta(2/\pi) ln|t| sgn(t) & \text{if } \alpha = 1, \end{cases}$$

with $K(\alpha) = \min(\alpha, 2-\alpha)$ and $\alpha, \beta, \lambda, \mu$ satisfy (3).

If a random variable Y has ch.f. ψ defined by (A) or (B) above, we say that it is (geo+)-stable with parameters $(\alpha, \beta, \lambda, \mu)$ (or $(\alpha, \beta, \sigma, \mu)$ if σ is used instead of λ). The parameters in forms (A) and (B) are related by the following formulas

$$\alpha_A = \alpha_B = \alpha, \qquad \mu_A = \mu_B = \mu.$$

If $\alpha \neq 1$, then

$$\beta_A = cot(\pi\alpha/2) tan(\pi/2K(\alpha)\beta_B),$$

and

$$\lambda_A = \lambda_B cos(\pi/2K(\alpha)\beta_B).$$

If $\alpha = 1$, then

$$\beta_A = \beta_B, \qquad \lambda_A = \lambda_B.$$

The most important parameter is α. We will call it the index of the (geo+)-stable law. The parameters μ and σ correspond to location and scale; β determines the skewness of the distribution. If in either form, μ=0 for $\alpha \neq 1$ or β=0 for α=1, the distribution is *strictly* stable. This corresponds to the strict stability of the α-stable distribution appearing in (2). If μ=0 and λ=1 (in either form), the distribution will be called *standard.*

The distribution function of a (geo+)-stable law will be denoted by $GS(x,\alpha,\beta,\lambda,\mu)$ (in form (A)) or $GS^*(x,\alpha,\beta,\lambda,\mu)$, (in form (B)). For the standard case we shall use $GS(x,\alpha,\beta)$ and $GS^*(x,\alpha,\beta)$ respectivelly. Similar expressions with "S" instead of "GS" will correspond to the distribution functions of α-stable laws. The lower-case letters will denote the densities (e.g. $gs(x,\alpha,\beta)$ is the density of the standard (geo+)-stable law in form (A)).

3 REPRESENTATION OF GEOMETRIC STABLE LAWS

Before giving representation of geometric stable laws, we need to establish two lemmas.

Lemma 3.1 *Let G, H be distribution functions on R and g, ψ be the corresponding characteristic functions. Let F(x,y) be a distribution function in x for every y, which is Borel measurable in y for every x. For all y, let $f_y(t)$ be the characteristic function corresponding to F(x,y), measurable in y for every t. Then*

$$H(x) = \int_{-\infty}^{\infty} F(x,y)dG(y) \tag{4}$$

if and only if

$$\psi(t) = \int_{-\infty}^{\infty} f_y(t)dG(y)\,. \tag{5}$$

Proof: Assume that (4) holds. In Robbins (1948) it was shown that under the above conditions

$$\int_{-\infty}^{\infty} f(x)d\,(\int_{-\infty}^{\infty} F(x,y)dG(y)) = \int_{-\infty}^{\infty}(\int_{-\infty}^{\infty} f(x)d_xF(x,y))dG(y)$$

for every bounded f. Therefore, since $f(x) = e^{itx}$ is bounded, we get

$$\psi(t) = \int_{-\infty}^{\infty} e^{itx}dH(x) = \int_{-\infty}^{\infty} e^{itx}d\,(\int_{-\infty}^{\infty} F(x,y)dG(y))$$

$$= \int_{-\infty}^{\infty}(\int_{-\infty}^{\infty} e^{itx}d_xF(x,y))dG(y) = \int_{-\infty}^{\infty} f_y(t)dG(y)\,,$$

where $f_y(t) = \int_{-\infty}^{\infty} e^{itx}d_xF(x,y)$ is the characteristic function corresponding to the distribution function F($\cdot$,y). Hence (5) holds.

Conversely, assume that (5) holds. Now (4) follows, since characteristic functions uniquely determine distributions; if

$$H_1(x) = \int_{-\infty}^{\infty} F(x,y)dG(y)\,,$$

then H_1 is a distribution function, as shown by Robbins (1948, Theorem 4). Let ψ_1 be the corresponding characteristic function. Then by the previous part we get

$$\psi_1(t) = \int_{-\infty}^{\infty} f_y(t)dG(y)\,.$$

Consequently, by (5) $\psi_1(t) = \psi(t)$ for all t, and thus $H_1(x) = H(x)$ for all x. Hence (4) follows.□

Lemma 3.2 *Let the assumptions of Lemma 3.1 hold and in addition assume that for all y $F(\cdot,y)$ is an absolutely continuous distribution function with density $f(\cdot,y)$, a measurable function on R^2. Then, H is also an absolutely continuous distribution function with a density $h(\cdot)$ given by*

$$h(x) = \int_{-\infty}^{\infty} f(x,y)dG(y) \qquad (a.e.) .$$

Proof: Again let ψ denote the characteristic function corresponding to H. Then

$$\psi(t) = \int_{-\infty}^{\infty} (\int_{-\infty}^{\infty} e^{itx} d_x F(x,y))dG(y) = \int_{-\infty}^{\infty}\int_{-\infty}^{\infty} e^{itx} f(x,y)dxdG(y)$$

$$\stackrel{1}{=} \int_{-\infty}^{\infty}\int_{-\infty}^{\infty} e^{itx} f(x,y)dG(y)dx = \int_{-\infty}^{\infty} e^{itx} h(x)dx ,$$

where

$$h(x) = \int_{-\infty}^{\infty} f(x,y)dG(y) .$$

To justify 1, we note that Fubini's theorem can be applied, since $S(x,y) = e^{itx} f(x,y)$ is a measurable function in R^2 and

$$\int_{-\infty}^{\infty}\int_{-\infty}^{\infty} |S(x,y)|dxdG(y) = \int_{-\infty}^{\infty}\int_{-\infty}^{\infty} f(x,y)dxdG(y) = \int_{-\infty}^{\infty} dG(y) = 1 < \infty .$$

It is a straightforward application of Fubini's theorem to show that the integral of $h(\cdot)$ defined above is 1, and so $h(\cdot)$ is valid probability density. Since $\psi(t) = \int_{-\infty}^{\infty} e^{itx} h(x)dx$, by the uniqueness theorem, h(·) corresponds to the characteristic function ψ and the distribution function H. Lemma 3.2 is thus proved. □

The application of the above lemmas to the case of geometric stable laws leads to the following Corollary, that provides the representation of densities and distribution functions of geometric stable laws.

Corollary 3.1 *Every (geo+)-stable distribution is absolutely continuous with the following representation of the distribution function GS and the density gs (in the case $\sigma \neq 0$)*

For $\alpha \neq 1$

$$GS(x,\alpha,\beta,\sigma,\mu) = \int_0^{\infty} S(\frac{x-\mu z}{\sigma z^{1/\alpha}},\alpha,\beta)e^{-z}dz , \tag{6}$$

$$gs(x,\alpha,\beta,\sigma,\mu) = \int_0^{\infty} \sigma^{-1} z^{-1/\alpha} s(\frac{x-\mu z}{\sigma z^{1/\alpha}},\alpha,\beta)e^{-z}dz . \tag{7}$$

For $\alpha = 1$

$$GS(x,\alpha,\beta,\sigma,\mu) = \int_0^{\infty} S(\frac{x - z\sigma\beta(2/\pi)ln(\sigma z) - \mu z}{\sigma z},\alpha,\beta)e^{-z}dz , \tag{8}$$

$$gs(x,\alpha,\beta,\sigma,\mu) = \int_0^{\infty} \sigma^{-1} z^{-1} s(\frac{x - z\sigma\beta(2/\pi)ln(\sigma z) - \mu z}{\sigma z},\alpha,\beta)e^{-z}dz . \tag{9}$$

Remark The above formulas use representation (A) of the (geo+)-stable laws where $\sigma = \lambda^{1/\alpha}$. Using the representation (B) we get exactly the same result, where GS is replaced by GS^*.

Proof: Let ψ be the ch.f. of a (geo+)-stable distribution given by form (A) and let ϕ be the ch.f. of the corresponding stable law, given by Theorem 2.1.
Consider $\alpha \neq 1$. Then

$$\phi(t) = exp(i\mu t - \sigma^\alpha |t|^\alpha w_A(t, \alpha, \beta))$$

and so

$$[\phi(t)]^z = exp(i\mu zt - (z^{1/\alpha}\sigma)^\alpha |t|^\alpha w_A(t, \alpha, \beta)) .$$

But the above corresponds to the stable distribution Y with parameters $(\alpha, \beta, z^{1/\alpha}\sigma, \mu z)$, where $z^{1/\alpha}\sigma$ is the scale and μz is the location parameter, so that Y can be written as

$$Y \stackrel{d}{=} \mu z + z^{1/\alpha}\sigma X ,$$

where X$\sim$ $S(x, \alpha, \beta)$ (standard stable random variable). Therefore, the distribution function of Y can be written as

$$F_Y(x) = F_X(\frac{x - \mu z}{\sigma z^{1/\alpha}}) = S(\frac{x - \mu z}{\sigma z^{1/\alpha}}, \alpha, \beta) ,$$

and we get (6) by the application of Lemma 3.1 with $G(z) = 1 - e^z$. Since all stable random variables have densities, the application of Lemma 3.2 yields (6).
Let α=1. Now

$$\phi(t) = exp(i\mu t - \sigma|t|w_A(t, \alpha, \beta)) ,$$

where $w_A(t, \alpha, \beta) = 1 + i\beta(2/\pi)ln|t|sgn(t)$. Let Y be a random variable with ch.f. ϕ. It easily follows by examining the above characteristic functions that Y is related to the standard stable random variable $X \sim S(x, 1, \beta)$ by the following formula

$$Y \stackrel{d}{=} \mu + \sigma X + \sigma\beta(2/\pi)ln(\sigma) .$$

Since

$$[\phi(t)]^z = exp(i\mu zt - z\sigma|t|^\alpha w_A(t, \alpha, \beta)) ,$$

then the above gives us

$$F_Y(x) = F_X(\frac{x - z\sigma\beta(2/\pi)ln(\sigma z) - \mu z}{\sigma z})$$

$$= S(\frac{x - z\sigma\beta(2/\pi)ln(\sigma z) - \mu z}{\sigma z}, \alpha, \beta) .$$

Hence, again by the application of Lemma 3.1 and Lemma 3.2 with $G(z) = 1 - e^z$ we get (8) and (9). □

Writing (6) - (9) in terms of random variables, we get the following representation of (geo+)-stable laws.

Corollary 3.2 *Let* $Y\sim$ $GS(x, \alpha, \beta, \sigma, \mu)$ *((geo+)-stable),* $X\sim$ $S(x, \alpha, \beta)$ *(standard α-stable),* $Z\sim$ $exp(1)$ *(exponential distribution with mean 1). Let X and Z be independent. Then*

$$\begin{cases} Y \stackrel{d}{=} \mu Z + Z^{1/\alpha}\sigma X, & \alpha \neq 1 , \\ Y \stackrel{d}{=} \mu Z + Z\sigma X + \sigma Z\beta(2/\pi)ln(Z\sigma), & \alpha = 1 . \end{cases}$$

Remark: The above Corollaries are also valid if ψ is the characteristic function of a (geo+)stable law in form (B). Using the representation (B) we get exactly the same results, where GS is replaced by GS^*.

4 PROPERTIES OF GEOMETRIC STABLE LAWS

The theory and properties of α-stable laws are well known (Zolotarev, 1986). Using the representation discussed in the previous section, we establish analogous properties of geometric stable laws.

Theorem 4.1 *For any admissible quadruple $(\alpha, \beta, \sigma, \mu)$, where $\sigma \neq 0$*

$$GS(-x, \alpha, \beta, \sigma, \mu) = 1 - GS(x, \alpha, -\beta, \sigma, -\mu),$$

and

$$gs(-x, \alpha, \beta, \sigma, \mu) = gs(x, \alpha, -\beta, \sigma, -\mu) .$$

Proof: As shown in Zolotarev (1986), the above is true for the α-stable laws

$$S(-x, \alpha, \beta) = 1 - S(x, \alpha, -\beta) \quad \text{and} \quad s(-x, \alpha, \beta) = s(x, \alpha, -\beta).$$

Now, using Corollary 3.1 we get for $\alpha \neq 1$

$$\begin{aligned} GS(-x, \alpha, \beta, \sigma, \mu) &= \int_0^\infty S(\frac{-x - \mu z}{\sigma z^{1/\alpha}}, \alpha, \beta) e^{-z} dz \\ &= \int_0^\infty [1 - S(\frac{x + \mu z}{\sigma z^{1/\alpha}}, \alpha, -\beta)] e^{-z} dz \\ &= 1 - \int_0^\infty S(\frac{x - (-\mu) z}{\sigma z^{1/\alpha}}, \alpha, -\beta) e^{-z} dz \\ &= 1 - GS(x, \alpha, -\beta, \sigma, -\mu) . \end{aligned}$$

The proof for the case α=1 is analogous. The equation for the densities immediately follows. □

Due to the above Theorem it is enough to consider the case x >0 when examining the properties of (geo+)-stable laws.

Theorem 4.2 *Suppose that Y is a (geo+)-stable random variable with parameters $(\alpha, \beta, \lambda, \mu)$ in form (A), where α >1 and $\lambda > 0$. Then the density of Y admits the following representation*

$$gs(x, \alpha, \beta, \lambda, \mu) = \frac{1}{\pi} \int_0^\infty \frac{(1 + \lambda t^\alpha) cos(tx) + (t\mu + \lambda \eta t^\alpha) sin(tx)}{(1 + \lambda t^\alpha)^2 + (t\mu + \lambda \eta t^\alpha)^2} dt,$$

where $\eta = \beta \tan(\pi\alpha/2)$.

Proof: Let ψ be the ch.f. of Y given in form (A). Then

$$|1/\psi(t)|^2 = |1 + \lambda |t|^\alpha (1 - i\beta \tan(\pi\alpha/2) sgn(t)) - i\mu t|$$

$$= (1 + \lambda|t|^\alpha)^2 + (\lambda |t|^\alpha \beta \tan(\pi\alpha/2) sgn(t) + \mu t)^2 \geq (1 + \lambda|t|^\alpha)^2 \geq 1.$$

Therefore

$$|\psi(t)| \leq \frac{1}{1 + \lambda|t|^\alpha} .$$

Now, for α >1 and $\lambda > 0$, $|\psi(t)| \in L_1(R)$ and so the inversion formula can be applied:

$$gs(x, \alpha, \beta, \lambda, \mu) = \frac{1}{2\pi} \int_{-\infty}^\infty e^{-itx} \psi(t) dt = \frac{1}{2\pi} \int_{-\infty}^\infty Re[e^{-itx} \psi(t)] dt.$$

Since

$$Re[e^{-itx}\psi(t)] = Re[(cos(tx) - isin(tx))(1 + \lambda|t|^{\alpha}(1 - i\eta\, sgn(t)) - i\mu\, t)^{-1}]$$

$$= \frac{(1 + \lambda|t|^{\alpha})cos(tx) + (t\mu + \lambda\eta|t|^{\alpha}sgn(t))sin(tx)}{(1 + \lambda|t|^{\alpha})^2 + (t\mu + \lambda\eta|t|^{\alpha}sgn(t))^2}$$

and the above is an even function of t, we get the result. □

It is a well known fact, that the densities of all α-stable laws are uniformly bounded, as shown in Zolotarev (1986). The next theorem shows that the similar property holds for (geo+)-stable laws when $\alpha > 1$.

Theorem 4.3 *Let Y be a (geo+)-stable random variable with density $gs(x,\alpha,\beta,\sigma,\mu)$, where $\alpha > 1$ and $\sigma > 0$. Then for any x*

$$gs(x,\alpha,\beta,\sigma,\mu) \le \frac{\Gamma(1/\alpha)\Gamma(1 - 1/\alpha)}{\alpha\sigma\pi} .$$

Proof: Let ϕ be the ch.f. of the stable random variable corresponding to Y. Now we can write

$$|\phi(t)| = |exp[i\mu\, t - \sigma^{\alpha}|t|^{\alpha}(1 - i\beta\, tan(\pi\alpha/2)sgn(t))]| = e^{-\sigma^{\alpha}|t|^{\alpha}} .$$

Combining the above with the inversion formula, we get the following inequality for the density of the stable law

$$s(x,\alpha,\beta,\sigma,\mu) = \frac{1}{2\pi}\int_{-\infty}^{\infty} e^{-itx}\phi(t)dt \le \frac{1}{2\pi}\int_{-\infty}^{\infty} |e^{-itx}\phi(t)|dt$$

$$\le \frac{1}{2\pi}\int_{-\infty}^{\infty} e^{-\sigma^{\alpha}|t|^{\alpha}}dt = \frac{1}{\pi}\int_{0}^{\infty} e^{-\sigma^{\alpha}t^{\alpha}}dt = \frac{1}{\pi}\int_{0}^{\infty} \frac{1}{\sigma\alpha}z^{1/\alpha-1}e^{-z}dz = \frac{\Gamma(1/\alpha)}{\alpha\sigma\pi} .$$

Using Corollary 3.1 we get

$$gs(x,\alpha,\beta,\sigma,\mu) = \int_{0}^{\infty} \sigma^{-1}z^{-1/\alpha}s(\frac{x - \mu\, z}{\sigma\, z^{1/\alpha}},\alpha,\beta)e^{-z}dz$$

$$\le \int_{0}^{\infty} \sigma^{-1}z^{-1/\alpha}\frac{\Gamma(1/\alpha)}{\alpha\pi}e^{-z}dz = \frac{\Gamma(1/\alpha)}{\alpha\sigma\pi}\int_{0}^{\infty} z^{(-1/\alpha)}e^{-z}dz = \frac{\Gamma(1/\alpha)\Gamma(1 - 1/\alpha)}{\alpha\sigma\pi}$$

as desired. □

Note: If we use the representation (B) of (geo+)-stable laws, then we get the same bound for $gs^*(x,\alpha,\beta,\sigma,\mu)$, but only in the case $x < 0$, $\mu \le 0$. Otherwise, the bound is slightly different

$$gs^*(x,\alpha,\beta,\sigma,\mu) \le \frac{\Gamma(1/\alpha)\Gamma(1 - 1/\alpha)}{\alpha\sigma\pi(cos(\gamma))^{1/\alpha}} ,$$

where $\gamma = (\pi/2)min(\alpha, 2 - \alpha)\beta$.

The previous Theorem described the bounds for densities of (geo+)-stable laws in the case $\alpha > 1$. For $\alpha < 1$, they are in fact unbounded.

Theorem 4.4 *Let Y be a standard (geo+)-stable distributed random variable with parameters (α,β) given in form (B). Then, for $\alpha < 1$ and $\alpha = 1, \beta = 0$, the density gs^* of Y is unbounded at 0.*

Proof:
(i) Let $\alpha < 1$, $|\beta| \neq 1$. Then, as shown in Zolotarev (1986), the corresponding stable random variable X with density s^*, is concentrated on the whole real line with $s^*(x,\alpha,\beta) \neq 0$ for any x. Consequently, using the representation of (geo+)-stable densities and Fatou's Lemma, we have

$$\liminf_{x\to 0} gs^*(x,\alpha,\beta) = \liminf_{x\to 0} \int_0^\infty z^{-1/\alpha} s^*(\frac{x}{z^{1/\alpha}},\alpha,\beta)e^{-z}dz$$

$$\geq \int_0^\infty z^{-1/\alpha} \liminf_{x\to 0} s^*(\frac{x}{z^{1/\alpha}},\alpha,\beta)e^{-z}dz\ .$$

By the continuity of s^*, the above equals

$$\int_0^\infty z^{-1/\alpha} s^*(0,\alpha,\beta)e^{-z}dz = s^*(0,\alpha,\beta)\int_0^\infty z^{-1/\alpha}e^{-z}dz = \infty.$$

If β=1, then Y is a positive random variable, since $Y \stackrel{d}{=} Z^{1/\alpha}X$ and X is positive for $\alpha < 1, \beta = 1$, as discussed in Zolotarev (1986). Such distributions were considered in Gnedenko (1970), where it was shown that the densities are unbounded. By symmetry, we get the result for $\alpha < 1, \beta = -1$.
(ii) Finally, in the case $\alpha = 1, \beta = 0$, we proceed in exactly the same way as in (i), using Fatou's Lemma. □

The next property gives another useful bound for the density gs^* of a standard (geo+)-stable random variable.

Theorem 4.5 *Let Y be a standard (geo+)-stable random variable with parameters (α,β) given in form (B). Then*

$$gs^*(x,\alpha,\beta) \leq \begin{cases} \Gamma(\alpha+1)/[\pi(x\cos\eta)^{\alpha+1}], & x>0, \\ \Gamma(\alpha+1)/[\pi(-x\cos\theta)^{\alpha+1}], & x<0, \end{cases}$$

where

$$\begin{aligned} \theta &= max(0,\pi/2+(-\gamma-\pi/2)/\alpha), \quad \eta = max(0,\pi/2+(\gamma-\pi/2)/\alpha)\ , \\ \gamma &= (\pi/2)K(\alpha)\beta. \end{aligned}$$

Proof: Since the density $s^*(x,\alpha,\beta)$ of the corresponding stable random variable also satisfies the above inequality (Devroye, 1986), taking $x > 0$ and using the the basic representation of (geo+)-stable densities, we have

$$\begin{aligned} gs^*(x,\alpha,\beta) &= \int_0^\infty z^{-1/\alpha} s^*(\frac{x}{z^{1/\alpha}},\alpha,\beta)e^{-z}dz \\ &\leq \int_0^\infty z^{-1/\alpha}\frac{\Gamma(\alpha+1)}{\pi(\cos\eta)^{\alpha+1}}(\frac{z^{1/\alpha}}{x})^{\alpha+1}e^{-z}dz \\ &= \frac{\Gamma(\alpha+1)}{\pi(\cos\eta)^{\alpha+1}}x^{-\alpha-1}\int_0^\infty ze^{-z}dz = \frac{\Gamma(\alpha+1)}{\pi(\cos\eta)^{\alpha+1}}x^{-\alpha-1}. \end{aligned}$$

The proof for $x < 0$ is analogous. □

Although we do not have explicit representations of the distribution function and density of (geo+)-stable laws, they can be expressed very simply at the origin.

Theorem 4.6 *Let Y be a standard (geo+)-stable random variable. Then the following is true*

$$gs^*(0,\alpha,\beta) = \Gamma(1+1/\alpha)\Gamma(1-1/\alpha)cos(\frac{\pi\beta\,K(\alpha)}{2\alpha})/\pi, \quad \alpha > 1,$$

$$GS^*(0,\alpha,\beta) = 1/2(1-\beta\,min(1,\frac{2-\alpha}{\alpha})), \qquad \alpha \neq 1\,.$$

Proof: If $\alpha \neq 1$, then it was shown in Zolotarev (1986) that

$$s^*(0,\alpha,\beta) = \Gamma(1+1/\alpha)cos(\frac{\pi\beta\,K(\alpha)}{2\alpha})/\pi\,,$$

and

$$S^*(0,\alpha,\beta) = 1/2(1-\beta\,min(1,\frac{2-\alpha}{\alpha}))\,.$$

Therefore

$$\begin{aligned} gs^*(0,\alpha,\beta) &= \int_0^\infty z^{-1/\alpha}s^*(0,\alpha,\beta)e^{-z}dz \\ &= \int_0^\infty z^{-1/\alpha}\Gamma(1+1/\alpha)cos(\frac{\pi\beta\,K(\alpha)}{2\alpha})/\pi\, e^{-z}dz \\ &= \Gamma(1+1/\alpha)\Gamma(1-1/\alpha)cos(\frac{\pi\beta\,K(\alpha)}{2\alpha})/\pi\,. \end{aligned}$$

For the distribution function, we proceed in exactly the same way

$$GS^*(0,\alpha,\beta) = \int_0^\infty S^*(0,\alpha,\beta)e^{-z}dz = S^*(0,\alpha,\beta)\int_0^\infty e^{-z}dz$$

$$= 1/2(1-\beta\,min(1,\frac{2-\alpha}{\alpha}))\int_0^\infty e^{-z}dz = 1/2(1-\beta\,min(1,\frac{2-\alpha}{\alpha}))\,.$$

The Theorem is thus proved. □

Theorem 4.7 *Let $\alpha \neq 1$. The (geo+)-stable density gs^* can be expanded for values $x > 0$ as follows*

$$\forall n > 0 \quad gs^*(x,\alpha,\beta) = \sum_j^n c_j x^{-j\alpha-1} + C^*_{n+1}(x)\,, \tag{10}$$

where

$$c_j = (-1)^{j-1}\Gamma(j\alpha+1)sin[j(\alpha\pi/2+\gamma)]/\pi\,,$$

$$|C^*_{n+1}(x)| \leq \frac{\Gamma(\alpha(n+1)+1)}{\pi(x\,cos\theta)^{\alpha(n+1)+1}}\,,$$

$$\gamma = \pi\,K(\alpha)\beta/2 = \pi\beta\,min(\alpha,2-\alpha)/2\,,$$

$$\theta = max(0,\pi/2+(\gamma-\pi/2)/\alpha)\,.$$

Proof: Under the same assumptions, we have the following representation of the α-stable density s for $x > 0$ (Devroye, 1986)

$$\forall n > 0 \quad s^*(x,\alpha,\beta) = \sum_j^n b_j x^{-j\alpha-1} + B^*_{n+1}(x)\,, \tag{11}$$

where

$$b_j = (-1)^{j-1}\Gamma(j\alpha+1)sin[j(\alpha\pi/2+\gamma)]/(n!\pi),$$

$$|B^*_{n+1}(x)| \leq \frac{\Gamma(\alpha(n+1)+1)}{\pi\,(n+1)!(x\,cos\theta)^{\alpha(n+1)+1}},$$

and γ, θ are the same as above. Using the basic representation of the (geo+)-stable densities given in Corollary 3.1 and the above, we have

$$\begin{aligned}|C^*_{n+1}(x)| &= |gs^*(x,\alpha,\beta) - \sum_j^n c_j x^{-j\alpha-1}| \\ &= |\int_0^\infty z^{-1/\alpha} s^*(\frac{x}{z^{1/\alpha}},\alpha,\beta)e^{-z}dz - \sum_j^n c_j x^{-j\alpha-1}|.\end{aligned}$$

Since $c^j = b^j j!$ and $j! = \Gamma(j+1) = \int_0^\infty z^{j+1}e^{-z}dz$, the above equals

$$\begin{aligned}&= |\int_0^\infty z^{-1/\alpha}s^*(\frac{x}{z^{1/\alpha}},\alpha,\beta)e^{-z}dz - \sum_j^n b_j x^{-j\alpha-1}\int_0^\infty z^{j+1}e^{-z}dz| \\ &= |\int_0^\infty z^{-1/\alpha}s^*(\frac{x}{z^{1/\alpha}},\alpha,\beta)e^{-z}dz - \int_0^\infty\sum_j^n b_j z^{j+1/\alpha}x^{-j\alpha-1}z^{-1/\alpha}e^{-z}dz| \\ &= |\int_0^\infty z^{-1/\alpha}e^{-z}(s^*(\frac{x}{z^{1/\alpha}},\alpha,\beta) - \sum_j^n b_j(\frac{x}{z^{-1/\alpha}})^{-j\alpha-1})dz| \\ &\leq \int_0^\infty z^{-1/\alpha}e^{-z}|B^*_{n+1}(\frac{x}{z^{1/\alpha}})|dz.\end{aligned}$$

By (11)

$$|B^*_{n+1}(\frac{x}{z^{1/\alpha}})| \leq Q(\frac{x}{z^{1/\alpha}})^{-\alpha(n+1)-1},$$

where

$$Q = \frac{\Gamma(\alpha(n+1)+1)}{\pi\,(n+1)!(cos\theta)^{\alpha(n+1)+1}}.$$

Therefore

$$\begin{aligned}|C^*_{n+1}(x)| &\leq \int_0^\infty z^{-1/\alpha}e^{-z}Q(\frac{x}{z^{1/\alpha}})^{-\alpha(n+1)-1}dz \\ &= Qx^{-\alpha(n+1)-1}\int_0^\infty z^{n+1}e^{-z}dz = \Gamma(n+2)Qx^{-\alpha(n+1)-1} \\ &= (n+1)!Qx^{-\alpha(n+1)-1} = \frac{\Gamma(\alpha(n+1)+1)}{\pi(x\,cos\theta)^{\alpha(n+1)+1}},\end{aligned}$$

as desired. □

Remarks:

(1) The above series is not convergent. It gives an asymptotic expansion at $x \to \infty$, since for fixed n, $|C^*_{n+1}(x)| \to 0$, as $x \to \infty$.
(2) The summation in (10) becomes identically 0 when $\alpha = 2$ or $\beta = -1$ and $\alpha < 1$ or $\beta = 1$ and $1 < \alpha < 2$. The right tails of these distributions decrease faster then the powers of x. In fact, for $\beta = -1$ and $\alpha < 1$, $gs^*(x,\alpha,\beta) \equiv 0$, since the distribution is concentrated on the negative half-line.
(3) Using the Theorem 4.1, we get the expansions for the densities for $x \to -\infty$, by replacing β by $-\beta$.

By integrating the series in the Theorem 4.7, we get the following theorem.

Theorem 4.8 *Let $\alpha \neq 1$. The (geo+)-stable distribution function GS^* can be expanded for values $x > 0$ as follows*

$$1 - GS^*(x, \alpha, \beta) = \sum_j^n d_j x^{-j\alpha} + D^*_{n+1}(x) , \qquad (12)$$

where

$$d_j = (-1)^{j-1}\Gamma(j\alpha) sin[j(\alpha\pi/2 + \gamma)]/\pi ,$$

$$|D^*_{n+1}(x)| \leq \frac{\Gamma(\alpha(n+1))}{\pi(x\, cos\theta)^{\alpha(n+1)}} ,$$

$$\gamma = \pi K(\alpha)\beta/2 = \pi\beta\, min(\alpha, 2-\alpha)/2 ,$$

$$\theta = max(0, \pi/2 + (\gamma - \pi/2)/\alpha) .$$

Proof: Using the notation from the previous Theorem, we have: $c_j = j\alpha\, d_j$. By representing GS^* according to the Corollary 3.1, we have

$$\begin{aligned} |D^*_{n+1}(y)| &= |1 - GS^*(y, \alpha, \beta) - \sum_j^n d_j x^{-j\alpha}| \\ &= |\int_y^\infty gs^*(x, \alpha, \beta) - \sum_j^n d_j(\alpha j) \int_y^\infty x^{-j\alpha-1} dx| \\ &= |\int_y^\infty (gs^*(x, \alpha, \beta) - \sum_j^n c_j x^{-j\alpha-1}) dx \\ &= \leq \int_y^\infty |gs^*(x, \alpha, \beta) - \sum_j^n c_j x^{-j\alpha-1}| dx = \int_y^\infty |C^*_{n+1}(x)| dx . \end{aligned}$$

Now, from the previous Theorem,

$$|C^*_{n+1}(x)| \leq P x^{-(n+1)\alpha-1}, \text{ where } P = \frac{\Gamma(\alpha(n+1)+1)}{\pi(cos\theta)^{\alpha(n+1)+1}} .$$

Consequently

$$\begin{aligned} |D^*_{n+1}(y)| &\leq \int_y^\infty P x^{-(n+1)\alpha-1} dx = \frac{P}{\alpha(n+1)} y^{-\alpha(n+1)} \\ &= \frac{\Gamma(\alpha(n+1)+1)}{\pi(cos\theta)^{\alpha(n+1)+1}} \frac{1}{\alpha(n+1)} y^{-\alpha(n+1)} = \frac{\Gamma(\alpha(n+1))}{\pi(y\, cos\theta)^{\alpha(n+1)}} , \end{aligned}$$

as desired. □

Corollary 4.1 *The distribution function of a standard (geo+)-stable random variable Y with parameters (α, β), where $\alpha \neq 1$, has the following asymptotic behavior at $x \to \infty$*

$$1 - GS^*(x, \alpha, \beta) = \begin{cases} o(x^{-\alpha}) & if\ \alpha = 2, \\ o(x^{-\alpha}) & if\ \alpha \in (1, 2), \beta = 1, \\ 0 & if\ \alpha < 1 \beta = 1, \\ d_1 x^{-\alpha} + o(x^{-\alpha}) & otherwise. \end{cases}$$

Proof: For $\alpha = 2$, Y has a Laplace distribution with exponential tails, so we get the first equality. If $\alpha < 1$, $\beta = -1$, the support of Y is the negative half-axis, so that $GS^*(x, \alpha, \beta)=0$ for any $x > 0$, and the third equality follows. The remaining equalities are a consequences of the previous Theorem. □

The next property identifies parameter β_A as a measure of the departure of (geo+)-stable distributions from symmetry.

Theorem 4.9 *Let Y be a standard (geo+)-stable distributed random variable, with $\alpha \neq 1$. Then the following is true*

$$\lim_{x\to\infty} \frac{1 - GS^*(x,\alpha,\beta_B) - GS^*(-x,\alpha,\beta_B)}{1 - GS^*(x,\alpha,\beta_B) + GS^*(-x,\alpha,\beta_B)} = \beta_A \,. \tag{13}$$

Proof: If $\alpha = 2$, then the parameter β is irrelevant, since the ch.f. of Y has the form: $\psi(t) = 1/(1+t^2)$. If we agree to set β=0 in this case (which is natural, since the distribution of Y is symmetric), the theorem will hold.
(i) Assume $|\beta| \neq 1$. Using Theorem 4.8 we have

$$1 - GS^*(x,\alpha,\beta_B) = d_1 x^{-\alpha} + o(x^{-\alpha}),\ x \to \infty \,,$$

$$1 - GS^*(-x,\alpha,\beta_B) = \tilde{d}_1 x^{-\alpha} + o(x^{-\alpha}),\ x \to \infty \,,$$

where

$$d_1 = \Gamma(\alpha) sin[(\alpha\pi/2 + \gamma)]/\pi \neq 0 \,,$$
$$\tilde{d}_1 = \Gamma(\alpha) sin[(\alpha\pi/2 - \gamma)]/\pi \neq 0 \,,$$
$$\gamma = (\pi/2)\beta_B min(\alpha, 2-\alpha) \,.$$

Using the above, we have

$$\begin{aligned} &\lim_{x\to\infty} \frac{1 - GS^*(x,\alpha,\beta_B) - GS^*(-x,\alpha,\beta_B)}{1 - GS^*(x,\alpha,\beta_B) + GS^*(-x,\alpha,\beta_B)} \\ = &\lim_{x\to\infty} \frac{1 - GS^*(x,\alpha,\beta_B) - GS^*(-x,\alpha,\beta_B)\, x^{-\alpha}}{1 - GS^*(x,\alpha,\beta_B) + GS^*(-x,\alpha,\beta_B)\, x^{-\alpha}} = \frac{d_1 - \tilde{d}_1}{d_1 + \tilde{d}_1} = Q \,. \end{aligned}$$

Substituting the expressions for d_1 and $\tilde{d}_1$ yields

$$\begin{aligned} Q &= \frac{\Gamma(\alpha) sin[(\alpha\pi/2+\gamma)]/\pi - \Gamma(\alpha) sin[(\alpha\pi/2-\gamma)]/\pi}{\Gamma(\alpha) sin[(\alpha\pi/2+\gamma)]/\pi + \Gamma(\alpha) sin[(\alpha\pi/2-\gamma)]/\pi} \\ &= \frac{sin[(\alpha\pi/2+\gamma)] - sin[(\alpha\pi/2-\gamma)]}{sin[(\alpha\pi/2+\gamma)] + sin[(\alpha\pi/2-\gamma)]} \\ &= \frac{cos(\alpha\pi/2) sin(\gamma)}{sin(\alpha\pi/2) cos(\gamma)} = \frac{tan(\gamma)}{tan(\alpha\pi/2)} \,. \end{aligned}$$

Since

$$\gamma = (\pi/2)\beta_B min(\alpha, 2-\alpha) \text{ and } \beta_A = cot(\alpha\pi/2) tan(\gamma) \,,$$

then $Q = \beta_A$, as desired.
(ii) The other cases where $|\beta_B|$=1 hold trivially. For example, if $\alpha > 1$ and $|\beta_B| = 1$, then using Corollary 4.1 we have

$$1 - GS^*(x,\alpha,\beta_B) = o(x^{-\alpha}), \quad x \to \infty \,,$$

$$GS^*(x,\alpha,\beta_B) = \tilde{d}_1 o(x^{-\alpha}) + o(x^{-\alpha}), \quad x \to \infty \,,$$

so that $Q = -1$. But $\beta_B = -1 \Leftrightarrow \beta_A = 1$, and the Theorem holds. □

Theorem 4.10 *Let Y be a (geo+)-stable random variable with parameters* $(\alpha, \beta, \lambda, \mu)$, *where* $\alpha < 2$ *and* $\lambda > 0$. *Then for any* $\tau > 0$

$$E|Y|^{\tau} < \infty \quad iff \quad \tau < \alpha .$$

The proof of the above Theorem is based on the following Lemmas given in Ramachandran (1969) and Ramachandran and Rao (1968), that establish the connection between the existence of moments and the behavior of the characteristic function in the neighborhood of the origin.

Lemma 4.1 (Ramachandran and Rao) *Let F be a distribution function and f the corresponding characteristic function. If* $ln|f(t)|/|t|^{\tau}$ *is bounded in the (deleted) neighborhood of the origin for some* $\tau < 2$, *then F has absolute moments of all orders* $< \tau$.

Lemma 4.2 (Ramachandran) *Let F be a distribution function and f the corresponding characteristic function. If* $ln|f(t)|/|t|^{\tau}$ *is bounded away from 0 for some* $\tau < 2$, *then F has no absolute moments of any order* $\geq \tau$.

Proof of Theorem 4.10: Let Y be a (geo+)-stable random variable with ch.f. ψ given in form (A). We examine the behavior of $ln|\psi(t)|/|t|^{\alpha}$ near the origin. First, note that for any real number x

$$\text{if } 1/2 \leq x \leq 1, \text{ then } 1-x \leq -ln(x) \leq 2(1-x) . \tag{14}$$

The above follows from the Taylor expansion of the logarithm. Applying (14) to $x = |\psi(t)|^2$ for small enough t, we get

$$0 \leq 1 - |\psi(t)|^2 \leq -2ln|\psi(t)| \leq 2(1 - |\psi(t)|^2) ,$$

which implies that

$$0 \leq \frac{1}{2}\left(\frac{1-|\psi(t)|^2}{|t|}\right) \leq \left|\frac{ln|\psi(t)|}{t}\right| \leq \frac{1-|\psi(t)|^2}{|t|} . \tag{15}$$

In the sequel, we show that

$$\lim_{t\to 0} \frac{1-|\psi(t)|^2}{|t|^{\alpha}} = 2\lambda > 0 , \tag{16}$$

which, together with (15), will imply that $ln|\psi(t)|/|t|^{\alpha}$ is not only bounded in the neighborhood of the origin, but is also bounded away from 0. The application of Lemma 4.1 and Lemma 4.2 with $\tau = \alpha$ will then give the result.

(i) Case α=1.

To establish (16), we use the representation of ψ in form A

$$|\psi(t)|^2 = \left|\frac{1}{1+\lambda|t| + i(\eta|t|\, ln|t| sgn(t) - \mu t)}\right|^2$$

$$= \frac{1}{(1+\lambda|t|)^2 + (\eta|t|\, ln|t| sgn(t) - \mu t)^2} ,$$

where $\eta = \lambda\beta(2/\pi)$. Now, using the above leads to

$$\frac{1-|\psi(t)|^2}{|t|} = \frac{1}{|t|}\frac{1+2\lambda|t|+\lambda^2|t|^2+((\eta|t|\,ln|t|sgn(t)-\mu\, t)^2-1}{(1+\lambda|t|)^2+(\eta|t|\,ln|t|sgn(t)-\mu\, t)^2}$$

$$=\frac{2\lambda+\lambda^2|t|+|t|\eta^2(ln|t|)^2+\mu^2|t|-2\eta\mu|t|\,ln|t|}{(1+\lambda|t|)^2+(\eta|t|\,ln|t|sgn(t)-\mu\, t)^2} \longrightarrow 2\lambda, \quad \text{as } t\to 0\,.$$

Therefore, (16) holds and so does the Theorem.
(ii) Case $\alpha \neq 1$ is very similar. Using the representation of ψ in form A and denoting $\eta = \beta\, tan(\pi\alpha/2)$, we have

$$\psi(t) = \frac{1}{1+\lambda|t|^\alpha(1-i\eta\, sgn(t))-i\mu\, t)}\,,$$

and so

$$|\psi(t)|^2 = \frac{1}{(1+\lambda|t|^\alpha)^2+(\lambda\eta|t|^\alpha\, sgn(t)+\mu\, t)^2}\,.$$

Consequently, for $\alpha < 2$ and $\lambda > 0$

$$\frac{1-|\psi(t)|^2}{|t|^\alpha} = \frac{1}{|t|^\alpha}\frac{1+2\lambda|t|^\alpha+\lambda^2|t|^{2\alpha}+\eta^2\lambda^2|t|^{2\alpha}+\mu^2t^2+2\eta\lambda\mu|t|^{\alpha+1}-1}{(1+\lambda|t|^\alpha)^2+(\lambda\eta|t|^\alpha\, sgn(t)+\mu\, t)^2}$$

$$=\frac{2\lambda+\lambda^2|t|^\alpha+\eta^2\lambda^2|t|^\alpha+\mu^2\, t^{2-\alpha}+2\eta\lambda\mu|t|}{(1+\lambda|t|^\alpha)^2+(\lambda\eta|t|^\alpha\, sgn(t)+\mu\, t)^2} \longrightarrow 2\lambda, \quad \text{as } t\to 0\,,$$

which proves (16), and so establishes the Theorem. □
Note that the above Theorem does not apply to the exponential distribution (λ=0) and to the Laplace distribution (α=2), which have moments of all orders.

The next property clarifies the meaning of μ as a location parameter.

Theorem 4.11 *Let Y be a (geo+)-stable random variable with parameters $(\alpha,\beta,\lambda,\mu)$, where $\alpha > 1$ and $\lambda > 0$. Then $E(Y)$ exists and equals μ.*

Proof: The existence of $E(Y)$ follows from Theorem 4.10.
But then $E(Y) = \frac{1}{i}\psi'(t)|_{t=0}$. Since

$$\psi(t) = \frac{1}{1+\lambda|t|^\alpha(1-i\eta\, sgn(t))-i\mu\, t)}\,,$$

where $\eta = \beta\, tan(\pi\alpha/2)$, we have

$$\lim_{t\to 0}\frac{\psi(t)-\psi(0)}{t} = \lim_{t\to 0}\frac{1}{t}[\frac{1}{1+\lambda|t|^\alpha(1-i\eta\, sgn(t))-i\mu\, t)}-1]$$

$$=\lim_{t\to 0}\frac{1}{t}\frac{i\mu\, t-\lambda|t|^\alpha(1-i\eta\, sgn(t))}{1+\lambda|t|^\alpha(1-i\eta\, sgn(t))-i\mu\, t)}$$

$$=\lim_{t\to 0}\frac{i\mu-\lambda|t|^{\alpha-1}(1-i\eta\, sgn(t))}{1+\lambda|t|^\alpha(1-i\eta\, sgn(t))-i\mu\, t)} = i\mu\,.$$

Therefore, $\psi'(t)|_{t=0} = i\mu$ and so $E(Y) = \frac{1}{i}i\mu = \mu$ as desired. □
Note that the above Theorem does not apply to the exponential distribution (λ=0), for which $E(Y) = 1/\mu$.

5 UNIMODALITY

A unimodal distribution on R is usually understood as one that has a density which has a maximum at a unique point x=m and decreases as x moves away from m in either direction. Since densities are not everywhere unique or may not exist, we would like to have a definition in terms of the distribution function. The following definition is due to Khintchine (1938).

Definition 5.1 *A real random variable X or its distribution function F is called unimodal about a mode m if F is convex for $x<m$ and concave for $x>m$. F is unimodal, if it is unimodal with mode m for some m.*

Unimodal distributions may have several modes. For example if F corresponds to the uniform distribution on (0,1), then F is unimodal about every m in (0,1). The properties of unimodal distributions are extensively discussed in Dharmadhikari and Joag-dev (1986). A useful characterization of unimodal distributions is given in Shepp (1962).

Theorem 5.1 (Shepp) *A distribution function F is unimodal about 0 if and only if there exist independent random variables U and Z such that U is uniform on (0,1) and the product UZ has distribution function F.*

Recall that a distribution function F is in class L if its corresponding characteristic function ϕ is self-decomposable, that is for each c $\in$ (0,1) there exists a ch.f. ϕ_c such that $\phi(t) = \phi(ct)\phi_c(t)$ for all t$\in$R (Lukacs, 1970). In particular, all α-stable distributions are in class L. It was shown in Yamazato (1978), that all distributions in class L are unimodal.

Theorem 5.2 *All strictly (geo+)-stable distribution functions are unimodal. Moreover, for $\alpha \leq 1$, the mode is equal to 0.*

Proof: Let Y be strictly (geo+)-stable random variable. Then, if its ch.f. is given by form (A), μ=0 for $\alpha \neq 1$ and β=0 for α=1.
(i) Assume α=1.
By Corollary 3.2, Y admits the following representation

$$Y \stackrel{d}{=} \mu Z + Z\sigma X \ , \tag{17}$$

where $Z \sim exp(1)$ and $X \sim S(x, \alpha, \beta)$, Z and X independent. Since the exponential distribution is unimodal with mode at 0, according to Theorem 5.1, Z can be represented as UW, where U has the uniform distribution on (0,1) and U and W are independent. In view of (17) we have the following

$$Y \stackrel{d}{=} \mu Z + Z\sigma X \stackrel{d}{=} Z(\sigma X + \mu) \stackrel{d}{=} U[W(\sigma X + \mu)] \ .$$

By Theorem 5.1 Y is unimodal with mode at 0.
(ii) Assume $\alpha \neq 1$.
If $\alpha < 1$ then we use similar arguments. By Corollary 3.2, Y admits the representation

$$Y \stackrel{d}{=} Z^{1/\alpha}\sigma X \ .$$

We conclude that Y is unimodal with mode at 0 if $W = Z^{1/\alpha}$ is unimodal with mode at 0. But W has the Weibull distribution with density f given by

$$f(w) = \alpha w^{\alpha-1} e^{-w^\alpha}, \quad w > 0 \ .$$

The unimodality of W about 0 will follow, if we can show that $f'(w)$ is negative for any $w > 0$. But this is immediate, since $f'(w) = \alpha w^{\alpha-2} e^{-w^\alpha}(\alpha - 1 - \alpha w^\alpha)$, which is negative for $\alpha < 1$.
Consider $\alpha > 1$. The ch.f. ψ of Y given in form (A) is equal to

$$\psi(t) = (1 + \lambda|t|^\alpha w)^{-1}, \ w = 1 - i\beta \tan(\pi\alpha/2) sgn(t) .$$

We show that ψ is in class L. Let $c \in (0,1)$. Note that

$$\frac{\psi(t)}{\psi(ct)} = \frac{1 + \lambda|t|^\alpha w c^\alpha}{1 + \lambda|t|^\alpha w} = 1/(\frac{1 + \lambda|t|^\alpha w}{1 + \lambda|t|^\alpha w c^\alpha}) = \frac{1}{\psi(ct) + \frac{\lambda|t|^\alpha w}{1+\lambda|t|^\alpha w c^\alpha}}$$

$$= \frac{c^\alpha}{c^\alpha \psi(ct) + (1 - \psi(ct)} = \frac{1/(1 - c^\alpha)}{1/(1 - c^\alpha) - \psi(ct)} = \psi_c(t) .$$

Thus, we have $\psi(t) = \psi(ct)\psi_c(t)$ for all $t \in$R. It remains to show that $\psi_c(t)$ is a characteristic function. As was shown by Lukacs (1970), if $p > 1$ and g is a characteristic function, then $(p-1)/(p-g(t))$ is also a characteristic function. But $\psi_c(t)$ defined above has precisely this form, where $p = 1/(1 - c^\alpha) > 1$ for $c \in (0,1)$ and $g(t) = \psi(ct)$. Therefore, for $\alpha \neq 1$, every (geo+)-stable law is in class L and so it is unimodal. This concludes the proof of the Theorem. □

REFERENCES

[1] Devroye, L., 1986, "Non-Uniform Random Variate Generation," Springer, New York.

[2] Dharmadhikari, S., and Joag-dev, K., 1986, "Unimodality, Convexity, and Applications," Academic Press.

[3] Gertsbakh, I.B., 1984, Asymptotic methods in reliability: a review, *Adv. in Appl. Probab.* 16, 147-175.

[4] Gnedenko, B.V., 1970, Limit theorems for sums of random number of positive independent random variables, *Proc. 6-th Berkeley Symp. on Math. Statist. Probab.* v.2, 537-549.

[5] Hall, P., 1981, A comedy of errors: the canonical form for a stable characteristic function, *Bull. London Math. Soc.* 13, 23-27.

[6] Jacobs, P.A., 1986, First passage times for combinations of random loads, *SIAM J. Appl. Math.* 46, 643-656.

[7] Khintchine, A.Y., 1938, On uni· ιoda distributions, *Izv. Nauchno-Isled. Inst. Mat. Mech. Tomsk. Gos. Uni*ι. ', 1-7.

[8] Klebanov, L.B., Maniya, G.M., and Melamed, I.A., 1984, A problem of Zolotarev and analogs of infinitely divisible and stable distributions in a scheme for summing a random number of random variables, *Theory Probab. Appl.* 29, 791-794.

[9] Kozubowski, T.J., and Rachev, S.T., 1992, The theory of geometric stable distributions and its use in modeling financial data, *European J. Oper. Res.* (to appear).

[10] Lukacs, E., 1970, " Characteristic Functions," Griffin, London.

[11] Mittnik, S., and Rachev, S.T., 1991, Alternative multivariate stable distributions and their applications to financial modeling, In: "Stable Processes and Related Topics," S. Cambanis et. al. Eds., Birkhäuser, Boston, 107-119.

[12] Ramachandran, B., 1969, On characteristic functions and moments, *Sankhyā* 31, 1-12.

[13] Ramachandran, B., and Rao, C.R., 1968, Some results on characteristic functions and characterizations of the normal and generalized stable laws, *Sankhyā* 30, 125-140.

[14] Robbins, H., 1948, Mixtures of distributions, *Ann. Statist.* 19, 360-369.

[15] Shepp, L.A., 1962, Symmetric random walk, *Trans. Amer. Math. Soc.* 104, 144-153.

[16] Yamazato, M., 1978, Unimodality of infinitely divisible distribution functions of class L, *Ann. Probab.* 6, 523-531.

[17] Zolotarev, V.M., 1986, "One-Dimensional Stable Distributions," Volume 65 of Translations of Mathematical Monographs, American Mathematical Society.

REARRANGEMENTS OF CONDITIONALLY INTEGRABLE FUNCTIONS

Thomas Kunkle

Department of Mathematics
University of Charleston, SC
Charleston, SC 29424
kunkle@math.cofc.edu

ABSTRACT

From an s-variate continuous function whose positive and negative parts both have infinite Lebesgue integral, we construct a continuous rearrangement whose generalized Riemann integral equals any prescribed value. The original function is assumed to be finite a.e., with domain an interval in $\mathbb{R}^s$. A similar result is obtained for functions defined on a ball.

1. INTRODUCTION

Recall the following definition.

Definition 1.1: *The Lebesgue measurable functions f and g are called* **identically distributed**, *denoted $f \sim g$, if they are finite almost everywhere, and if, for every real α, the sets $\{x : f(x) < \alpha\}$ and $\{x : g(x) < \alpha\}$ have the same measure. If f and g have the same domain, one is said to be a* **rearrangement** *of the other.*

An elementary fact from measure theory states that a Lebesgue integrable function and its rearrangements have the same Lebesgue integral.

In his survey lecture [4], Shisha compares the generalized Riemann integral (**GRI**) to the Lebesgue integral (**LI**) and makes the analogy that the GR-integrability of a function is to its L-integrability as the convergence of a series is to its absolute convergence. Hearing this, and noticing that L-integrable functions and absolutely convergent series behave alike under rearrangements, the listener might wonder if rearrangements effect a "conditionally integrable" function in the same way that they effect a conditionally convergent series. It turns out that they do. We'll show that if a (continuous) function is GR- but not L-integrable in $\mathbb{R}^s$, then it possesses a (continuous) rearrangement whose GRI takes any prescribed value.

(The analogous result for series is due to Riemann, whose name therefore arises twice here in connection with two different areas).

A brief description of the GRI appears in Section 3. Over a large class of func-

Approximation, Probability, and Related Fields, Edited by
G. Anastassiou and S.T. Rachev, Plenum Press, New York 1994

tions, the GRI is identical to the LI. For instance, if $f \geq 0$, then the GRI and LI of f over the Lebesgue measurable set $E \subseteq \mathbb{R}^s$ are the same, in the sense that the existence of one implies the existence of the other and their equality. If $|f|$ is integrable in either sense, then the GRI and LI of f exist and are equal.

However, there are functions which are not absolutely integrable in either sense, but which are GR-integrable. Such a function has positive and negative parts with infinite GRI. Obviously, then, if f GR-integrable over the measurable set E, then we obviously cannot conclude that that it is GR-integrable on every subset of E. It is true, however, that if f is GR-integrable over the closed interval I in $\mathbb{R}^s$ (defined in Section 2), then it is GR-integrable over every closed subinterval of I. Fubini's theorem for the GRI states that if a multivariate function is GR-integrable over a product of intervals, then its GRI is computable via iterated integration; absolute integrability is not required.

An interesting property of the GRI is its ability to give meaning to integrals that are improper in the L-sense. Consider, for example, the following result ([2], §1.5).

Theorem 1.2: *Let $f : [a,b] \to \mathbb{R}$ have a GRI on $[a,s]$ for every $s \in (a,b)$. Then the GRI $\int_a^b f\,dx$ exists if and only if the limit of GRIs $\lim_{s\to b^-} \int_a^s f\,dx$ exists. Moreover, when they exist, the two are equal.*

Thus, some singular integrals, i.e., integrals that exist only improperly in the L-sense, exist as proper GRIs—some, that is, but not all. For example, for every positive ε, $\left(\int_{-1}^{-\varepsilon} + \int_{\varepsilon}^{1}\right) x^{-1}\,dx$ is zero, but x^{-1} is not GR-integrable on $(-1,1)$. The following definition is made to include functions that are integrable in the GR- but not L-sense, as well as functions like $f(x) = x^{-1}$ on $(-1,1)$.

Definition 1.3: *A Lebesgue measurable function f of s variables is* **conditionally integrable** *on the measurable set E if f is finite almost everywhere and if both Lebesgue integrals $\int_E f^+$ and $\int_E f^-$ are infinite.*

Here $f^+ := \max(f,0)$ and $f^- := \max(-f,0)$, the **positive part** and **negative part** of f. The integrability of f is said to be conditional since, as we shall show, the existence and value of the GRI varies among f's rearrangements.

2. NOTATION

We use := to denote definition, as in $f(x) := x^2$.

A sequence is said to converge if it has a limit in $[-\infty,\infty]$, and diverge otherwise.

We denote the domain of a function f by $\operatorname{dom} f$; if $M \subseteq \operatorname{dom} f$, then $f|_M$ is the restriction of f to M. Denote the greatest integer, or floor, function by $\lfloor\cdot\rfloor$, and the identity function by $(\cdot)$. By **monotone**, we mean either nondecreasing or nonincreasing.

For $x \in \mathbb{R}^s$, we refer to the ith coordinate of x by $x(i)$; subscripts are reserved to denote sequences. Define $r(x)$ to be the Euclidean norm of x. Let $B(R)$ denote $\{x \in \mathbb{R}^s : r(x) < R\}$, the open ball of radius R centered at the origin. $B^c(R)$ shall denote the closed ball. $A(R,S)$ shall denote the open annulus $\{x \in \mathbb{R}^s : R < r(x) < S\}$, and $A^c(R,S)$ the closed annulus. In this context, R and S will always stand for positive, real numbers. By **radial** function, we mean a function of the form $g \circ r$, where $\operatorname{dom} g \subseteq [0,\infty)$. We refer to $g \circ r$ as the **radial extension** of g, and say that $g \circ r$ is **nondecreasing** if g is.

Define π_s to be $m_s(B(1))$, the volume of the unit ball.

In spherical coordinates, the integral of $g \circ r$ over $B(R)$ is $\int_0^R g(r)\pi_s \, dr^s$, where dr^s is $sr^{s-1} \, dr$, provided $g \circ r$ is absolutely integrable on $B(R)$.

By **measurable**, we will always mean Lebesgue measurable. We'll denote the LI of f over the E by $\int_E f \, dm$, and its GRI by $\int_E f \, dx$, with the usual convention when $E = (a, b)$. We may refer simply to the integral of f, denoted $\int_E f$, when it is unambiguous to do so. The s-dimensional measure of E is written $m(E)$, or $m_s(E)$, if there exists a chance for confusion over the dimension.

The set $I \subseteq \mathbb{R}^s$ is called an **interval** if it is the Cartesian product of s intervals in $\mathbb{R}$. Unless otherwise specified, all intervals are bounded and nondegenerate; i.e., $0 < m_s(I) < \infty$. The interior of an interval I is denoted I°.

The **essential supremum** of f, written ess sup f, is the infimum of the collection of real β for which $m(f^{-1}(\beta, \infty]) = 0$. The **essential infimum** of f is defined as $\operatorname{ess\,inf} f = -\operatorname{ess\,sup} -f$.

3. THE GENERALIZED RIEMANN INTEGRAL

See [1] for an overview of the GRI and its current status; for a more detailed introduction to the GRI, see [2].

Developed by Henstock and Kurzweil in the mid 1950's, the generalized Riemann integral of a function over the closed and bounded interval $I \subset \mathbb{R}^s$, is defined as a generalized limit of Riemann sums, as follows.

A **gauge** γ is a function which associates to every z in I an s-dimensional open interval $\gamma(z)$ containing z. It is not required that $\gamma(z) \subseteq I$, that $\gamma(z)$ be bounded, or that γ be in any way continuous.

A **division** $\mathcal{D}$ of I is a finite collection of closed intervals J and **tags** $z_J \in J$ such that $I = \bigcup J$ and $J \cap H$ has measure zero for every $J \neq H$ in $\mathcal{D}$. A division $\mathcal{D}$ is said to be γ-**fine** if $J \subseteq \gamma(z_J)$ for every $J \in \mathcal{D}$. It can be shown that for any γ there exists a γ-fine division.

The gauge γ' is said to be **finer** than γ if $\gamma'(z) \subseteq \gamma(z)$ for all $z \in I$. Clearly, any γ'-fine division is also γ-fine.

For f defined on I, and for $\mathcal{D}$ a division of I, the Riemann sum $fm(\mathcal{D})$ is defined as $\sum_{\mathcal{D}} f(z_J)m(J)$. The GRI $\int_I f \, dx$ is said to exist if, for every positive ε, there exists a gauge γ such that

$$\left| \int_I f \, dx - fm(\mathcal{D}) \right| < \varepsilon \tag{3.1}$$

for any γ-fine division $\mathcal{D}$ of I.

For example, take $s = 1$. If, for every $\varepsilon > 0$, there is a $\delta > 0$ so that, when $\gamma(z) = (z - \delta, z + \delta)$, any γ-fine $\mathcal{D}$ satisfies (3.1), then f is Riemann integrable on I.

The concept of GRI exists for unbounded intervals but is not relevant to our discussion. The GRI of f over a set $E \subseteq I$ is defined to be $\int_I g \, dx$, where $g = f$ on E and $g = 0$ on $I \setminus E$.

Specializing the arguments in [3], we state and prove a multivariate generalization by McLeod of Theorem (1.2).

Let I' be an s-dimensional interval containing all but one of its $(s-1)$-dimensional faces. Let I be the closure of I'. Let $f : I \to \mathbb{R}$ be identically zero on $I \setminus I'$ and

GR-integrable on every closed subinterval J of I'. For $\mathcal{D}$ a division of I, define

$$\nu(\mathcal{D}) := \sum_{\substack{J\in\mathcal{D}\\ J\subseteq I'}} \int_J f\,dx.$$

If G is a real-valued function whose domain is the set of all divisions on I, we say that $A = \lim_{\mathcal{D}} G(\mathcal{D})$ if, for every positive ε, there exists a gauge γ on I such that $|A - \nu(\mathcal{D})| < \varepsilon$ for every γ-fine division $\mathcal{D}$ of I. For example, the GRI of f on I is $\lim_{\mathcal{D}} fm(\mathcal{D})$.

In contrast to the standard limit theorems, McLeod's theorem below does not require that f be absolutely integrable on I.

Theorem 3.2: *Under the above assumptions, the GRI of f over I is the same as $\lim_{\mathcal{D}} \nu(\mathcal{D})$, in the sense if either exists, the two are equal.*

Proof: It will suffice to prove $\lim_{\mathcal{D}}\big(\nu(\mathcal{D}) - fm(\mathcal{D})\big) = 0$.

Let $A : \mathbb{R}^s \to \mathbb{R}^s$ be an invertible map of the form $A(x) = A(0) + PDx$, where P is a permutation matrix and D is a diagonal matrix. It is not difficult to show via Riemann sums that f is GR-integrable on I if and only if $f \circ A$ is GR-integrable on $A^{-1}I$. It is without loss of generality, then, to let $I = [0,1]^s$ and $I' = [0,1) \times [0,1]^{s-1}$.

For $n \geq 1$, set $I_n := [0, n/(n+1)] \times [0,1]^{s-1}$, and let γ_n be a gauge on I_n that satisfies

- a: $\left|\int_{I_n} f\,dx - fm(\mathcal{D})\right| < \varepsilon 2^{-n}$ for every γ_n-fine division $\mathcal{D}$ on I_n;
- b: if $z(1) < n/(n+1)$, then $I^\circ \cap \gamma_n(z) \subseteq I_n^\circ$; and
- c: if $z(1) = n/(n+1)$ for some $n \geq 1$, then $\gamma_{n+1}(z) \subseteq \gamma_n(z)$.

That there exists a γ_n to satisfy a follows from the integrability of f on I_n. Having such a γ_n, one can replace γ_n by a finer gauge that satisfies a and b, and then by an even finer gauge that satisfies a, b, and c.

Define γ on I as follows. If $z(1) = 1$, set $\gamma(z) = \mathbb{R}^s$. Otherwise, set $\gamma(z) = \gamma_n(z)$ for the unique n that satisfies $(n-1)/n \leq z(1) < n/(n+1)$. Then

- b': if $z(1) < n/(n+1)$, then $I^\circ \cap \gamma(z) \subseteq I_n^\circ$.

Because I_n° is an increasing sequence of sets, it is sufficient to prove b' for the minimal n such that $z(1) < n/(n+1)$. But then $\gamma(z) = \gamma_n(z)$ and b' reduces to b.

Let $\mathcal{D}$ be any γ-fine division of I. We claim that

$$\big|\nu(\mathcal{D}) - fm(\mathcal{D})\big| < \varepsilon, \tag{3.3}$$

which will complete the proof.

Partition $\mathcal{D}$ as follows. Let $\mathcal{D}_0$ consist of all $J \in \mathcal{D}$ that intersect $I \setminus I'$. Let $\mathcal{D}_n$ consist of all $J \in \mathcal{D}$ for which n is the smallest integer satisfying $J \subseteq I_n$. By definition, $\nu(\mathcal{D}_0) = 0$, and since $J \in \mathcal{D}_0$ implies $z_J(1) = 0$, the Riemann sum $fm(\mathcal{D}_0) = 0$ also. Therefore

$$\nu(\mathcal{D}) - fm(\mathcal{D}) = \sum_{n=1}^{\infty} \nu(\mathcal{D}_n) - fm(\mathcal{D}_n). \tag{3.4}$$

The summand is zero when $\mathcal{D}_n$ is empty, which occurs for all but finitely many n.

Let $n \geq 1$ and consider $J \in \mathcal{D}_n$. The corresponding tag satisfies $z_J \in J \subseteq I_n$, so $z_J(1) \leq n/(n+1)$. Furthermore, $z_J(1) < (n-1)/n$ implies that $n > 1$, and, with b', that $I^\circ \cap \gamma(z_J) \subseteq I_{n-1}^\circ$, contradicting $J \not\subseteq I_{n-1}$. Therefore

$$\frac{n-1}{n} \leq z_J(1) \leq \frac{n}{n+1}.$$

If $z_J(1) < n/(n+1)$, then $\gamma(z_J) = \gamma_n(z_J)$, and if $z_J(1) = n/(n+1)$, then $\gamma(z_J) = \gamma_{n+1}(z_J) \subseteq \gamma_n(z_J)$ by c. Thus $J \in \mathcal{D}_n$ implies that $\gamma(z_J) \subseteq \gamma_n(z_J)$. Because $J \subseteq \gamma(z_J)$, there exists a γ_n-fine division of I_n, say C_n, which contains $\mathcal{D}_n$. Henstock's Lemma ([2]) implies that $fm(\mathcal{D}_n)$ approximates $\nu(\mathcal{D}_n)$ almost as well as $fm(C_n)$ approximates $\int_{I_n} f\,dx$. Specifically,

$$\left|\nu(\mathcal{D}_n) - fm(\mathcal{D}_n)\right| \leq \varepsilon 2^{-n}$$

(compare with a). Since the left side is zero when $\mathcal{D}_n = \emptyset$,

$$\sum_{n=1}^{\infty} |\nu(\mathcal{D}_n) - fm(\mathcal{D}_n)| < \varepsilon.$$

Combined with (3.4), this proves (3.3). □

4. A UNIVARIATE REARRANGEMENT

We begin this section with two simple but useful lemmas.

Lemma 4.1: *If g is nondecreasing on (R, S), and if the radial function $\rho \circ r$ is identically distributed to $r|_{A(R,S)}$ then $g \circ \rho \circ r \sim g \circ r$.*

Proof: To see this, let α be a real number, and define S_α to be $\sup\{x > R : g(x) < \alpha\}$. Then

$$(g \circ \rho \circ r)^{-1}(-\infty, \alpha) = (\rho \circ r)^{-1}(R, S_\alpha).$$

By hypothesis, this set has the same measure as $r^{-1}(R, S_\alpha) = (g \circ r)^{-1}(-\infty, \alpha)$, completing the proof. □

Note that when $s = 1$, the hypothesis on ρ is equivalent to $\rho \sim (\cdot)|_{(R,S)}$, and the conclusion is equivalent to $g \circ \rho \sim g$.

Often, as in the proof of the next lemma, it is useful to note that identically distributed functions have domains of equal measure.

Lemma 4.2: *Let $\{f_n\}$ and $\{g_n\}$ be sequences of functions satisfying $f_n \sim g_n$ and $\operatorname{ess\,sup} f_n \leq \operatorname{ess\,inf} f_{n+1}$. Assume that, for $n \neq m$, both $\operatorname{dom} f_n \cap \operatorname{dom} f_m$ and $\operatorname{dom} g_n \cap \operatorname{dom} g_m$ are of measure zero. Define f a.e. on $\bigcup \operatorname{dom} f_n$ by the rule $f|_{\operatorname{dom} f_n} := f_n$; define g similarly. Then $f \sim g$.*

Proof: Since $f_n \sim g_n$, these two functions have the same essential supremum and infimum. Consider the increasing sequence

$$\cdots \leq \operatorname{ess\,inf} f_{n-1} \leq \operatorname{ess\,sup} f_{n-1} \leq \operatorname{ess\,inf} f_n \leq \operatorname{ess\,sup} f_n \leq \cdots. \tag{4.3}$$

If (4.3) has a largest member $\leq \alpha$, then $\operatorname{ess\,sup} f_{n-1} \leq \alpha < \operatorname{ess\,sup} f_n$ for some n, and

$$m(f^{-1}(-\infty, \alpha)) = m(f_n^{-1}(-\infty, \alpha)) + \sum_{k<n} m(\operatorname{dom} f_k).$$

The same is true when f is replaced by g, and the right-hand sides of each are equal, as desired.

On the other hand, if there is no largest member of (4.3) $\leq \alpha$, then either $\alpha >$ ess sup f_n for every n, or $\alpha <$ ess inf f_n for every n. In either case, $m(f^{-1}(-\infty,\alpha)) = m(g^{-1}(-\infty,\alpha))$. □

Consider now a conditionally integrable function f on $(0,1)$, and a number $\xi \in [-\infty,\infty]$. It follows from Riemann's result on rearrangements of conditionally convergent series that f has a rearrangement whose GRI equals ξ, as we now show.

Replace f by its nondecreasing rearrangement ([5], p. 29). Since $\int_0^1 f^+\,dx$ and $\int_0^1 f^-\,dx$ are infinite,

i) there exists $x_0 \in (0,1)$ such that $f \geq 0$ on $(x_0,1)$ $f \leq 0$ on $(0,x_0)$;

ii) one can define $x_m \in (0,1)$ for $m = \pm1,\pm2,\ldots$ by the rule

$$I_m := \int_{x_{m-1}}^{x_m} f\,dx = \begin{cases} 2m^{-1} & \text{if } m > 0, \text{ and} \\ 2m+1^{-1} & \text{if } m < 0; \end{cases}$$

and *iii)* $\lim_{m\to\infty} x_m = 1$ and $\lim_{m\to-\infty} x_m = 0$.

Take a one-to-one function σ from $\mathbb{Z}$ onto $\mathbb{N}$ for which $\sum_{m=1}^{\infty} I_{\sigma(m)} = \xi$. We'll take a corresponding rearrangement g for which

$$\lim_{\alpha\to1^-} \int_0^\alpha g\,dx = \lim_{N\to\infty} \sum_{m=1}^{N} I_{\sigma(m)}. \tag{4.4}$$

Then, by Theorem (1.2), the GRI of g exists and also equals ξ.

To construct g, begin with the infinite partition of $(0,1)$ consisting of the intervals (x_{m-1},x_m). Set $\Delta_m := x_m - x_{m-1}$ for $m \in \mathbb{Z}$ and $s_m := \sum_{i=1}^{m} \Delta_{\sigma(i)}$ for $m \geq 0$. On (s_{m-1},s_m), define $\rho_m(x) := x - s_m + x_{\sigma(m)}$. Clearly, $\rho_m \sim (\cdot)|_{(x_{\sigma(m)-1},x_{\sigma(m)})}$. Define ρ on $\bigcup \operatorname{dom}\rho_m$ as in Lemma (4.2). Then $\rho \sim (\cdot)|_{(0,1)}$; consequently, $m(0,1) = m(\operatorname{dom}\rho)$, so $\operatorname{dom}\rho = [0,1)$. Therefore ρ is a rearrangement of $(\cdot)|_{(0,1)}$. By Lemma (4.1), $g := f\circ\rho$ is a rearrangement of f.

One can check directly that

$$\int_{s_{m-1}}^{s_m} g\,dx = \int_{x_{\sigma(m)-1}}^{x_{\sigma(m)}} f\,dx.$$

Viewed as a function of its upper limit, the integral $\int_0^\alpha g\,dx$ is monotone on (s_{m-1},s_m), since g has constant sign on each such interval. Equation (4.4) follows, as desired.

Finally, we note that there exist σ such that the partial sums of $\sum I_{\sigma(m)}$ diverge. By Theorem (1.2), the corresponding rearrangement g will fail to be GR-integrable.

We summarize the results of this section in the following theorem, making the obvious generalization from $(0,1)$ to (a,b).

Theorem 4.5: *Let f be a conditionally integrable function on the finite interval (a,b), and let $\xi \in [-\infty,\infty]$. Then f has a rearrangement whose GRI equals ξ. There also exists a rearrangement of f that is not GR-integrable.*

5. CONTINUOUS REARRANGEMENTS ON DISKS

In all likelihood, the rearrangement constructed in Theorem (4.5) will be discontinuous. In this section and the next, we deal with the more challenging problem of finding continuous rearrangements of continuous functions.

We begin with the following lemma.

Lemma 5.1: *Let f be a s-variate measurable function whose domain* $\operatorname{dom} f$ *has finite measure. Let f be finite almost everywhere. Then there exists a nondecreasing function g, with* $\operatorname{dom} g = (0, A)$ *(for some $A < \infty$), whose radial extension is identically distributed to f. The function $g \circ r$ is finite on a punctured open ball with volume equal to* $m_s(\operatorname{dom} f)$.

Proof: To construct g, we modify a construction appearing in [5].

For real x, set $p(x) := m_s\big(f^{-1}(-\infty, x)\big)$. Then p is left-continuous and nondecreasing. If $x < \inf f$ or $x > \sup f$, then $p(x) = 0$ or $p(x) = m_s(\operatorname{dom} f)$, respectively. Define f_* to be the inverse function p^{-1}, with the understanding that the jump discontinuities of p will correspond to intervals of constancy of f_* and at the jump discontinuities of f_* (corresponding to intervals of constancy of p), f_* will be defined to left continuous. Then f_* has domain $[0, m_s(\operatorname{dom} f)]$ and range $[\inf f, \sup f]$. Furthermore, f_* is nondecreasing, and, for $\alpha \in \mathbb{R}$,

$$m_1\big(f_*^{-1}(\infty, \alpha)\big) = m_s\big(f^{-1}(-\infty, \alpha)\big), \tag{5.2}$$

since both sides equal $p(\alpha)$.

If $s = 1$ and $\operatorname{dom} f = [0, c]$ for some real c, then f_* is the nondecreasing rearrangement used in the proof of Theorem (4.5).

Set $g(r) := f_*(\pi_d r^d)$ for all allowable $r \geq 0$. Then g is nondecreasing and its domain is of the required form. For any real α, the set $m_s\big((g \circ r)^{-1}(-\infty, \alpha)\big)$ is empty if $\alpha < \inf f$, and otherwise it is a ball with radius

$$\sup\big\{r : f_*(\pi_s r^s) < \alpha\big\} = \sup\big\{r : \pi_s r^s < m_s\big(f^{-1}(-\infty, \alpha)\big)\big\},$$

so that its volume is $m_s\big(f^{-1}(-\infty, \alpha)\big)$. Hence $g \circ r$ is identically distributed to f. Since $\operatorname{dom} f$ and $\operatorname{dom} g \circ r$ have the same measure, the proof is complete. □

From the proof of Lemma (5.1), one observes the following corollary.

Corollary 5.3: *If f is bounded below, then $g \circ r$ is defined at the origin and equals* $\inf f$ *there. If, in addition, f_* is continuous on* $\big[0, m_s(\operatorname{dom} f)\big)$ *then $g \circ r$ is continuous.*

We next construct a radial function $\rho \circ r$ that is identically distributed to $r|_{A(R,S)}$.

Let $X := \{x_0, x_1, \ldots\}$ be an increasing sequence of nonnegative reals with the property that

$$x_n^s + (S^s - R^s)2^{-\lfloor n/2 \rfloor - 2} \leq x_{n+1}^s \quad (n \geq 0). \tag{5.4}$$

This condition ensures that the intersection of any two of the intervals

$$\left[x_n, \sqrt[s]{x_n^s + (S^s - R^s)2^{-\lfloor n/2 \rfloor - 2}}\right] \tag{5.5}$$

has measure zero. Define $\rho := \rho(\cdot \mid X, R, S)$ on the union of all such intervals as follows. For n even, set ρ on (5.5) to be

$$\rho(x) := \sqrt[s]{2^{\lfloor n/2 \rfloor + 2}(x^s - x_n^s) + R^s}$$

and for n odd,

$$\rho(x) := \sqrt[s]{2^{\lfloor n/2 \rfloor + 2}(x_n^s - x^s) + S^s}$$

Note that on the even intervals, ρ increases from R to S, and on the odd intervals, it decreases from S to R.

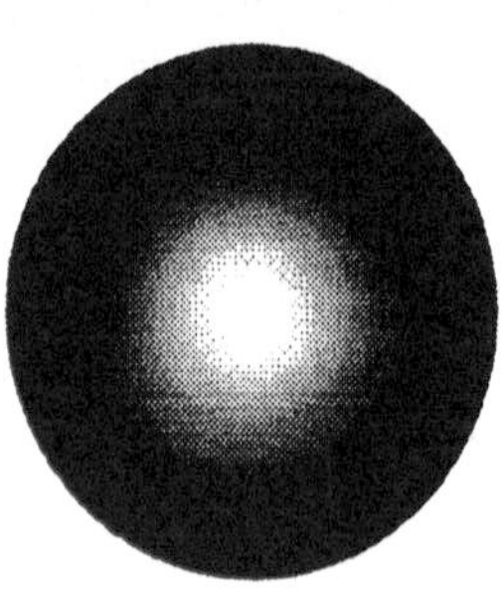

Figure 1. $r|_{A(R,S)}$

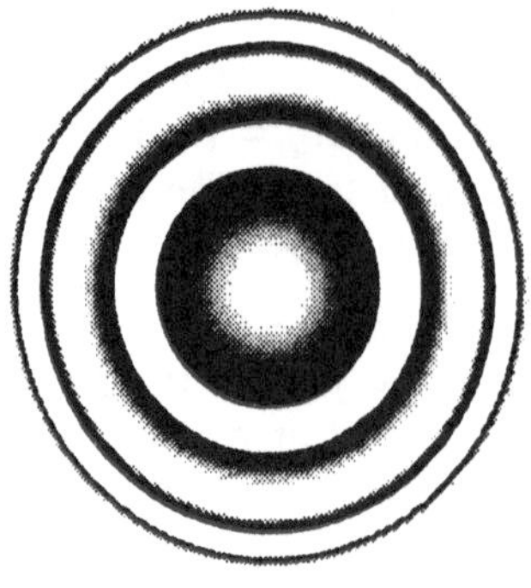

Figure 2. $\rho(r(\cdot) \mid X, R, S)$

Lemma 5.6: *For X satisfying (5.4) and for $0 \le R < S < \infty$, the function $r|_{A(R,S)}$ is identically distributed to $\rho(r(\cdot) \mid X, R, S)$.*

Proof: The domain of $\rho \circ r$ is the union of closed annuli of the form

$$A^c\left(x_n, \sqrt[s]{x_n^s + (S^s - R^s)2^{-\lfloor n/2 \rfloor - 2}}\right).$$

To see that $\rho \circ r$ is identically distributed to r's restriction to $A(R, S)$, note first that if $\alpha \le R$, then both $(\rho \circ r)^{-1}(-\infty, \alpha)$ and $r^{-1}(-\infty, \alpha)$ are empty, and if $\alpha \ge S$,

$$\begin{aligned} m_s\big((\rho \circ r)^{-1}(-\infty,\alpha)\big) &= m_s\big(\operatorname{dom} \rho \circ r\big) \\ &= \sum_{n \ge 0} \pi_s\big((x_n^s + (S^s - R^s)2^{-\lfloor n/2 \rfloor - 2}) - x_n^s\big) \qquad (5.7) \\ &= \pi_s(S^s - R^s) = m_s\big(r^{-1}(-\infty,\alpha)\big). \end{aligned}$$

Finally, if $R < \alpha < S$, then it is not hard to see that $(\rho \circ r)^{-1}(-\infty, \alpha)$ is a union of annuli whose volumes sum geometrically to $\pi_s(\alpha^s - R^s) = m_s\big(r^{-1}(-\infty, \alpha)\big)$. □

Figures 1 and 2 show the density plots of $r|_{A(R,S)}$ and $\rho\big(r(\cdot) \mid X, R, S\big)$ (for a particular X, R, and S). The domain of $\rho \circ r$ is an infinite sequence of annuli, and $\rho \circ r$ alternates between increasing and decreasing on these. If one interprets r and $\rho \circ r$ as point densities, then the objects in figures 1 and 2 have the same total mass. In figure 2, this mass is divided into parts $\frac{1}{4}, \frac{1}{4}, \frac{1}{8}, \frac{1}{8}, \ldots$, which are summed in (5.7).

We now state and prove main result of this section.

Theorem 5.8: *Let f be a conditionally integrable continuous function on $B(R) \subset \mathbb{R}^s$, and let $\xi \in [-\infty, \infty]$. Then f has a continuous radial rearrangement $g \circ r$ such that the singular integral*

$$\lim_{\alpha \to R^-} \int_{B(\alpha)} g \circ r = \xi.$$

Furthermore, f has a continuous rearrangement $g \circ r$ whose singular integral over $B(R)$ does not exist.

Since g is continuous, $\int_{B(\alpha)} g \circ r = \pi_s \int_0^\alpha g(r)\, dr^s$ for $\alpha < R$. Hence Theorem (1.2) implies that the GRI $\pi_s \int_0^R g(r)\, dr^s$ equals ξ in the first case, and in the second does not exist. When $s = 1$, the radial functions are simply the even functions, and the existence of $\int_0^R g(r)\, dr$ is equivalent to the existence of $\int_{-R}^R g(|x|)\, dx$. Hence Theorem (5.8) has the following corollary.

Corollary 5.9: *Let f be a univariate function that is continuous and conditionally integrable on $(-R, R)$. Let $\xi \in [-\infty, \infty]$. Then f has a continuous even rearrangement $g(|\cdot|)$ whose GRI on $(-R, R)$ equals ξ. There also exists a continuous rearrangement of f which is not GR-integrable on $(-R, R)$.*

For $s > 1$, it would be interesting to know whether or not the GRIs $\int_{B(R)} g \circ r \, dx$ and $\pi_s \int_0^R g(r)\, dr^s$ are the same for g continuous on $[0, R)$. If so, one could restate the conclusion of Theorem (5.8) in terms of the GRI of the rearrangement, as we have in Corollary (5.9) and will do in the next section. However, the multivariate change-of-variables rule requires absolute integrability, as do most limit theorems that could be used to prove $\lim_{\alpha \to R^-} \int_{B(\alpha)} = \int_{B(R)}$.

The proof is similar to that of Theorem (4.5). We'll assume, without loss of generality, that $R = 1$, then choose a series which converges conditionally to ξ/π_s (or fails to converge), and then construct a (radial) rearrangement $g \circ r$ such that, as α increases to 1, $\int_0^\alpha g(r)\, dr^s$ behaves like the partial sums of the series.

Proof of (5.8): Without loss of generality, $R = 1$. Furthermore, it will suffice to prove Theorem (5.8) in case $\xi \geq 0$. Then, if $\xi < 0$, we can obtain a rearrangement of $-f$ whose GRI equals $-\xi$, and therefore a rearrangement of f whose GRI is ξ.

Apply Lemma (5.1) to the restriction $f|_{f^{-1}[0,\infty)}$, obtaining a nondecreasing function g_+ whose radial extension is identically distributed to $f|_{f^{-1}[0,\infty)}$ and has domain $B(R_+)$, where $\pi_s R_+^s = m_s(f^{-1}[0, \infty))$. Likewise, let $g_- \circ r$ be a nonincreasing radial rearrangement of $f|_{f^{-1}(-\infty,0)}$ with domain $B(R_-)$. We claim that g_+ and g_- are continuous.

Consider g_+ first.

Corollary (5.3) implies that, because $f|_{f^{-1}[0,\infty)}$ is bounded below, we need only show that $\left(f|_{f^{-1}[0,\infty)}\right)^*$, or f^* for short, is continuous on $[0, R_+)$. Since f^* is nondecreasing, it will suffice to show that it has no jump discontinuities. These exist if and only if there are intervals on which $m_s(f^{-1}(0, x))$ is constant, or, equivalently, if $m_s(f^{-1}(a, b)) = 0$ for some nontrivial (a, b). But because f is continuous and is unbounded above and below, $f^{-1}(a, b)$ is nonempty and open, and hence has positive measure. Therefore g_+ is continuous. The proof of g_-'s continuity is similar.

Choose $\sum_{m\geq 2} a_m$ to be a positive series summing to ∞ as follows.

If $\xi = 0$, take $a_{2m} = a_{2m+1} = (m+3)^{-1}$; if $\xi = \infty$ take $a_{2m} \equiv 2$ and $a_{2m+1} \equiv 1$; and if ξ is finite and positive, take $a_m = c/(m+1)$, where $c > 0$ is chosen so that, as in the other cases,

$$\sum_{m=2}^{\infty} (-1)^m a_m = \frac{\xi}{\pi_s}. \tag{5.10}$$

To construct a rearrangement whose GRI does not exist, take $a_m \equiv 1$, so that

$$\sum_{m=2}^{\infty} (-1)^m a_m \text{ diverges.} \tag{5.11}$$

With $\sum a_m$ so chosen, both $\sum_{m=1}^{\infty} 2a_{2m} - a_{2m-2}$ and $\sum_{m=1}^{\infty} 2a_{2m+1} - a_{2m-1}$ are positive series summing to infinity. (We adopt the convention that $a_m = 0$ for $m < 2$.) Thus, it is possible to define the sequences (r_m) and (s_m) by

$$\begin{aligned} r_0 = s_0 &= 0; \\ \int_{r_{m-1}}^{r_m} g_+(r)\, dr^s &= 2a_{2m} - a_{2m-2} \quad m \geq 1; \quad \text{and} \\ \int_{s_{m-1}}^{s_m} g_-(r)\, dr^s &= -2a_{2m+1} + a_{2m-1} \quad m \geq 1. \end{aligned} \tag{5.12}$$

Furthermore, r_m (or s_m) increases to R_+ (or R_-) as $m \to \infty$.

Now define the functions g_m for $m \in \mathbb{Z} \setminus 0$ by

$$g_m := \begin{cases} g_+|_{[r_{m-1}, r_m]} & \text{if } m > 0; \text{ and} \\ g_-|_{[s_{-m-1}, s_{-m}]} & \text{if } m < 0. \end{cases} \tag{5.13}$$

Next we construct, for every $m \geq 1$, a sequence $X_m = \{x_{m,0}, x_{m,1}, \ldots\}$ of non-negative reals satisfying (5.4). The associated

$$\rho_m := \rho(\cdot \mid X_m, r_{m-1}, r_m)$$

is defined on closed intervals with left endpoint $x_{m,n}$ and right endpoint

$$\mathrm{rep}^x_{m,n} := \sqrt[s]{x^s_{m,n} + (r^s_m - r^s_{m-1})2^{-\lfloor n/2 \rfloor -2}}.$$

We also construct $Y_m = \{y_{m,0}, y_{m,1}, \ldots\}$ also to satisfy (5.4). For $m > 0$, define

$$\rho_{-m} := \rho(\cdot \mid Y_m, s_{m-1}, s_m),$$

whose domain consists of closed intervals from $y_{m,n}$ to

$$\mathrm{rep}^y_{m,n} := \sqrt[s]{y^s_{m,n} + (s^s_m - s^s_{m-1})2^{-\lfloor n/2 \rfloor -2}}.$$

We define the $x_{m,n}$s and $y_{m,n}$s by passing through the following list and assigning each entry to be the rep of its predecessor.

$$\begin{array}{llllll} 0 =: x_{1,0} & x_{1,1} & & & & \\ y_{1,0} & y_{1,1} & & & & \\ x_{1,2} & x_{2,0} & x_{2,1} & x_{1,3} & & \\ y_{1,2} & y_{2,0} & y_{2,1} & y_{1,3} & & \\ x_{1,4} & x_{2,2} & x_{3,0} & x_{3,1} & x_{2,3} & x_{1,5} \\ y_{1,4} & y_{2,2} & y_{3,0} & y_{3,1} & y_{2,3} & y_{1,5} \quad \cdots \end{array} \tag{5.14}$$

That is, we start by setting $x_{1,0}$ equal to 0, and then set $x_{1,1} = \mathrm{rep}^x_{1,0}$, then $y_{1,0} = \mathrm{rep}^x_{1,1}$, then $y_{1,1} = \mathrm{rep}^y_{1,0}$, and $x_{1,2} = \mathrm{rep}^y_{1,1}$, etc.. The typical mth row of xs in this list is

$$\begin{array}{ccccccccc} x_{1,2m-2} & x_{2,2m-4} & \cdots & x_{m-1,2} & x_{m,0} & & & & \\ & x_{m,1} & x_{m-1,3} & \cdots & x_{2,2m-3} & x_{1,2m-1} \end{array} \tag{5.14.m}$$

(to be followed by $y_{1,2m-2}, y_{2,2m-4}$, etc.), in which the first two elements of X_m are defined. In the $(m+1)$th row, the next two members of X_m are computed, and so on, so that every $x_{m,n}$ (and $y_{m,n}$) is eventually given a value. The condition (5.4) is satisfied by each X_m (and Y_m), since X_m (and Y_m) appears, in order, as a subsequence of (5.14). Thus the function ρ_m is well-defined as in Lemma (5.6) for every $m \neq 0$.

For $m \neq n$, the sets $\operatorname{dom} \rho_n$ and $\operatorname{dom} \rho_m$ intersect at most on a set of measure zero, since any two intervals of the form

$$[x_{m,n}, \mathrm{rep}^x_{m,n}] \quad \text{or} \quad [y_{m,n}, \mathrm{rep}^y_{m,n}]$$

intersect at most at an endpoint. Thus the sets $\{\operatorname{dom} \rho_m \circ r\}$ are mutually almost disjoint, intersecting on a the surface of a ball if at all. As in Lemma (4.2), we define the radial function $g \circ r$ on their union. Since $g_m \circ \rho_m \circ r \sim g_m \circ r$ for every m, the functions f and $g \circ r$ are identically distributed.

We now show $\operatorname{dom} g \circ r = \operatorname{dom} f$, so that $g \circ r$ is a rearrangement of f. Since $\operatorname{dom} g \circ r = \bigcup \operatorname{dom} g_m \circ \rho_m \circ r = \bigcup \operatorname{dom} \rho_m \circ r$, it will suffice to show that $\bigcup \operatorname{dom} \rho_m$ is $[0,1)$.

Since every $x_{m,n}$ and $y_{m,n}$ is nonnegative, each $\operatorname{dom} \rho_m$ lies entirely in $[0,\infty)$. By construction, $\bigcup \operatorname{dom} \rho_m$ is connected, its smallest member is 0, and it contains no largest member; therefore it is of the form $[0,c)$ for some $c \leq \infty$. From $g \circ r \sim f$, it follows that $m_s(\operatorname{dom} f) = m_s(\operatorname{dom} g \circ r) = \pi_s c^s$, so $c = 1$, as desired.

To see that $g \circ r$ is continuous on $B(1)$, or, equivalently, that g is continuous on $[0,1)$, first note that g_m is continuous, and that ρ_m is continuous on each interval of its domain. Hence, one need only verify that g is continuous on any nonempty $\operatorname{dom} \rho_m \cap \operatorname{dom} \rho_n$. To this end, we list all possible intersections between intervals of the forms

$$[x_{m,n}, \operatorname{rep}^x_{m,n}] \quad \text{or} \quad [y_{m,n}, \operatorname{rep}^y_{m,n}]$$

below. The intersection point is the left endpoint of the second interval and right endpoint of the first as listed; the right endpoint of each interval is simply rep of its left. We therefore suppress the argument of rep.

$$\begin{array}{rc}
a: & [x_{l,2(m-l)}, \operatorname{rep}] \cap [x_{l+1,2(m-l)-2}, \operatorname{rep}] \\
b: & [x_{m,0}, \operatorname{rep}] \cap [x_{m,1}, \operatorname{rep}] \\
c: & [x_{l+1,2(m-l)-1}, \operatorname{rep}] \cap [x_{l,2(m-l)+1}, \operatorname{rep}] \\
d: & [x_{1,2m-1}, \operatorname{rep}] \cap [y_{1,2m-2}, \operatorname{rep}] \\
a', b', c': & \text{(Replace } x \text{ by } y \text{ throughout } b,\ c,\ d.) \\
d': & [y_{1,2m-1}, \operatorname{rep}] \cap [x_{1,2m}, \operatorname{rep}]
\end{array}$$

Recall that, when $m \geq 1$, the function ρ_m increases on $[x_{m,2k}, \operatorname{rep}]$ and decreases on $[x_{m,2k+1}, \operatorname{rep}]$. Specifically,

$$\begin{aligned}
\rho_m(x_{m,2k}) &= r_{m-1} = \rho_m(\operatorname{rep}^x_{m,2k+1}), \quad \text{and} \\
\rho_m(x_{m,2k+1}) &= r_m \ = \rho_m(\operatorname{rep}^x_{m,2k}).
\end{aligned} \tag{5.15}$$

Likewise, ρ_{-m} takes the extreme values

$$\begin{aligned}
\rho_{-m}(y_{m,2k}) &= s_{m-1} = \rho_{-m}(\operatorname{rep}^y_{m,2k+1}) \quad \text{and} \\
\rho_{-m}(y_{m,2k+1}) &= s_m \ = \rho_{-m}(\operatorname{rep}^y_{m,2k}).
\end{aligned} \tag{5.16}$$

Using (5.15), (5.16), and $g_+(0) = g_-(0) = 0$, it is not hard to check continuity at each intersection point. For example, g is continuous at point a, since from the left g approaches

$$g_l(\rho_l(\operatorname{rep}^x_{l,2k})) = g_l(r_l)$$

and from the right

$$g_{l+1}(\rho_{l+1}(x_{l+1,2\bar{k}})) = g_{l+1}(r_l)$$

and these are equal by (5.13). The proof of continuity at c, a', and c' are similar. At b, the left-hand limit is $g_m(\rho_m(\operatorname{rep}^x_{m,0}))$ and the right hand limit is $g_m(\rho_m(x_{m,1}))$. Both of these equal $g_m(r_m)$ by (5.15); continuity at b' follows similarly. The left-hand limit at d is

$$g_1(\rho_1(\operatorname{rep}^x_{1,2m-1})) = g_1(r_0) = g_+(0) = 0,$$

while the right-hand limit is

$$g_{-1}(\rho_{-1}(y_{1,2m-2})) = g_{-1}(s_0) = g_-(0) = 0.$$

Hence g is continuous at d; continuity at d' is proved similarly.

Thus g is continuous on $[0,1)$, and $g \circ r$ is continuous on $B(1)$.

Finally, we consider the integral of $g(r)\,dr^s$ on $(0,1)$.

We claim

$$\int_{x_{1,2k-2}}^{\mathrm{rep}^x_{1,2k-1}} g(r)\,dr^s = a_{2k} \quad (k \geq 1) \tag{5.17}$$

and

$$\int_{y_{1,2k-2}}^{\mathrm{rep}^y_{1,2k-1}} g(r)\,dr^s = -a_{2k+1} \quad (k \geq 1). \tag{5.18}$$

By a simple change of variables, one can check that

$$\int_{x_{m,n}}^{\mathrm{rep}^x_{m,n}} g(r)\,dr^s = 2^{-\lfloor n/2 \rfloor -2} \int_{r_{m-1}}^{r_m} g_+(u)\,du^s \qquad (m \geq 1, \quad n \geq 0) \tag{5.19}$$

and

$$\int_{y_{m,n}}^{\mathrm{rep}^y_{m,n}} g(r)\,dr^s = 2^{-\lfloor n/2 \rfloor -2} \int_{s_{m-1}}^{s_m} g_-(u)\,du^s \qquad (m \geq 1, \quad n \geq 0) \tag{5.20}$$

As a consequence,

$$\int_{x_{m,n}}^{\mathrm{rep}^x_{m,n}} g(r)\,dr^s = \frac{1}{2}\int_{x_{m,n-2}}^{\mathrm{rep}^x_{m,n-2}} g(r)\,dr^s \tag{5.21}$$

and

$$\int_{y_{m,n}}^{\mathrm{rep}^y_{m,n}} g(r)\,dr^s = \frac{1}{2}\int_{y_{m,n-2}}^{\mathrm{rep}^y_{m,n-2}} g(r)\,dr^s. \tag{5.22}$$

By (5.19) and (5.12), the integral of g from $x_{m,0}$ to $\mathrm{rep}^x_{m,0} = x_{m,1}$ and from there to $\mathrm{rep}^x_{m,1}$ is

$$\int_{x_{m,0}}^{\mathrm{rep}^x_{m,1}} g(r)\,dr^s = \frac{1}{2}\int_{r_{m-1}}^{r_m} g_+(u)\,du^s = a_{2m} - \frac{1}{2}a_{2m-2}. \tag{5.23}$$

Similarly, (5.20) and (5.12) imply

$$\int_{y_{m,0}}^{\mathrm{rep}^y_{m,1}} g(r)\,dr^s = \frac{1}{2}\int_{s_{m-1}}^{s_m} g_-(u)\,du^s = -a_{2m+1} + \frac{1}{2}a_{2m-1}. \tag{5.24}$$

Letting $m = 1$ in (5.23) and (5.24) verifies the $k = 1$ cases of (5.17) and (5.18). We now induct on k. By (5.14.m), (suppressing $g(r)\,dr^s$)

$$\int_{x_{1,2k-2}}^{\mathrm{rep}^x_{1,2k-1}} = \int_{x_{1,2k-2}}^{\mathrm{rep}^x_{k-1,2}} + \int_{x_{k-1,3}}^{\mathrm{rep}^x_{1,2k-1}} + \int_{x_{k,0}}^{\mathrm{rep}^x_{k,1}}.$$

By (5.23), the last of these is $a_{2k} - \frac{1}{2}a_{2k-2}$, and, by (5.21) and (5.14.m), the first two equal

$$\frac{1}{2}\left(\int_{x_{1,2k-4}}^{\mathrm{rep}^x_{k-1,0}} + \int_{x_{k-1,1}}^{\mathrm{rep}^x_{1,2k-3}}\right),$$

which by the induction hypothesis is $\frac{1}{2}a_{2k-2}$, proving (5.17). The proof of (5.18) is similar.

Keeping the order of (5.14) in mind, we can rewrite

$$\int_0^{\mathrm{rep}^y_{1,2k-1}} g(r)\,dr^s = \sum_{n=1}^{k} \int_{x_{1,2n-2}}^{\mathrm{rep}^x_{1,2n-1}} g(r)\,dr^s + \sum_{n=1}^{k} \int_{y_{1,2n-2}}^{\mathrm{rep}^y_{1,2n-1}} g(r)\,dr^s.$$

By virtue of (5.17) and (5.18),

$$\int_0^{\mathrm{rep}^y_{1,2k-1}} g(r)\,dr^s = \sum_{m=2}^{2k+1} (-1)^m a_m,$$

and similarly,

$$\int_0^{\mathrm{rep}^x_{1,2k-1}} g(r)\,dr^s = \sum_{m=2}^{2k} (-1)^m a_m.$$

Viewed as a function of its upper limit, $\int_0^\alpha g(r)\,dr^s$ is monotone for α between $\mathrm{rep}^x_{1,2k-1} = y_{1,2k-2}$ and $\mathrm{rep}^y_{1,2k-1}$, and for α between $\mathrm{rep}^y_{1,2k-1} = x_{1,2k}$ and $\mathrm{rep}^x_{1,2k+1}$, because g is of constant sign on those intervals. Since

$$\lim_{k\to\infty} x_{1,k} = \lim_{k\to\infty} y_{1,k} = R,$$

this implies

$$\lim_{\alpha\to R^-} \int_0^\alpha g(r)\,dr^s = \sum_{m=2}^{\infty} (-1)^m a_m.$$

Whether the series converges to ξ/π_s (as in (5.10)), or fails to converge (as in (5.11)), this finishes the proof of Theorem (5.8). □

The appendix shows an example of this construction in one variable.

We note that if $\mathrm{dom}\, f = \mathbb{R}^s$, then the same construction yields a rearrangement whose GRI equals ξ or fails to exist, as desired, provided that both f^+ and f^- have nondecreasing radial rearrangements. For this, it is necessary that $m\big(f^{-1}(a,b)\big) < \infty$ whenever $0 < b - a < \infty$.

6. CONTINUOUS REARRANGEMENTS ON INTERVALS

In this section, we construct a rearrangement on an interval in $\mathbb{R}^s$ and investigate the resulting GRI.

Theorem 6.1: *Let f be a conditionally integrable continuous function on H, an open and bounded interval in $\mathbb{R}^s$. Then f has a continuous rearrangement which is not GR-integrable on H. Furthermore, for every $\xi \in [-\infty,\infty]$, there exists a continuous rearrangement of f whose GRI over H equals ξ.*

Proof: Without loss of generality, we take $\xi \geq 0$ and $H = [-1,1] \times [0,1]^{s-1}$.

Define f_* as in Lemma (5.1); replace f_* by $f_*(\cdot+1)$, so that f_* is nondecreasing with domain $(-1,1)$. For every real α, (5.2) is satisfied, and, because $m_s(f^{-1}(a,b)) > 0$ for (a,b) any nontrivial interval, f_* is continuous.

Define the linear projector P on H by $P(x) = x(1)$. Then $f_* \circ P$ is a continuous rearrangement of f, since

$$m_s(P^{-1}f_*^{-1}(-\infty,\alpha)) = m_s(f_*^{-1}(-\infty,\alpha) \times [0,1]^{s-1})$$

and this is $m_s(f^{-1}(-\infty,\alpha))$ by (5.2). Choose $x_0 \in (-1,1)$ so that $f_* \le 0$ on $[-1,x_0]$ and $f_* \ge 0$ on $[x_0,1]$. The integral of $f_* \circ P$ over $[-1,x_0] \times [0,1]^{s-1}$ in infinite, because $(f_* \circ P)^- \sim f^-$. Since $f_* \circ P$ is not GR-integrable on a subinterval of H, it cannot be integrable on H, proving the first conclusion of Theorem (6.1).

To construct a rearrangement whose GRI is ξ, apply Corollary (5.9) to obtain an even rearrangement $g(|\cdot|)$ of f_* such that $\int_{-1}^{1} g(|x|)\,dx = \xi$. Then

$$G := g(|P(\cdot)|)$$

is defined on H and, for every real α,

$$\begin{aligned} m_s(G^{-1}(-\infty,\alpha)) &= m_s(g(|\cdot|)^{-1}(-\infty,\alpha) \times [0,1]^{s-1}) \\ &= m_1(f_*^{-1}(-\infty,\alpha)) \end{aligned}$$

By (5.2), $G \sim f$.

Let $I := [0,1]^s$, and $I' = [0,1) \times [0,1]^{s-1}$. By Theorem (3.2), if

$$\lim_{\mathcal{D}} \nu(\mathcal{D}) = \frac{\xi}{2}, \tag{6.2}$$

where

$$\nu(\mathcal{D}) := \sum_{\substack{J \in \mathcal{D} \\ J \subseteq I'}} \int_J G(x)\,dx$$

for $\mathcal{D}$ a division of I, then $\int_I G\,dx = \xi/2$. It will then follow that $\int_H G\,dx = \xi$, as desired.

From the construction of g, there exists $a_n > 0$ such that $\sum_{n=2}^{\infty}(-1)^n a_n = \xi/2$, and there exists a sequence $0 = b_1 < b_2 < \cdots$ converging to 1 such that

$$\int_{b_{n-1}}^{b_n} g = (-1)^n a_n.$$

Furthermore, $(-1)^n g \ge 0$ on (b_{n-1}, b_n). Define $S_m := \sum_{n=2}^{m}(-1)^n a_n$.

For every $\varepsilon > 0$, there exists N such that $|S_m - \xi/2| < \varepsilon$ for all $m \ge N$. Define $\gamma = \gamma_\varepsilon$ on I by

$$\gamma(z) = \begin{cases} (b_N, \infty) \times \mathbb{R}^{s-1} & \text{if } z(1) = 1; \text{ and} \\ (-\infty, \frac{1}{2}(z(1)+1)) \times \mathbb{R}^{s-1} & \text{if } z(1) < 1. \end{cases}$$

Let $\mathcal{D} = \{(z_J, J)\}$ be any γ-fine division on I. We'll show that

$$\left|\nu(\mathcal{D}) - \frac{\xi}{2}\right| < \varepsilon, \tag{6.3}$$

proving (6.2).

For $J \in \mathcal{D}$, we let $[j_1, j_2]$ denote the image of J under the projection P, and define $A_J := m_s(J)/(j_2 - j_1)$, the $(s-1)$-dimensional area of J's cross-section. Then, provided $j_2 < 1$, Fubini's theorem allows $\int_J G = A_J \int_{j_1}^{j_2} g$.

Let $\bigcup' J$ denote the union of all J in $\mathcal{D}$ with $J \not\subseteq I'$. Let $\mathcal{D}^n$ denote the collection of those $J \not\subseteq I'$ for which $j_1 \le b_n$, and let $\bigcup^n$ and $\sum^n$ stand for a union and a sum over $\mathcal{D}^n$.

Define $I_n := [b_{n-1}, b_n] \times [0,1]^{s-1}$. Then $I_n \setminus \bigcup' J = I_n \setminus \bigcup^n J$. Since $(-1)^n G \ge 0$ on I_n,

$$\int_{I_n \setminus \bigcup^n J} G \le \int_{I_n} G = a_n.$$

Define

$$w_n := a_n^{-1}(-1)^n \int_{I_n \setminus \bigcup^n J} G.$$

Then $w_n \in [0,1]$.

Fix n. Every $J \in \mathcal{D}^n$ has the form $[j_1, 1] \times K$ for $j_1 \le b_n$ and K a closed interval in $\mathbb{R}^{s-1}$. For $J \in \mathcal{D}^n$ define $J^* := [b_{n-1}, 1] \times K$. Then $A_J = A_{J^*}$ and $I_n \setminus \bigcup^n J \supset I_n \setminus \bigcup^n J^*$ together imply

$$(-1)^n \int_{I_n \setminus \bigcup^n J} G \ge (-1)^n \int_{I_n \setminus \bigcup^n J^*} G = a_n(1 - \sum\nolimits^n A_J).$$

Hence $w_n \ge 1 - \sum^n A_J$. Likewise,

$$(-1)^{n+1} \int_{I_{n+1} \setminus \bigcup^{n+1} J} G \le (-1)^{n+1} \int_{I_{n+1} \setminus \bigcup^n J} G = a_{n+1}(1 - \sum\nolimits^n A_J),$$

so that $w_{n+1} \le 1 - \sum^n A_J$. Therefore, for every n,

$$w_n \ge w_{n+1}.$$

Since $\mathcal{D}$ is γ-fine, if $J \not\subseteq I'$, then $j_1 > b_N$, proving $\mathcal{D}^n = \emptyset$ for $n \le N$. Also, for each of the (finitely many) $J \subseteq I'$, we have $j_2 < 1$, so there exists $M > N$ for which $j_2 < b_M$ for all $J \subseteq I'$. In terms of w_n, these two facts translate to

$$w_n = \begin{cases} 1 & \text{if } n \le N, \text{ and} \\ 0 & \text{if } n > M. \end{cases}$$

We now consider $\nu(\mathcal{D})$, by definition $\sum_{J \subseteq I'} \int_J G$. We choose to write this as

$$\sum_{n=2}^{\infty} \int_{I_n \setminus \bigcup' J} G = \sum_{n=2}^{\infty} \int_{I_n \setminus \bigcup^n J} G.$$

The summand above is zero for all but finitely many terms. In fact,

$$\nu(\mathcal{D}) = \sum_{n=2}^{N} (-1)^n a_n + \sum_{n=N+1}^{M} (-1)^n a_n w_n.$$

Summation by parts gives

$$S_N + \sum_{n=N+1}^{M} (-1)^n a_n w_n = \sum_{n=N}^{M} S_n(w_n - w_{n+1}),$$

which, on the one hand, is

$$\begin{aligned} &\leq \left(\frac{\xi}{2} + \varepsilon\right) \sum_{n=N}^{M} (w_n - w_{n+1}) \\ &= \frac{\xi}{2} + \varepsilon, \end{aligned}$$

and, on the other, is $\geq \xi/2 - \varepsilon$. Thus (6.3) is true, completing the proof. □

APPENDIX

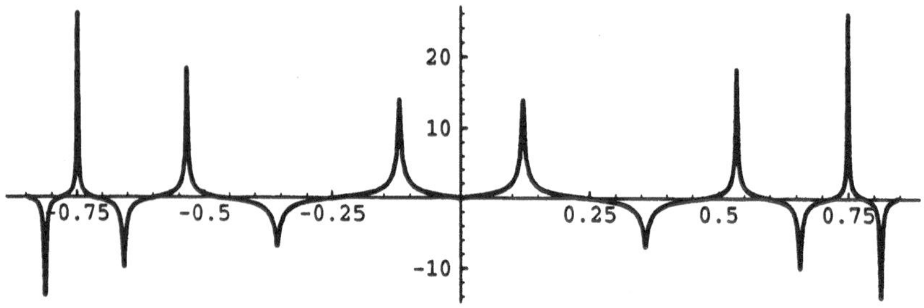

Figure 3. A rearrangement $g(|x|)$ of $f(x) = x/(1 - x^2)$.

Figure 3 shows a rearrangement of $f(x) = x/(1 - x^2)$ on the interval $(-1, 1)$, constructed as in the proof of Theorem (5.8). Since f is monotone, the radial (i.e., even) rearrangements of f^+ and $-f^-$ satisfy $g_+(|x|) = -g_-(|x|) = f(2|x|)$ on $(-\frac{1}{2}, \frac{1}{2})$. Calculations were made using $a_m = (m+1)^{-1}$, so that the GRI equals $\ln 4 - 1$.

Each ρ_m is piecewise linear, with its derivative taking the values $\pm 2^k$ for $k \geq 2$. Thus the graph of $g(|x|)$ is composed of horizontally scaled segments of the original graph, each appearing four times with a scaling factor $\pm 2^k$ for every $k \geq 3$. By (5.17) and (5.18), the integral of the rearrangement between any two successive x-intercepts equals one term of the series $\sum_{m \geq 2} (-1)^m a_m$.

ACKNOWLEDGMENTS

The author wishes to thank to Professor Robert McLeod for generous and helpful clarifications of his remarks in [2].

BIBLIOGRAPHY

[1] R.G. Bartle, Review of *The General Theory of Integration*, by Ralph Henstock, *Bull. Amer. Math. Soc. (New Series)* **29** (1993), 136–139.

[3] R.M. McLeod, private communication, 1993.

[2] R.M. McLeod, *The Generalized Riemann Integral*, Carus Mathematical Monographs, Number 20, Math. Assoc. of America., 1980.

[4] O. Shisha, The generalized Riemann integral (lecture), International Conference on Approximation, Probability, and Related Fields, Santa Barbara, CA, May 22, 1993.

[5] A. Zygmund, *Trigonometric Series*, Vol. 1, Cambridge University Press, Cambridge, 1977.

WAVELET TRANSFORMS AND WAVELET APPROXIMATIONS

E. B. Lin

Department of Mathematics
University of Toledo
Toledo, OH 43606

Abstract We summarize properties of classical wavelet transforms and Wavelet Stieltjes transforms. Wavelet approximation problems are also considered for Wavelet Stieltjes transforms. This will give rise to some characterizations of general signals.

I. Introduction

We study and summarize some properties of wavelet transforms and Wavelet Stieltjes transforms which are recently introduced and obtained in [2,3]. Further observations and analysis on the space of general signals are given in this paper. To apply wavelet approximation methods to Wavelet Stieltjes transforms, we provide some useful ideas and approaches for approximation problems in general.

Notation

Z and R denote the set of integers and real numbers respectively. $L^2(R)$ denotes the vector space of measurable, square-integrable one-dimensional functions. For $f(x),\ g(x) \in L^2(R)$, the inner product of $f(x)$ with $g(x)$ is

$$< g(u), f(u) > = \int_{-\infty}^{\infty} g(u)f(u)du.$$

The norm of $f(x)$ in $L^2(R)$ is given by

$$\|f\|^2 = \int_{-\infty}^{\infty} |f(u)|^2 du.$$

We will use the same inner product, norm notation for other Hilbert spaces as well.

We denote the convolution of two functions $f(x) \in L^2(R)$ and $g(x) \in L^2(R)$ by

$$f * g(x) \ = \ (f(u) * g(u))(x)$$

Approximation, Probability, and Related Fields, Edited by
G. Anastassiou and S.T. Rachev, Plenum Press, New York 1994

$$= \int_{-\infty}^{\infty} f(u)g(x-u)du.$$

The Fourier transform of $f(x) \in L^2(R)$ is written $\hat{f}(w)$ and is defined by

$$\hat{f}(w) = \int_{-\infty}^{\infty} f(x)e^{-iwx}\, dx.$$

For any function $\psi : R \to R$ and $a, b \in R$, the following notations are being used

$$\psi^{a,b}(t) := \begin{cases} \psi(\frac{t-b}{a}), & a \neq 0 \\ 0, & a = 0 \end{cases}$$

$$\psi_{kj}(x) := 2^{\frac{k}{2}}\psi(2^k x - j) \quad for\ k, j \in Z$$

$$c_\psi \ := \int_{-\infty}^{\infty} \frac{|\hat{\psi}(\omega)|^2}{|\omega|} d\omega.$$

II. Wavelet Transforms

§2.1 Classical wavelet transforms

We call the following wavelet transform as a classical wavelet transform:

$$W_\psi\ f(a,b) \ := \ |a|^{-\frac{1}{2}} \int_{-\infty}^{\infty} f(x)\overline{\psi(\frac{x-b}{a})}\, dx, \quad f \in L^2(R), \quad a, b \in R \qquad (2.1.1)$$

Properties of (2.1.1) are summarized as follows:

Let $f,\ g \in L^2(R)$

(i) $\forall \alpha,\ \beta\ \in\ R$

$W_\psi\ (\alpha f + \beta g)(a,b) \ = \ \alpha W_\psi f(a,b) + \beta W_\psi g(a,b)$

(ii) Let $g(x) = f(-x)$ Then

$W_\psi g(a,b) = W_\psi f(-a,-b)$

(iii) Let $g(x) = f(x - x_0)$ Then

$W_\psi\ g(a,b) = W_\psi\ f(a, b - x_0)$

(iv) $\int_{-\infty}^{\infty} \int_{-\infty}^{\infty} |W_\psi\ f(a,b)|^2\ \frac{da db}{a^2} \ = \ c_\psi \|f\|^2$

(v) Let $g(x) = f(\frac{x}{\lambda}) \quad \lambda > 0$ Then

$W_\psi\ g(a,b) \ = \ \sqrt{\lambda}\, W_\psi\ f(a\lambda^{-1},\ b\lambda^{-1})$

(vi) Let $g(x) = f'(x)$ Then

$\frac{\partial W_\psi f}{\partial b}\ (a,b) \ = \ W_\psi\ g(a,b)$

(vii) Let $g(x) = f * h(x) = \int_{-\infty}^{\infty}\ f(u)h(x-u)du$

Then $W_\psi\ g(a,b) \ = \ f * W_\psi h(a,\cdot)(b) \ = \ h * W_\psi f(a,\cdot)(b)$

(viii) $\int_{-\infty}^{\infty} \int_{-\infty}^{\infty} [(W - \psi f)(b,a) W_\psi g(b,a)] \frac{da db}{a^2} = c_\psi < f, g >$ (Parseval Identity)

(ix) $f(x) = \frac{1}{c_\psi} \int_{-\infty}^{\infty} \int_{-\infty}^{\infty} [(W_\psi f)(b,a)]\psi_{b,a}(x) \frac{da db}{a^2}$ for any $f \in L^2(R)$ and $x \in R$ at which f is continuous.

Remark 2.1.1

Substitute (2.1.2) into (2.1.3) below. Equating both sides of (2.1.3), in general, one can choose l, m in the following settings:

$$W_\psi\ f(a,b) \ = \ |a|^l \int f(t)\psi(\frac{t-b}{a})\, dt \qquad (2.1.2)$$

and

$$\int_{-\infty}^{\infty}\int_{-\infty}^{\infty}[W_\psi f(b,a)\ \overline{(W_\psi g)}(b,a)]a^m\,dadb = c_\psi < f,g > \tag{2.1.3}$$

where l, m can be shown to satisfy $2l + m + 3 = 0$.

For simplicity, we usually choose $l = -\frac{1}{2}, m = -2$.

Remark 2.1.2

Covolution theorem only holds for Laplace transform, in particular, for Fourier transform. In general, it is not true for any wavelet transform.

§2.2 Wavelet Stieltjes Transforms (WST)

A. Definition of WST

Definition 2.2.1

$$BV(1,2) = \{F|F : R \to R, F(\bullet) = \int_{-\infty}^{\bullet} f(t)dt + \sum_{s\leq\bullet} \rho(s),$$

$$f \in L^1(R) \cap L^2(R), \quad \rho \in l_1 \cap l_2 \ \} \tag{2.2.1}$$

where summation convention is summed over countable discontinuities of F.

Note that any function F in $BV(1,2)$ is of bounded first variation, where the integral part is the absolutely continuous component of F and the sum part is the jump component of F. Of course, the singular component is absent from any function F in $BV(1,2)$. We shall write $\mathbf{F} \sim (\mathbf{f}, \rho)$ to associate F with its absolutely continuous and jump components.

Without loss of generality, we shall be assuming that

$$interior\ (supp\ f)\ \cap\ J_F\ = \emptyset \tag{2.2.2}$$

where $F \sim (f, \rho)$ and J_F is the support of jump component of F, that is $J_F = supp\ \rho$. Consequently, we have $f(t)\rho(t) = 0,\ t \in R$.

Definition 2.2.2

For $F \in BV(1,2)$, define

$$\|F\|_1 := \int_{-\infty}^{\infty} |f(t)|dt + \sum_{-\infty<s<\infty} |\rho(s)| \tag{2.2.3}$$

and

$$\|F\|_2 := [\int_{-\infty}^{\infty} |f(t)|^2 dt + \sum_{-\infty<s<\infty} |\rho(s)|^2]^{\frac{1}{2}} \tag{2.2.4}$$

In fact, $\|\cdot\|_1$ and $\|\cdot\|_2$ are norms on $BV(1,2)$.

Proposition 2.2.1 Let $F \in BV(1,2)$ and $V^1(F)$ be its first variation. Then

$$V^1(F) = \|F\|_1 \tag{2.2.5}$$

Let $BV(2)$ denote the completion of $BV(1,2)$ with respect to the $\|\cdot\|_2$ norm given in (2.2.4). We endow the Hilbert space $BV(2)$ with an inner product defined in (2.2.6) below:

Let $F, G \in BV(2)$ and $F \sim (f, \rho)$, $G \sim (g, \gamma)$ then

$$< F, G >:= \int_{-\infty}^{\infty} f(t)g(t)dt + \frac{1}{2}\sum_{s\in J_F} \rho(s)\gamma(s)+$$

$$\frac{1}{2}\sum_{s\in J_G} \rho(s)\gamma(s) + \frac{1}{2}\sum_{s\in J_F} \rho(s)g(s) + \frac{1}{2}\sum_{s\in J_G} f(s)\gamma(s) \tag{2.2.6}$$

Remark 2.2.1

In the above definition, we have slightly abused notation by writing $F \sim (f, \rho)$ for $F \in BV(2)$. The inner product introduced in (2.2.6) is compatible with norm (2.2.4).

In order to define the Wavelet-Stieltjes Transform, we need to introduce the following class of functions.

$$\Psi := \{\psi | \psi : R \to R, (\psi, F) < \infty,\ F,\ \psi \in BV(2)\} \tag{2.2.7}$$

where

$$(\psi, F) := < \psi, F > \tag{2.2.8}$$

that is $(\psi, F) = \int_{-\infty}^{\infty} \psi(t)dF(t)$, the last integral being the Lebesgue-Stieltjes integral.

Definition 2.2.3

The integral transform of $F \in BV(1,2)$ given by

$$(WS_\psi F)(a, b) := (\psi_{a,b}, F) \tag{2.2.9}$$

for $\psi \in \Psi$, is called the Wavelet-Stieltjes transform (WST) of F with kernel $\psi_{a,b}$.

Remark 2.2.2

In fact, $WS_\psi F \in L^2(R^2, \mu)$ where $\mu(da, db) = \frac{da\,db}{|a|^3}$, for $F \in BV(1,2)$, $\psi \in \Psi$. Using this result we naturally extend the definition of WST to all functions in BV(2).

B. Uniqueness of the WST

The following proposition shows the uniqueness of the WST for functions in $BV(1,2)$. Uniqueness for functions in $BV(2)$ will follow from the results of reference 2, 3.

Theorem 2.2.1 Let $F_1, F_2 \in BV(1,2)$ be such that

$$WS_\psi F_1 \equiv WS_\psi F_2$$

for some $\psi \in \Psi$ and

$$0 < c_\psi < \infty\ . \tag{2.2.10}$$

Then

$$F_1 \equiv F_2.$$

Example

Let us consider the following example, similar to the one given on page 4 of reference 5. Let $F \sim (f, \rho)$. Where for some $T > 0$ and $t_1, t_2, \in [-T, T]$,

$$f(t) = \begin{cases} \sin(2\pi\nu_1 t), & t \neq t_1, t_2, t \in [-T, T] \\ 0, & t = t_1, t_2 \text{ or } t \notin [-T, T] \end{cases} \tag{2.2.11}$$

$$\rho(t_1) = \rho(t_2) = \gamma + \sin(2\pi\nu_1 t_1) \tag{2.2.12}$$

Take any $\psi \in \Psi$. Note that the classical wavelet transform does not distinguish between the two signals f and $f + \rho$ in the sense that

$$W_\psi f \equiv W_\psi(f + \rho) \tag{2.2.13}$$

However, consider $F_1 \sim (f, 0)$ and $F_2 \sim (f, \rho)$ with f and ρ as above. The Wavelet-Stieltjes transform does discriminate between F_1 and F_2 which is due to the fact that

$$WS_\psi F_1 \not\equiv WS_\psi F_2 \tag{2.2.14}$$

C. Parseval Identity and Inversion Formula

Theorem 2.2.2 (Parseval Identity)

Let $F, G \in BV(2)$ and $\widetilde{WS}[F](a,b) = \int_{-\infty}^{\infty} \psi_{a,b}\, dF(x)$
Then

$$\int_{-\infty}^{\infty} \int_{-\infty}^{\infty} \widetilde{WS}[F](a,b)\widetilde{WS}[G](a,b)\mu = c_\psi < F, G >$$

where $\mu = \frac{da\,db}{a^3}$
Apply the above theorem, we have the following inverse WST formula.

Theorem 2.2.3

Let $F \in BV(2)$ and $\tilde{F}(x) = \frac{F(x+) + F(x-)}{2}$
Then

$$\tilde{F}(x_1, x_2] = \frac{1}{c_\psi} \int_{-\infty}^{\infty} \int_{-\infty}^{\infty} [\widetilde{WS}[F](a,b) \int_{x_1}^{x_2} \psi_{a,b}(x)dx]\mu$$

The proofs of theorems of this section can be found in reference 2, 3.

Properties of WST are listed in the following theorem which is analogous to properties of classical wavelet transforms in §2.1.

Theorem 2.2.4

Let $F \sim (f, \rho)$, $G \sim (g, \sigma)$. Then
(i) $\forall \alpha, \beta \in R$
$\alpha F + \beta G \sim (\alpha f + \beta g,\ \alpha\rho + \beta\sigma)$
and hence

$$WS_\psi(\alpha F + \beta G)(a,b) =$$

$$\alpha WS_\psi[F](a,b) + \beta WS_\psi[G](a,b)$$

(ii) If $G(x) = F(-x)$ then $f(x) = g(-x)$, $\rho(x) = \sigma(-x)$ and

$$WS_\psi[G](a,b) = WS_\psi[F](-a,-b)$$

(iii) If $G(x) = F(x - x_0)$

then $g(x) = f(x - x_0), \sigma(x) = \rho(x - x_0)$ and

$$WS_\psi[G](a,b) = WS_\psi[F](a, b - x_0)$$

(iv) Let $G(x) = F(\frac{x}{\lambda}),\ \ \lambda > 0$
then $g(x) = f(\frac{x}{\lambda}),\ \ \sigma(x) = \rho(\frac{x}{\lambda})$

$$WS_\psi[G](a,b) =$$

$$\lambda \int_{-\infty}^{\infty} f(t)\psi^{\lambda^{-1}a,\lambda^{-1}b}\,(t)dt + \sum \rho(s)\psi^{\lambda^{-1}a,\lambda^{-1}b}\,(s)$$

(v) $\int_{-\infty}^{\infty} \int_{-\infty}^{\infty}\ |\tilde{W}S[F](a,b)|^2 \mu(da,db) = c_\psi \parallel F \parallel^2$

D. Wavelet Stieltjes Decompositions

Let $F \in BV(2),\ F \sim (f, \rho)$ where $f \in L^2,\ \rho \in l_2$ and let ν be counting measure on R. Since $f \in L^2$, we have discrete wavelet decomposition

$$f = \sum c_{m,n}\psi_{m,n}$$

where $\psi_{m,n}$ is generated by a mother wavelet ψ in multiresolution analysis, and

$$c_{m,n} = < f, \psi_{m,n} > .$$

Hence the continuous part has the same features as MRA in L^2, e.g. fast algorithm, decomposition and reconstruction algorithms.

For the discrete part, we consider the following translation and dilation of measure,

$$\nu_{m,\varphi} = a_0^{-\frac{m}{2}}\nu(\frac{x - \varphi(n)b_0 a_0^m}{a_0^m})$$

where m, n range over Z, $\varphi : Z \to R$ and $a_0 > 1,\ b_0 > 1$ are fixed.

We have Wavelet Stieltjes Decomposition

$$F = \sum c_{m,n} \int \psi_{m,n}\,(x)dx + \sum_{\Gamma} d_\Gamma \nu_\Gamma$$

where $d_\Gamma = < \rho, \nu_\Gamma >$.

The topology in discrete part is standard product topology.[7]

Remark 2.2.3

Since the jump part is randomly located, there is, in general, no explicit expression for $\varphi(n)$. More precisely, the decomposition in the jump part is generally uncountable.

Remark 2.2.4

If the locations of discontinuity are fixed, i.e. time scale of jump are fixed, say on integers, then we can write $\varphi(n) = n$, whose representation is actually similar to the analysis of representation in L^2 or l_2 MRA. A detail analysis is in the next paragraph. We will denote this subspace as $BV(2, Z)$, i.e., a subspace of $BV(2)$ whose jump parts are fixed at integer points.

E. Multiresolution Analysis (MRA)

As mentioned in the previous remarks, there is, in general, no explicit expression for Wavelet Stieltjes decomposition. However, in the case of remark 2.2.4 stated, we can apply the following construction for each interval $[n, n+1]$ on the real line and l^2 MRA for integer points where discontinuities occur.

Multiresolution analysis on $[0,1]$ [4]

Consider approximation subspaces

$$V_{j_0} \subset V_{j_0+1} \subset V_{j_0+2} \cdots toL^2[0,1]$$

$V_j (j \geq j_0)$ is generated by

$$\{\varphi_{j,k}\}_{1 \leq k \leq 2^j - 2N} \quad \{\tilde{\varphi}^{l,i}(2^j x)\}_{0 \leq i \leq N-1}$$

and

$$\{\tilde{\varphi}^{r,i}(2^j(x-1)\}_{0 \leq i \leq N-1}$$

where φ is scaling function in $L^2(R)$ MRA with $supp\,(\varphi_{j,0}) = [0, 2^{-j}(2N-1)]$ included in $[0, \frac{1}{2}]$.

$$\tilde{\varphi}^{l,i}(x) = (P_i(x) - \sum_{k>0} k^i \varphi(x-k))\chi_{[0,+\infty]}$$

$$= \sum_{k \leq 0} k^i \varphi(x-k)\chi_{[0,+\infty]}$$

$$\tilde{\varphi}^{r,i}(x) = (P_i(x) - \sum_{k<(1-2N)} k^i \varphi(x-k))\chi_{[-\infty,0]} = \sum_{k=-(2N-1)}^{-1} k^i \varphi(x-k)\chi_{[-\infty,0]}$$

and

$$P_i(x) = \sum_k k^i \varphi(x-k)$$

In fact, for all $j \geq j_0$, $V_j \subset V_{j+1}$ and the set of functions

$$\{\tilde{\varphi}^{l,i}(2^j x), \tilde{\varphi}^{r,i}(2^j x)\}_{0 \leq i \leq N-1} \ \cup \ \{\varphi_{j,k}\}_{1 \leq k \leq 2^j - 2N}$$

form a basis of V_j. Hence, MRA on $L^2[0,1]$ and l^2 give rise to MRA on $BV(2,Z)$.

III. Wavelet Approximations

§3.1 Classical Wavelet Approximations

For $f \in C(R)$, define

$$A_k(f)(x) := \sum_{j=\infty}^{\infty} < f, \varphi_{kj} > \varphi_{kj}(x), \qquad k \in Z$$

where $< f, \varphi_{kj} > := \int_{-\infty}^{\infty} f(t)\varphi_{kj}(t)dt \quad \varphi$ is compactly supported. We recall an approximation theorem in [1].

Assumption 3.1.1

Let $\varphi(x)$ be a bounded function with $supp\ \varphi(x) \leq [-a, a]$, $0 < a < +\infty$ and satisfies the following conditions

(i) $\sum_{j=-\infty}^{\infty} \varphi(x-j) \equiv 1 \ \ on\ R$;

(ii) there is a number b such that $\varphi(x)$ is non-decreasing if $x \leq b$ and is non-increasing if $x \geq b$.

Theorem 3.1.1

Suppose that $\varphi(x)$ satisfies assumption 3.1.1 above.

Then, for $f \in C(R)$, if f is a non-decreasing function, the linear wavelet operator $A_k(f)$ are also non-decreasing on R and satisfy

$$|A_k(f)(x) - f(x)| \leq \omega(f, 2^{-k+1}a), \quad x \in R, \ k \in Z$$

where $\omega(f,h) := sup_{|t|\leq h} |f(x+t) - f(x)|$. Moreover, the inequalities are sharp.

For $F \in BV(2)$, F is not in $C(R)$. We can not apply the above theorem directly. However, we can proceed by modifying F with a help of a nontrivial continuous function l such that $F^l = lF$ belongs to $C(R)$. Then estimate $|A_k(F)(x) - F(x)|$ by comparing to $|A_k(F^l)(x) - F^l(x)|$. In $BV(2, Z)$, alternatively, one can apply the above theorem in each subinterval $x \in (n, n+1)$, $n \in Z$. We therefore have the following propositions.

Proposition 3.1.1

Suppose that $\varphi(x)$ satisfies assumption 3.1.1 above. In addition, let l be a continuous function such that $F^l \in C(R)$ for $F \in BV(2)$.

Then, if F^l is a non-decreasing function, the linear wavelet operator $A_k(F^l)$ are also non-decreasing on R and satisfy

$$|A_k(F^l)(x) \ - \ F^l(x)| \leq \omega(F^l, \ 2^{-k+1}a), \quad x \in R, \ k \in Z.$$

Proposition 3.1.2

Suppose that $\varphi(x)$ satisfies assumption 3.1.1 above. Let $F \in BV(2, Z)$ and $f_n = F\lceil_{(n,n+1)}$. Then if f_n is a non-decreasing function, the linear wavelet operator $A_k(f_n)$ are also non-decreasing on R and satisfy

$$|A_k(f_n)(x) - f_n(x)| \leq \omega(f_n, 2^{-k+1}a), \quad x \in R, \ k \in Z.$$

§3.2 Compression

As it is mention in paragraph D of section 2.2, Wavelet Stieltjes decomposition in jump part is usually uncountable. It is therefore interesting in the following nonlinear approximation problem: how shall we pick N coefficients b_I in order to minimize the error

$$\sigma_N(f)_S \ := \ inf_{\sharp\Gamma \leq N} \ \| f - \sum_{I \in \Gamma} b_I \psi_I \|_S, \ N \leq 1$$

when S is $BV(2)$?

Here $\{\psi_I(x) = 2^{\frac{i}{2}}\psi(2^i x - k) : I = I_{ik} = [\frac{k}{2^i}, \frac{k+1}{2^i}) \ i, k \in Z\}$ is an orthonormal wavelet basis of a subspace in S. To answer this question, we need the following definitions and theorem

Let e_I be the Kronecker sequence

$$(e_I)_J = \begin{cases} 1 & \text{if J=I} \\ 0 & \text{otherwise} \end{cases}$$

Given a finite set Γ we let $1_{\Gamma,s}$ be the normalized characteristic sequence

$$1_{\Gamma,s} = \sum_{I \in \Gamma} \frac{e_I}{\| e_I \|_s}$$

where s is a sequence space which is isomorphic to the corresponding function space S.

We call the normed space S a p-space $1 \leq p < \infty$ if there are constants c_1 and c_2 such that

$$c_1(\sharp\Gamma)^{\frac{1}{p}} \leq \| 1_{\Gamma,s} \|_s \leq c_2(\sharp\Gamma)^{\frac{1}{p}}$$

for every finite set Γ.

Theorem 3.2.1 [6]

Let $1 \leq p \leq \infty$. If S is a p-space then

$$\frac{1}{c} \| f^N \|_S \leq \sigma_N(f)_S \leq \| f^N \|_S \quad N \geq 1,$$

where f^N is obtained from the decomposition of f by taking N terms with $| < f, \psi_I > | \, \| \psi_I \|_s$ as large as possible.

Apply this theorem, we have

Proposition 3.2.1

$$\frac{1}{c} \| F^N \|_B \leq \sigma_N(F)_B \leq \| F^N \|_B \quad N \geq 1$$

where $B = BV(2, Z)$ and F^N is obtained from the decomposition of F by taking N terms with $| < F, \psi_I > | \, \| \psi_I \|$ as large as possible.

proof

Since $BV(2, Z) \cong L^2 \times l^2$ as described in paragraph E of section 2.2. It is easy to see $BV(2, Z)$ is a p-space. Therefore, this implies the result.

Remark 3.2.1

It is interesting to know whether one can have a similar result for the space $BV(2)$. We finally remark that one need to understand MRA of $BV(2)$ clearly before answer questions in approximation problems.

References

[1] G. A. Anastassiou and X. M. Yu, "Monotone and Probabilistic Wavelet Approximation", Stochastic Analysis and Applications, 10(3), pp. 251-264, 1992.

[2] T. Bielecki, J. Chen, E. B. Lin and Stephen Yau, "Some Remarks on Wavelet Transforms", *Proceeding of first IEEE Conference on Aerospace Control Systems*, pp. 148-150, 1993.

[3] T. Bielecki, J. Chen, E. B. Lin and Stephen Yau, "Wavelet Representation of General Signals", preprint, 1993.

[4] A. Cohen, I. Daubechies, B. Jawerth and P. Vial, "Multiresolution Analysis, Wavelets and Fast Algorithms on the Interval", to appear.

[5] I. Daubechies, *Ten Lectures on Wavelets*, SIAM. Philadelphia, 1992.

[6] B. Jawerth, C.C. Hsiao, B. J. Lucier and X. M. Yu, "Near Optimal Compression of Orthonormal Wavelet Expansiona", to appear.

[7] H. L. Royden, *Real Analysism*, 3rd ed., Macmillan, New York, 1986.

APPROXIMATIONS AND DERIVATIVES OF PROBABILITY FUNCTIONS

K. Marti

Universität der Bundeswehr München, Fak. LRT
Fakultät Luft- und Raumfahrttechnik
D-85577 Neubiberg / München

1. INTRODUCTION

A very important tool in reliability based systems design and optimization are, see [1],[6],[9], probability functions of the type

$$P(x) := P(y_{1i}<(\leq)y_i(a(\omega),x)<(\leq)y_{2i},\ i=1,\dots,m),\ x\in \mathbb{R}^n \tag{1}$$

$$P_f(x) := P(\min_{1\leq i\leq m} g_i(a(\omega),x)<0),\ x\in \mathbb{R}^n. \tag{2}$$

Here, $x=(x_1,\dots,x_n)'$ denotes the vector of (nominal) decision/design/input variables x_k, e.g. sizing variables, geometric variables (nodal coordinates), degree of refinement of the material and/or manufacturing process in structural design, factors of production in production problems. Furthermore, $a=a(\omega)$ is the ν-vector of random parameters, e.g. material coefficients (elastic moduli,...), manufacturing errors, load parameters in structural design, demands, technological coefficients, cost factors in production problems.

In (1), $y=y(a,x)=(y_1(a,x),\dots,y_m(a,x))'$ is the vector of basic response/output/performance/error variables, e.g. certain displacement and/or stress variables, differences between demand and production output, where y must fulfill the basic behavioral constraints

$$y_{1i} <(\leq)\ y_i <(\leq)\ y_{2i},\ i=1,\dots,m \tag{1.1}$$

with given upper and lower bounds $y_{1i}<y_{2i}, i=1,\dots,m$. Moreover, in structural design the functions $g_i=g_i(a,x)$, $i=1,\dots,m$, in (2) denote the so-called "limit state functions" of certain elements, or parts of the structure, where the "failure domain" is described by

$$\bigcap_{i=1}^{m} \{a\in \mathbb{R}^\nu:\ g_i(a,x)<0\}. \tag{2.1}$$

Since the failure probabilities P_f can be described by means of proba-

Approximation, Probability, and Related Fields, Edited by
G. Anastassiou and S.T. Rachev, Plenum Press, New York 1994

bility functions of the type (1), we may concentrate also for structural design problems on functions of type (1).

In systems safety and reliability, in parameter sensitivity and in optimal systems design, the following mathematical problems occur:

A) Find sufficiently accurate approximations $\hat{P}(x)$ of the probability $P(x)$; determine estimates of the approximation error

$$\epsilon=\epsilon(x):=P(x)-\hat{P}(x).$$

While there are already several very efficient techniques for the approximative calculation of $P(x)$, see [1-4], the mathematical estimation of $\epsilon(x)$ is still an open problem, cf. [5].

B) Find efficient formulas for the (approximative) calculation of the first and higher order partial derivatives $\nabla P(x), \nabla^k P(x)$ of the probability function $P(x)$. Looking for first order differentiation formulas for probability functions in the literature, see [10], [11], one observes that their mathematical derivation is incomplete, and - as a major disadvantage - the numerical evaluation in systems reliability and design is very difficult.

Since we also need derivatives of $P(x)$ in the estimation of the approximation error $\epsilon(x)$, based on [6], we start with the presentation of differentiation formulas for probability functions of the type (1).

2. DIFFERENTIATION FORMULAS FOR PROBABILITY FUNCTIONS

As can be seen from an input-output analysis of the underlying technological system, e.g., from a FE-analysis of a mechanical structure, the basic response variables $y_i=y_i(a,x)$, $i=1,\dots,m$, of the system have often the following functional form:

$$y_i(a,x) = \eta_i(b_o, Q^{(1)}(x)b_1, \dots, Q^{(r)}(x)b_r). \tag{3}$$

Here,

$$b_j = (a_k)_{k\in I_j} \text{ with } I_j \subset \{1,\dots,\nu\},\ j=0,1,\dots,r, \tag{3.1}$$

denote certain $|I_j|$-subvectors of a, where

$$I_0 \cap I_j = \emptyset,\ j=1,\dots,r, \tag{3.2}$$

but I_j, I_ℓ, $j,\ell\geq 1, j\neq\ell$, are not necessarily disjoint,

$$Q^{(j)}=Q^{(j)}(x),\ j=1,\dots,r, \text{ are positive definite } |I_j|\times|I_j| \text{ matrix functions on a set } D\subset\mathbb{R}^n; \tag{3.3}$$

in many practical applications $Q^j(x)$ is a very simple (e.g. a linear) function of the design vector x. Furthermore,

$$\eta_i = \eta_i(b_o, q^{(1)}, \dots, q^{(r)}),\ i=1,\dots,m, \tag{3.4}$$

are given functions on $\mathbb{R}^{|I_o|}\times\mathbb{R}^{|I_1|}\times\dots\times\mathbb{R}^{|I_r|}$ having important analytical properties such as continuity, differentiability.

Considering -as an important practical application- a planar rigid-jointed frame, the vector of displacements u of the frame reads [8]

$$u = u(a,x) = \tilde{K}(a_{II},x)^{-1}p(a_I), \tag{4}$$

where $a_I(= b_o), a_{II}(= b_1)$ is a certain partition of a, and $p=p(a_I)$ denotes the load vector depending on random parameters $a_I=a_I(\omega)$. Moreover, the stiffness matrix $\tilde{K}$ can be represented [8] by

$$\tilde{K} = \tilde{K}(a_{II},x) = \sum_{k=1}^{\kappa} (q_k^R \tilde{K}_R^{(k)} + q_k^B \tilde{K}_B^{(k)}), \tag{4.1}$$

where $\tilde{K}_R^{(k)}, \tilde{K}_B^{(k)}$ are fixed matrices corresponding to the behavior of the k-th element of the structure as a rod, a beam, resp., and therefore

$$q_k^R = \frac{E_k A_k(x)}{L_k}, \quad q_k^B = \frac{E_k I_k(x)}{L_k}, \quad k=1,\dots,\kappa. \tag{4.2}$$

Obviously, E_k, $A_k=A_k(x)$, $I_k=I_k(x)$, L_k denote the elastic modulus, the cross-sectional area, the second moment of area, the length, resp., of the k-th element of the frame. Consequently, u can be represented by

$$u = U(a_I,q^R,q^B) \tag{4.3}$$

where

$$q^R = Q^R(x)a_{II}, \quad q^B = Q^B(x)a_{II} \tag{4.3.1}$$

$$a_{II} := (E_1(\omega),\dots,E_\kappa(\omega))' \tag{4.3.2}$$

$$Q^R(x) := \left(\frac{A_k(x)}{L_k}\delta_{\ell k}\right), \quad Q^B(x) := \left(\frac{I_k(x)}{L_k}\delta_{\ell k}\right), \tag{4.3.3}$$

$\delta_{\ell k}$ denotes the Kronecker symbol. If only pin-jointed trusses are considered, then (4.3) is reduced to

$$u = U(a_I,q^R), \quad q^R = Q^R(x)a_{II}. \tag{4.3)'$$

Note that this important special case can be represented in (3) by setting r:=1, i.e.

$$y_i(a,x) = \eta_i(a_I,Q^{(1)}(x)a_{II}), \quad i=1,\dots,m. \tag{5}$$

2.1. The case r=1

If (5) holds, and the random vector $a=\binom{a_I}{a_{II}}$ has the probability density

$$f(a) = f(a_I,a_{II}) = \psi(a_{II}|a_I)\varphi(a_I),$$

then by the integral transformation

$$a_{II} := Q^{(1)}(x)^{-1}q^{(1)} \tag{5.1}$$

we find

$$P(x) = P(y_{1i} < \eta_i(a_I(\omega),Q^{(1)}(x)a_{II}(\omega)) < y_{2i}, \ i=1,\dots,m)$$

$$= \int\int_{\substack{y_{1i}<\eta_i(a_I,q^{(1)})<y_{2i},\\ i=1,\dots,m}} \psi(Q^{(1)}(x)^{-1}q^{(1)}|a_I)\varphi(a_I)\frac{dq^{(1)}}{\det Q^{(1)}(x)}da_I. \tag{6}$$

Since the domain of integration in (6) is now **independent of the design**

vector x, by differentiation under the integrals we obtain - after the back-transformation $q^{(1)}:=Q^{(1)}(x)a_{II}$ - the first **differentiation formula**

$$\frac{\partial P}{\partial x_t}(x) = - \iint_{B(x)} \left(tr(Q^{(1)}(x)^{-1}\frac{\partial}{\partial x_t}Q^{(1)}(x)) + (\frac{\partial}{\partial x_t}Q^{(1)}(x)Q^{(1)}(x)^{-1}\frac{\nabla\psi(a_{II}|a_I)}{\psi(a_{II}|a_I)})'a_{II}\right)\psi(a_{II}|a_I)\varphi(a_I)da_{II}da_I, \quad (7)$$

where trM denotes the trace of a matrix M, $\nabla\psi(a_{II}|a_I)$ is the gradient of $\psi(a_{II}|a_I)$ with respect to a_{II}, x_t is the t-th component of x, and the domain of integration B(x) is given by

$$B(x) = \{(a_I,a_{II}):\ y_{1i}<\eta_i(a_I,Q^{(i)}(x)a_{II})<y_{2i},\ i=1,\dots,m\}. \quad (7.1)$$

2.2. The case r>1

The above transformation-based method can be generalized immediately to the case r>1, provided that also, cf. (3.1),(3.2),

$$I_j \cap I_\ell = \emptyset \text{ for all } j\neq\ell,\ j,\ell=1,\dots,r, \quad (8)$$

which simply means that $b_1,\dots,b_r$ are subvectors of $(a_k)_{k\notin I_o}$ containing **different** components of a.

If (8) does not hold, then approximating differentiation formulas for (1) can be obtained by the following generalized transformation-based method which was suggested first in [6]:

Step 1: Using random **completion** vectors $\delta_1(\omega),\dots,\delta_r(\omega)$, the probability function P(x) defined by (1), (3) is approximated by

$$\tilde{P}(x):=P(y_{1i} < \eta_i(b_o(\omega),Q^{(1)}(x)(b_1(\omega)+\delta_1(\omega)),\dots,Q^{(r)}(x)(b_r(\omega)+\delta_r(\omega))) < y_{2i},\ 1\le i\le m), \quad (9.1)$$

where the random $\nu_j(:=|I_j|)$-vectors $\delta_j(\omega)$, j=1,...,r, e.g. zero mean normal distributed random vectors, are selected such that the derivative $\frac{\partial\tilde{P}}{\partial x_t}(x)$ can be obtained by using a similar integral transformation as described in Section 2.1 and in the beginning of this section.

Step 2: The derivative $\frac{\partial P}{\partial x_t}$ is obtained then by the limit process

$$\frac{\partial P}{\partial x_t}(x) := \lim_{\substack{\delta_j(\cdot)\to 0,\\ 1\le j\le r}} \frac{\partial\tilde{P}}{\partial x_t}(x). \quad (9.2)$$

For simplicity of notation, this method (Step 1 and Step 2) is worked out here for the important special case, cf. (4),

$$y_i(a,x) = \eta_i(a_I,Q^{(1)}(x)a_{II},Q^{(2)}(x)a_{II}),\ i=1,\dots,m. \quad (10)$$

According to (9.1) we replace now P(x) by

$$\tilde{P}(x):=P(y_{1i}<\eta_i(a_I(\omega),Q^{(1)}(x)a_{II}(\omega),Q^{(2)}(x)(a_{II}(\omega)+\delta(\omega)))<y_{2i},1\le i\le m), \quad (11)$$

where $\delta=\delta(\omega)$ is a random ν_2-vector having stochastically independent

$N(0,\sigma_k^2)$-distributed components $\delta_k(\omega)$, $k=1,\dots,\nu_2$, which are also independent of $a(\omega)=(a_I(\omega),a_{II}(\omega))$. Defining $q^{(1)}:=Q^{(1)}(x)a_{II}$, $q^{(2)}:=(x)(a_{II}+\delta)$, hence, using the integral transformation, cf. (5.1),

$$a_{II}:= Q^{(1)}(x)^{-1}q^{(1)},\quad \delta:= -Q^{(1)-1}q^{(1)}+Q^{(2)-1}q^{(2)}, \tag{11.1}$$

corresponding to (6) we get

$$\tilde{P}(x)= \int\int\int_{\substack{y_{1i}<\eta_i(a_I,q^{(1)},q^{(2)})<y_{2i},\\ 1\le i\le m}} \psi(Q^{(1)}(x)^{-1}q^{(1)}|a_I)\gamma(-Q^{(1)}(x)^{-1}q^{(1)} +Q^{(2)}(x)^{-1}q^{(2)})\ \frac{dq^{(1)}}{\det Q^{(1)}(x)}\ \frac{dq^{(2)}}{\det Q^{(2)}(x)}\ da_I, \tag{11.2}$$

where $\gamma=\gamma(\delta)$ denotes the density of $\delta(\omega)$. Since the transformed domain of integration is now independent of x, by differentiation under the integrals and back-transformation, corresponding to (7) we have the following first result:

Theorem 2.1. Approximative differentiation formula. If the above assumptions hold, then

$$\frac{\partial\tilde{P}}{\partial x_t}(x) = -\iiint_{\tilde{B}(x)} \Big(\mathrm{tr}(Q^{(1)}(x)^{-1}\frac{\partial}{\partial x_t}Q^{(1)}(x)) + \mathrm{tr}(Q^{(2)}(x)^{-1}\frac{\partial}{\partial x_t}Q^{(2)}(x)) + \frac{\nabla\gamma(\delta)'}{\gamma(\delta)}(-Q^{(1)}(x)^{-1}\ \frac{\partial Q^{(1)}}{\partial x_t}(x)a_{II} + Q^{(2)}(x)^{-1}\ \frac{\partial Q^{(2)}}{\partial x_t}(x)(a_{II}+\delta)) + \frac{\nabla\psi(a_{II}|a_I)'}{\psi(a_{II}|a_I)}Q^{(1)}(x)^{-1}\ \frac{\partial Q^{(1)}}{\partial x_t}(x)a_{II}\Big)\psi(a_{II}|a_I)\gamma(\delta)\varphi(a_I)da_{II}d\delta da_I, \tag{12}$$

where

$$\tilde{B}(x):=\{(a_I,a_{II},\delta):y_{1i}<\eta_i(a_I,Q^{(1)}(x)a_{II},Q^{(2)}(x)(a_{II}+\delta))<y_{2i},i=1,\ \dots,m\}. \tag{12.1}$$

Remark 2.1

i) $\nabla\gamma(\delta)= -\gamma(\delta)\Sigma\delta$, where Σ denotes the diagonal matrix having the diagonal elements $\frac{1}{\sigma_k^2}$, $k=1,\dots,\nu_2$.

ii) Stochastic quasigradients, i.e. approximations of stochastic gradients, can be easily obtained from (12).

iii) The exact differentiation formula for $\frac{\partial P}{\partial x_t}(x)$ is obtained from (12) by the limit process $\sigma_k \longrightarrow 0$ for all $k=1,\dots,\nu_2$.

iv) Note that the **higher order partial derivatives** of $\tilde{P}(x)$ can be obtained by the same procedure.

In the important special case

$$Q^{(1)}(x):=(q_{kk}^{(1)}(x_k)\delta_{\ell k}),\quad Q^{(2)}(x):=(q_{kk}^{(2)}(x_k)\delta_{\ell k}), \tag{13}$$

cf. (4.3.3), we find

$$\frac{\partial \tilde{P}}{\partial x_t}(x) = - \iiint_{\tilde{B}(x)} \Big(\frac{\dot{q}_{tt}^{(1)}(x_t)}{q_{tt}^{(1)}(x_t)} + \frac{\dot{q}_{tt}^{(2)}(x_t)}{q_{tt}^{(2)}(x_t)} \tag{13.1}$$

$$+ \frac{\dot{\gamma}_t(\delta_t)}{\gamma_t(\delta_t)} \Big(- \frac{\dot{q}_{tt}^{(1)}(x_t)}{q_{tt}^{(1)}(x_t)} a_{\nu_1+t} + \frac{\dot{q}_{tt}^{(2)}(x_t)}{q_{tt}^{(2)}(x_t)} (a_{\nu_1+t}+\delta_t)\Big)$$

$$+ \frac{\frac{\partial \psi}{\partial a_{\nu_1+t}}(a_{II}|a_I)}{\psi(a_{II}|a_I)} \frac{\dot{q}_{tt}^{(1)}(x_t)}{q_{tt}^{(1)}(x_t)} a_{\nu_1+t} \Big) \psi(a_{II}|a_I)\varphi(a_I)\gamma(\delta) da_{II} d\delta da_I .$$

2.2.1. The limit $\sigma_k \longrightarrow 0$, $k=1,\dots,\nu_2$

For sake of simplicity we consider here only the case (13). Using the theory of distributions, by taking $\sigma_k \longrightarrow 0$, $k=1,\dots,\nu_2$, for the first and third term in (13.1) we find the limit

$$J_I^{\infty} := - \iint_{B(x)} \Big(\Big(\frac{\dot{q}_{tt}^{(1)}(x_t)}{q_{tt}^{(1)}(x_t)} + \frac{\dot{q}_{tt}^{(2)}(x_t)}{q_{tt}^{(2)}(x_t)} \Big) \psi(a_{II}|a_I)\varphi(a_I) \tag{14}$$

$$+ \frac{\partial \psi}{\partial a_{\nu_1+t}}(a_{II}|a_I) \frac{\dot{q}_{tt}^{(1)}(x_t)}{q_{tt}^{(1)}(x_t)} a_{\nu_1+t}\varphi(a_I) \Big) da_{II} da_I$$

$$= - \iint_{B(x)} \Big(\frac{\dot{q}_{tt}^{(2)}(x_t)}{q_{tt}^{(2)}(x_t)} \psi(a_{II}|a_I)\varphi(a_I) + \frac{\dot{q}_{tt}^{(1)}(x_t)}{q_{tt}^{(1)}(x_t)} \frac{\partial}{a_{\nu_1+t}} (a_{\nu_1+t}\psi(a_{II}|a_I))\varphi(a_I) \Big)$$

$$\times da_{II} da_I ,$$

where, cf. (12.1),

$$B(x) := \{(a_I, a_{II}) : y_{1i} < \eta_i(a_I, Q^{(1)}(x)a_{II}, Q^{(2)}(x)a_{II}) < y_{2i}, i=1,\dots,m\}. \tag{14.1}$$

For the second term in (13.1) we have

$$J_{II}(\sigma_k, 1 \le k \le \nu_2)$$

$$= - \int_{-\infty}^{+\infty} \dots \int \prod_{k \neq t} \gamma_k(\delta_k) d\delta_k \int_{-\infty}^{+\infty} g_t(\delta_1,\dots,\delta_{t-1},\delta_t,\delta_{t+1},\dots,\delta_{\nu_2}) d\delta_t, \tag{15}$$

where $g_t = g_t(\delta)$ is given by

$$g_t(\delta) := \dot{\gamma}_t(\delta_t) h_t(\delta) \tag{15.1}$$

with

$$h_t(\delta) := \iint_{\tilde{B}(x,\delta)} \Big(- \frac{\dot{q}_{tt}^{(1)}(x_t)}{q_{tt}^{(1)}(x_t)} a_{\nu_1+t}$$

$$+ \frac{\dot{q}_{tt}^{(2)}(x_t)}{q_{tt}^{(2)}(x_t)} (a_{\nu_1+t}+\delta_t)\Big) \psi(a_{II}|a_I)\varphi(a_I) da_{II} da_I , \tag{15.2}$$

where

$$\tilde{B}(x,\delta):=\{(a_I,a_{II}):y_{i1}<\eta_i(a_I,Q^{(1)}(x)a_{II},Q^{(2)}(x)(a_{II}+\delta))<y_{2i},1\le i\le m\}. \tag{15.3}$$

Thus, by partial integration with respect to δ_t we find

$$\int_{-\infty}^{+\infty} g_t(\delta_1,\ldots,\delta_{t-1},\delta_t,\delta_{t+1},\ldots,\delta_{\nu_2})d\delta_t = -\int_{-\infty}^{+\infty}\frac{\partial h_t}{\partial\delta_t}(\delta)\gamma_t(\delta_t)d\delta_t,$$

assuming that $\gamma_t(\delta_t)h_t(\delta)\longrightarrow 0$, as $\delta_t\longrightarrow\pm\infty$, and therefore

$$J_{II}(\sigma_k,1\le k\le\nu_2) = \int_{-\infty}^{+\infty}\!\!\ldots\int \frac{\partial h_t}{\partial\delta_t}(\delta)\gamma(\delta)d\delta. \tag{16}$$

Taking in (16) now the limit $\sigma_k\longrightarrow 0$ for all $1\le k\le\nu_2$, we get

$$J_{II}^{\infty} = \frac{\partial h_t}{\partial\delta_t}(0). \tag{16.1}$$

By partial differentiation of h_t at $\delta=0$ with respect to δ_t we obtain, see (14.1),

$$\begin{aligned}\frac{\partial h_t}{\partial\delta_t}(0) = &\iint_{B(x)}\frac{\dot q_{tt}^{(2)}(x_t)}{q_{tt}^{(2)}(x_t)}\psi(a_{II}|a_I)\varphi(a_I)da_{II}da_I\\ &+\int\ldots\int da_I\prod_{k\neq t}da_{\nu_1+k}\varphi(a_I)\\ &\times\left[\frac{\partial}{\partial\delta_t}\int_{\tilde{B}_t(x,(a_{\nu_1+k})_{k\neq t},\delta_t)}\tilde{\psi}(a_{\nu_1+t}|(a_{\nu_1+k})_{k\neq t},a_I)da_{\nu_1+t}\right]_{\delta_t=0},\end{aligned} \tag{17}$$

where $k\ge 1$ and

$$\tilde{\psi}(a_{\nu_1+t}|(a_{\nu_1+k})_{k\neq t},a_I):=\left(\frac{\dot q_{tt}^{(2)}(x_t)}{q_{tt}^{(2)}(x_t)}-\frac{\dot q_{tt}^{(1)}(x_t)}{q_{tt}^{(1)}(x_t)}\right)a_{\nu_1+t}\psi(a_{II}|a_I), \tag{17.1}$$

$$\begin{aligned}\tilde{B}_t(x,(a_{\nu_1+k})_{k\neq t},\delta_t):=\{a_{\nu_1+t}:\ &y_{i1}<\eta_i(a_I,(q_{kk}^{(1)}(x_k)a_{\nu_1+k})_{k\neq t},\\ &q_{tt}^{(1)}(x_t)a_{\nu_1+t},(q_{kk}^{(2)}(x_k)a_{\nu_1+k})_{k\neq t},q_{tt}^{(2)}(x_t)(a_{\nu_1+t}+\delta_t))<y_{i2},1\le i\le m\}.\end{aligned} \tag{17.2}$$

Supposing now for simplicity that $m=1$, for the inner integral $I_t(x,(a_{\nu_1+k})_{k\neq t},\delta_t)$ of the second term in (17) we have

$$I_t(x,(a_{\nu_1+k})_{k\neq t},\delta_t) = \sum_{\ell=1}^{\pi}\int_{A_{\ell 1}}^{A_{\ell 2}}\tilde{\psi}(a_{\nu_1+t}|(a_{\nu_1+k})_{k\neq t},a_I)da_{\nu_1+t}, \tag{18}$$

assuming a locally fixed integer π, where the bounds

$$A_{\ell\tau} = A_{\ell\tau}(x,(a_{\nu_1+k})_{k\neq t},\delta_t),\ \ell=1,\ldots,\pi,\ \tau=1,2, \tag{18.1}$$

are defined by the equations

$$\begin{aligned}\eta_1(a_I,(q_{kk}^{(1)}(x_k)a_{\nu_1+k})_{k\neq t},q_{tt}^{(1)}(x_t)A_{\ell\tau},(q_{kk}^{(2)}(x_k)a_{\nu_1+k})_{k\neq t},\\ q_{tt}^{(2)}(x_t)(A_{\ell\tau}+\delta_t)) = Y_{\ell\tau},\end{aligned} \tag{18.2}$$

where $Y_{\ell\tau}\in\{y_{11},y_{12}\}$, $\ell=1,\ldots,\pi$, $\tau=1,2$. By partial differentiation of I_t with respect to δ_t, from (18) we get

$$\frac{\partial I_t}{\partial\delta_t} = \sum_{\ell=1}^{\pi}\left(\tilde{\psi}(A_{\ell 2}|(a_{\nu_1+k})_{k\neq t},a_I)\frac{\partial A_{\ell 2}}{\partial\delta_t} - \tilde{\psi}(A_{\ell 1}|(a_{\nu_1+k})_{k\neq t},a_I)\frac{\partial A_{\ell 1}}{\partial\delta_t}\right), \quad (19.1)$$

where by partial differentiation of (18.2) with respect to δ_t, for $\frac{\partial A_{\ell\tau}}{\partial\delta_t}$ we find

$$\frac{\partial A_{\ell\tau}}{\partial\delta_t} = \quad (19.2)$$

$$-\frac{\frac{\partial\eta_1}{\partial q_t^{(2)}}(a_I,(q_{kk}^{(1)}a_{\nu_1+k})_{k\neq t},q_{tt}^{(1)}A_{\ell\tau},(q_{kk}^{(2)}a_{\nu_1+k})_{k\neq t},q_{tt}^{(2)}(A_{\ell\tau}+\delta_t))\cdot q_{tt}^{(2)}}{\frac{\partial\eta_1}{\partial a_{\nu_1+t}}(\ldots)}.$$

Hence, (19.1) and (19.2) yield

$$\left.\frac{\partial I_t}{\partial\delta_t}\right|_{\delta_t=0} = -\left(\frac{\dot{q}_{tt}^{(2)}}{q_{tt}^{(2)}} - \frac{\dot{q}^{(1)}}{q_{tt}^{(1)}}\right)$$

$$\times\int_{\tilde{B}_t(x,(a_{\nu_1+k})_{k\neq t},0)}\frac{\partial}{\partial a_{\nu_1+t}}\left(a_{\nu_1+t}\psi(a_{II}|a_I)\frac{\frac{\partial\eta_1}{\partial q_t^{(2)}}}{\frac{\partial\eta_1}{\partial a_{\nu_1+t}}}\,q_{tt}^{(2)}\right)da_{\nu_1+t}. \quad (19.3)$$

From (19.3), (17) and (16.1) we find

$$J_{II}^{\infty} = \iint_{B(x)}\frac{\dot{q}_{tt}^{(2)}}{q_{tt}^{(2)}}\,\psi(a_{II}|a_I)\varphi(a_I)da_{II}da_I \quad (20)$$

$$-\left(\frac{\dot{q}_{tt}^{(2)}}{q_{tt}^{(2)}} - \frac{\dot{q}_{tt}^{(1)}}{q_{tt}^{(1)}}\right)\iint_{B(x)}\frac{\partial}{\partial a_{\nu_1+t}}\left(a_{\nu_1+t}\psi(a_{II}|a_I)\frac{\frac{\partial\eta_1}{\partial q_t^{(2)}}}{\frac{\partial\eta_1}{\partial a_{\nu_1+t}}}\,q_{tt}^{(2)}\right)\varphi(a_I)da_{II}da_I.$$

Adding then J_I^{∞}, see (14), and J_{II}^{∞}, we obtain, cf. (13.1),

$$\frac{\partial P}{\partial x_t} = \lim_{\substack{\sigma_k\to 0\\ 1\le k\le\nu_2}}\frac{\partial\tilde{P}}{\partial x_t} = J_I^{\infty} + J_{II}^{\infty} = -\iint_{B(x)}\frac{\partial}{\partial a_{\nu_1+t}}\Bigg(\Bigg(\frac{\dot{q}_{tt}^{(1)}}{q_{tt}^{(1)}}$$

$$+\left(\frac{\dot{q}_{tt}^{(2)}}{q_{tt}^{(2)}} - \frac{\dot{q}_{tt}^{(1)}}{q_{tt}^{(1)}}\right)\frac{\frac{\partial\eta_1}{\partial q_t^{(2)}}}{\frac{\partial\eta_1}{\partial a_{\nu_1+t}}}\,q_{tt}^{(2)}\Bigg)a_{\nu_1+t}\psi(a_{II}|a_I)\Bigg)\varphi(a_I)da_{II}da_I$$

$$= - \iint_{B(x)} \frac{\partial}{\partial a_{\nu_1+t}} (a_{\nu_1+t} \psi(a_{II}|a_I) \frac{\frac{\partial \eta_1}{\partial q_t^{(1)}} \dot{q}_{tt}^{(1)} + \frac{\partial \eta_1}{\partial q_t^{(2)}} \dot{q}_{tt}^{(2)}}{\frac{\partial \eta_1}{\partial a_{\nu_1+t}}}) \varphi(a_I) da_{II} da_I. \tag{21}$$

Thus, we have the following differentiation formula:

Theorem 2.2. If m=1 and B(x) is given by (14.1), then

$$\frac{\partial P}{\partial x_t}(x) = - \iint_{B(x)} \frac{\partial}{\partial a_{\nu_1+t}} (\psi(a_{II}|a_1) \frac{\frac{\partial \eta_1}{\partial x_t}(a_I, Q^{(1)}(x)a_{II}, Q^2(x)a_{II})}{\frac{\partial \eta_1}{\partial a_{\nu_1+t}}(a_I, Q^{(1)}(x)a_{II}, Q^{(2)}(x)a_{II})}) \varphi(a_I) da_{II} da_I. \tag{22}$$

Using the Gaussian divergence theorem for volume integrals in $\mathbb{R}^{\nu}$, we may write $\frac{\partial P}{\partial x_t}$ also in the following form:

Theorem 2.3. If m=1 and B(x) is given by (14.1), then

$$\frac{\partial P}{\partial x_t}(x) = - \int_{B(x)} \mathrm{div}_a (f(a) \frac{\partial \eta_1}{\partial x_t} \frac{\nabla_a \eta_1}{||\nabla_a \eta_1||^2}) da. \tag{23}$$

Remark 2.2

Using a further smoothing process, cf. (9), the case m>1 can be reduced to the case m=1.

3. ERROR ESTIMATION

In systems reliability theory, approximation $\ddot{P}(x)$ of

$P(x) = P(y_1 < y(a(\omega),x) < y_2)$,

where $y(a,x):=(y_1(a,x),\dots,y_m(a,x))'$, $y_1:=(y_{11},\dots,y_{1m})'$, $y_2:=(y_{21},\dots,y_{2m})'$, see (1), are defined [1-4] mostly by

$$\hat{P}(x) := P(y_1 < \hat{y}(a(\omega),x) < y_2), \tag{24}$$

where $\hat{y}=\hat{y}(a,x)$ is an appropriate approximation, e.g. a linearization, of y(a,x) with respect to a.

For a given, **fixed** design vector x, we suppose that y(a,x) can be represented by the **parametric model**

$$y(a,x) = Y(\theta_o,a) := \hat{y}(a,x) + R(\theta_o,a) \tag{25}$$

with the remainder

$$R(\theta,a) := \sum_{j=1}^{q} \theta_j R_j(a), \theta \in \Theta (\subset \mathbb{R}^q, 1 \le q \le \infty), \tag{25.1}$$

where $R_j=R_j(a;x)$, j=1,...,q, are given functions, e.g. certain polynomials, of the variables $a_1,\dots,a_{\nu}$, and $\theta_j=\theta_j(x)$, j=1,...,q, are certain real parameters, e.g. partial derivatives of y(a,x) with respect to $a_1,\dots,a_\nu$ at $\bar{a}:=Ea(\omega)$, see [7]. Defining now for a given, **fixed** vector x

$$W(\theta):= P(y_1 < Y(\theta,a(\omega)) < y_2), \tag{25.2}$$

the error ϵ can be represented by

$$\epsilon = W(\theta_o)-W(0) = \nabla W(\lambda\theta_o)\cdot\theta_o \text{ (with } 0<\lambda<1)$$
$$= \nabla W(0)\cdot\theta_o+\dots\ .$$

By the differentiation formula (23), we find - replacing there x_t by θ_j, and using that $R(\theta,a)$ is linear in θ - the following **error representation**:

Theorem 3.1. If m=1 and the above assumptions hold, then

$$\epsilon = - \int_{y_1<Y(\lambda\theta_o,a)<y_2} \mathrm{div}_a(f(a)R(\theta_o,a) \frac{\nabla_a Y(\lambda\theta_o,a)}{||\nabla_a Y(\lambda\theta_o,a)||^2})da$$
$$= - \int_{y_1<\hat{y}(a,x)<y_2} \mathrm{div}_a(f(a)R(\theta_o,a) \frac{\nabla_a \hat{y}(a,x)}{||\nabla_a \hat{y}(a,x)||^2})da + \dots\ . \tag{26}$$

Note. Upper bounds for $|\epsilon|$ can be obtained now from (26).

REFERENCES

[1] Abdo, T., Rackwitz, R.: Reliability of Uncertain Structural Systems. Finite Elements in Engineering Applications, p. 161-176. Stuttgart, INTES GmbH 1990

[2] Bjerager, P.: On Computation Methods for Structural Reliability Analysis. Structural Safety 9, 79-96 (1990)

[3] Breitung, K.: Asymptotic Approximation for Multinomial Integrals. ASCE J. of the Eng. Mechanics Division 110, 357-367 (1984)

[4] Breitung, K.: Asymptotische Approximation für Wahrscheinlichkeitsintegrale. Habilitationsschrift, Fakultät für Philosophie, Wissenschaftstheorie und Statistik der Universität München, 1990

[5] Breitung, K.: Parameter Sensitivity of Failure Probabilities. In: A. Der Kiureghian, P. Thoft-Christensen (eds.): Reliability and Optimization of Structural Systems '90. Lecture Notes in Engineering, Vol. 61, 43-51 (1990)

[6] Marti, K.: Stochastic Optimization Methods in Structural Mechanics. ZAMM 70, T742-T745 (1990)

[7] Marti, K.: Approximations and Derivatives in Structural Design. ZAMM 72, T575-T578 (1992)

[8] McGuire, W., Gallagher, R.H.: Matrix Structural Analysis. New York-London, Wiley 1979

[9] Schueller, G.I.: A critical appraisal of methods to determine failure probabilities. J. Structural Safety 4, No.4, 293-309 (1987)

[10] Streeter, V.L., Wylie, E.B.: Fluid Mechanics. New York, McGraw Hill 1979

[11] Uryas'ev. St.: A differentiation formula for integrals over sets given by inclusion. Numer. Funct. Anal. and Optimiz. 10 (7.u.8), 827-841 (1989)

RATE OF CONVERGENCE IN THE CENTRAL LIMIT THEOREM

FOR GENERALIZED CONVOLUTIONS

Anna K. Panorska[1]

Department of Mathematics
University of Tennessee at Chattanooga
615 McCallie Ave
Chattanooga, TN 37403

1. INTRODUCTION

The theory of Generalized Convolutions represents an unifying approach to many limit schemes in the probability theory, like Central Limit Theorems for i.i.d. random variables, extreme value theory, Kingman convolution and others. The main idea is, that in contrast to the usual summation and maxima scheme, the operation between random variables may itself be random. This random effect of the operation first appeared in a seminal work of Kingman (1963). The structure of Kingman convolution attracted attention of many specialists, among them Urbanik (1964, 1973, 1988), Bingham (1971, 1984), Volkovich (1980, 1984). It was Urbanik (1964), who developed a theory of generalized convolutions studying binary operations on probability measures on the positive half-line, that possess analogues of the most important properties of ordinary convolution. Kingman's work provides only an example of such an operation. We shall refer to generalized convolutions of probability measures on R^+ as to *Urbanik convolutions*. The theory of Urbanik convolutions is very extensive now. We summarize it in Section 2. The main open problem in this area is the rate of convergence in the Central Limit Theorem. In Section 3 we deal with sharp estimates of the rate of convergence for normalized Urbanik convolutions to generalized stable laws. In Section 4 we present an example of a generalized convolution of random vectors and provide an estimate of the rate of convergence of an n-fold generalized convolution to a generalized stable vector.

[1]Research partially supported by 1993 CCI grant from CECA, University of Tennessee at Chattanooga.

Approximation, Probability, and Related Fields, Edited by
G. Anastassiou and S.T. Rachev, Plenum Press, New York 1994

2. URBANIK CONVOLUTIONS

In this section, we give a brief summary of the extensive theory of Urbanik convolutions. Let $\mathcal{P}$ be a class of probability measures on $[0,\infty)$. By E_a $(a \geq 0)$ we shall denote the probability measure concentrated at the point a and by T_a the mapping from $\mathcal{P}$ into itself by means of the formula $(T_aP)(B) = P(a^{-1}B)$ for all Borel subsets of $[0,\infty)$. If $a = 0$ we put $T_0P = E_0$ for all $P \in \mathcal{P}$.
Let $\circ$ be a $\mathcal{P}$-valued commutative and associative binary operation on $\mathcal{P}$ which is linear, continuous with respect to the weak convergence, homogeneous under the maps T_x, (i.e. $T_aP \circ T_aQ = T_a(P \circ Q)$, for all $P, Q \in \mathcal{P}$, has E_0 as identity element , and satisfies the following axiom:
There exists a sequence of positive numbers c_n such that the sequence of measures $T_{c_n}E_1^{\circ n}$ is weakly convergent to some measure other than E_0.
Here $P^{\circ n}$ denotes the nth power of P under the operation $\circ$. Such a map $\circ$ is called here an *Urbanik convolution.* $(\mathcal{P}, \circ)$ is then an *Urbanik convolution algebra.*
A real valued, continuous map h defined on $\mathcal{P}$ which is linear on convex combinations and multiplicative with respect to $\circ$ is called a *homomorphism* of $\mathcal{P}$. If $(\mathcal{P}, \circ)$ possesses a non-trivial homomorphism it is called *regular.* We restrict our attention to regular Urbanik convolution algebras. We refer to Urbanik's papers (see Urbanik, 1964, 1985) for examples of such algebras.
We say that an algebra $(\mathcal{P}, \circ)$ admits a characteristic function if there exists a one-to-one correspondence $P \leftrightarrow \varphi_P$ between $P \in \mathcal{P}$ and real valued, continuous function φ_P defined on $[0,\infty)$ such that $\varphi_{aP+bQ} = a\varphi_P + b\varphi_Q$ $(a, b \geq 0,\ \ a+b=1)$, $\varphi_{P\circ Q} = \varphi_P\varphi_Q$, $\varphi_{T_aP}(t) = \varphi_P(at)$ $(a, t \geq 0)$ and the uniform convergence in every finite interval of φ_{P_n} is equivalent to weak convergence of P_n. The function φ_P is called *characteristic function* of P. Urbanik (1964) proves that $(\mathcal{P}, \circ)$ admits characteristic function if and only if it is regular. Moreover each characteristic function is an integral transform

$$\varphi_P(t) = \int_0^\infty \Omega(tx)P(dx),$$

where kernel Ω satisfies the inequality $\Omega(x) < 1$ in the deleted neighborhood of the origin and

$$\lim_{x\to 0} \frac{1-\Omega(tx)}{1-\Omega(x)} = t^\kappa$$

uniformly in every finite interval. The positive constant κ does not depend upon a choice of characteristic function and is called a *characteristic exponent* of the algebra $(\mathcal{P}, \circ)$. Further, there exists a probability measure M called *characteristic measure* of the algebra for which

$$\varphi_M = e^{-ct^\kappa}.$$

Throughout this paper we assume that φ is a fixed characteristic function in $(\mathcal{P}, \circ)$.
Urbanik defines P to be *stable* if and only if to each $a, b > 0$ there corresponds $c > 0$ such that

$$(T_aP) \circ (T_bP) = T_cP.$$

He proves that a measure is stable if and only if, for some $c_n > 0$ and some $Q \in \mathcal{P}$, we have

$$T_{c_n}Q^{\circ n} \Longrightarrow P.$$

The stable measures have characteristic functions of the form

$$\varphi_P(t) = e^{-ct^\alpha} \quad (c \geq 0),\ \ 0 < \alpha \leq \kappa, \tag{1}$$

(see Urbanik, 1964, Bingham, 1971). We will say that measure P is *α-stable* if P has characteristic function of the form (1). In this case, we shall call α the *index of stability.* Urbanik (1984) shows, that it is enough to consider *standard* α- stable measures, that is α- stable measures with $c = 1$ in (1).

2.1 Uniqueness of Characteristic Functions

By C_b we denote the space of all continuous and bounded functions defined on the positive half-line. We define transformation T_a on $f \in C_b$ by $T_a f(x) = f(ax)$ for all $x \geq 0$. We say that two functions f and g from C_b (two measures P and Q from $\mathcal{P}$) are *similar*, in symbols $f \sim g$ ($P \sim Q$), if $f = T_a g$ ($P = T_a Q$) for a certain positive number a. Note that stable measures with the same index of stability and their characteristic functions are similar. It is thus enough to consider *standard* α- stable measures, that is α- stable measures with $c = 1$ in (1). We can now formulate a result (Urbanik 1984) on uniqueness of the characteristic function.
All kernels Ω *corresponding to characteristic functions of an Urbanik convolution algebra are similar.*
This theorem enables us to associate with every Urbanik convolution $\circ$ the set

$$\mathcal{C}(\circ) = \{\varphi_P : P \in \mathcal{P}\},$$

which does not depend upon the choice of a characteristic function. Urbanik (1984) shows that: $\mathcal{C}(\circ_1) = \mathcal{C}(\circ_2)$ *if and only if* $\circ_1 = \circ_2$. He also gives a description of the set $\mathcal{C}(\circ)$ in terms of stable measures. We say, that a real valued function g defined on $[0, \infty)$ is *completely monotone* if it has derivatives of all orders and $(-1)^n d^{(n)}(x) \geq 0$ for all $x > 0$, where $g^{(n)}$ denotes the nth derivative of g.
Let P_α $(0 < \alpha < \kappa)$ *be an α- stable measure in* $(\mathcal{P}, \circ)$. *Let* $f \in C_b$ *and* $f(0) = 1$. *Then* $f \in \mathcal{C}(\circ)$ *if and only if the function* $\int_0^\infty f(t^{1/\alpha} x) P_\alpha(dx)$ *is completely monotone.*

2.2 Properties of Stable Measures and Their Connection with Ordinary Strictly Stable Measures on R^+

Throught this section we shall use the following notation: Y_α is a standard, strictly stable random variable with respect to the ordinary summation scheme, that is characteristic function φ_{Y_α} of Y_α satisfies:

$$\ln \varphi_{Y_\alpha}(t) = -|t|^\alpha \omega(x, \alpha, \beta),$$

where

$$\omega(x, \alpha, \beta) = \begin{cases} exp(-i(\pi/2)\beta K(\alpha) sgn(t) & \text{if } \alpha \neq 0 \\ (\pi/2) + i\beta \ln |t| sgn(t) & \text{if } \alpha = 1, \end{cases}$$

and $K(\alpha) = \alpha - 1 + sgn(1 - \alpha)$, $0 < \alpha \leq 2$, $-1 \leq \beta \leq 1$; $g(\cdot, \alpha)$ is the density function of Y_α, $G(\cdot, \alpha)$ is the distribution function of Y_α, and X_α is α-stable in the Urbanik's sense random variable with distribution P_α.
Smirnov (1985) and Urbanik (1984, 1985, 1988) discuss the existence of moments of stable measures in $(\mathcal{P}, \circ)$.
Let P_α be a standard α-stable measurc, M characteristic measure, and κ characteristic exponent of $(\mathcal{P}, \circ)$.
For all $0 < \alpha \leq \kappa$ *and* $0 < s < \alpha$

$$\int_0^\infty x^s P_\alpha(dx) = \frac{\Gamma(1 - s/\alpha)}{sK(s)},$$

where $K(s) = \int_0^\infty (1 - \Omega(x)) x^{-s-1} dx < \infty$.

The following result (Smirnov, 1985, Urbanik, 1984) shows connection between measures stable in the Urbanik's sense and ordinary strictly stable measures on R_+.
Let Y_α be independent of X_λ. For any $0 < \lambda \leq \kappa$ and $0 < \alpha \leq 1$

$$X_{\alpha\lambda} \stackrel{d}{=} (Y_\alpha)^{1/\lambda} X_\lambda,$$

where $\stackrel{d}{=}$ denotes equality in distribution.
It follows that *for any $0 < \alpha \leq \kappa$*

$$X_\alpha \stackrel{d}{=} (Y_{\alpha/\kappa})^{1/\kappa} X_\kappa,$$

and distribution function F_α of X_α is given by

$$F_\alpha(x) = \int_0^\infty G((x/y)^\kappa, \lambda/\kappa) M(dy).$$

Using these results, Smirnov (1985) proves, that all stable laws in Urbanik algebras have densities.
For all $0 < \alpha < \kappa$, measures P_α are absolutely continuous with densities

$$f_\alpha(x) = \kappa \int_0^\infty g[(x/y)^\kappa, \alpha/\kappa] \frac{x^{\kappa-1}}{y^\kappa} M(dy).$$

Smirnov also studies asymptotic behavior of the distribution function of α-stable measures in Urbanik algebras.

2.3 Domains of Attraction for Urbanik Convolutions

We shall say that measure $Q \in \mathcal{P}$ *belongs to the domain of attraction* of a stable (in the Urbanik sense) measure P, in symbols $Q \in DOA(P)$, if there exists a *normalizing sequence* of positive numbers $\{c_n\}_{n=1}^\infty$, such that P is a weak limit of the sequence $T_{c_n} Q^{\circ n}$.
Bingham (1971) obtained a characterization of the domains of attraction of stable measures in terms of characteristic functions: *Let P_α be an α-stable measure in an algebra $(\mathcal{P}, \circ)$. A measure $Q \in DOA(P_\alpha)$ if and only if its characteristic function satisfies*

$$\varphi_Q(t) = 1 - t^\alpha L(t),$$

for some function L varying slowly at zero. Bingham also proves that *if $Q \in DOA(P_\alpha)$ then the normalizing sequence c_n varies slowly with exponent $1/\alpha$, that is*

$$\lim_{n\to\infty} \frac{c_{nk}}{c_n} = k^{1/\alpha}.$$

Klosowska (1977) gives a characterization of the domains of attraction in Urbanik convolution algebras in terms of distributions only but under the assumption that

$$\int_{[0,\infty)} x^\kappa M(dx) < \infty,$$

where κ is the characteristic exponent of the algebra and M characteristic measure. For any $Q \in \mathcal{P}$ let us define the measure Q^κ on $[0,\infty)$ by the formula:

$$Q^\kappa(B) = Q\{x^{\kappa/2} : \ x \in B\},$$

for every Borel subset B of $[0,\infty)$. Klosowska proves the following result:
The measure $Q \in \mathcal{P}$ is attracted in the Urbanik's sense by an α-stable measure $P_\alpha \in \mathcal{P}$ if and only if the measure Q^κ is attracted in the ordinary sense by a stable measure on the real line with index of stability $\frac{2\alpha}{\kappa} \in (0,2]$.

2.4 Other Works

Analogues of classical theorems on convergence to infinitely divisible measures (in the Urbanik sense) are given in Urbanik (1964, 1973), Bingham (1971), Klosowska (1977). Convergence to selfdecomposable measures is discussed in Urbanik (1964, 1973). Representation of the characteristic function of quasi-stable measures and characterization of measures which are simultanously self-decomposable and quasi-stable are given by Jajte (1975, 1976). Characterizations of class $\mathcal{L}$ in Urbanik convolution algebras are presented by Jurek (1980). Multiply self-decomposable measures and their characteristic functions are treated by Nguyen Van Thu (1979). Simple characterization of α convolutions and $(\alpha, 1)$ convolutions in terms of the cardinal number of the support of convolution of two point measures is given by Urbanik (1967). Connection between delphic semigroups (Kendall, 1968, Davidson, 1968) and Urbanik convolutions discussed Gilewski (1972). Analogues of classical factorization theorems of Khinchin are due to Bingham (1971) and Urbanik (1964). Connection between commutative semigroups (with 0 as neutral element) on nonnegative integers and Urbanik convolutions of measures supported by subsets of natural numbers discussed Gilewski and Urbanik (1968). Normed rings given by Urbanik convolutions and connection between Urbanik convolutions and β-stable functions were studied by Volkovich (1980) and Kucharczak and Urbanik (1974).

3. RATE OF CONVERGENCE IN THE CENTRAL LIMIT THEOREM FOR URBANIK CONVOLUTIONS

3.1 Introduction

Since the operation of Urbanik convolution is nontrivial, standard methods of proving rate of convergence results using characteristic functions are not applicable. For example, we do not have Berry - Esseen type inequalities available to work with. To solve this problem, we shall employ probability metrics. The concept of ideal metrics for summation of independent random variables was introduced by Zolotarev (1976) and developed by Rachev (1991). Let $(U, \|\cdot\|)$ be a separable Banach space, $\mathcal{B}(U)$ Borel σ-algebra of subsets of U, and $\mathcal{X}(U)$ be the vector space of all U-valued random variables defined on a probability space $(\Omega, \mathcal{F}, P)$.

Definition (Zolotarev, Rachev). A probability semimetric $\mu : \mathcal{X} \times \mathcal{X} \to [0, \infty]$ is called an **ideal** (probability) **metric** of order $r \in R$ if for any random variables $X_1, X_2, X_3 \in \mathcal{X}$, X_3 independent of X_1 and X_2 and any nonzero constant c the following two properties are satisfied

(z1) *Regularity:* $\mu(X_1 + X_3, X_2 + X_3) \leq \mu(X_1, X_2)$, and

(z2) *Homogeneity of order r:* $\mu(cX_1, cX_2) = |c|^r \mu(X_1, X_2)$.

Zolotarev (1976, 1983), Rachev and Ignatov (1984), Rachev and Yukich (1989) showed the existence of an ideal metric of a given order $r \geq 0$.

We shall investigate the rate of convergence in the central limit theorem for Urbanik convolutions. Let $(P, \circ)$ be a generalized convolution algebra. P_α, $0 < \alpha \leq \kappa$, $\circ$-stable measure with index of stability α, κ characteristic exponent of the algebra. Then, as shown in Urbanik (1964) P_α satisfies

$$T_{n^{-1/\alpha}} P_\alpha^{\circ n} = P_\alpha. \tag{2}$$

Central Limit Theorem (CLT). Measure P belongs to the domain of normal attraction of P_α if and only if

$$T_{n^{-1/\alpha}} P^{\circ n} \Longrightarrow P_\alpha. \tag{3}$$

We reserve letter P_α to denote o-stable measure with index of stability $\alpha \in (0, \kappa]$. Next, we extend Zolotarev's definition of ideal metric (with respect to the summation of random variables) to ideal metric with respect to Urbanik convolutions. These ideal metrics will be used to provide exact convergence rates for convergence to α-stable measures in Urbanik convolution algebras. Moreover, the rates will hold with respect to uniform and L_2- type metrics on $\mathcal{P}$.

3.2 Ideal Metrics for Urbanik Convolutions

Definition. A probability semimetric $\mu : \mathcal{P} \times \mathcal{P} \to [0, \infty]$ is called an **ideal** (probability) metric of order $r \in \mathbf{R}$ if for any probability measures $P_1, P_2, P_3 \in \mathcal{P}$ and any positive constant c the following properties are satisfied

(i) *Regularity*: $\mu(P_1 \circ P_3, P_2 \circ P_3) \le \mu(P_1, P_2)$, and

(ii) *Homogeneity of order* r : $\mu(T_c P_1, T_c P_2) = |c|^r \mu(P_1, P_2)$.

We shall study rates of convergence in the Central Limit Theorem with respect to the following metrics on $\mathcal{P}$:

Uniform metric between characteristic functions

$$\chi(P, Q) := \sup_{t \ge 0} |\varphi_P(t) - \varphi_Q(t)|,$$

where $P, Q \in \mathcal{P}$, and φ_P denotes the generalized characteristic function of P.

Weighted χ_r-metric

$$\chi_r(P, Q) := \sup_{t \ge 0} t^{-r} |\varphi_P(t) - \varphi_Q(t)|.$$

l_2 - metric between characteristic functions

$$l_2(P, Q) = \left(\int_0^\infty |\varphi_P(t) - \varphi_Q(t)|^2 dt\right)^{1/2}.$$

Generalized Convolution metric $\mu_{r,\theta}$

Let $\Theta \in \mathcal{P}$

$$\mu_{r,\Theta}(P, Q) = \sup_{h \ge 0} h^r l_2(P \circ T_h \Theta, Q \circ T_h \Theta).$$

Here each $\Theta \in \mathcal{P}$, and $r > 0$ generate a metric $\mu_{r,\Theta}$. If $\Theta = P_\alpha$ for some α-stable measure P_α, then we denote μ_{r,P_α} by $\mu_{r,\alpha}$.

Note: All introduced metrics satisfy the *triangle inequality*, that is:

$$\chi(P, Q) \le \chi(P, R) + \chi(R, Q) \tag{4}$$

$$\chi_r(P, Q) \le \chi_r(P, R) + \chi_r(R, Q) \tag{5}$$

$$l_2(P, Q) \le l_2(P, R) + l_2(R, Q) \tag{6}$$

$$\mu_{r,\Theta}(P, Q) \le \mu_{r,\Theta}(P, R) + \mu_{r,\Theta}(R, Q) \tag{7}$$

The proofs of the above inequalities are easy to see. We will study ideality properties of all introduced ideal metrics.

Lemma 3.1 *χ is an ideal metric of order $r = 0$.*

Proof. *Regularity* of χ

$$\begin{aligned}\chi(P \circ R, Q \circ R) &= \sup_{t \geq 0} |\varphi_{P \circ R}(t) - \varphi_{Q \circ R}(t)| \\ &= \sup_{t \geq 0} [|\varphi_P(t) - \varphi_Q(t)||\varphi_R(t)|] \leq \chi(P, Q),\end{aligned}$$

because $|\varphi_P(t)| \leq 1$ for all $t \geq 0$, (see Urbanik (1964)).
Zero - homogeneity of χ

$$\begin{aligned}\chi(T_c P, T_c Q) &= \sup_{t \geq 0} |\varphi_{T_c P}(t) - \varphi_{T_c Q}(t)| \\ &= \sup_{t \geq 0} |\varphi_P(ct) - \varphi_Q(ct)| = \chi(P, Q). \quad \square\end{aligned}$$

Lemma 3.2 *For all $r > 0$, χ_r is an ideal metric of order r.*

Proof. *Regularity* of χ_r.

$$\begin{aligned}\chi_r(P_1 \circ P_3, P_2 \circ P_3) &= \sup_{t \geq 0} t^{-r} |\varphi_{P_1 \circ P_3}(t) - \varphi_{P_2 \circ P_3}(t)| \\ &= \sup_{t \geq 0} t^{-r} |\varphi_{P_1}(t) - \varphi_{P_2}(t)||\varphi_{P_3}(t)| \\ &\leq \sup_{t \geq 0} t^{-r} |\varphi_{P_1}(t) - \varphi_{P_2}(t)| = \chi_r(P_1, P_2).\end{aligned}$$

Homogeneity of order r, $c > 0$.

$$\begin{aligned}\chi_r(T_c P_1, T_c P_2) &= \sup_{t \geq 0} t^{-r} |\varphi_{P_1}(ct) - \varphi_{P_2}(ct)| \\ &= c^r \sup_{t \geq 0} (ct)^{-r} |\varphi_{P_1}(ct) - \varphi_{P_2}(ct)| = c^r \chi_r(P_1, P_2). \qquad (8)\end{aligned}$$

Metric χ_r is therefore ideal of order r. $\square$

Lemma 3.3 *For any $P_1, P_2 \in \mathcal{P}$ and $\sigma > 0$*

$$\chi(P_1 \circ T_\sigma P_\alpha, P_2 \circ T_\sigma P_\alpha) \leq \sigma^{-r} C_{r,\alpha} \chi_r(P_1, P_2),$$

where $C_{r,\alpha} = (\frac{r}{\alpha e})^{r/\alpha}$

Proof.

$$\begin{aligned}\chi(P_1 \circ T_\sigma P_\alpha, P_2 \circ T_\sigma P_\alpha) &= = \sup_{t \geq 0} (|\varphi_{P_1}(t) - \varphi_{P_2}(t)||\varphi_{T_\sigma P_\alpha}(t)|) \\ &= \sup_{t \geq 0} |\varphi_{P_1}(t) - \varphi_{P_2}(t)| \exp(-\sigma^\alpha t^\alpha) \\ &\leq \sigma^{-r} \chi_r(P_1, P_2) \sup_{t \geq 0} (\sigma t)^r \exp(-\sigma^\alpha t^\alpha).\end{aligned}$$

Let $C_{r,\alpha} = \sup_{t \geq 0} (\sigma t)^r e^{-(t\sigma)^\alpha}$. Simple calculation gives $C_{r,\alpha} = (\frac{r}{\alpha e})^{r/\alpha}$. $\square$

Lemma 3.4 *For any $P_1, P_2, P, Q \in \mathcal{P}$ the following inequality holds:*

$$\chi(P_1 \circ Q, P_2 \circ Q) \leq \chi(P_1, P_2) \chi(Q, P) + \chi(P_1 \circ P, P_2 \circ P).$$

Proof.

$$\begin{aligned}
\chi(P_1\circ Q, P_2\circ Q) &= \sup_{t\geq 0}|\varphi_{P_1}(t)-\varphi_{P_2}(t)||\varphi_Q(t)| \\
&\leq \sup_{t\geq 0}\{|\varphi_{P_1}(t)-\varphi_{P_2}(t)|(|\varphi_Q(t)-\varphi_P(t)|+|\varphi_P(t)|)\} \\
&\leq \sup_{t\geq 0}|\varphi_{P_1}(t)-\varphi_{P_2}(t)||\varphi_Q(t)-\varphi_P(t)| + \sup_{t\geq 0}|\varphi_{P_1}(t)-\varphi_{P_2}(t)||\varphi_P(t)| \\
&\leq \chi(P_1,P_2)\chi(Q,P)+\chi(P_1\circ P, P_2\circ P). \quad \square
\end{aligned}$$

Lemma 3.5 *l_2 is an ideal metric of order $r = 1/2$.*

Proof. *Regularity*

$$\begin{aligned}
l_2(P\circ R, Q\circ R) &= \left(\int_0^\infty |\varphi_P(t)-\varphi_Q(t)|^2|\varphi_R(t)|^2dt\right)^{1/2} \\
&\leq \left(\int_0^\infty |\varphi_P(t)-\varphi_Q(t)|^2dt\right)^{1/2} = l_2(P\circ Q).
\end{aligned}$$

Homogeneity of order 1/2

$$\begin{aligned}
l_2(T_cP, T_cQ) &= \left(\int_0^\infty |\varphi_P(ct)-\varphi_Q(ct)|^2dt\right)^{1/2} \\
&= c^{-1/2}\left(\int_0^\infty |\varphi_P(t)-\varphi_Q(t)|^2dt\right)^{1/2} = c^{-1/2}l_2(P,Q). \quad \square
\end{aligned}$$

Lemma 3.6 *For all $\Theta\in\mathcal{P}$ and all $r>0$, $\mu_{r,\Theta}$ is an ideal metric of order $r-\frac{1}{2}$.*

Proof. *Regularity*

$$\begin{aligned}
\mu_{r,\Theta}(P\circ R, Q\circ R) &= \sup_{h\geq 0} h^r l_2(P\circ R\circ T_h\Theta, Q\circ R\circ T_h\Theta) \\
&\leq \sup_{h\geq 0} h^r l_2(P\circ T_h\Theta, Q\circ T_h\Theta) = \mu_{r,\Theta}(P,Q).
\end{aligned}$$

Homogeneity of order $r-\frac{1}{2}$
Let $c>0$. Using homogeneity of order $-\frac{1}{2}$ of l_2 we get

$$\begin{aligned}
\mu_{r,\Theta}(T_cP, T_cQ) &= \sup_{h\geq 0} h^r l_2(T_cP\circ T_h\Theta, T_cQ\circ T_h\Theta) \\
&= c^r \sup_{h\geq 0}(h/c)^r l_2(T_c(P\circ T_{h/c}\Theta), T_c(Q\circ T_{h/c}\Theta)) \\
&= c^{r-\frac{1}{2}} \sup_{h\geq 0}(h/c)^r l_2(P\circ T_{h/c}\Theta, Q\circ T_{h/c}\Theta) \\
&= c^{r-\frac{1}{2}}\mu_{r,\Theta}(P,Q).
\end{aligned}$$

Therefore $\mu_{r,\Theta}$ is regular of order $r-\frac{1}{2}$. $\square$

Lemma 3.7 *(i) For any $P, Q\in\mathcal{P}$ and $\sigma>0$*

$$l_2(P\circ T_\sigma P_\alpha, Q\circ T_\sigma P_\alpha)\leq \sigma^{-r}\mu_{r,\alpha}(P,Q).$$

(ii) For any $P, Q, Z, Y\in\mathcal{P}$

$$l_2(P\circ Z, Q\circ Z)\leq l_2(P,Q)\chi(Z,Y)+l_2(P\circ Y, Q\circ Y).$$

Proof. (i)

$$l_2(P \circ T_\sigma P_\alpha, Q \circ T_\sigma P_\alpha) = \sigma^{-r}\sigma^r l_2(P \circ T_\sigma P_\alpha, Q \circ T_\sigma P_\alpha) \le \sigma^{-r}\mu_{r,\alpha}(P, Q).$$

(ii) By Minkowski's inequality we have

$$\begin{aligned} l_2(P \circ Z, Q \circ Z) &= \Big(\int_0^\infty |\varphi_P(t) - \varphi_Q(t)|^2 |\varphi_Z(t) + \varphi_Y(t) - \varphi_Y(t)|^2 dt\Big)^{1/2} \\ &\le \Big(\int_0^\infty |\varphi_P(t) - \varphi_Q(t)|^2 |\varphi_Z(t) - \varphi_Y(t)|^2 dt\Big)^{1/2} \\ &\quad + \Big(\int_0^\infty |\varphi_P(t) - \varphi_Q(t)|^2 |\varphi_Y(t)|^2 dt\Big)^{1/2} \\ &\le l_2(P, Q)\chi(Z, Y) + l_2(P \circ Y, Q \circ Y), \end{aligned}$$

as required. □

3.3 Rate of Convergence Results

Theorem 3.1 *Let P_α be an α-stable measure in $(\mathcal{P}, \circ)$ generalized convolution algebra. Let $r > \alpha$, $a = 2^{-r/\alpha}A^{-1}$ and*
$A := max(3^{r/\alpha}, 2C_{r,\alpha}(2^{\frac{r}{\alpha}-1} + 3^{\frac{r}{\alpha}}))$, where $C_{r,\alpha} = (\frac{r}{\alpha e})^{\frac{r}{\alpha}}$.
If $P \in \mathcal{P}$ satisfies

$$t_0 := t_0(P, P_\alpha) := max\{\chi(P, P_\alpha), \chi_r(P, P_\alpha)\} \le a, \tag{9}$$

then for all $n \ge 1$

$$\chi(T_{n^{-1/\alpha}} P^{\circ n}, P_\alpha) \le A t_0 n^{1-\frac{r}{\alpha}} \le 2^{\frac{-r}{\alpha}} n^{1-\frac{r}{\alpha}}.$$

Proof. We proceed by induction; for $n = 1$ the result is trivial. For $n = 2$ we have

$$\begin{aligned} \chi(T_{2^{-1/\alpha}} P^{\circ 2}, P_\alpha) &= \chi(T_{2^{-1/\alpha}}(P \circ P), T_{2^{-1/\alpha}}(P_\alpha \circ P_\alpha)) = \chi(P^{\circ 2}, P_\alpha^{\circ 2}) \\ &\le \chi(P^{\circ 2}, P_\alpha \circ P) + \chi(P \circ P_\alpha, P_\alpha^{\circ 2}) \\ &\le \chi(P, P_\alpha) + \chi(P, P_\alpha) = 2\chi(P, P_\alpha) \\ &\le 2t_0 \le A t_0 2^{1-\frac{r}{\alpha}}, \end{aligned}$$

since $A \ge 2^{\frac{r}{\alpha}}$. Similar calculation holds for $n = 3$. Suppose now, that the estimate

$$\chi(T_{j^{-1/\alpha}} P^{\circ j}, P_\alpha) \le A t_0 j^{1-\frac{r}{\alpha}} \tag{10}$$

holds for all $j < n$. To complete the proof, we need to show that (10) holds for $j = n$. By (10) and (9) we have

$$\chi(T_{j^{-1/\alpha}} P^{\circ j}, P_\alpha) \le A t_0 j^{1-\frac{r}{\alpha}} \le A t_0 \le 2^{\frac{-r}{\alpha}}. \tag{11}$$

For any $n \ge 4$ and $m = [n/2]$, where $[\cdot]$ denotes integer part, the triangle inequality gives

$$\begin{aligned} &K := \chi(T_{n^{-1/\alpha}} P^{\circ n}, P_\alpha) = \chi(T_{n^{-1/\alpha}} P^{\circ n}, T_{n^{-1/\alpha}} P_\alpha^{\circ n}) \le \\ &\le \chi(T_{n^{-1/\alpha}} P^{\circ m} \circ T_{n^{-1/\alpha}} P^{\circ(n-m)}, T_{n^{-1/\alpha}} P_\alpha^{\circ m} \circ T_{n^{-1/\alpha}} P^{\circ(n-m)}) + \\ &\quad + \chi(T_{n^{-1/\alpha}} P_\alpha^{\circ m} \circ T_{n^{-1/\alpha}} P^{\circ(n-m)}, T_{n^{-1/\alpha}} P_\alpha^{\circ m} \circ T_{n^{-1/\alpha}} T_{n^{-1/\alpha}} P_\alpha^{\circ(n-m)}). \end{aligned}$$

Hence, by Lemma 3.4

$$K \leq I_1 + I_2 + I_3,$$

where

$$I_1 = \chi(T_{n^{-1/\alpha}}P^{om}, T_{n^{-1/\alpha}}P_\alpha^{om})\chi(T_{n^{-1/\alpha}}P^{o(n-m)}, T_{n^{-1/\alpha}}P_\alpha^{o(n-m)}),$$

$$I_2 = \chi(T_{n^{-1/\alpha}}P^{om} \circ T_{n^{-1/\alpha}}P_\alpha^{o(n-m)}, T_{n^{-1/\alpha}}P_\alpha^{om} \circ T_{n^{-1/\alpha}}P_\alpha^{o(n-m)}),$$

and

$$I_3 = \chi(T_{n^{-1/\alpha}}P_\alpha^{om} \circ T_{n^{-1/\alpha}}P^{o(n-m)}, T_{n^{-1/\alpha}}P_\alpha^{om} \circ T_{n^{-1/\alpha}}P_\alpha^{o(n-m)}).$$

Let us first estimate I_1. By (10), (11) and Lemma 3.2

$$\begin{aligned}
I_1 &= k(T_{n^{-1/\alpha}}P^{om}, T_{n^{-1/\alpha}}P_\alpha^{om})\chi(T_{n^{-1/\alpha}}P^{o(n-m)}, T_{n^{-1/\alpha}}P_\alpha^{o(n-m)}) \\
&= \chi(T_{m^{-1/\alpha}}P^{om}, T_{m^{-1/\alpha}}P_\alpha^{om})\chi(T_{(n-m)^{-1/\alpha}}P^{o(n-m)}, T_{(n-m)^{-1/\alpha}}P_\alpha^{o(n-m)}) \\
&= \chi(T_{m^{-1/\alpha}}P^{om}, P_\alpha)\chi(T_{(n-m)^{-1/\alpha}}P_\alpha^{o(n-m)}, P_\alpha) \leq 2^{\frac{-r}{\alpha}}At_0(n-m)^{1-\frac{r}{\alpha}} \\
&\leq (1/2)At_0n^{1-\frac{r}{\alpha}}.
\end{aligned}$$

To estimate I_2 and I_3 we use Lemma 3.2 and relation (2). By Lemma 3.3, (2) and ideality of χ_r we have

$$\begin{aligned}
I_2 &= \chi(T_{n^{-1/\alpha}}P^{om} \circ T_{(\frac{n-m}{n})^{1/\alpha}}T_{(n-m)^{1/\alpha}}P_\alpha^{o(n-m)}, \\
&\qquad T_{n^{-1/\alpha}}P_\alpha^{om} \circ T_{(\frac{n-m}{n})^{1/\alpha}}T_{(n-m)^{1/\alpha}}P_\alpha^{o(n-m)}) \\
&= \chi(T_{n^{-1/\alpha}}P^{om} \circ T_{(\frac{n-m}{n})^{1/\alpha}}P_\alpha, T_{n^{-1/\alpha}}P_\alpha^{om} \circ T_{(\frac{n-m}{n})^{1/\alpha}}P_\alpha) \\
&\leq (\frac{n-m}{n})^{-\frac{r}{\alpha}}\chi_r(T_{n^{-1/\alpha}}P^{om}, T_{n^{-1/\alpha}}P_\alpha^{om})C_{r,\alpha} \\
&\leq (\frac{n-m}{n})^{-\frac{r}{\alpha}}\sum_{i=1}^{m}\chi_r(T_{n^{-1/\alpha}}P, T_{n^{-1/\alpha}}P_\alpha)C_{r,\alpha} \\
&= (\frac{n-m}{n})^{-\frac{r}{\alpha}}m\chi_r(T_{n^{-1/\alpha}}P, T_{n^{-1/\alpha}}P_\alpha)C_{r,\alpha} \\
&\leq (\frac{n-m}{n})^{-\frac{r}{\alpha}}m(n^{-\frac{r}{\alpha}})\chi_r(P, P_\alpha)C_{r,\alpha} \\
&\leq 2^{\frac{r}{\alpha}}m(n^{-\frac{r}{\alpha}})\chi_r(P, P_\alpha)C_{r,\alpha} \leq 2^{\frac{r}{\alpha}-1}n^{1-\frac{r}{\alpha}}\chi_r(P, P_\alpha)C_{r,\alpha} \\
&\leq 2^{\frac{r}{\alpha}-1}n^{1-\frac{r}{\alpha}}t_0C_{r,\alpha}.
\end{aligned}$$

Finally

$$\begin{aligned}
I_3 &= \chi(T_{n^{-1/\alpha}}P^{o(n-m)} \circ T_{(m/n)^\alpha}T_{m^{-1/\alpha}}P_\alpha^{om}, \\
&\qquad T_{n^{-1/\alpha}}P_\alpha^{o(n-m)} \circ T_{(m/n)^{1/\alpha}}T_{m^{-1/\alpha}}P_\alpha^{om}) \\
&= \chi(T_{n^{-1/\alpha}}P_\alpha^{o(n-m)} \circ T_{(m/n)^{1/\alpha}}P_\alpha, T_{n^{-1/\alpha}}P_\alpha^{o(n-m)} \circ T_{(m/n)^{1/\alpha}}P_\alpha) \\
&\leq (m/n)^{-\frac{r}{\alpha}}\chi_r(T_{n^{-1/\alpha}}P^{o(n-m)}, T_{n^{-1/\alpha}}P_\alpha^{o(n-m)})C_{r,\alpha} \\
&\leq (m/n)^{-\frac{r}{\alpha}}(n-m)n^{-\frac{r}{\alpha}}\chi_r(P, P_\alpha)C_{r,\alpha} \\
&\leq (3)^{\frac{r}{\alpha}}n^{1-\frac{r}{\alpha}}\chi_r(P, P_\alpha)C_{r,\alpha} \leq (3)^{\frac{r}{\alpha}}n^{1-\frac{r}{\alpha}}t_0C_{r,\alpha}.
\end{aligned}$$

Therefore

$$\begin{aligned}
K &\leq I_1 + I_2 + I_3 \\
&\leq (1/2)At_0n^{1-\frac{r}{\alpha}} + 2^{\frac{r}{\alpha}-1}n^{1-\frac{r}{\alpha}}t_0C_{r,\alpha} + 3^{\frac{r}{\alpha}}n^{1-\frac{r}{\alpha}}t_0C_{r,\alpha} \\
&= ((1/2)A + 2^{\frac{r}{\alpha}-1}C_{r,\alpha} + 3^{\frac{r}{\alpha}}C_{r,\alpha})t_0n^{1-\frac{r}{\alpha}} = At_0n^{1-\frac{r}{\alpha}}.
\end{aligned}$$

This completes the proof of Theorem 3.1. □

Theorem 3.2 *Let P_α be an α- stable measure in $(\mathcal{P}, \circ)$ generalized convolution algebra. Let $r > \alpha$, $C_{r,\alpha} := (\frac{r}{c\alpha e})^{\frac{r}{\alpha}}$, $A_1 := max(3^{\frac{r}{\alpha}}, 2C_{r,\alpha}(2^{\frac{r}{\alpha}-1} + 3^{\frac{r}{\alpha}}))$, $D := 3^{1/\alpha}2^{\frac{r}{\alpha}}$, $A := 2(2^{\frac{r}{\alpha}-1} + 3^{\frac{r}{\alpha}})$, $a := 2^{\frac{-r}{\alpha}}A^{-1}$. If $P \in \mathcal{P}$ satisfies*

$$t_0 := t_0(P, P_\alpha) := max(\chi(P, P_\alpha), \chi_r(P, P_\alpha)) \leq \frac{1}{A_1 D},$$

and

$$\tau := \tau(P, P_\alpha) := max(l_2(P, P_\alpha), \mu_{r,\alpha}(P, P_\alpha)) \leq a, \tag{12}$$

then for all $n \geq 1$

$$l_2(T_{n^{-1/\alpha}}P^{\circ n}, P_\alpha) \leq A\tau n^{1-\frac{r}{\alpha}+1/2\alpha} \leq 2^{\frac{-r}{\alpha}} n^{1-\frac{r}{\alpha}+1/2\alpha}. \tag{13}$$

Proof. Like in the proof of Theorem 3.1 we proceed by induction. For $n = 1$ the result is trivial, since $A > 1$. For $n = 2$, using (2}, (9), and (6) we get

$$\begin{aligned}
l_2(T_{2^{-1/\alpha}}P^{\circ 2}, P_\alpha) &= l_2(T_{2^{-1/\alpha}}P^{\circ 2}, P_\alpha^2) \\
&= 2^{1/2\alpha} l_2(P^{\circ 2}, P_\alpha^{\circ 2}) \\
&\leq 2^{1/2\alpha}(l_2(P^{\circ 2}, P \circ P_\alpha) + l_2(P \circ P_\alpha, P_\alpha^{\circ 2})) \\
&\leq 2^{1/2\alpha} 2 l_2(P, P_\alpha) \leq 2^{\frac{1}{2\alpha}+1}\tau \leq A\tau 2^{1-\frac{r}{\alpha}+\frac{1}{2\alpha}}.
\end{aligned}$$

A similar calculation holds for $n = 3$. Let us assume now that

$$l_2(T_{j^{-1/\alpha}}P^{\circ j}, P_\alpha) \leq A\tau j^{1-\frac{r}{\alpha}+\frac{1}{2\alpha}} \tag{14}$$

holds for all $0 \leq j < n$. To complete the induction we need to show that (14) holds for $j = n$. Let $m := [n/2]$. By (2) and (6), we have

$$\begin{aligned}
l_2(T_{n^{-1/\alpha}}P^{\circ n}, P_\alpha) &= l_2(T_{n^{-1/\alpha}}P^{\circ n}, T_{n^{-1/\alpha}}P_\alpha^{\circ n}) \\
&\leq l_2(T_{n^{-1/\alpha}}P^{\circ m} \circ T_{n^{-1/\alpha}}P^{\circ(n-m)}, \\
&\qquad T_{n^{-1/\alpha}}P_\alpha^{\circ m} \circ T_{n^{-1/\alpha}}P^{\circ(n-m)}) \\
&\quad + l_2(T_{n^{-1/\alpha}}P_\alpha^{\circ m} \circ T_{n^{-1/\alpha}}P^{\circ(n-m)}, \\
&\qquad T_{n^{-1/\alpha}}P_\alpha^{\circ m} \circ T_{n^{-1/\alpha}}P_\alpha^{\circ(n-m)}).
\end{aligned}$$

Let

$$\begin{aligned}
I_1 &:= l_2(T_{n^{-1/\alpha}}P^{\circ m}, T_{n^{-1/\alpha}}P_\alpha^{\circ m})\chi(T_{n^{-1/\alpha}}P^{\circ(n-m)}, T_{n^{-1/\alpha}}P_\alpha^{\circ(n-m)}), \\
I_2 &:= l_2(T_{n^{-1/\alpha}}P^{\circ m} \circ T_{n^{-1/\alpha}}P_\alpha^{\circ(n-m)}, T_{n^{-1/\alpha}}P_\alpha^{\circ m} \circ T_{n^{-1/\alpha}}P_\alpha^{\circ(n-m)}), \\
I_3 &:= l_2(T_{n^{-1/\alpha}}P_\alpha^{\circ m} \circ T_{n^{-1/\alpha}}P^{\circ(n-m)}, T_{n^{-1/\alpha}}P_\alpha^{\circ m} \circ T_{n^{-1/\alpha}}P_\alpha^{\circ(n-m)}).
\end{aligned}$$

Then by Lemma 3.7 (ii), $l_2(T_{n^{-1/\alpha}}P^{\circ n}, P_\alpha) \leq I_1 + I_2 + I_3$.
By (2), (9) and (14)

$$\begin{aligned}
l_2(T_{n^{-1/\alpha}}P^{\circ m}, T_{n^{-1/\alpha}}P_\alpha^{\circ m}) &= (m/n)^{-\frac{1}{2\alpha}} l_2(T_{m^{-1/\alpha}}P^{\circ m}, P_\alpha) \\
&\leq (m/n)^{-\frac{1}{2\alpha}} A\tau m^{1-\frac{r}{\alpha}+\frac{1}{2\alpha}} \\
&\leq (3)^{\frac{1}{2\alpha}} A\tau n^{1-\frac{r}{\alpha}+\frac{1}{2\alpha}} 2^{-(1-\frac{r}{\alpha}+\frac{1}{2\alpha})}.
\end{aligned}$$

By homogeneity of χ and Theorem 3.1

$$\begin{aligned}\chi(T_{n^{-1/\alpha}}P^{\circ(n-m)}, T_{n^{-1/\alpha}}P_\alpha^{\circ(n-m)}) &= \chi(T_{(n-m)^{-1/\alpha}}P^{\circ(n-m)}, P_\alpha)\\ &\leq A_1 t_0 (n-m)^{1-\frac{r}{\alpha}} \leq A_1 t_0 n^{1-\frac{r}{\alpha}} \leq 1/D.\end{aligned}$$

Therefore

$$I_1 \leq 3^{\frac{1}{2\alpha}} \frac{A}{2D} \tau n^{1-\frac{r}{\alpha}+\frac{1}{2\alpha}} 2^{\frac{r}{\alpha}-\frac{1}{2\alpha}} \leq \frac{A}{2}\tau n^{1-\frac{r}{\alpha}+\frac{1}{2\alpha}}.$$

To estimate I_2 we use (2), Lemma 3.7 (i), (7), (9), (9) and (12)

$$\begin{aligned}I_2 &= l_2(T_{n^{-1/\alpha}}P^{\circ m} \circ T_{(\frac{n-m}{n})^{1/\alpha}}P_\alpha, T_{n^{-1/\alpha}}P_\alpha^{\circ m} \circ T_{(\frac{n-m}{n})^{1/\alpha}}P_\alpha)\\ &\leq (\frac{n-m}{n})^{-\frac{r}{\alpha}} \mu_{r,\alpha}(T_{n^{-1/\alpha}}P^{\circ m}, T_{n^{-1/\alpha}}P_\alpha^{\circ m})\\ &\leq (\frac{n-m}{n})^{-\frac{r}{\alpha}} m \mu_{r,\alpha}(T_{n^{-1/\alpha}}P, T_{n^{-1/\alpha}}P_\alpha)\\ &\leq (\frac{n-m}{n})^{-\frac{r}{\alpha}} m (n^{-1/\alpha})^{r-1/2} \mu_{r,\alpha}(P, P_\alpha)\\ &\leq 2^{\frac{r}{\alpha}} m n^{-\frac{r}{\alpha}+\frac{1}{2\alpha}} \mu_{r,\alpha}(P, P_\alpha) \leq 2^{\frac{r}{\alpha}-1} n^{1-\frac{r}{\alpha}+\frac{1}{2\alpha}}\tau.\end{aligned}$$

By (9), Lemma 3.7 (i), (7), (9) and (12)

$$\begin{aligned}I_3 &= l_2(T_{(m/n)^{1/\alpha}}P_\alpha \circ T_{n^{-1/\alpha}}P^{\circ(n-m)}, T_{(m/n)^{1/\alpha}}P_\alpha^{\circ m} \circ T_{n^{-1/\alpha}}P_\alpha^{\circ(n-m)})\\ &\leq (m/n)^{-\frac{r}{\alpha}} \mu_{r,\alpha}(T_{n^{-1/\alpha}}P^{\circ(n-m)}, T_{n^{-1/\alpha}}P_\alpha^{\circ(n-m)})\\ &\leq (m/n)^{-\frac{r}{\alpha}} (n-m) \mu_{r,\alpha}(T_{n^{-1/\alpha}}P, T_{n^{-1/\alpha}}P_\alpha)\\ &= (m/n)^{-\frac{r}{\alpha}} (n-m)(n^{-1/\alpha})^{r-1/2} \mu_{r,\alpha}(P, P_\alpha)\\ &\leq 3^{\frac{r}{\alpha}} n^{1-\frac{r}{\alpha}+\frac{1}{2\alpha}} \mu_{r,\alpha}(P, P_\alpha) \leq 3^{\frac{r}{\alpha}} n^{1-\frac{r}{\alpha}+\frac{1}{2\alpha}}\tau.\end{aligned}$$

Finally

$$\begin{aligned}K &\leq I_1 + I_2 + I_3\\ &\leq \frac{A}{2}\tau n^{1-\frac{r}{\alpha}+\frac{1}{2\alpha}} + 2^{\frac{r}{\alpha}-1} n^{1-\frac{r}{\alpha}+\frac{1}{2\alpha}}\tau + 3^{\frac{r}{\alpha}} n^{1-\frac{r}{\alpha}+\frac{1}{2\alpha}}\tau\\ &\leq \tau n^{1-\frac{r}{\alpha}+\frac{1}{2\alpha}}(\frac{A}{2} + 2^{\frac{r}{\alpha}-1} + 3^{\frac{r}{\alpha}}) \leq A\tau n^{1-\frac{r}{\alpha}+\frac{1}{2\alpha}},\end{aligned}$$

as requested. □

4. GENERALIZED SUM OF RANDOM VECTORS

The difficulties in the description of generalized convolutions on R^d or in the Banach space setting increase dramatically. In order to have some idea of generalized stable distributions in more general spaces we adopt the Central Limit Theorem approach. Namely, we start with "simple" distributions on R^d, apply *normalized convolution* to these distributions, and approximate the unknown stable law, by corresponding n-fold convolution. To secure necessary accuracy of this approximation we need to solve the rate of convergence problems. Initial steps towards this line of research are given in this section.

Let $X_1, X_2, \ldots$ be i.i.d. d-dimensional random vectors, and $\|\cdot\|$ be the Euclidean norm in R^d. Define the generalized sum in $(R^d, \|\cdot\|)$ by

$$X \circ Y = (X^{\uparrow p} + Y^{\uparrow p})^{\uparrow 1/p}, \qquad (15)$$

where $X^{\uparrow p} = X \parallel X \parallel^{p-1}$, $p > 0$. It seems that Vatan (1984) first pointed out the importance of investigating the generalized sum (15). For the reminder of this section, by generalized sum of random vectors, we will understand operation defined in (15). Let Y_α be a symmetric α-stable random vector, that is

$$Y_\alpha \stackrel{d}{=} n^{-1/\alpha}(Y_\alpha^{(1)} \circ \cdots \circ Y_\alpha^{(n)}),$$

where $Y_\alpha^{(i)}$, $i = 1, \ldots, n$ are i.i.d. copies of Y_α. Then

$$Y_\alpha^{(1)\uparrow p} + \cdots + Y_\alpha^{(n)\uparrow p} \stackrel{d}{=} n^{p/\alpha} Y_\alpha^{\uparrow p}$$

which means, that $Y_\alpha^{\uparrow p}$ is an ordinary α/p-symmetric stable random vector, with condition $\alpha \leq 2p$. Since

$$P(|Y_\alpha| > x) = P(|Y_\alpha^{\uparrow p}| > x^{\uparrow p}),$$

then the Y_α is in the domain of attraction of an ordinary symmetric α-stable law, for any p. It follows, that generalized stable distributions will have "heavy tails". This property may prove to be useful for modelling. Therefore we need to answer the question of the rate of convergence to generalized stable distributions.

Consider the space $\mathcal{X}(R^d)$ of R^d valued random variables with *uniform (Kolmogorov)* metric

$$\rho(X,Y) := \sup_{C \in \mathcal{C}} |P(X \in C) - P(Y \in C)|,$$

where $\mathcal{C}$ is the family of all convex Borel sets in R^d.

We will investigate the rate of convergence in the following limit relation:

$$\lim_{n \to \infty} \rho(n^{-1/\alpha}(X_1 \circ \cdots \circ X_n), Y_\alpha) = 0. \tag{16}$$

We will use the total variation metric **Var** and its *smoothed* version $\tilde{\nu}_{r,\alpha}$:

$$\mathbf{Var}(X,Y) := 2 \sup_{A \in \mathcal{B}(R^d)} |P(X \in A) - P(Y \in A)|,$$

and

$$\tilde{\nu}_{r,\alpha}(X,Y) := \sup_{h \in R} |h|^r \mathbf{Var}(X \circ hY_\alpha, Y \circ hY_\alpha),$$

where X, $Y \in \mathcal{X}(R^d)$ and Y_α is an symmetric α-stable random vector (with respect to generalized convolution "$\circ$"). Let

$$\tilde{\tau}_{rp,\alpha} := \max\{\rho(X,Y_\alpha), \tilde{\nu}_{p,\alpha}(X,Y_\alpha), [\tilde{\nu}_{rp,\alpha}(X,Y_\alpha)]^{\frac{1}{r-\alpha/p}}\}.$$

The following result provides a sharp estimate of the rate of convergence in relation (16).

Theorem 4.1 *Let $rp > \alpha$, then for some constant C*

$$\rho(n^{-1/\alpha} X^{\circ n}, Y_\alpha) \leq C \tilde{\nu}_{rp,\alpha}(X,Y_\alpha) n^{1-\frac{rp}{\alpha}} + C \tilde{\tau}_{rp,\alpha} n^{-p/\alpha}.$$

Remarks: Constant C depends on r and α, but it does not depend on the distributions of X and Y_α. The proof relies heavily on the ideality properties of $\tilde{\nu}_{r,\alpha}$ and **Var** metrics and follows the main idea of Senatov (1980) and Rachev (1991).

Future Research. One direction of the future investigations is to extend generalized convolutions to the probability measures on the entire real line and study their limiting properties. The initial steps towards this line of research are presented in Panorska (1991). Another interesting problem is that of statistical modeling (and computer simulations) with generalized stable laws, where the rate of convergence estimates will be of great use.

REFERENCES

[1] Bingham, N.H., 1971, Factorization theory and domains of attraction for generalized convolutions algebras, *Proc. London Math. Soc.* 23, 16-30.

[2] Bingham, N.H., 1984, On a theorem of Klosowska about generalized convolutions, *Colloq. Math.* 48, 117 -125.

[3] Gilewski, J., 1972, Generalized convolutions and delphic semigroups, *Colloq. Math.* 25, 281 - 289.

[4] Gilewski, J., and Urbanik, K., 1968, Generalized convolutions and generating functions, *Bull. Acad. Polon. Sci. Sér. Sci. Math. Astronom. Phys.* 16, No 6, 481-487.

[5] Jajte, R., 1976, Quasi-stable measures in generalized convolution algebras, *Bull. Acad. Polon. Sci. Sér. Sci. Math. Astronom. Phys.* 14, 503-511.

[6] Jajte, R., 1977, Quasi-stable measures in generalized convolution algebras, II, *Bull. Acad. Polon. Sci. Sér. Sci. Math. Astronom. Phys.* 25, 67 - 72.

[7] Jurek, Z.J., 1985, Limit distributions in generalized convolution algebras, *Probab. Math. Statist.* 5, 113-135.

[8] Kingman, J.F.C., 1963, Random walks with spherical symmetry, *Acta Math.* 109, 11-53.

[9] Klosowska, M., 1977, On the domain of attraction for generalized convolution algebras, *Rev. Roumaine Math. Pures Appl.* 22, 669-677.

[10] Kucharczak, J., and Urbanik, K., 1974, Quasi-stable functions, *Bull. Acad. Polon. Sci. Sér. Sci. Math. Astronom. Phys.* 22, 263 - 268.

[11] Nguyen Van Thu., 1979, Multiply self-decomposable measures in generalized convolution algebras, *Studia Math.* 66, 177-184.

[12] Rachev, S.T., and Yukich, J.E., 1989, Rates for the CLT via new ideal metrics, *Ann. Probab.* 17, 775 - 778.

[13] Rachev, S.T., and Ignatov, Zv., 1984, Ideal quadratic metrics, *Stability Problems for Stoch. Models, Proceedings*, Moscow, VNIISI, 119 - 128. (In Russian). (Engl. transl. (1989) *J. Soviet Math.,* 35, 2386 - 2394.)

[14] Rachev, S. T., 1991, "Probability Metrics and the Stability of Stochastic Models", Wiley, New York.

[15] Senatov, V.V., 1980, Uniform estimates of the rate of convergence in the multidimensional central limit theorem, *Theory Probab. Appl.* 25, 745 - 759.

[16] Smirnov, A.K., 1984, On some properties of measures in Urbanik's algebras, *Stability Problems for Stoch. Models, Proceedings, Moscow, VNIISI.* 136 - 139. (in Russian).

[17] Smirnov, A. K., 1985, Properties of measures in Urbanik's convolutions algebras, *Stability Problems for Stoch. Models, Proceedings, Moscow, VNIISI.* 126 - 136. (in Russian).

[18] Urbanik, K., 1964, Generalized convolutions, *Studia Math.* 23, 217-245.

[19] Urbanik, K., 1967, A characterization of a class of convolutions. *Colloq. Math.* 18, 239-249.

[20] Urbanik, K., 1973, Generalized convolutions II, *Studia Math.* 45,57-70.

[21] Urbanik, K., 1984, Generalized convolutions III, *Studia Math.* 80, 167 - 189.

[22] Urbanik, K., 1985, Moments and generalized convolutions, *Probability and Math. Statist.* 6, 173 - 185.

[23] Urbanik, K., 1986, Generalized convolutions IV, *Studia Math.* 83, 57 - 95.

[24] Urbanik, K., 1987, Domains of attraction and moments, *Probab. Math. Statist.* 8, 89 - 101.

[25] Urbanik, K., 1988, Generalized convolutions V, *Studia Math.* 91, 153 - 178.

[26] Vatan, P., 1984, Max-infinite divisibility and max-stability in infinite dimensions, *Lecture Notes in Math.*, v 1153, 1985, *Probability in Banach Spaces*, Proceedings. 1984, 400 - 425.

[27] Volkovich, V. E., 1980, Normed rings, generated by generalized convolutions, *Stability Problems for Stoch. Models, Proceedings, Moscow, VNIISI.* 12 - 18. (in Russian).

[28] Volkovich, V. E., 1984, Multivariate β-stable measures and generalized convolutions, *Stability Problems for Stoch. Models, Proceedings, Moscow, VNIISI.* 40 -54. (in Russian).

[29] Zolotarev, V. M., 1976, Metric distances in spaces of random variables and their distributions, *Math. USSR-Sb.* 30, 373 - 401.

[30] Zolotarev, V. M., 1983, Probability metrics, *Theory Probab. Appl.* 28, 278 - 302.

[31] Zolotarev, V. M., 1986, "One-Dimensional Stable Distributions", Transl. Math. Monographs. 65. Society.

ORDER COMPLETENESS OF L_1 WITH APPLICATIONS TO STOCHASTICS

D. Plachky

Institute of Math. Statistics
Einsteinstr. 62
D-48149 Münster

Abstract

A short and straightforward proof of the order completeness of $L_1(\Omega, \mathcal{A}, \mu)$ for arbitrary positive measure spaces $(\Omega, \mathcal{A}, \mu)$ is given including the fact that the corresponding least upper bound coincides with the least upper bound of a countable subset. As an application a characterization of atomless probability measures is rederived, a refinement of the Halmos-Savage result concerning families of probability measures dominated by a σ-finite measure is treated, some basic properties concerning least upper bounds of bounded, finitely additive set functions are presented, and a Riesz type decomposition for finitely additive measures, which includes the Hammer-Sobczyk and the Hewitt-Yosida decomposition, is proved.

1. INTRODUCTION

Let $L_1(\Omega, \mathcal{A}, \mu)$ denote the set of all μ-integrable functions $f : \Omega \to \mathbb{R}$, where $(\Omega, \mathcal{A}, \mu)$ denotes an arbitrary positive measure space. Furthermore, let $f_i \in L_1(\Omega, \mathcal{A}, \mu)$, $i \in I$, be bounded from above, i.e. there exists some $g \in L_1(\Omega, \mathcal{A}, \mu)$ satisfying $f_i \leq g$ μ-a.e., $i \in I$. Then it will be shown that there exists a least upper bound of $\{f_i : i \in I\}$ with respect to the partial order "$\leq$ μ-a.e." of $L_1(\Omega, \mathcal{A}, \mu)$, i.e. there exists some $h \in L_1(\Omega, \mathcal{A}, \mu)$ such that $f_i \leq h$ μ-a.e., $i \in I$, is valid and $f_i \leq k$ μ-a.e., $i \in I$, for some $k \in L_1(\Omega, \mathcal{A}, \mu)$ implies $h \leq k$ μ-a.e. Furthermore, it will be shown that h might be chosen as the least upper bound $\sup_{i \in J} f_i$ for some countable subset J of I. This result is already known for positive and σ-finite measure spaces, where the corresponding proof is not elementary (cf. Dunford and Schwartz (1964), IV.11.7).

Approximation, Probability, and Related Fields, Edited by
G. Anastassiou and S.T. Rachev, Plenum Press, New York 1994

2. MAIN RESULT

THEOREM. For any subset $\{f_i : i \in I\}$ of $L_1(\Omega, \mathcal{A}, \mu)$ bounded from above by some $g \in L_1(\Omega, \mathcal{A}, \mu)$, where $(\Omega, \mathcal{A}, \mu)$ is an arbitrary positive measure space, there exists some countable subset J of I such that $\sup_{i\in J} f_i$ stands for some least upper bound of $\{f_i : i \in I\}$ with respect to the partial order "$\leq \mu$-a.e." of $L_1(\Omega, \mathcal{A}, \mu)$. The countable subset J of I might be determined by $J = \bigcup_{n=1}^{\infty} J_n$, where J_n, $n \in \mathbb{N}$, are countable subsets of I such that $\int \inf_{j\in J_n}(g - f_j)d\mu \to \inf\{\int \inf_{k\in K}(g - f_k)d\mu : K$ countable subset of $\Gamma\}$ holds true for $n \to \infty$.

Proof: The assertion of the theorem follows from $\inf_{j\in J}(g-f_j) = g - \sup_{j\in J} f_j$. □

Remarks. *(Concerning least upper bounds of bounded, finitely additive set functions on some algebra of subsets of a set)*

(i) Let $ca(\Omega, \mathcal{A})$ denote the set consisting of all finite signed and σ-additive (countably additive) set functions λ defined on the σ-algebra $\mathcal{A}$ of subsets of the set Ω. Then the theorem of Radon-Nikodym implies for any non-empty subset Λ of $ca(\Omega, \mathcal{A})$ bounded from above by some λ_o, i.e. $\lambda \leq \lambda_o$, $\lambda \in \Lambda$, is valid, the existence of some countable subset Λ_o of Λ such that the least upper bounded $\sup_{\lambda\in\Lambda_o} \lambda$ of Λ_o coincides with the least upper bound $\sup_{\lambda\in\Lambda} \lambda$ of Λ with respect to the order $\leq$ (setwise), where $\sup_{\lambda\in\Lambda} \lambda$ might be determined by $\sup_{\lambda\in\Lambda} \lambda(S) = \sup\{\lambda_1(S_1) + \ldots + \lambda_n(S_n) : S_j \in \mathcal{A}, j = 1, \ldots, n$, pairwise disjoint, $\bigcup_{j=1}^{n} S_j = S$, $\lambda_j \in \Lambda$, $j = 1, \ldots, n$, $n \in \mathbb{N}\}$, $S \in \mathcal{A}$ (cf. Dunford and Schwartz (1964), III.7.5). The result $\sup_{\lambda\in\Lambda} \lambda = \sup_{\lambda\in\Lambda_o} \lambda$ for some countable subset of $\Lambda \subset ca(\Omega, \mathcal{A})$ satisfying $\lambda \leq \lambda_o$, $\lambda \in \Lambda$, for some $\lambda_o \in ca(\Omega, \mathcal{A})$, can be carried over to subsets of $ba(\Omega, \mathcal{A}_o)$ introduced as the set consisting of all bounded, finitely additive set functions defined on $\mathcal{A}_o$, where $\mathcal{A}_o$ is merely an algebra of subsets of Ω. For this purpose one should observe that for any subset $\Lambda \subset ca(\Omega, \mathcal{A})$, which is bounded from above by some $\lambda_o \in ca(\Omega, \mathcal{A})$, the relation $(\sup_{\lambda\in\Lambda} \lambda)|\mathcal{A}_o = \sup_{\lambda\in\Lambda}(\lambda|\mathcal{A}_o)$ holds true, where $\mathcal{A}_o$ stands for some algebra of subsets of Ω generating the σ-algebra $\mathcal{A}$, since $\sup_{\lambda\in\Lambda}(\lambda|\mathcal{A}_o) \leq (\sup_{\lambda\in\Lambda} \lambda)|\mathcal{A}_o$ follows from $(\sup_{\lambda\in\Lambda}(\lambda|\mathcal{A}_o))(A) = \sup\{\lambda_1(A_1) + \ldots + \lambda_n(A_n) : A_j \in \mathcal{A}_o$, $j = 1, \ldots, n$ pairwise disjoint, $\bigcup_{j=1}^{n} A_j = A$, $\lambda_j \in \Lambda$, $j = 1, \ldots, n$, $n \in \mathbb{N}\}$, $A \in \mathcal{A}_o$, whereas $\sup_{\lambda\in\Lambda}(\lambda|\mathcal{A}_o) \geq (\sup_{\lambda\in\Lambda} \lambda)|\mathcal{A}_o$ follows from $(\inf_{\lambda\in\Lambda}(\lambda_o - \lambda))|\mathcal{A}_o \geq \inf_{\lambda\in\Lambda}((\lambda_o - \lambda)|\mathcal{A}_o)$, since the finite, σ-additive measure $(\lambda_o - \lambda)|\mathcal{A}_o$ can be extended uniquely to $\lambda_o - \lambda$ on $\mathcal{A}$ as a finite σ-additive measure. Finally, a Stonian representation argument (cf. Dunford and Schwartz (1964), IV.9.11) yields the existence of some countable subset Λ_o of $\Lambda \subset ba(\Omega, \mathcal{A}_o)$ bounded from above by some $\lambda_o \in ba(\Omega, \mathcal{A}_o)$ such that $\sup_{\lambda\in\Lambda} \lambda = \sup_{\lambda\in\Lambda_o} \lambda$ is valid.

(ii) Let $\mathcal{A}$ resp. $\mathcal{A}'$ denote σ-algebras of subsets of some set Ω resp. Ω' and let $\lambda : \mathcal{A} \times \Omega' \to \mathbb{R}$, where $A \to \lambda(A, \omega')$, $A \in \mathcal{A}$ is a finite, σ-additive measure on $\mathcal{A}$ for any $\omega' \in \Omega'$, and $\omega' \to \lambda(A, \omega')$, $\omega' \in \Omega'$, is $\mathcal{A}'$-measurable for any

$A \in \mathcal{A}$, introduce as a finite transition measure. If now Λ stands for some set of finite transition measures λ satisfying $\lambda(\ ,\omega') \leq \lambda_{\omega'}$ for some finite, σ-additive measure $\lambda_{\omega'}$ on $\mathcal{A}$, $\omega' \in \Omega'$, $\sup_{\lambda \in \Lambda} \lambda(\ ,\omega')$ defines for any $\omega' \in \Omega'$ some finite, σ-additive measure on $\mathcal{A}$. It will now be shown that the last expression defines even a finite transition measure, if the σ-algebra $\mathcal{A}$ is countably generated and $\lambda(\ ,\omega') \ll \lambda_o$, $\lambda \in \Lambda$, $\omega' \in \Omega'$, holds true for some finite measure λ_o on $\mathcal{A}$, i.e. $\lambda_o(N) = 0$ for some $N \in \mathcal{A}$ implies $\lambda(N,\omega') = 0$, $\lambda \in \Lambda$, $\omega' \in \Omega'$. The property of $\mathcal{A}$ to be countably generated implies that $L_1(\Omega, \mathcal{A}, \lambda_o)$ is separable with respect to the L_1-norm and, therefore, $\sup_{\lambda \in \Lambda} \lambda(\ ,\omega') = \sup_{n \in \mathbb{N}} \lambda_n(\ ,\omega')$, $\omega' \in \Omega'$, is valid for some countable subset $\{\lambda_n : n \in \mathbb{N}\}$ of Λ (independent of $\omega' \in \Omega'$). Furthermore, $\sup_{n \in \mathbb{N}} \lambda_n(A,\omega') = \sup\{\lambda_{j_1}(A_1) + \ldots + \lambda_{j_k}(A_k) : A_j \in \mathcal{A}_o,\ j = 1,\ldots,k$, pairwise disjoint, $\lambda_{j_\nu} \in \Lambda$, $\nu = 1,\ldots,k$, $k \in \mathbb{N}\}$, $A \in \mathcal{A}_o$, where $\mathcal{A}_o$ stands for some countable algebra of subsets of Ω generating $\mathcal{A}$, shows that $\omega' \to \sup_{\lambda \in \Lambda}(A,\omega')$, $\omega' \in \Omega'$, is $\mathcal{A}'$-measurable for any $A \in \mathcal{A}_o$. Finally, $\sup_{\lambda \in \Lambda}(\lambda(\ ,\omega')|\mathcal{A}_o) = (\sup_{\lambda \in \Lambda} \lambda(\ ,\omega'))|\mathcal{A}_o$, $\omega' \in \Omega'$, proves that $\omega' \to \sup_{\lambda \in \Lambda} \lambda(S,\omega')$, $\omega' \in \Omega'$, is $\mathcal{A}'$-measurable for any $S \in \mathcal{A}$.

3. APPLICATIONS

3.1 Characterization of Atomless Probability Measures

Let P denote some probability measure on some σ-algebra $\mathcal{A}$ of subsets of some set Ω. Then $\mathcal{F} = \{I_A : A \in \mathcal{A},\ A \subset S,\ P(A) \leq \alpha\}$, $S \in \mathcal{A}$, $\alpha \in (0, P(S))(P(S) > 0)$, has some maximal element I_A with respect to the order $\leq$ P-a.e., since the order completeness of $L_1(\Omega, \mathcal{A}, P)$ implies that some least upper bound of some totally complete subset of $\mathcal{F}$ might be represented according to the theorem above as some element belonging to $\mathcal{F}$. Now the maximality of $I_A \in \mathcal{F}$ yields $P(A) = \alpha$ if P is in addition atomless. Otherwise, one might choose some $B \in \mathcal{A}$ satisfying $B \subset S \backslash A$ and $0 < P(B) \leq \alpha - P(A)$. Therefore, $I_{A \cup B} \in \mathcal{F}$ would be a contradiction to the maximality of $I_A \in \mathcal{F}$.

3.2 Refinement and Generalization of the Halmos-Savage Criterion for Families of Probability Measures Dominated by Some σ-Finite Measure

Let $\mathcal{P}$ denote a set of probability measures on some σ-algebra $\mathcal{A}$ of subsets of some set Ω satisfying $P \ll \mu$ for any $P \in \mathcal{P}$, for some σ-finite measure μ on $\mathcal{A}$, i.e. $\mu(N) = 0$ for some $A \in \mathcal{A}$ implies $P(A) = 0$, $P \in \mathcal{P}$. Then the probability measure ν on $\mathcal{A}$ defined by $\nu(A) = \sum_{n=1}^{\infty} \frac{1}{2^n} \frac{\mu(A \cap S_n)}{\mu(S_n)}$, $A \in \mathcal{A}$, where $S_n \in \mathcal{A}$, $n \in \mathbb{N}$, pairwise disjoint, $\sum_{n=1}^{\infty} S_n = \Omega$, $0 < \mu(S_n) < \infty$, $n \in \mathbb{N}$, satisfies $P \ll \nu$, $P \in \mathcal{P}$, too. Let $P_n \in \mathcal{P}$, $n \in \mathbb{N}$, denote a countable subset of $\mathcal{P}$ such that $\sup_{n \in \mathbb{N}} \frac{dP_n}{d\nu} I_{\{\frac{dP_n}{d\nu} \leq N\}}$ stands for any $N \in \mathbb{N}$ for some least upper bound of $\{\frac{dP}{d\nu} I_{\{\frac{dP}{d\nu} \leq N\}} : P \in \mathcal{P}\}$ with respect to the order $\leq \nu$ a.e. according to the theorem above. This implies for any $P \in \mathcal{P}$ and $A \in \mathcal{A}$ the inequality $P(A) = \int_A \frac{dP}{d\nu} d\nu \leq \sup_{N \in \mathbb{N}} \int_A \sup_{n \in \mathbb{N}} \frac{dP_n}{d\nu} I_{\{\frac{dP_n}{d\nu} \leq N\}} d\nu = \int_A \sup_{n \in \mathbb{N}} \frac{dP_n}{d\nu} d\nu = \sup_{m \in \mathbb{N}} \int_A \sup_{1 \leq n \leq m} \frac{dP_n}{d\nu} d\nu$ i.e. $\sup_{P \in \mathcal{P}} P(A) \leq \sup_{m \in \mathbb{N}} \int_A \sup_{1 \leq n \leq m} \frac{dP_m}{d\nu} d\nu$, $A \in \mathcal{A}$, holds true, from which for $P_n(A) = 0$, $n \in \mathbb{N}$, for some $A \in \mathcal{A}$, the equation $P(A) = 0$, $P \in \mathcal{P}$, can be derived, since $\int_A \sup_{1 \leq n \leq m} \frac{dP_n}{d\nu} d\nu \leq \sum_{n=1}^{m} \int_{A \cap A_n} \frac{dP_n}{d\nu} d\nu$, $A \in \mathcal{A}$, is valid, where A_n stands for $\{\frac{dP_n}{d\nu} \geq \sup_{1 \leq n \leq m} \frac{dP_n}{d\nu}\}$, $n = 1, \ldots, m$. Therefore, $P \ll \mu$, $P \in \mathcal{P}$, for some σ-finite measure μ, is equivalent to the existence of $P_n \in \mathcal{P}$, $n = 1, 2, \ldots$, such that $P \ll \sum_{n=1}^{\infty} \frac{1}{2^n} P_n$, $P \in \mathcal{P}$, is valid (cf. Halmos and Savage (1949)). This characterization admits the following finitely additive version: Let $\mathcal{M}$ denote a set of finite, finitely additive measures ν on an algebra $\mathcal{A}$ of subsets of a set Ω, such that there exists some finite, finitely additve measure μ on $\mathcal{A}$ having the property that any $\nu \in \mathcal{M}$ is absolutely continuous with respect to μ (i.e. for any $\varepsilon > 0$ there exists some $\delta > 0$ satisfying $\nu(A) \leq \varepsilon$ for all $A \in \mathcal{A}$ with the property $\mu(A) \leq \delta$). Then a Stonian representation argument (cf. Dunford and Schwartz (1964), IV.9.11) together with the fact that absolute continuity of finite, σ-additive measures on some algebras $\mathcal{A}$ of subsets of some set Ω is preserved under (the uniquely determined) extensions as finite, σ-additive measures on the σ-algebra generated by $\mathcal{A}$ (cf. Dunford and Schwartz (1964), IV.9.13) yields the existence of $\nu_n \in \mathcal{M}$, $n \in \mathbb{N}$, such that ν is absolutely continuous with respect to the finite, finitely additive measure $\sum_{n=1}^{\infty} \frac{1}{2^n} \frac{\nu_n}{\nu_n(\Omega)}$ on $\mathcal{A}$ for every $\nu \in \mathcal{M}$.

3.3 A Riesz Type Decomposition for Finite, Finitely Additive Measures on an Algebra of Subsets of Some Set

Let M denote some set of finite, finitely additive measures ν on an algebra $\mathcal{A}$ of subsets of a set Ω, i.e. ν is non-negative, finitely additive and satisfies $\nu(\Omega) < \infty$, where $o \in M$ holds true. Then any finite, finitely additive measure μ on $\mathcal{A}$ can be written as $\mu = \mu_1 + \mu_2$, where $\mu_1 = \sum_{j \in \mathbb{N}} \nu_j$, $\nu_j \in M$, $j \in \mathbb{N}$, is valid, and $\mu_2 \geq \nu$ for some $\nu \in M$ implies $\nu = 0$. In this connection $\sum_{k \in K} \nu_k$ denotes the least upper bound of all finite sums $\sum_{i \in F} \nu_i$, where F is some finite subset of K, with respect to the natural order of finite, finitely additive measures on $\mathcal{A}$. This decomposition follows by introducing the set of all classes of finite, finitely additive measures $\nu_i \in M$, $i \in I$, satisfying $\sum_{i \in I} \nu_i \leq \mu$, where $\sum_{i \in I} \nu_i$ is defined as the least upper bound of all finite sums $\sum_{i \in F} \nu_i$, F being some finite subset of K. This set is inductively ordered with respect to inclusion and a maximal element yields the first part $\mu_1 = \sum_{i \in I} \nu_i$ resp. the second part $\mu_2 = \mu - \mu_1$ of this decomposition, if one takes into consideration that according to the theorem above together with a Stonian representation argument (cf. Dunford and Schwartz (1964), IV.9.11) that $\sum_{i \in I} \nu_i$ can be represented as an ordinary sum $\sum_{i=1}^{\infty} \nu_i$. Furthermore, this decomposition is uniquely determined, if M is solid i.e. $\nu \geq \lambda$ for some $\nu \in M$ and some finite, finitely additive measure λ on $\mathcal{A}$ implies $\lambda \in M$. Finally, the first part $\mu_1 = \sum_{i=1}^{\infty} \nu_i$, $\nu_i \in M$, $i \in \mathbb{N}$, of the decomposition of μ might be represented simply by some $\mu_1 \in M$, if M is in addition Σ-stable, i.e. $\sum_{i=1}^{\infty} \nu_i \in M$ is valid for any countable subset $\{\nu_i : \nu_i \in M,\ i \in \mathbb{N}\}$ of M such that $\sum_{i=1}^{\infty} \nu_i(\Omega)$ is finite. Special cases of this decomposition are the decomposition of Hammer-Sobczyk, if one chooses for M the family of all finite, finitely additive measures on $\mathcal{A}$, which are atmost two-valued (cf. Bhaskara Rao and Bhaskara Rao (1983), p. 146) and the decomposition of Hewitt-Yosida, if one chooses for M the set of all finite and σ-additive measures on $\mathcal{A}$ (cf. Bhaskara Rao and Bhaskara Rao (1983), p. 240).

REFERENCES

Bhaskara Rao, K. P. S. and Bhaskara Rao, M., 1983, "Theory of Charges", Academic Press, London.

Dunford, N. and Schwartz, J., 1964, "Linear Operators I", Interscience Publishers, New York.

Halmos, P. R. and Savage, L. J., 1949, Application of the Radon-Nikodym theorem to the theory of sufficient statistics, *Ann. Math. Stat.* 20, 225 - 241.

ON RELIABILITY ANALYSIS OF CONSECUTIVE-k-OUT-OF-n: F SYSTEMS AND THEIR GENERALIZATIONS - A SURVEY

Wolfgang W. Preuss[1] and Thomas K. Boehme[2]

[1]Hochschule für Technik & Wirtschaft Dresden, Germany
[2]University of California Santa Barbara, CA

INTRODUCTION

A special type of system has attracted considerable attention during the past decade. This is the so-called consecutive-k-out-of-n:F system. A consecutive-k-out-of-n:F system consists of a sequence of n ordered components which either fail or operate. The system fails whenever k consecutive components are failed, $1<k<n$. In case of $k=1$ and $k=n$ the system is the series and the parallel system, respectively. There are two topologies for the system considered: a line or a circle. Therefore we have to distinguish between the *linear* and the *circular* consecutive-k-out-of-n:F system.

Konteleon[23] seems to be the first who dealt with the reliability determination of the consecutive-k-out-of-n:F system and established a computer algorithm for it. Then Chiang and Niu[9] introduced the denotation "consecutive-k-out-of-n:F system" and described a telecommunication system with n relay stations (either satelites or ground stations) and an oil pipeline system with n pump stations (that both fail if any two stations in the system fail) as consecutive-2-out-of-n:F systems. Assuming that the components fail independently and that their lifetimes are identically distributed they established a closed formula to compute the exact reliability of a (linear) consecutive-2-out-of-n:F system, presented a recursive formula in the general case of a (linear) consecutive-k-out-of-n:F system and gave bounds for the system reliability. Bollinger and Salvia[3] developed another recursive formula and gave another practical example for the system considered from the field of integrated circuits. Independently of these investigations, the research team of the first author of the present paper started a research project supported by the mining industry in 1981. One of the topics of the project was the reliability analysis of extremely long belt conveyors as they are used in open-cast mining. The failure policy used by the engineers was the following. If a roll station of the belt conveyor's upper range (which has to carry the heavy material) fails then, by use of a special equipment, it can be removed without to stop the conveyor, i.e. the system does not fail. The belt conveyor must be stoped, i.e. the system is considered to be failed, if a hole of a certain number k of consecutive failed roll stations occurs. Obviously, the upper range of the belt conveyor (and similarly the

Approximation, Probability, and Related Fields, Edited by
G. Anastassiou and S.T. Rachev, Plenum Press, New York 1994

lower range too) can be considered as a linear consecutive-k-out-of-n:F system (with lifetimes of the components that are dependent). The reliability of the belt conveyor was approximated by a closed formula (see Dietl, Preuss, Kossow and Kirchner[12]) established under the assumption that the roll stations fail independently. Another interesting practical example for linear consecutive-k-out-of-n:F systems is given by Chiang and Chiang[10] in connection with questions that came up when the Pioneer 11 got lost in space. In order to explore distant stars by spacecrafts, they considered the following model. After a first spacecraft is launched and before the ground stations lose the contact with it a second spacecraft is to launch which will relay signals transmitted by the first one to ground stations. If necessary, then additional spacecrafts will be launched to relay signals from the first spacecraft to ground stations. The question was, "What is the mean number of spacecrafts needed to explore the distant star for a given reliability of spacecraft?"

The circular consecutive-k-out-of-n:F system was first studied by Derman, Lieberman and Ross[11] and its reliability is expressed in terms of the reliability of linear systems. Closed formulas for circular (and linear) consecutive-k-out-of-n:F systems are established by Lambiris and Papastavridis[29].

Until now, a great number of closed formulas, recursive and direct algorithms, limit formulas and bounds for the reliability of linear and circular consecutive-k-out-of-n:F systems are established by use of methods from combinatorics, probability theory, switching-algebra, graph theory and so on (see References).

There are also many papers dealing with questions from (e.g.) optimization, renewal and maintenance theory that are not listed in our References.

In the meantime, some papers dealt with certain generalizations of the consecutive-k-out-of-n:F systems (see References[2, 17, 18, 22, 28, 32, 33, 36, 39, 42]).

SOME RESULTS ON CONSECUTIVE-k-OUT-OF-n:F SYSTEMS

First, let us introduce some symbols. We denote the reliability of a linear and a circular consecutive-k-out-of-n:F system by

$$R_L(p_1,\dots,p_n;k) \quad , \quad R_C(p_1,\dots,p_n;k)$$

where p_i $(i = 1, \dots, n)$ represents the reliability of component (i). The lifetimes of the components are assumed to be independent random variables. In case of $p_1 = \dots = p_n = p$ we will use the symbols

$$R_L(p;k;n) \quad , \quad R_C(p;k;n)$$

$q_i = 1 - p_i$ and $q = 1 - p$ stand for the unreliability of a component.

The case $p_i = p$ $(i = 1, \dots, n)$

Lambiris and Papastavridis[29] established the formulas

$$R_L(p;k;n) = \sum_{\lambda=0}^{n} \begin{bmatrix} n-\lambda k \\ \lambda \end{bmatrix} (-1)^\lambda (pq^k)^\lambda - q^k \sum_{\lambda=0}^{n} \begin{bmatrix} n-\lambda k-k \\ \lambda \end{bmatrix} (-1)^\lambda (pq^k)^\lambda \tag{1}$$

and

$$R_C(p;k;n) = \sum_{\lambda=0}^{n} \begin{bmatrix} n-\lambda k \\ \lambda \end{bmatrix} (-1)^\lambda (pq^k)^\lambda - k\sum_{\lambda=0}^{n} \begin{bmatrix} n-\lambda k-k-1 \\ \lambda \end{bmatrix} (-1)^k (pq^k)^{\lambda+1} - q^n \qquad (2)$$

($n \geq k$). Another formula (derived from paper[12])

$$R_L(p;k;n) = \sum_{j=1}^{n} g_k(n,j) p^{n-j} q^j \qquad (3)$$

with numbers

$$g_k(n,j) = \sum_{\nu=0}^{[j/k]} (-1)^\nu \begin{bmatrix} n-\nu k \\ j-\nu k \end{bmatrix} \begin{bmatrix} n-j+1 \\ \nu \end{bmatrix} \qquad (0 \leq j \leq n-[n/k])$$

and

$$g_k(n,j) = 0 \qquad (n-[n/k] < j \leq n)$$

can be found in Kossow and Preuss[26] (see also Goulden[16]). In case of $k = 2$ Preuss, Kossow, Kirchner and Dietl[34] established the reliability expression

$$R_C(p;2;n) = \sum_{j=0}^{[n/2]} \begin{bmatrix} n-j \\ j \end{bmatrix} \frac{n}{n-j} p^{n-j} q^j \qquad (4)$$

($[x]$ stands for the largest integer less than or equal to x).
The reliability formulas (1), (2) and (3), (4) were found independently of each other.

Derman, Lieberman and Ross[11] expressed the reliability of a circular system by terms of the reliability of a linear one. That is

$$R_C(p;k;n) = p^2 \sum_{j=0}^{k-1} (j+1) q^j R_L(p;k;n-j-2) \qquad (5)$$

with

$$R_L(p;k;i) = 1 \qquad (i < k)$$

A further closed formula for the circular system similar to (3) is given by Goulden[16].

Some Recursion Formulas

A closed reliability formula is nice to have for many purposes, but it is not automatically the best and the fastest way to compute the reliability of the system. Therefore many papers provide recursion algorithms, especially if the component reliabilities are different.

Some of the many results should be given here. The recursive procedure

$$R_L(p_1,\dots,p_n;k) = R_L(p_1,\dots,p_{n-1};k) - p_{n-k} \left[\prod_{r=0}^{k-1} q_{n-r} \right] R_L(p_1,\dots,p_{n-k-1};k) \qquad (6)$$

with initial conditions

$$R_L(p_1,\dots,p_i;k) = 1 \qquad (i < k)$$

and

$$R_L(p_1,\dots,p_k;k) = 1 - \prod_{r=1}^{k} q_r$$

is given by Shanthikumar[40]. In case of $p_i = p$, $q_i = q$ $(i = 1, \dots , n)$ this procedure lead to the recursion formula

$$R_L(p;k;n) = R_L(p;k;n-1) - pq^k R_L(p;k;n-k-1) \tag{7}$$

with initial conditions

$$R_L(p;k;i) = 1 \qquad (i < k)$$

and

$$R_L(p;k;k) = 1 - q^k$$

One of the fastest recursive algorithm to compute the reliability of a circular consecutive-k-out-of-n:F system is established by Antonopoulou and Papastavridis[1]. That is

$$\left.\begin{aligned} R_C(p_1,\dots,p_n;k) = {}& p_n R_L(p_1,\dots,p_{n-1};k) + q_n R_C(p_1,\dots,p_{n-1};k) \\ & - \sum_{i=0}^{k-1} \left[\prod_{\nu=1}^{i} q_\nu\right] p_{i+1} \left[\prod_{\nu=n-k+i+1}^{n} q_\nu\right] p_{n-k+i} R_L(p_{i+2},\dots,p_{n-k+i-1};k) \end{aligned}\right\} \tag{8}$$

where

$$R_L(p_{i+2},\dots,p_{n-k+i-1};k) = R_L(p_1^*,\dots,p_{n-k-2}^*;k)$$

is the reliability of a linear consecutive-k-out-of-$(n$-k-2):F system having different component reliabilities $p_\nu^* = p_{i+\nu+1}$ $(\nu = 1, \dots , n\text{-}k\text{-}2)$. If $p_i = p$, $q_i = q$ $(i = 1, \dots , n)$, then (from formula (8)) one can derive immediately the algorithm

$$R_C(p;k;n) = pR_L(p;k;n-1) + qR_C(p;k;n-1) - kp^2q^k R_L(p;k;n-k-2) \tag{9}$$

Another recursion algorithm which does not need the reliabilities of linear systems and that corresponds to formula (7) is given by Lambiris and Papastavridis[29] too. That is

$$R_C(p;k;n) = R_C(p;k;n-1) - pq^k R_C(p;k;n-k-1) \qquad (n \geq 2k+1) \tag{10}$$

with initial conditions

$$R_C(p;k;n) = 0 \qquad (n < 0)$$

$$R_C(p;k;n) = 1 \qquad (0 \leq n < k)$$

$$R_C(p;k;k) = 1 - q^k$$

$$R_C(p;k;n) = 1 - npq^k - q^n \qquad (k < n \leq 2k)$$

Remark on Dependent Component Lifetimes

The majority of the published papers deals with consecutive-k-out-of-n:F systems consisting of components whose lifetimes are independent. The class of consecutive-k-out-of-n:F systems with exchangeable lifetimes studied by Shanthikumar[41] includes both systems with independent component lifetimes and with a special type of dependency of component failures. By use of the inclusion-exclusion formula and methods from graph theory, Kossow and Preuss[26] established topological formulas for exact reliability evaluation of linear and circular consecutive-k-out-of-n:F systems whose component lifetimes may be dependent.

SOME GENERALIZATIONS OF THE CONSECUTIVE SYSTEMS

During the past eight years some papers introduced certain generalizations of the consecutive systems that should be mentioned here.

The Consecutive k-within-m-out-of-n:F system

The consecutive k-within-m-out-of-n:F system introduced by Griffith[17] consists of n linearly or cyclically ordered components and fails iff there are m consecutive components which include at least k failed components. This system is a consecutive-k-out-of-n:F system if $m = k$ holds.

Systems like that may play a role in radar detection and other contexts, namely in quality control and inspection procedures (see Saperstein[37,38] and Papastavridis and Koutras[33]).

Until now, some authors dealt with the consecutive k-within-m-out-of-n:F system (see papers[22,32,33,39]) and established limit and recursion formulas and bounds for its reliability. But many problems are still unsolved.

Consecutively Connected Systems

Shanthikumar[42] introduced the consecutively connected system as a system consisting of a source (0), n components $\{1, \dots, n\}$ and a sink $(n+1)$ such that the source is connected to $\{1, \dots, k(0)\}$ and component j $(1 \leq j \leq n)$ is connected to $\{j+1, \dots, j+k(j)\}$ (for some $k(j) \geq 1$, $0 \leq j \leq n$) by arcs. The source, sink and arcs are perfect and the n components are failure prone. The system functions if there is a connection from the source to the sink trough functioning components.

This system is a generalization of the (linear) consecutive-k-out-of-n:F system because the latter is the special case of Shanthikumar's system if $k(j) = \min(k\,;\, n+1-j)$ $(0 \leq j \leq n)$.

Shanthikumar considered a consecutively connected system that consists of a source, a sink and 6 components $\{1, \dots, 6\}$ having the parameters

$$\left.\begin{array}{l} k(0)=k(5)=2\,;\,k(1)=4\,;\,k(2)=k(3)=k(6)=1\,;\,k(4)=3 \\ p_1=p_2=p_3=0{,}98\,;p_4=p_5=p_6=0{,}99 \end{array}\right\} \tag{11}$$

and computed the value 0,99891792 for its reliability.

A further generalization of Shanthikumar's consecutively connected system (and therefore of the linear consecutive-k-out-of-n:F system too) is defined by Kossow and Preuss[28]. That is the consecutively connected system with multistate components (c.c.s. m.c.). A c.c.s.m.c. consists of $n+2$ linarly ordered components $\{0, 1, \dots, n, n+1\}$ where the sink $(n+1)$ is absolutely reliable and the unreliability of the system is caused by the system having the components $\{0, 1, \dots, n\}$ with the states

$$Z_i : \begin{bmatrix} 0 & 1 & \dots & j & \dots & k(i) \\ p_{i,0} & p_{i,1} & \cdots & p_{i,j} & \cdots & p_{i,k(i)} \end{bmatrix} \quad , \quad \sum_{j=0}^{k(i)} p_{i,j} = 1 \qquad (i=0,1,\dots,n)$$

where $Z_i = j$ ($j = 1, \dots, k(i)$) means that the component (i) is connected to $\{i+1, i+2, \dots, m\}$ ($m = \min(i+j, n+1)$) by absolutely reliable arcs. $Z_i = 0$ represents the failure state of component (i). The Z_i are assumed to be independent and the system is considered to be failed if there is no path from the source (0) to the sink $(n+1)$. In case of the c.c.s.m.c. the source is not absolutely reliable.

As an practical example take a set of relay stations $\{1, \dots, n\}$, a transmitter station (0) and a receiver $(n+1)$. Each of the stations (i) , $0 \le i \le n$, may consist of $k(i)$ amplifiers. If all of the amplifiers of station (i) are operating then a signal from (i) reaches the next $k(i)$ stations (or the receiver). The failure of an amplifier reduces the range of station (i) by one station, and so on.

In case of

$$p_{i,j}=0 \qquad (j=1,\dots,k(i)-1) \quad ; \quad p_{i,k(i)} = p_i \quad ; \quad p_{i,0} = q_i \ ,$$

i.e.

$$Z_i : \begin{bmatrix} 0 & 1 & \dots & k(i)-1 & k(i) \\ q_i & 0 & \dots & 0 & p_i \end{bmatrix} \quad , \quad q_i + p_i = 1 \ , \quad (i = 0,1,\dots,n)$$

and $q_0 = 0$, $p_0 = 1$ the c.c.s.m.c. is Shanthikumar's system. The linear consecutive-k-out-of-n:F system follows if $1 \le k(\mathrm{i}) = k$ ($i = 0, 1, \dots, n$).

The paper[28] provides recursion algorithms to compute the reliability of a c.c.s.m.c. and of Shanthikumar's system. Let $R(k(\mathrm{i});n)$ denote the reliability of the c.c.s.m.c. and define

$$k = \max(k(i) \mid 0 \le i \le n)$$

Then the reliability of the c.c.s.m.c. can be computed as follows:

$$R(k(i);0) = 1 - p_{0,0}$$

$$R(k(i);n) = \sum_{\nu=1}^{n} a_\nu(n) R(k(i);n-\nu) + a_{n+1}(n) \qquad (1 \le n \le k(0)-1)$$

$$R(k(i);n) = \sum_{\nu=1}^{\min(k,n)} a_\nu(n) R(k(i);n-\nu) \qquad (n \geq k(0))$$

$$a_1(n) = \sum_{j=1}^{k(n)} p_{n,j} = 1 - p_{n,0}$$

$$a_2(n) = p_{n,0} \sum_{j=2}^{k(n-1)} p_{n-1,j}$$

$$a_\nu(n) = p_{n,0} \left[\prod_{r=1}^{\nu-2} \sum_{s=0}^{\min(r,k(n-r))} p_{n-r,s} \right] \sum_{j=\nu}^{k(n-\nu+1)} p_{n-\nu+1,j} \qquad (3 \leq \nu \leq k)$$

(agreement: $\sum_{j=u}^{v} (\ldots) = 0$ if $v < u$).

In case of Shanthikumar's system, this procedure can be specified by

$$R(k(i),n) = \sum_{\nu=1}^{\min(k,n)} a_\nu(n) R(k(i),n-\nu) \qquad (n \geq k(0)) \tag{12}$$

where

$$R(k(i),n) = 1 \qquad (0 \leq n \leq k(0)-1)$$

$$a_1(n) = 1 - q_n = p_n$$

$$a_2(n) = q_n p_{n-1} \quad (k(n-1) \geq 2) \ , \quad a_2(n) = 0 \quad (k(n-1) = 1)$$

$$a_\nu(n) = q_n p_{n-\nu+1} \prod_{r=1}^{\nu-2} f(k(n-r)) \qquad (3 \leq \nu \leq k(n-\nu+1)) \ ,$$

$$a_\nu(n) = 0 \qquad (k(n-\nu+1) < \nu \leq k)$$

$$f(k(n-r)) = 1 \quad (k(n-r) \leq r) \ , \quad f(k(n-r)) = q_{n-r} \quad (k(n-r) > r).$$

In case of a (linear) consecutive-k-out-of-n:F system (i.e. $1 \leq k(i) = k \;\; i = 0, 1, \ldots ,$ n), the procedure (12) with $R(k(i);n) = R(k;n)$ leads to

$$R(k,n) = 1 \qquad (0 \leq n \leq k-1)$$

$$R(k,k) = 1 - \prod_{r=1}^{k} q_r$$

$$R(k,n) = R(k,n-1) - p_{n-k} \left[\prod_{r=0}^{k-1} q_{n-r} \right] R(k,n-k-1)$$

which is equivalent to the procedure (6).

Let us consider, as an example, Shanthikumar's system with parameters (11). Then, by use of the procedure (12), we obtain:

$$k = \max(k(i) \mid 0 \le i \le n) = 4 ;$$

$$R(k(i);0) = R(k(i);1) = 1 ; \quad a_1(n) = p_n \quad (n = 2, \dots, 6) ;$$

$$a_2(2) = q_2 p_1 ; \quad R(k(i);2) = p_2 + q_2 p_1 ;$$

$$a_2(3) = 0 ; \quad a_3(3) = q_3 p_1 ; \quad R(k(i);3) = p_3 R(k(i);2) + q_3 p_1 ;$$

$$a_2(4) = a_3(4) = 0 ; \quad a_4(4) = q_4 p_1 ; \quad R(k(i);4) = p_4 R(k(i);3) + q_4 p_1 ;$$

$$a_2(5) = q_5 p_4 ; \quad a_3(5) = a_4(5) = 0 ; \quad R(k(i);5) = p_5 R(k(i);4) + q_5 p_4 R(k(i);3) ;$$

$$a_2(6) = q_6 p_3 ; \quad a_3(6) = q_6 p_4 q_5 ; \quad a_4(6) = 0 .$$

Then the system's reliability is

$$R(k(i);6) = p_6 R(k(i);5) + q_6 p_5 R(k(i);4) + q_6 p_4 q_5 R(k(i);3)$$

which could be expressed by the p_i , q_i . If we take the given values for the p_i , q_i then we obtain

$$R(k(i);2) = 0{,}9996 , \quad R(k(i);3) = 0{,}999208 ;$$

$$R(k(i);4) = 0{,}99901592 ; \quad R(k(i);5) = 0{,}99891792$$

and the required reliability

$$R(k(i);6) = 0{,}99891792$$

which is the same value as calculated by Shanthikumar.

Let us remark that the c.c.s.m.c. could be considered in a circular form too.

Lattice Systems

Two dimensional generalizations of the linear and circular consecutive-k-out-of-n:F systems are given by Salvia and Lasher[36] and Boehme, Kossow and Preuss[2]. Because the system k^2/n^2 in paper[36] is a special case of the lattice systems defined in paper[2] let us introduce the lattice systems[2].

Consider a system whose components are ordered like the elements of a (m,n)-matrix and assume that these components are 1 (operating) or 0 (failed). A system like that is called a linear (m,n)-lattice system.

The circular (m,n)-lattice system is similar. Take m circles centred at the same point and n beams starting from that point and crossing the circles. The junctions of the beams and the circles represent the components of the system (the circles and the beams are not necessarily physical objects). The components have also only the states 1 (operating) or 0 (failed).

A (linear or circular) (m,n)-lattice system is called a (linear or circular) connected-X-out-of-(m,n):F lattice system if it fails whenever (at least) one subset of connected failed components occurs which includes failed components in the meaning of connected-X (for each context one must define what connected-X means).

For example, a (linear or circular) connected-(r,s)-out-of-(m,n):F lattice system fails if at least one connected (r,s)-submatrix of failed components occurs. A (linear or circular) connected-(r,s)-or-(s,r)-out-of-(m,n):F lattice system fails if at least one connected (r,s)-submatrix ore one connected (s,r)-submatrix of failed components occurs.

A (linear or circular) connected-$(1,k)$-out-of-$(1,n)$:F lattice system is obviously a (linear or circular) consecutive-k-out-of-n:F system.

Connected-X-out-of-(m,n):F lattice systems can be used to model (e.g.) supervision systems.

In the field of reliability analysis of these systems there are many open problems. But some results are already known.

For example, if the reliability of all system components is p $(q = 1 - p$ is the unreliability) and the components fail independently then the reliabilities

$$R_L(p;(r,s);(m,n)) \quad , \quad R_C(p;(r,s);(m,n))$$

of the linear and the circular connected-(r,s)-out-of-(m,n):F lattice systems are known for simple cases (see paper[2]):

$$R_L(p;(1,k);(m,n)) = \left[R_L(p;k;n)\right]^m$$

$$R_L(p;(k,1);(m,n)) = \left[R_L(p;k;m)\right]^n$$

$$R_C(p;(1,k);(m,n)) = \left[R_C(p;k;n)\right]^m$$

$$R_C(p;(k,1);(m,n)) = \left[R_L(p;k;m)\right]^n$$

$$R_L(p;(m,k);(m,n)) = \sum_{j=0}^{n-[n/k]} g_k(n,j)(1-q^m)^{n-j}q^{mj}$$

$$R_C(p;(m,k);(m,n)) = R_C(1-q^m;k;n)$$

where $R_L(p;k;n)$, $R_C(p;k;n)$, $g_k(n,j)$ can be calculated by the formulas (1) to (5) or by use of the procedures (7), (9) or (10).

One possibility to find the reliability of a linear connected-(r,s)-out-of-(m,n):F lattice system whose components fail independently and have all the same reliability p is to calculate

$$R_L(p;(r,s);(m,n)) = \sum_{k=0}^{mn} g_{(r,s)}((m,n),k)p^{mn-k}q^k$$

where $g_{(r,s)}((m,n),k)$ denotes the number of all (m,n)-matrices including exactly k failed and mn-k operating elements but no connected (r,s)-submatrix of failed elements. In the circular case the reliability could be expressed similarly. But these problems are still unsolved.

Finally, let us remark that one could generalize the lattice systems to three dimensions. Also one could consider two or three dimensional extensions of the consecutive k-within-m-out-of-n:F system.

REFERENCES

1. J. Antonopoulou and S. Papastavridis, Fast recursive algorithm to evaluate the reliability of a circular consecutive-k-out-of-n:F system, *IEEE Transactions on Reliability*, vol. R-36, pp. 83-84 (1987).
2. T.K. Boehme, A. Kossow and W. Preuss, A generalization of consecutive-k-out-of-n:F systems, *IEEE Transactions on Reliability*, vol. R-41, 3, pp. 451-457 (1992).
3. R.C. Bollinger and A.A. Salvia, Consecutive-k-out-of-n:F networks, *IEEE Transactions on Reliability*, vol. R-31, pp. 53-56 (1982).
4. R.C. Bollinger, Direct computation for consecutive-k-out-of-n:F systems, *IEEE Transactions on Reliability*, vol. R-31, 5, pp. 444-446 (1982).
5. R.C. Bollinger and A.A. Salvia, Consecutive-k-out-of-n:F systems with sequential failures, *IEEE Transactions on Reliability*, vol. R-34, pp. 43-45 (1985).
6. R.C. Bollinger, An algorithm for direct computation in consecutive-k-out-of-n:F systems, *IEEE Transactions on Reliability*, vol. R-35, pp. 611-612 (1986).
7. F.Y. Chan, L.K. Chan and G.D. Lin, On consecutive-k-out-of-n:F systems, *Europian Journal of Operational Research* 36, pp. 207-216 (1988).
8. R.W. Chen and F.K. Hwang, Failure distribution of consecutive-k-out-of-n:F systems, *IEEE Transactions on Reliability*, vol. R-34, pp. 338-341 (1985).
9. D.T. Chiang and S.C. Niu, Reliability of consecutive-k-out-of-n:F systems, *IEEE Transactions on Reliability*, vol. R-30, pp. 87-89 (1981).
10. D.T. Chiang and R.F. Chiang, Relayed communication via consecutive-k-out-of-n:F systems, *IEEE Transactions on Reliability*, vol. R-35, pp. 65-67 (1986).
11. C. Derman, G.J. Lieberman and S.M. Ross, On the consecutive-k-out-of-n:F system, *IEEE Transactions on Reliability*, vol. R-31, pp. 57-63 (1982).
12. W. Dietl, W. Preuss, H. Kirchner und A. Kossow, Mathematische Modellierung der Zuverlässigkeit von Bandstrassen, *Hebezeuge/Fördermittel* 25, pp. 292-295 (1985).
13. D.Z. Du and F.K. Hwang, A direct algorithm for computing reliability of a consecutive-k cycle, *IEEE Transactions on Reliability*, vol. 37, pp. 70-72 (1988).
14. J.C. Fu, Reliability of a large consecutive-k-out-of-n:F system, *IEEE Transactions on Reliability*, vol. R-34, pp. 127-130 (1985).
15. J.C. Fu, Bounds for reliability of large consecutive-k-out-of-n:F systems with unequal component reliability, *IEEE Transactions on Reliability*, vol. R-35, pp. 316-319 (1986).
16. I.P. Goulden, Generating functions and reliabilities for consecutive-k-out-of-n:F systems, *Utilitas Mathematica* 32, pp. 141-147 (1987).
17. W.S. Griffith, On consecutive-k-out-of-n failure systems and their generalizations, A.P. Basu (ed), *Reliability and quality control*, pp. 157-165 (1986); Elsevier (North Holland).
18. W. Griffith and Z. Govindarajulu, Consecutive-k-out-of-n failure systems: reliability and availability analysis with some multistate extensions, *Amer. J. Math. Management Sci* 5, pp. 125-160 (1985).
19. F.K. Hwang, Fast solutions for consecutive-k-out-of-n:F system, *IEEE Transactions on Reliability*, vol. R-31, pp. 447-448 (1982).
20. F.K. Hwang, An $O(kn)$-time algorithm for computing the reliability of a circular consecutive-k-out-of-n:F system, *IEEE Transactions on Reliability*, vol. 42, 1, pp. 161-162 (1993).
21. S.N. Iyer, Distribution of time to failure of consecutive-k-out-of-n:F systems, *IEEE Transactions on Reliability*, vol. 39, pp. 97-100 (1990).
22. S.N. Iyer, Distribution of the lifetime of consecutive-k-within-m-out-of-n:F systems, *IEEE Transactions on Reliability*, vol. 41, 3, pp. 448-450 (1992).
23. J.M. Konteleon, Reliability determination of a r-successive-out-of-n:F system, *IEEE Transactions on Reliability*, vol. 29, p. 437 (1980).
24. A. Kossow und W. Preuss, Zuverlässigkeitsanalyse spez. Netzwerke am Beispiel konsekutiver-k-aus-n:F Systeme, *Nachrichtentechnik Elektronik* 36, pp. 299-301 (1986).
25. A. Kossow und W. Preuss, Zuverlässigkeitsanalyse konsekutiver Systeme - eine Übersicht, *OR Spektrum* 11, pp. 121-130 (1989).
26. A. Kossow and W. Preuss, Reliability of consecutive-k-out-of-n:F systems with nonidentical component

reliabilities, *IEEE Transactions on Reliability*, vol. 38, pp. 229-233 (1989).

27. A. Kossow and W. Preuss, Mean time-to-failure for a linear consecutive-*k*-out-of-*n*:F system, *IEEE Transactions on Reliability*, vol. 40, 3, pp. 271-272 (1991).
28. A. Kossow and W. Preuss, Reliability analysis of consecutively connected systems with multistate components, *IEEE Transactions on Reliability* (TR 91-215).
29. M. Lambiris and S. Papastavridis, Exact reliability formulas for linear & circular consecutive-*k*-out-of-*n*:F systems, *IEEE Transactions on Reliability*, vol. R-34, 2, pp. 124-126 (1985).
30. S. Papastavridis, Upper and lower bounds for the reliability of a consecutive-*k*-out-of-*n*:F systems, *IEEE Transactions on Reliability*, vol. R-35, 5, pp. 607-610 (1986).
31. S. Papastavridis and J. Hadjichristos, Formulas for the reliability of a consecutive-*k*-out-of-*n*:F system, *J. Appl. Prob.* 26, pp. 772-779 (1988).
32. S. Papastavridis, A Weibull limit for the reliability of a consecutive-*k*-within-*m*-out-of-*n* system, *Adv. Applied Probability*, vol. 20, pp. 690-692 (1988).
33. S. Papastavridis and Koutras, Bounds for reliability of consecutive-*k*-within-*m*-out-of-*n*:F systems, *IEEE Transactions on Reliability*, vol. 42, 1, pp. 150-160 (1993).
34. W. Preuss, A. Kossow, H. Kirchner und W. Dietl, Bestimmung der Zuverlässigkeit von konsekutiven-*k*-aus-*n*:F Netzwerken, *Wiss. Beiträge IH Wismar* 3, pp. 10-16 (1985).
35. A.M. Rushdi, A switching-algebraic analysis of circular consecutive-*k*-out-of-*n*:F systems, *Reliability Engineering and System Safety* 21, pp. 119-127 (1988).
36. A.A. Salvia and W.C. Lasher, 2-dimensional consecutive-*k*-out-of-*n*:F models, *IEEE Transactions on Reliability*, vol. 39, pp. 382-385 (1990).
37. B. Saperstein, On the occurance of *n* success within *N* Bernoulli trials, *Technometrics* 15, pp. 809-818 (1973).
38. B. Saperstein, Note on a clustering problem, *J. Appl. Prob.* 12, pp. 629-632 (1975).
39. M. Sfakianakis, S. Kounias and A. Hillaris, Reliability of a consecutive *k*-out-of-*r*-from-*n*:F system, *IEEE Transactions on Reliability*, vol. 41, 3, pp. 442-446 (1992).
40. J.G. Shanthikumar, A recursive algorithm to evaluate the reliability of a consecutive-*k*-out-of-*n*:F system, *IEEE Transactions on Reliability*, vol. R-31, pp. 57-63 (1982).
41. J.G. Shanthikumar, Lifetime distribution of consecutive-*k*-out-of-*n*:F systems with exchangeable lifetimes, *IEEE Transactions on Reliability*, vol. R-34, 5, pp. 480-483 (1985).
42. J.G. Shanthikumar, Reliability of systems with consecutive minimal cutset, *IEEE Transactions on Reliability*, vol. R-36, 5, pp. 546-550 (1987).
43. Y.L. Tong, A rearrangement inequality for the longest run with an application to networks reliability, *J. Appl. Prob.* 22, pp. 386-393 (1985).
44. J.S. Wu and R.J. Chen, Efficient algorithm for reliability of a circular consecutive-*k*-out-of-*n*:F system, *IEEE Transactions on Reliability*, vol. 42, 1, pp. 163-164 (1993).

THE UNIFORM CLOSURE OF CONVEX SEMI-LATTICES

João B. Prolla

IMECC - UNICAMP
Caixa Postal 6065
Campinas, Brazil

ABSTRACT

Let X be a compact Hausdorff space, and let $C(X;\mathbb{R})$ be the Banach lattice of all continuous real-valued functions on X with the sup-norm and pointwise ordering. We describe the uniform closure of convex (resp. convex conic) semi-lattices, i.e. inf-lattices and sup-lattices. Our result is an improvement on the classical Choquet-Deny Theorem.

INTRODUCTION

Let X be a compact Hausdorff space and let $C(X;\mathbb{R})$ be the Banach lattice of all continuous real-valued functions defined on X, equipped with the sup-norm

$$\|f\| = \sup\{|f(x)|; x \in X\}$$

and the lattice structure defined by the usual ordering $f \leq g$ if and only if $f(x) \leq g(x)$ for all $x \in X$, and $f \wedge g$ and $f \vee g$ defined by

$$\begin{aligned}(f \wedge g)(x) &= \inf\{f(x), g(x)\} \\ (f \vee g)(x) &= \sup\{f(x), g(x)\}\end{aligned}$$

for every $x \in X$, and $f, g \in C(X;\mathbb{R})$. A subset S of $C(X;\mathbb{R})$ is called an **inf-lattice**, or is said to be **infimum-stable**, if $f \wedge g \in S$ for every pair, f and g, of elements of S. Similarly, S is called a **sup-lattice**, or is said to be **supremum-stable**, if $f \vee g \in S$ for every pair, f and g, of elements of S.

The classical Choquet-Deny Theorem describes the uniform closure of an inf-lattice (resp. a sup-lattice) which is a convex set or a convex cone. In [4] we presented an improved version of this result for the case of convex cones, and applied it to give a simpler proof of a result of Mc Afee and Reny [2]. Our aim is to show that an improved version of the Choquet-Deny Theorem for convex sets is likewise true and derive from it the case of convex cones.

We first present two lemmas, essentially due to Nachbin [3] and already used in our paper [4]. For the sake of completeness we state and prove them.

Lemma 1. *Let $S \subset C(X;\mathbb{R})$ be an infimum-stable subset and, for each point $x \in X$, let $P_x = \{f \in C(X;\mathbb{R});\ f \geq 0,\ f(x) = 0\}$. Then*

$$\overline{S} = \cap\{\overline{S - P_x};\ x \in X\}.$$

Proof. For each $x \in X$, we have $0 \in P_x$. Hence $\overline{S} \subset \overline{S - P_x}$, for each $x \in X$. Conversely, assume that $f \in \overline{S - P_x}$, for each $x \in X$. Let $\varepsilon > 0$ be given. For each $x \in X$, there are $g_x \in S$ and $h_x \in P_x$ such that $\|g_x - h_x - f\| < \varepsilon/2$. Let $V_x = \{t \in X;\ h_x(t) < \varepsilon/2\}$. Then V_x is open and contains x. By compactness, there are $x_1, \ldots, x_n \in X$ such that $X = V_{x_1} \cup \ldots \cup V_{x_n}$. Let $g = \inf\{g_{x_1}, \ldots, g_{x_n}\}$. Then $g \in S$. Let $t \in X$. Then, for each $j = 1, \ldots, n$, we have

$$g_{x_j}(t) \geq g_{x_j}(t) - h_{x_j}(t) > f(t) - \varepsilon/2.$$

Hence $g(t) > f(t) - \varepsilon$. On the order hand, there is some index i such that $t \in V_{x_i}$, and then $h_{x_i}(t) < \varepsilon/2$ and $g(t) \leq g_{x_i}(t)$ imply $g(t) - \varepsilon/2 < g_{x_i}(t) - h_{x_i}(t) < f(t) + \varepsilon/2$. Hence $g(t) < f(t) + \varepsilon$. Therefore $\|f - g\| < \varepsilon$ and $f \in \overline{S}$. □

Lemma 2. *Let φ be a non-zero continuous linear form on $C(X;\mathbb{R})$ and let $x \in X$. If φ is positive on $P_x = \{f \in C(X;\mathbb{R});\ f \geq 0, f(x) = 0\}$, there is $r \in \mathbb{R}$ such that $r < 0$ and $\varphi \geq r\delta_x$.*

Proof. (Nachbin [3], §21). Assume that $\varphi(f) \geq 0$ for all $f \geq 0$ such that $f(x) = 0$. Let

$$B = \{f \in C(X;\mathbb{R}); f \geq 0, f(x) = 1\}.$$

For any $f \in B$, notice that $g = f - \inf(\mathbf{1},\ f)$ belongs to P_x. Hence $\varphi(g) \geq 0$ and so $\varphi(f) \geq \varphi(\inf(\mathbf{1},\ f))$. Now $0 \leq \inf(\mathbf{1},\ f) \leq \mathbf{1}$ and therefore $|\varphi(\inf(\mathbf{1},\ f))| \leq \|\varphi\|$. Hence $\varphi(f) \geq -\|\varphi\|$, for all $f \in B$. Let $r = -\|\varphi\|$.

Let $f \geq 0$ be given in $C(X;\mathbb{R})$. If $f(x) > 0$, Then $\varphi(f/f(x)) \geq r$ and so $\varphi(f) \geq rf(x)$. If $f(x) = 0$ then $f \in P_x$ and so $\varphi(f) \geq 0 = rf(x)$. Hence $\varphi \geq r\delta_x$. □

Theorem 1. *Let S be a non-empty subset of $C(X;\mathbb{R})$. Denote by $\Lambda(S)$ the set of all triples (φ, x, α) such that*

$$\varphi(g) \leq g(x) + \alpha$$

for all $g \in S$, where φ is a positive linear form on $C(X;\mathbb{R})$, $x \in X$ and $\alpha \in \mathbb{R}$. If S is a convex inf-lattice and $f \in C(X;\mathbb{R})$, then f belongs to the uniform closure of S if, and only if,

$$\varphi(f) \le f(x) + \alpha$$

for all triples $(\varphi, x, \alpha) \in \Lambda(S)$.

Proof. Clearly, $\varphi(f) \le f(x) + \alpha$, for all triples $(\varphi, x, \alpha) \in \Lambda(S)$, whenever $f \in \overline{S}$.

Conversely, let $f \in C(X;\mathbb{R})$ be such that $f \notin \overline{S}$. By Lemma 1, there exists some $x \in X$ such that $f \notin \overline{S - P_x}$. Since $S - P_x$ is convex, by the Hahn-Banach Theorem there is a non-zero continuous linear form ψ on $C(X;\mathbb{R})$ and $c \in \mathbb{R}$ such that $\psi(g - h) \le c < \psi(f)$ for all $g \in S$ and $h \in P_x$. Fix some $g \in S$. Then for each $h \in P_x$ and $\lambda > 0$ we have $\psi(g - \lambda h) \le c$. Dividing by λ and letting $\lambda \to \infty$ we get $\psi(h) \ge 0$, for all $h \in P_x$. By Lemma 2, there is $r \in \mathbb{R}$ such that $r < 0$ and $\psi \ge r\delta_x$. Then $\varphi = \delta_x - r^{-1}\psi$ is positive. Since $0 \in P_x$, we have $\psi(g) \le c < \psi(f)$, for every $g \in S$. Hence

$$(*) \qquad \varphi(g) - g(x) = -r^{-1}\psi(g) \le -r^{-1}c < -r^{-1}\psi(f) = \varphi(f) - f(x).$$

Let $\alpha = -r^{-1}c$. Then (*) shows that $(\varphi, x, \alpha) \in \Lambda(S)$, and $\alpha + f(x) < \varphi(f)$, and Theorem 1 is proved. □

Corollary 1. *Let $S \subset C(X;\mathbb{R})$ be an infimum-stable convex non-empty set. Then S is uniformly dense if, and only if, for every $f \in C(X;\mathbb{R})$, every positive linear form φ on $C(X;\mathbb{R})$, every $x \in X$ and every $\alpha \in \mathbb{R}$, one has $\varphi(f) \le f(x) + \alpha$, whenever $\varphi(g) \le g(x) + \alpha$ for all $g \in S$.*

Remark 1. For any non-empty subset $S \subset C(X;\mathbb{R})$ and any $f \in C(X;\mathbb{R})$, the conditions

(1) $\varphi(f) - f(x) \le \sup\{\varphi(g) - g(x); g \in S\}$
(2) $(\varphi, x, \alpha) \in \Lambda(S)$ implies $\varphi(f) \le f(x) + \alpha$

are easily seen to be equivalent. Hence Theorem 1 can be rewritten in the following form:

THE MAIN RESULTS

Theorem 1a. *Let $S \subset C(X;\mathbb{R})$ be a convex inf-lattice, and let $f \in C(X;\mathbb{R})$. Then f belongs to the uniform closure of S if, and only if, for every positive linear form φ*

on $C(X;\mathbb{R})$ and every $x \in X$ we have

$$\varphi(f) - f(x) \leq \sup\{\varphi(g) - g(x);\ g \in S\}.$$

Let us now turn to the case of convex conic inf-lattices.

Theorem 2. *Let S be a non-empty subset of $C(X;\mathbb{R})$. Denote by $\Lambda_0(S)$ the set of all pairs (φ, x) such that*

$$\varphi(g) \leq g(x)$$

for all $g \in S$, where φ is a positive linear form on $C(X;\mathbb{R})$ and $x \in X$. If S is a convex conic inf-lattice and $f \in C(X;\mathbb{R})$, then f belongs to the uniform closure of S if, and only if,

$$\varphi(f) \leq f(x)$$

for all pairs $(\varphi, x) \in \Lambda_0(S)$.

Proof. Clearly, $\varphi(f) \leq f(x)$ for all pairs $(\varphi, x) \in \Lambda_0(S)$, whenever $f \in \overline{S}$.

Conversely, assume that $\varphi(f) \leq f(x)$ for all pairs $(\varphi, x) \in \Lambda_0(S)$. We claim that $f \in \overline{S}$. Let $(\varphi, x, \alpha) \in \Lambda(S)$ be given. Let $g \in S$ be given. For each $\lambda > 0$, we have $\lambda g \in S$ and therefore $\varphi(\lambda g) \leq \lambda g(x) + \alpha$. Dividing by λ and letting $\lambda \to \infty$ we get $\varphi(g) \leq g(x)$. Hence $(\varphi, x) \in \Lambda_0(S)$. Therefore $\varphi(f) \leq f(x)$. On the other hand, $0 \in S$ implies $\alpha \geq 0$. Hence $\varphi(f) \leq f(x) + \alpha$. Since S is convex, by Theorem 1 we get $f \in \overline{S}$.

□

Corollary 2. *let $S \subset C(X;\mathbb{R})$ be an infimum-stable convex cone. Then S is uniformly dense if, and only if for every $f \in C(X;\mathbb{R})$, every positive linear form φ on $C(X;\mathbb{R})$ and every $x \in X$, one has $\varphi(f) \leq f(x)$ whenever $\varphi(g) \leq g(x)$ for all $g \in S$.*

Remark 2. A positive linear form φ on $C(X;\mathbb{R})$ is said to be a **probability measure** if $\varphi(\mathbf{1}) = 1$, where $\mathbf{1}$ is the constant function whose value is 1.

Corollary 3. *Let $S \subset C(X;\mathbb{R})$ be a convex conic inf-lattice containing the constants and let $f \in C(X;\mathbb{R})$. Then f belongs to the uniform closure of S if and only if $\mu(f) \leq f(x)$, for every probability measure μ and every point $x \in X$, such that $\mu(g) \leq g(x)$ for all $g \in S$.*

Proof. The condition is clearly necessary. Conversely, assume that f satisfies the above condition and let $(\varphi, x) \in \Lambda_0(S)$ be given. Hence $\varphi(g) \leq g(x)$ for all $g \in S$. Since $\mathbf{1}$ and $-\mathbf{1}$ belong to S, we see that $\varphi(\mathbf{1}) = 1$, i.e., φ is a probability measure. Hence $\varphi(f) \leq f(x)$, and by Theorem 2, $f \in \overline{S}$. □

Let us now turn to the case of convex sup-lattices and convex conic sup-lattices.

Theorem 3. *Let S be a non-empty subset of $C(X;\mathbb{R})$. Denote by $V(S)$ the set of all triples (φ, x, α) such that*

$$\varphi(g) \geq g(x) + \alpha$$

for all $g \in S$, where φ is a positive linear form on $C(X;\mathbb{R}), x \in X$ and $\alpha \in \mathbb{R}$. If S is a convex sup-lattice and $f \in C(X;\mathbb{R})$, then f belongs to the uniform closure of S if, and only if,

$$\varphi(f) \geq f(x) + \alpha$$

for all triples $(\varphi, x, \alpha) \in V(S)$.

Proof. Notice that $f \in \overline{S}$ if and only if $-f \in \overline{-S}$, and $(\varphi, x, \alpha) \in V(S)$ if and only if $(\varphi, x, -\alpha) \in \Lambda(-S)$. It remains to apply Theorem 1 to $-f$ and $-S$, since $-S$ is a convex inf-lattice. □

Corollary 4. *Let $S \subset C(X;\mathbb{R})$ be a supremum-stable convex non-empty set. Then S is uniformly dense if, and only if, for every $f \in C(X;\mathbb{R})$, every positive linear form φ on $C(X;\mathbb{R})$, every $x \in X$ and every $\alpha \in \mathbb{R}$, one has $\varphi(f) \geq f(x) + \alpha$, whenever $\varphi(g) \geq g(x) + \alpha$ for all $g \in S$.*

Remark 3. For any non-empty subset $S \subset C(X;\mathbb{R})$ and any $f \in C(X;\mathbb{R})$, the conditions

(1) $\varphi(f) - f(x) \geq \inf\{\varphi(g) - g(x); g \in S\}$

(2) $(\varphi, x, \alpha) \in V(S)$ implies $\varphi(f) \geq f(x) + \alpha$

are easily seen to be equivalent. Hence Theorem 3 can be rewritten in the following form (cf. Nachbin [3], §21)

Theorem 3a. *Let $S \subset C(X;\mathbb{R})$ be a convex sup-lattice and let $f \in C(X;\mathbb{R})$. Then f belongs to the uniform closure of S if, and only if, for every positive linear form φ on $C(X;\mathbb{R})$ and every $x \in X$ we have*

$$\varphi(f) - f(x) \geq \inf\{\varphi(g) - g(x); g \in S\}.$$

Theorem 4. *Let S be a non-empty subset of $C(X;\mathbb{R})$. Denote by $V_0(S)$ the set of all pairs (φ, x) such that*

$$\varphi(g) \geq g(x)$$

for all $g \in S$, where φ is a positive linear form on $C(X;\mathbb{R})$ and $x \in X$. If S is a convex conic sup-lattice and $f \in C(X;\mathbb{R})$, then f belongs to the uniform closure of S if, and only if,

$$\varphi(f) \geq f(x)$$

for all pairs $(\varphi, x) \in V_0(S)$.

Proof. Theorem 4 follows from Theorem 3 in the same way that Theorem 2 follows from Theorem 1. □

Corollary 5. *Let $S \subset C(X;\mathbb{R})$ be a supremum-stable convex cone. Then S is uniformly dense if, and if, for every $f \in C(X;\mathbb{R})$, every positive linear form φ on $C(X;\mathbb{R})$ and every $x \in X$, one has $\varphi(f) \geq f(x)$ whenever $\varphi(g) \geq g(x)$ for all $g \in S$.*

Corollary 6. *Let $S \subset C(X;\mathbb{R})$ be a convex conic sup-lattice containing the constants and let $f \in C(X;\mathbb{R})$. Then f belongs to the uniform closure of S if and only if $\mu(f) \geq f(x)$ for every probability measure μ on $C(X;\mathbb{R})$ and every point $x \in X$, such that $\mu(g) \geq g(x)$ for all $g \in S$.*

Proof. Apply Corollary 3 to $-f$ and $-S$. □

Remark 4. If φ is a continuous linear form on $C(X;\mathbb{R})$ and β is any real number the functional $\gamma : C(X;\mathbb{R}) \to \mathbb{R}$ defined by $\gamma(f) = \varphi(f) + \beta$, for every $f \in C(X;\mathbb{R})$ is called a continuous **affine** functional. When φ is positive, then γ is **monotone**: if $f \leq g$ then $\gamma(f) \leq \gamma(g)$. With this notion we can restate Theorems 1 and 3. For example, Theorem 1 would read now as follows:

Theorem 1'. *Let S be a non-empty subset of $C(X;\mathbb{R})$. Denote by $\Lambda(S)$ the set of all pairs (γ, x) such that*

$$\gamma(g) \leq g(x)$$

for all $g \in S$, where γ is a monotone affine functional on $C(X;\mathbb{R})$ and $x \in X$. If S is a convex inf-lattice and $f \in C(X;\mathbb{R})$, then f belongs to the uniform closure of S if, and only if,

$$\gamma(f) \leq f(x)$$

for all pairs $(\gamma, x) \in \Lambda(S)$.

Definition 1. Let S be a non-empty subset of $C(X;\mathbb{R})$. We say that S **separates isotonically points from monotone affine functionals**, if for every $x \in X$ and every monotone affine functional γ on $C(X;\mathbb{R})$, the following holds for every $f \in C(X;\mathbb{R})$:
(1) if $\gamma(f) > f(x)$, then $\gamma(g) > g(x)$ for some $g \in S$
(2) if $\gamma(f) < f(x)$, then $\gamma(h) < h(x)$ for some $h \in S$.

We say that S **separates isotonically points from positive linear forms**, if for every $x \in X$ and every positive linear form φ on $C(X;\mathbb{R})$, the following holds for every $f \in C(X;\mathbb{R})$:
(1) if $\varphi(f) > f(x)$, then $\varphi(g) > g(x)$ for some $g \in S$
(2) if $\varphi(f) < f(x)$, then $\varphi(h) < h(x)$ for some $h \in S$.

Similarly, we say that S **separates isotonically points from probability measures** when the above holds for positive linear forms that are probability measures, i.e., $\varphi(\mathbf{1}) = 1$.

Let us denote by S_m and S_M, respectively, the inf-lattice and the sup-lattice generated by a non-empty subset $S \subset C(X; \mathbb{R})$, i.e.,

$$S_m = \{\inf(f_1, \ldots, f_n);\ f_i \in S,\ 1 \le i \le n,\ n \in \mathbb{N}\}$$
$$S_M = \{\sup(f_1, \ldots, f_n);\ f_i \in S,\ 1 \le i \le n,\ n \in \mathbb{N}\}$$

With these notations we have the following result.

Theorem 5. *Let $S \subset C(X; \mathbb{R})$ be a non-empty subset. If S is convex, the following are equivalent:*
(1) $\overline{S_m} = \overline{S_M} = C(X; \mathbb{R})$
(2) *S separates isotonically points from monotone affine functionals.*

If S is a convex cone, the following are equivalent:
(1) $\overline{S_m} = \overline{S_M} = C(X; \mathbb{R})$
(3) *S separates isotonically points from positive linear forms.*

If S is a convex cone containing the constants, the following are equivalent:
(1) $\overline{S_m} = \overline{S_M} = C(X; \mathbb{R})$
(4) *S separates isotonically points from probability measures.*

Proof. (1) $\Rightarrow$ (2). Since $\overline{S_m} = C(X; \mathbb{R})$ it follows from Corollary 1 that for every monotone affine functional ψ on $C(X; \mathbb{R})$ and for every $x \in X$, if $\psi(f) > f(x)$ then $\psi(g) > g(x)$ for some $g \in S_m$. Notice that $g = \inf(g_1, \ldots g_n)$, where $g_i \in S, i = 1, \ldots, n$. Now $g(x) = g_i(x)$ for some $i \in \{1, \ldots, n\}$. Hence $\psi(g_i) \ge \psi(g) > g(x) = g_i(x)$. Similarly, $\overline{S_M} = C(X; \mathbb{R})$ and Corollary 4 imply that, if $\psi(f) < f(x)$ then $\psi(h) < h(x)$, for some $h \in S_M$. Now $h = \sup(h_1, \ldots, h_m)$ where $h_j \in S, j = 1, \ldots, m$. Notice that $h(x) = h_j(x)$, for some index $j \in \{1, \ldots, m\}$. Hence $\psi(h_j) \le \psi(h) < h(x) = h_j(x)$. Since $g_i \in S$ and $h_j \in S$, we see that S separates isotonically points from monotone affine functionals.

(2) $\Rightarrow$ (1). Since $S \subset S_m$ and $S \subset S_M$, we get by Corollaries 1 and 4 that $\overline{S_m} = C(X; \mathbb{R})$ and $\overline{S_M} = C(X; \mathbb{R})$.

The proof of (1) $\Leftrightarrow$ (3), when S is a convex cone is similar and based on Corollaries 2 and 5. When S contains the constants, (1) $\Leftrightarrow$ (4) follows from Corollaries 3 and 6.□

Definition 2. Let us say that a subset A **separates points from positive linear forms** if for every $x \in X$ and every positive linear form φ on $C(X; \mathbb{R})$, and every $f \in C(X; \mathbb{R})$ such that $\varphi(f) \ne f(x)$, there exists $g \in A$ such that $\varphi(g) \ne g(x)$.

Theorem 6. *Let $A \subset C(X; \mathbb{R})$ be a linear subspace. The following are equivalent*

(1) $\overline{A_m} = \overline{A_M} = C(X; \mathbb{R})$

(2) *A separates points from positive linear forms.*

Proof. By Theorem 5, it suffices to show that (2) is equivalent to say that A separates isotonically points from positive linear forms. Assume (2). Let $\varphi(f) > f(x)$. If $\varphi(g) \leq g(x)$ for all $g \in A$, then $A = -A$ implies $\varphi(g) = g(x)$ for all $g \in A$. By (2), $\varphi(f) = f(x)$, a contradiction. Hence $\varphi(g) > g(x)$ for some $g \in A$. Similarly, if $\varphi(f) < f(x)$, then $\varphi(g) < g(x)$ for some $g \in A$. Therefore A separates isotonically points from positive linear forms. Conversely, if A separates isotonically points from positive linear forms, then a fortiori A separates points from positive linear forms. □

Definition 3. A subset $A \subset C(X; \mathbb{R})$ **separates points from probability measures** if for every $x \in X$, every probability measure μ on $C(X; \mathbb{R})$, and every $f \in C(X; \mathbb{R})$ such that $\mu(f) \neq f(x)$, there exits $g \in A$ such that $\mu(g) \neq g(x)$.

The following result was proved in [4]. It generalizes to the non-metric case a result of McAfee and Reny [2].

Theorem 7. *Let A be a linear subspace of $C(X; \mathbb{R})$ containing the constants. The following are equivalent:*

(1) $\overline{A_m} = \overline{A_M} = C(X; \mathbb{R})$

(2) *A separates points from probability measures.*

Proof. By Theorem 6, it suffices to show that (2) is equivalent to say that A separates points from positive linear forms. Clearly, if A separates points from positive linear forms, then A separates points from probability measures.

Conversely assume (2). Let φ be a positive linear form. If $\varphi(g) = g(x)$ for all $g \in A$, then $\mathbf{1} \in A$ implies $\varphi(\mathbf{1}) = 1$ and so φ is a probability measure. By (2), $\varphi(f) = f(x)$ for every $f \in C(X; \mathbb{R})$. Hence A separates points from positive linear forms. □

Remark 4. If X is a completely regular Hausdorff space and $C(X; \mathbb{R})$ is endowed with the compact-open topology given by the directed family of continuous seminorms $f \mapsto \sup\{|f(x)|; x \in K\}$ where K is a compact subset of X, then all results presented here remain true, with the obvious modifications. For example, where one has a positive linear form on $C(X; \mathbb{R})$ in the compact case, now one must take a positive **continuous** linear form.

REFERENCES

1. G. Choquet et J. Deny, Ensembles semi-réticulés et ensembles réticulés de fonctions continues, *J. Math. Pures Appl.* **36** (1957), 179-189.

2. R. P. McAfee and P. Reny, A Stone-Weierstrass theorem without closure under suprema, *Proc. Amer. Math. Soc.* **114** (1992), 61-67.

3. L. Nachbin, "Elements of Approximation Theory", van Nostrand, New York, 1967. Reprinted by R. E. Krieger Publ. Co., Huntington NY, 1976.

4. J. B. Prolla, Density of infimum-stable convex cones, to appear in *Proc. Amer. Math. Soc.*

ANOTHER PAIRWISE INDEPENDENT, STATIONARY CHAIN

James B. Robertson [1,2]

Department of Statistics and Applied Probability
University of California
Santa Barbara, CA 93106–3110

ABSTRACT

We construct a pairwise independent, stationary chain $\{X_n : n = 1, 2, \ldots\}$ such that $P\{X_n = 1\} = P\{X_n = -1\} = \frac{1}{2}$ and $E[X_1X_2X_3] = -E[X_1X_3X_5] = \frac{1}{2}$. The properties of this chain are also studied. In particular it is shown that the chain is ergodic but not weakly mixing.

INTRODUCTION

In recent years several authors have been interested in stationary pairwise independent chains $\{X_n : n = 1, 2, \ldots\}$ such that $P\{X_n = 1\} = P\{X_n = -1\} = \frac{1}{2}$. For such chains pairwise independence is equivalent to the condition that $\{X_n\} \cup \{1\}$ is an orthonormal sequence in L_2 of the underlying probability space. Bradley (1989) and Jason (1988) investigated the central limit theorem for such sequences. Robertson and Womack (1985), Robertson and Simons (1988) and Robertson (1985) (1988) were interested in the possible dependence among triples. Mathew and Nadkarni (1984) have examples of such chains that have zero entropy (Their chains have the property that $\{X_n : n = 1, 2, \ldots\}$ and $\{-X_n : n = 1, 2, \ldots\}$ have the same distribution and so all third moments are zero). We are interested in the nonlinear structure of such chains and in particular in the open question of whether or not there is a chain such that the

[1] *AMS 1991 subject classifications*: Primary, 60G10; secondary, 58F11
[2] *Key words and phrases.* ergodic, mixing, third moments, embedding

constant function 1 and the set $\{X_n : n = 0, \pm 1, \pm 2, \ldots\}$ is a complete orthonormal set for L_2. The answer to this question depends on having sufficiently large third moments (cf. Robertson (1988)).

Robertson and Womack (1985) constructed a stationary, pairwise independent chain (Chain I) such that $E[X_1X_2X_3] = \frac{1}{2}$ and also showed that this is the maximum value for this moment. Robertson (1988) produced another such chain (Chain II) which in addition to having $E[X_1X_2X_3] = \frac{1}{2}$ also had $E[X_1X_3X_5] = -\frac{1}{4}$. However it was not clear in that paper what was the maximum value of $|E[X_1X_3X_5]|$ for such a chain. Since $\{X_n : n = 1, 3, 5, \ldots\}$ is also such a chain, it follows that $|E[X_1X_3X_5]| \leq \frac{1}{2}$. The purpose of the present paper is to construct an example of such a chain (Chain III) with maximal absolute value: $E[X_1X_3X_5] = -\frac{1}{2}$.

All three of these chains are finitary chains. In the next section we shall give the finitary system for the new chain and develop its properties. In the final section we will discuss a related chain that acts as a sufficient statistic for the pairwise independent chain and has a somewhat simpler structure. These sections are logically independent.

THE FINITARY SYSTEM

In this section we develop the properties of the following finitary system which we will refer to as $\mathcal{S}$. This finitary system can be viewed as the transitions on a stationary Markov chain with 14 states. In the diagram below a double arrow is a positive transition and a single arrow is a negative transition.

$$\begin{array}{ccccccccc}
2 & \longrightarrow & 3 & & & & & & \\
\Uparrow & & \downarrow & & & & & & \\
1 & \longleftarrow & 4 & \Longrightarrow & 5 & \Longrightarrow & 6 & & \\
 & & \Uparrow & & & & \Downarrow & & \\
 & & 11 & & & & 7 & & \\
 & & \uparrow & & & & \Downarrow & & \\
 & & 10 & \longleftarrow & 9 & \Longleftarrow & 8 & \longrightarrow & 12 \\
 & & & & & & \uparrow & & \downarrow \\
 & & & & & & 14 & \Longleftarrow & 13
\end{array} \qquad (1)$$

States 4 and 8 have two arrows coming out from them. Each of these transitions has probability $\frac{1}{2}$. All other transitions are deterministic.

Formally the system consist of $\mathcal{S} = \{\pi, \xi, A_-, A_+\}$ given as follows.

$$\begin{aligned}
\pi &= \frac{1}{16}(1,1,1,2,1,1,1,2,1,1,1,1,1,1) \\
\xi &= (1,1,1,1,1,1,1,1,1,1,1,1,1,1)^t
\end{aligned}$$

$$A_- = \begin{pmatrix} 0&0&0&0&0&0&0&0&0&0&0&0&0&0 \\ 0&0&1&0&0&0&0&0&0&0&0&0&0&0 \\ 0&0&0&1&0&0&0&0&0&0&0&0&0&0 \\ \frac{1}{2}&0&0&0&0&0&0&0&0&0&0&0&0&0 \\ 0&0&0&0&0&0&0&0&0&0&0&0&0&0 \\ 0&0&0&0&0&0&0&0&0&0&0&0&0&0 \\ 0&0&0&0&0&0&0&0&0&0&0&0&0&0 \\ 0&0&0&0&0&0&0&0&0&0&0&\frac{1}{2}&0&0 \\ 0&0&0&0&0&0&0&0&0&1&0&0&0&0 \\ 0&0&0&0&0&0&0&0&0&0&1&0&0&0 \\ 0&0&0&0&0&0&0&0&0&0&0&0&0&0 \\ 0&0&0&0&0&0&0&0&0&0&0&0&1&0 \\ 0&0&0&0&0&0&0&0&0&0&0&0&0&0 \\ 0&0&0&0&0&0&0&1&0&0&0&0&0&0 \end{pmatrix}$$

$$A_+ = \begin{pmatrix} 0&1&0&0&0&0&0&0&0&0&0&0&0&0 \\ 0&0&0&0&0&0&0&0&0&0&0&0&0&0 \\ 0&0&0&0&0&0&0&0&0&0&0&0&0&0 \\ 0&0&0&0&\frac{1}{2}&0&0&0&0&0&0&0&0&0 \\ 0&0&0&0&0&1&0&0&0&0&0&0&0&0 \\ 0&0&0&0&0&0&1&0&0&0&0&0&0&0 \\ 0&0&0&0&0&0&0&1&0&0&0&0&0&0 \\ 0&0&0&0&0&0&0&0&\frac{1}{2}&0&0&0&0&0 \\ 0&0&0&0&0&0&0&0&0&0&0&0&0&0 \\ 0&0&0&0&0&0&0&0&0&0&0&0&0&0 \\ 0&0&0&1&0&0&0&0&0&0&0&0&0&0 \\ 0&0&0&0&0&0&0&0&0&0&0&0&0&0 \\ 0&0&0&0&0&0&0&0&0&0&0&0&0&1 \\ 0&0&0&0&0&0&0&0&0&0&0&0&0&0 \end{pmatrix}$$

That $\mathcal{S}$ is a finitary system is shown by verifying the following properties.

$$\begin{aligned} \pi \cdot \xi &= 1 \\ (A_- + A_+) \cdot \xi &= \xi \\ \pi \cdot (A_- + A_+) &= \pi \\ \pi \cdot A_{x_1} \cdots A_{x_n} \cdot \xi &\geq 0 \end{aligned}$$

where $x_1, \ldots, x_n$ is any sequence of +'s and −'s. All of these are easily verified. This implies that the formula

$$P\{X_1 = x_1, \ldots, X_n = x_n\} = \pi \cdot A_{x_1} \cdots A_{x_n} \cdot \xi \qquad (2)$$

defines the distribution of a stationary sequence of +'s and −'s.

In order to determine the ergodic properties of the chain $\{X_n : n = 1, 2, \ldots\}$, we need to know that the system $\mathcal{S}$ is reduced (cf. Robertson (1973)). This means that both the sets of vectors $\{A_{x_1} \cdots A_{x_n}\xi : n \geq 0, x_i \in \{-,+\}\}$ and $\{\pi A_{x_1} \cdots A_{x_n} : n \geq 0, x_i \in \{-,+\}\}$ span all of $\mathbf{R}^{14}$. This is rather tedious. It consist of using finite induction on the n in $A_{x_1} \cdots A_{x_n}$. The value $n = 6$ gives a spanning set for both sets. We did these calculation by computer and used row operations to eliminate redundant vectors. Given a finitary chain, a reduced finitary system is unique up to isomorphism. In particular the dimension 14 is a property of the chain. The corresponding dimensions for Chain I and II are 4 and 5 respectively. This number is in some sense a measure of the complexity of the distribution function.

We now list the properties of the chain $\{X_n : n = 1, 2, \ldots\}$, each of which is followed by an indications of its proof.

I. *The chain* $\{X_n : n = 1, 2, \ldots\}$ *is ergodic but not weakly mixing.* The results in Robertson (1973) show that the eigenvalues (and their multiplicities) of the shift operator for the chain $\{X_n : n = 1, 2, \ldots\}$ are the same as the eigenvalues (and their multiplicities) of absolute value one of the matrix $A_- + A_+$. This matrix is just the transition matrix of the above Markov chain. That $\{X_n : n = 1, 2, \ldots\}$ is ergodic says that 1 is a simple eigenvalue which is equivalent to the Markov chain being irreducible which is evident from the diagram. That $\{X_n : n = 1, 2, \ldots\}$ is weakly mixing is equivalent to saying that 1 is the only eigenvalue of the shift operator which says that the Markov chain is aperiodic. However the above Markov chain has period 4 (the periodic subclasses are $\{1, 5, 9, 12\}, \{2, 6, 10, 13\}, \{3, 7, 11, 14\}, \{4, 8\}$).

II. *For the chain* $\{X_n : n = 1, 2, \ldots\}$ *the random variables* $\{X_1, \ldots, X_6\}$ *are pairwise independent and* $E[X_1X_2X_3] = -E[X_1X_3X_5] = \frac{1}{2}$. The distribution of the finite sequence $\{X_1, \ldots, X_5\}$ is calculated from formula (2) to be the uniform distribution on the following 16 points.

$+ + + + +$	$- + + + +$
$+ + + + -$	$- + - + -$
$+ + + - -$	$- + - - +$
$+ + - - +$	$- + - - -$
$+ - + - -$	$- - + + +$
$+ - - + +$	$- - + - +$
$+ - - + -$	$- - + - -$
$+ - - - +$	$- - - + -$

In fact, this is the only distribution of $\{X_1, \ldots, X_5\}$ such that it is stationary, pairwise independent, $P\{X_n = +\} = P\{X_n = -\} = \frac{1}{2}$, and $E[X_1X_2X_3] = -E[X_1X_3X_5] = \frac{1}{2}$. The distribution of $\{X_1, \ldots, X_6\}$ is given by:

x	$64p(x)$	x	$64p(x)$
$+ + + + + +$	1	$- + + + + +$	3
$+ + + + + -$	3	$- + + + + -$	1
$+ + + + - -$	4	$- + - + - -$	4
$+ + + - - +$	4	$- + - - + +$	3
$+ + - - + +$	1	$- + - - + -$	1
$+ + - - + -$	3	$- + - - - +$	4
$+ - + - - +$	3	$- - + + + +$	4
$+ - + - - -$	1	$- - + - + -$	4
$+ - - + + +$	4	$- - + - - +$	1
$+ - - + - +$	3	$- - + - - -$	3
$+ - - + - -$	1	$- - - + - +$	1
$+ - - - + -$	4	$- - - + - -$	3

From this the above properties can be verified by direct calculation.

III. *The chain* $\{X_n : n = 1, 2, \ldots\}$ *is* $(5, 1)$ *independent, i.e. for any sequence* $x_1, \ldots, x_n$, $P\{X_1 = x_1, \ldots, X_n = x_n, X_{n+6} = +\} = \frac{1}{2}P\{X_1 = x_1, \ldots, X_n = x_n\}$ ($\forall n \{X_1, \ldots, X_n\} \perp\!\!\!\perp X_{n+6}$). This probability can be calculated from the finitary system by the formula

$$\pi \cdot A_{x_1} \cdots A_{x_n} \cdot (A_- + A_+)^5 \cdot A_+ \xi.$$

The desired result follows from the fact that

$$(A_- + A_+)^5 \cdot A_+\xi = \frac{1}{2}\xi$$

which can be verified directly.

The last two results together clearly imply that

IV. *The chain* $\{X_n : n = 1,2,\ldots\}$ *is pairwise independent.*

V. *The chain* $\{X_n : n = 1,2,\ldots\}$ *is reversible.* If $\{Z_n : n = 1,2,\ldots\}$ is the stationary Markov chain with transition matrix $A_- + A_+$, then the distribution of $\{X_n : n = 1,2,\ldots\}$ is given by $X_n = g(Z_n, Z_{n+1})$ where g is the function given in diagram 1 (for example $g(4,1) = -$ and $g(4,5) = +$). Thus the reverse chain has the distribution of $g(Z_n, Z_{n-1})$. If the arrows in Diagram 1 are reversed and if the states are relabeled (14, 13, 12, 8, 7, 6, 5, 4, 11, 10, 9, 3, 2, 1) respectively, then one obtains exactly the same transition matrix. This shows that $g(Z_n, Z_{n+1})$ and $g(Z_n, Z_{n-1})$ have the same distribution. Hence $\{X_n : n = 1,2,\ldots\}$ is reversible.

VI. *The third moments.* Since, by **III**, the factors of nonzero third moments cannot be separated by more than four time units, $X_nX_{n+5}X_{n+10}$ is the longest possible span. The distribution of $\{X_n,\ldots,X_{n+10}\}$ is calculated from equation (2) and is given below.

The nonzero third moments are:

$$E[X_nX_{n+1}X_{n+2}] = \frac{1}{2} \qquad E[X_nX_{n+2}X_{n+4}] = -\frac{1}{2}$$
$$E[X_nX_{n+1}X_{n+5}] = -\frac{1}{4} \qquad E[X_nX_{n+4}X_{n+5}] = -\frac{1}{4}$$
$$E[X_nX_{n+2}X_{n+7}] = \frac{1}{4} \qquad E[X_nX_{n+5}X_{n+7}] = \frac{1}{4}$$

Let $: X_nX_m :$ denote the orthogonal projection of X_nX_m onto the closed linear space spanned by $\{X_n : n = 0,\pm 1,\ldots\}$. Then the nonzero projections and their norms are easily calculated from the above third moments.

$$\begin{aligned}
: X_nX_{n+1} : &= -\frac{1}{4}X_{n-4} + \frac{1}{2}X_{n-1} + \frac{1}{2}X_{n+2} - \frac{1}{4}X_{n+5} \\
\| : X_nX_{n+1} : \|^2 &= \frac{5}{8} \\
: X_nX_{n+2} : &= \frac{1}{4}X_{n-5} - \frac{1}{2}X_{n-2} + \frac{1}{2}X_{n+1} - \frac{1}{2}X_{n+4} + \frac{1}{4}X_{n+7} \\
\| : X_nX_{n+2} : \|^2 &= \frac{7}{8} \\
: X_nX_{n+4} : &= -\frac{1}{4}X_{n-1} - \frac{1}{2}X_{n+2} - \frac{1}{4}X_{n+5} \\
\| : X_nX_{n+4} : \|^2 &= \frac{3}{8} \\
: X_nX_{n+5} : &= \frac{1}{4}X_{n-2} - \frac{1}{4}X_{n+1} - \frac{1}{4}X_{n+4} + \frac{1}{4}X_{n+7} \\
\| : X_nX_{n+5} : \|^2 &= \frac{1}{4} \\
: X_nX_{n+7} : &= \frac{1}{4}X_{n+2} + \frac{1}{4}X_{n+5} \\
\| : X_nX_{n+7} : \|^2 &= \frac{1}{8}
\end{aligned}$$

x	$128p(x)$	x	$128p(x)$
++++++−−+++	1	−++++++−−++	1
++++++−−+−+	1	−++++++−−+−	1
+++++−−++++	2	−+++++−−+++	1
+++++−−+−+−	3	−+++++−−+−+	2
+++++−−+−−−	1	−+++++−−+−−	1
++++−−+++++	2	−++++−−+−+−	1
++++−−+−+−−	4	−++++−−+−−−	1
++++−−+−−−+	2	−+−+−−+++++	3
+++−−++++++	1	−+−+−−++++−	1
+++−−+++++−	1	−+−+−−+−+−−	2
+++−−+−+−−+	3	−+−+−−−+−−+	1
+++−−+−+−−−	1	−+−+−−−+−−−	1
+++−−+−−−+−	2	−+−−++++++−	1
++−−++++++−	1	−+−−+++++−−	3
++−−+++++−−	1	−+−−++++−−+	2
++−−+−+−−++	2	−+−−+−+−−++	1
++−−+−+−−+−	1	−+−−+−+−−−+	1
++−−+−+−−−+	1	−+−−−+−+−−+	2
++−−+−−−+−+	1	−+−−−+−−+++	2
++−−+−−−+−−	1	−+−−−+−−−+−	4
+−+−−++++++	1	−−++++++−−+	2
+−+−−+++++−	2	−−+++++−−++	1
+−+−−++++−−	1	−−+++++−−+−	3
+−+−−+−+−−+	1	−−++++−−+−+	1
+−+−−+−+−−−	1	−−++++−−+−−	1
+−+−−−+−−++	1	−−+−+−−++++	4
+−+−−−+−−−+	1	−−+−+−−+−+−	2
+−−++++++−−	2	−−+−+−−−+−−	2
+−−+++++−−+	4	−−+−−+++++−	1
+−−++++−−+−	2	−−+−−++++−−	1
+−−+−+−−+++	3	−−+−−−+−+−−	2
+−−+−+−−+−+	1	−−+−−−+−−++	1
+−−+−+−−−+−	2	−−+−−−+−−−+	3
+−−+−−−+−+−	1	−−−+−+−−+++	1
+−−+−−−+−−−	1	−−−+−+−−+−+	1
+−−−+−+−−++	1	−−−+−−+++++	1
+−−−+−+−−+−	1	−−−+−−++++−	1
+−−−+−−++++	2	−−−+−−−+−+−	1
+−−−+−−−+−+	1	−−−+−−−+−−+	1
+−−−+−−−+−−	3	−−−+−−−+−−−	2

The Distribution of $X_1, \ldots, X_{11}$

VII. *The distribution of* $\{-X_{2n}\}$ *is the same as the distribution of chain I.* $\{-X_{2n}\}$ is automatically a finitary chain with the same π and ξ and operators $B_\pm$ given by $B_\pm = (A_- + A_+)A_\mp$. This system is not reduced. The space spanned by $B_{x_1} \cdots B_{x_n} \cdot \xi$ is 6 dimensional. If $\hat{B}_\pm$ is the restriction of $B_\pm$ to this space, and $\hat{\pi}$ is the projection of π onto this space, then the space spanned by $\hat{\pi} \cdot \hat{B}_{x_1} \cdots \hat{B}_{x_n}$ is four dimensional and isomorphic to chain I:

$$\tilde{\pi} = \frac{1}{4}(1,1,1,1)$$
$$\tilde{\xi} = (1,1,1,1)^t$$

$$\tilde{A}_- = \begin{pmatrix} 0 & 0 & 0 & \frac{1}{2} \\ 0 & 0 & \frac{1}{2} & 0 \\ 0 & 0 & 0 & 0 \\ 0 & 1 & 0 & 0 \end{pmatrix}$$

$$\tilde{A}_+ = \begin{pmatrix} 0 & 0 & \frac{1}{2} & 0 \\ 0 & 0 & 0 & \frac{1}{2} \\ 1 & 0 & 0 & 0 \\ 0 & 0 & 0 & 0 \end{pmatrix}$$

This illustrates the distinction between embedding a chain and square roots of measure preserving transformations. Since chain I is ergodic and has -1 in its point spectrum, there does not exists a measure preserving transformation T such that T^2 is isomorphic to the measure preserving transformation, S, of chain I (cf. Halmos (1956) Square Root Theorem, page 49). However, if T is the measure preserving transformation of chain II, then the *factor* of T^2 obtained by restricting it to the σ–algebra generated by $\{X_{2n}\}$ is isomorphic to S.

There are alternative (not reduced) representations of the finitary systems for chains I and III that exhibit the similarity of these chains. For chain I we have:

$$\begin{array}{ccc} 1(+) & \longrightarrow & 2(+) \\ 3(+) & \longrightarrow & 4(-) \\ 5(-) & \longrightarrow & 6(+) \\ 7(-) & \longrightarrow & 8(-) \end{array}$$

State 2 (resp. 8) goes to state 1 and state 3 with equal probability. State 4 (resp. 6) goes to state 5 and state 7 with equal probability.

For chain III we have:

$$\begin{array}{ccccccc} 1(+) & \longrightarrow & 2(+) & \longrightarrow & 3(+) & \longrightarrow & 4(+) \\ 5(+) & \longrightarrow & 6(-) & \longrightarrow & 7(-) & \longrightarrow & 8(+) \\ 9(-) & \longrightarrow & 10(+) & \longrightarrow & 11(-) & \longrightarrow & 12(-) \\ 13(-) & \longrightarrow & 14(-) & \longrightarrow & 15(+) & \longrightarrow & 16(-) \end{array}$$

State 4 (resp. 16) goes to state 5 and state 13 with equal probability. State 8 (resp. 12) goes to state 1 and state 9 with equal probability.

In these examples we have given the transition matrices of stationary Markov chains with 8 and 16 states respectively. The chains are the functions of the chains where the value of the functions are given by the sign in parentheses. Every function of a finite Markov chain is finitary, they can be seen to be equivalent to the previous definitions by reducing these systems (cf. Robertson (1973)).

VIII. *Sufficient Statistics.* We will give a simple description of the distribution of $\{X_n : n = 1, 2, \ldots\}$. For a given sequence $y_1, \cdots, y_n$ of +'s and −'s there are exactly four sequences $x_1, \cdots, x_{n+2}$ of +'s and −'s such that $y_i = x_i x_{i+1} x_{i+2}$ for all $1 \leq i \leq n$. For each of these four sequences $P\{X_1 = x_1, \cdots, X_{n+2} = x_{n+2}\}$ is the same. With probability one no two consecutive y's are −. Sequences of y's that are all +'s with −'s on each end can only have 2, 3, or 4 +'s: − + + −, − + + + −, or − + + + + −. Finally a sequence of two (four) +'s will be followed by any number of sequences of three +'s, followed by a string of four (two) +'s (e.g. −++−++++−, −++−+++−++++−, ...) at each step a string of three +'s and four (two) +'s occur with equal probability. This means that strings like − + + − + + − have zero probability.

The above assertions can be seen by chasing around Diagram 1. The only branching occurs at states 4 and 8. The x sequences between these states are deterministic and consist of y sequences of two +'s.

from	*to*	x's
4	4	− + − −
4	8	+ + ++
8	4	+ − −+
8	8	− − +−

It takes three transitions for the y sequence to contain two minuses.

$z_n z_{n+4} z_{n+8} z_{n+12}$	$x_n \cdots x_{n+11}$
4444	− + −[− − + − −] − + − −
4448	− + −[− − + − −+] + ++
4484	− + − − [+ + + + +] − −+
4488	− + − − [+ + ++] − − + −
4844	+ + + + [+ − − + −] + −−
4848	+ + + + [+ − −+] + + + +
4884	+ + +[+ − − + −] + − − +
4888	+ + +[+ − − + −−] − +−
8444	+ − − + [− + −−] − + − −
8448	+ − − + [− + − − +] + ++
8484	+ − −[+ + + + ++] − −+
8488	+ − −[+ + + + +] − − + −
8844	− − +[− + − − +−] + −−
8848	− − +[− + − − +] + + + +
8884	− − + − [− − +−] + − − +
8888	− − + − [− − + − −] − +−

The brackets inclose the values of $x_i, \ldots, x_{j+2}$ for which $y_i, \ldots, y_j$ are + with y_{i-1} and y_{j+1} being −. Each of the four strings of 4 (2) +'s occurs once. Following these through further cycles one sees that once a $-++++-$ $(-++-)$ occurs, it cannot occur again until after a $-++-$ $(-++++-)$ occurs. The rest of the assertion should be apparent.

IX. *The automorphism.* Any stationary chain is given by a probability space $(\Omega, \mathcal{F}, P)$, a measure preserving transformation T, and a measurable partition $\pi = \{P_1, \ldots, P_k\}$ (k=2). $X_n(\omega) = a_k$ if $T^n(\omega) \in P_k$ where $\{a_1, \ldots, a_k\}$ is a given alphabet $(a_1 = -, a_2 = +)$. The above analysis suggest the following model for our chain. $(\Omega, \mathcal{F}, P, T)$ is the disjoint union of four copies of an ordinary Bernoulli chain $(\Omega', \mathcal{F}', P', T')$. Thus $\Omega = \{(\omega', i) : \omega' \in \Omega', i \in \{1,2,3,4\}\}$. We define T by $T(\omega', i) = (\omega', i+1)$ if $i = 1,2,3$ and $T(\omega', 4) = (T'(\omega'), 1)$. The set $P_2 = (A \Delta TA, TA, A^c, A \Delta TA)$, where (A_1, A_2, A_3, A_4) denotes the union of the sets $A_i \times \{i\}$, and $A \subset \Omega$ is a set such that $P(A) = \frac{1}{2}$ and $A, TA, \ldots$ are mutually independent and identically distributed. P_1 is the complement of P_2. It is easily verified that this has the properties listed above. It follows immediately from this that the spectral type of the unitary operator $Vf = f \circ T, f \in L_2$ is a point spectrum consisting of the fourth roots of unity with simple multiplicity and Lebesgue spectrum with countable multiplicity.

SUFFICIENT STATISTICS

Each of the three chains that we have studied can be obtained as an *extension* of a simpler chain. In this section we develop this connection.

Let $\{X_n : n = 1, 2, \ldots\}$ be a stationary chain with values in the multiplicative group $\{1, -1\}$ which we shall abbreviate as $\{+, -\}$. As in **VIII**, we define a new stationary chain $\{Y_n : n = 1, 2, \ldots\}$ by the formula $Y_n = X_n X_{n+1} X_{n+2}$ (multiplication, not concatenation). This map on sequences is four to one: i.e. for each of the four values of (x_1, x_2) and for each sequence $(y_1, y_2, \ldots, y_n)$ there is a unique sequence $(x_1, x_2, \ldots, x_{n+2})$ that gets mapped to $(y_1, y_2, \ldots, y_n)$. We shall study chains for which each of these four $(x_1, x_2, \ldots, x_{n+2})$ sequences has the same probability, i.e.

$$P[X_1 = x_1, \ldots, X_{n+2} = x_{n+2}] = \frac{1}{4} P[Y_1 = y_1, \ldots, Y_n = y_n] \tag{3}$$

where $y_i = x_i x_{i+1} x_{i+2}$ for $i = 1, 2, \ldots, n$. Given any stationary chain $\{Y_n : n = 1, 2, \ldots\}$ with state space $\{+, -\}$ it can easily be verified that the above formula defines the distribution of a stationary chain $\{X_n : n = 1, 2, \ldots\}$. It follows from this that X_n and X_{n+1} are independent and that $P[X_n = +] = P[X_n = -] = \frac{1}{2}$ (take $n = 0$ in (3)). We also have

Proposition 3.1 If $m \not\equiv 0 \bmod(3)$, then X_n and X_{n+m} are independent.

Proof. X_n and X_{n+m} are independent if and only if $P[X_1 = X_{m+1} = +] = \frac{1}{4}$. Fix a $(y_1, y_2, \ldots, y_{m-1})$. We will calculate x_{m+1} in terms of $x_1, x_2, y_1, \ldots, y_{m-1}$. In general $x_k = x_{k-2} x_{k-1} y_{k-2}$. Using this and induction we obtain

$$x_{3k} = x_1 x_2 y_1 \prod_{i=1}^{k-1} (y_{3i} y_{3i+1}) \tag{4}$$

$$x_{3k+1} = x_1 \prod_{i=1}^{k} (y_{3i-2} y_{3i-1}) \tag{5}$$

$$x_{3k+2} = x_2 \prod_{i=1}^{k} (y_{3i-1} y_{3i}). \tag{6}$$

The expressions for x_{3k} and x_{3k+2} have a factor of x_2. Thus for fixed y vector the x sequences starting with $++$ and $+-$ end in opposite signs. Since these sequences have the same probabilities, for $m \not\equiv 0 \bmod(3)$ we must have

$$P[X_1 = X_2 = X_{m+1} = +] = P[X_1 = +, X_2 = X_{m+1} = -]$$

and

$$P[X_1 = X_2 = +, X_{m+1} = -] = P[X_1 = +, X_2 = -, X_{m+1} = +].$$

Thus

$$P[X_1 = +, X_{m+1} = +] = P[X_1 = +, X_{m+1} = -]$$

and hence X_1 and X_{m+1} are independent. □

Proposition 3.2 X_n and X_{n+3m} are independent if and only if

$$E\left[\prod_{i=0}^{m-1} (Y_{3i+1} Y_{3i+2})\right] = 0.$$

Proof. Set $n = 1$ and observe that

$$X_1 X_{3n+1} = \prod_{i=0}^{m-1} (Y_{3i+1} Y_{3i+2}).$$

□

Chains I and II where finitary chains which means that there exists vectors π and ξ and matrices A_+ and A_- such that for any sequence of states $x_1, x_2, \ldots, x_n$ we have

$$P[X_1 = x_1, X_2 = x_2, \ldots, X_n = x_n] = \pi A_{x_1} A_{x_2} \cdots A_{x_n} \xi.$$

We remark, without proof, that in fact $\{Y_n : n = 1, 2, \ldots\}$ is also finitary. For Chain I we may take

$$\pi = \frac{1}{2}(1,1), \xi = \begin{pmatrix} 1 \\ 1 \end{pmatrix}, A_- = \begin{pmatrix} 0 & \frac{1}{2} \\ 0 & 0 \end{pmatrix}, A_+ = \begin{pmatrix} 0 & \frac{1}{2} \\ 1 & 0 \end{pmatrix}.$$

For Chain II we may take

$$\pi = \frac{1}{2}(1,1), \xi = \begin{pmatrix} 1 \\ 1 \end{pmatrix}, A_- = \begin{pmatrix} 0 & \frac{1}{2} \\ 0 & 0 \end{pmatrix}, A_+ = \begin{pmatrix} \frac{1}{2} & 0 \\ \frac{1}{2} & \frac{1}{2} \end{pmatrix}.$$

This illustrates that it is much easier working with the $\{Y_n : n = 1, 2, \ldots\}$ chain. The condition of pairwise independence and $E[X_1 X_2 X_3] = \frac{1}{2}$ completely determine the distribution of $\{X_1, \ldots, X_4\}$. The distribution of $\{Y_1, Y_2\}$ are determined by this to be

$$p(++) = \frac{1}{2}, p(+-) = p(-+) = \frac{1}{4}, p(--) = 0. \tag{7}$$

The additional condition $E[X_1 X_3 X_5] = -\frac{1}{2}$ can also be expressed in terms of the $\{Y_n : n = 1, 2, \ldots\}$ chain as

$$p(-+-) = p(+++++) = 0. \tag{8}$$

Thus the sequences consist of runs of 2, 3, or 4 +'s separated by a single −.

In trying to construct a chain with $E[X_1 X_3 X_5] = -\frac{1}{2}$, we first ask what are the possible distributions of the chain $\{X_1, X_3, \ldots\}$. Natural candidates are the negatives of Chains I and II. We calculated, on the computer, the distributions of chains $\{X_n : n = 1, 2, \ldots\}$ that were stationary, pairwise independent, reversible, and such that the distribution of $\{-X_1, -X_3, \ldots - X_{2n-1}\}$ was the same as $\{X_1, X_2, \ldots, X_n\}$ for Chain I (Chain II). When Chain II was used, it turned out X_1 and X_{16} could not be independent and so no such chain exists. However with Chain I (up to dimension 20) there appeared to be many such possibilities. To reduce the possibilities we imposed the additional restriction that $\{X_1, \ldots, X_n\}$ and X_{n+6} were independent (6 was the smallest such possible number). The results suggested that the resulting distribution was unique and that for the Y chain a run of 4 (2) +'s followed by any number of runs of 3 +'s would by followed, with equal probability, by a run of 3 or 2 (4) +'s.

This suggest that the chain $\{Y_n : n = 1, 2, \ldots\}$ will be a function of a Markov chain. The states of the chain and the function can be viewed as arrange in the following diagram.

1	5	6			−	+	+			
2	7	8	9		−	+	+	+		
3	10	11	12	13	−	+	+	+	+	
4	14	15	16		−	+	+	+		

In each row the chain moves to the right. Thus

$$\begin{aligned} p_{1,5} = p_{5,6} = p_{2,7} = p_{7,8} = p_{8,9} = p_{3,10} &= \\ p_{10,11} = p_{11,12} = p_{12,13} = p_{4,14} = p_{14,15} = p_{15,16} &= 1. \end{aligned}$$

At the end of a line one goes to beginning of another line according to the following probabilities.

$$\begin{aligned} p_{6,2} = p_{6,3} = p_{9,2} = p_{9,3} &= \\ p_{13,1} = p_{13,4} = p_{16,1} = p_{16,4} &= \frac{1}{2}. \end{aligned}$$

All other transition probabilities are zero. This is an irreducible, doubly stochastic transition matrix with period 4. Its invariant vectors are

$$\pi = \frac{1}{16}\xi^t, \quad \xi^t = (1,1,1,1,1,1,1,1,1,1,1,1,1,1,1,1).$$

The matrix A_- is the 16×16 matrix whose first four rows are the same as P and the last 12 rows are zero. The first four rows of A_+ are zero while the last 12 rows are the same as P. The 4-tuple π, ξ, A_+, A_- is clearly a finitary system for some stationary chain $\{Y_n : n = 1, 2, \ldots\}$, and it is not too difficult to verify that it has the desired properties.

The above Markov chain $\{Z_n : n = 1, 2, \ldots\}$ is perhaps the most natural one satisfying the conditions at the end of the last section. However it is not the only one. Another possibility is given by the six state chain below.

$$\begin{array}{ccccc} + & \longrightarrow & + & \longleftarrow & - \\ \uparrow & & \downarrow & & \uparrow \\ - & \longleftarrow & + & \longrightarrow & + \end{array}$$

The middle + on the bottom row has two arrows leaving it. Each of these are with probability $\frac{1}{2}$. All other arrows have probability one. Both of these Markov chains have the properties that only runs of 2, 3, or 4 +'s are possible, only isolated −'s can occur, and once a run of 4 (2) +'s has occurred the next runs of +'s will be either 2 (4) or 3 with equal probability. The $\{Y_n : n = 1, 2, \ldots\}$ chain for both of these have the same distributions. In working with finitary systems it is necessary, in order to determine, the mixing properties to work with reduced systems (see Robertson (1973)). It is standard, but tedious to derive a reduced system and the reduction itself is not necessary as all the properties can be derived from the the following reduced system. This system can be viewed as the transitions of a 5–state Markov chain. The system is as follows:

$$\pi = \frac{1}{8}(1,2,2,2,1), \quad \xi = (1,1,1,1,1)^t,$$

$$A_+ = \begin{pmatrix} 0 & 1 & 0 & 0 & 0 \\ 0 & 0 & 1 & 0 & 0 \\ 0 & 0 & 0 & 1 & 0 \\ 0 & 0 & 0 & 0 & \frac{1}{2} \\ 0 & 0 & 0 & 0 & 0 \end{pmatrix}, \quad A_- = \begin{pmatrix} 0 & 0 & 0 & 0 & 0 \\ 0 & 0 & 0 & 0 & 0 \\ 0 & 0 & 0 & 0 & 0 \\ \frac{1}{2} & 0 & 0 & 0 & 0 \\ 0 & 1 & 0 & 0 & 0 \end{pmatrix}.$$

The system studied in section 2 was obtained by lifting this to the X chain and then reducing that system.

REFERENCES

Bradley, R. (1989), A stationary, pairwise independent, absolutely regular sequence for which the central limit theorem fails, *Probab. Th. Rel. Fields*, **81**, 1–10.

Halmos, P. (1956), *Ergodic Theory*, Chelsea, New York.

Janson, S. (1988), Some pairwise independent sequences for which the central limit theorem fails, *Stochastics*, **23**, 439–448.

Mathew, J. and Nadkarni, M. (1984), A measure preserving transformation whose spectrum has Lebesgue component of multiplicity two, *Bull. London Math. Soc.* **16**, 402–406.

Robertson, J. (1973), A spectral representation of the states of a measure preserving transformation, *Z. Wahr. verw. Geb.* **27**, 185–194.

Robertson, J. and Womack, J. (1985), A pairwise independent stationary stochastic process, *Stats. and Prob. Letters*, **3**, 195–199.

Robertson, J. and Simons, S. (1988), A De Finetti theorem for a class of pairwise independent stationary processes, *Ann. of Prob.* **16**, 344–354.

Robertson, J. (1985), Independence of fair coin tossing, *The Math. Sci.* **10**, 109–117.

Robertson, J. (1988), A two state pairwise independent stationary process for which X_1, X_3, X_5 is dependent, *Sankhya: The Indian Stat. J.* **50**, 171–183.

THE LEBESGUE INTEGRAL AS AN IMPROPER RIEMANN INTEGRAL

Oved Shisha

Department of Mathematics
University of Rhode Island
Kingston, RI 02881, USA

1. The purpose of this note is to point out that the Lebesgue integral of a real function, summable over a measurable set in some real Euclidean n-space, can be defined as a (one-dimensional) improper Riemann integral. This is in line (though in a different direction) with recent literature exhibiting the Lebesgue integral as a special case of the generalized Riemann integral (cf., e.g., [**1**, p.289]). A treatment of improper Riemann integral can be found, e.g., in [**4**], Chapters 28, 29.

2. *Theorem. Let f be a real function, defined and measurable on a measurable set S in some real Euclidean n−space. For every real y, let*

$$\mu(y) = \begin{cases} m\{P\epsilon S:\ f(P) \geq y\} & \text{if } y \geq 0; \\ -m\{P\epsilon S:\ f(P) \leq y\} & \text{if } y < 0, \end{cases}$$

m denoting Lebesgue measure.

Then f is summable over S iff, for every real $y \neq 0, \mu(y)$ is finite and the improper Riemann integral $(IR) \int_{-\infty}^{\infty} \mu(y)dy$ converges, in which case it equals the Lebesgue integral $(L) \int_S f(P)dP$.

Observe that $\mu(y)$ is monotone on both $[0, \infty)$ and $(-\infty, 0)$.

Please draw a figure describing the Theorem (say, for $n = 1$) and observe how plausible the Theorem is.

3. *Proof.* We may suppose S is nonempty as otherwise the theorem is trivial.

a) Assume, first, that f is bounded and ≥ 0 on S, and $m(S) < \infty$.

Let

$$s = 1 + sup\{f(P) : P\epsilon S\}.$$

Then

$$\int_0^s \mu(y)dy + \int_0^s yd\mu(y) = s\mu(s) = 0, \tag{1}$$

the second integral being Riemann-Stieltjes, by virtue of a well known identity [**3**, p.255, Theorem 323].

If

$$0 = y_0 < y_1 \ldots < y_N = s,$$

Approximation, Probability, and Related Fields, Edited by
G. Anastassiou and S.T. Rachev, Plenum Press, New York 1994

then, for $k = 1, 2, \ldots, N$,

$$\mu(y_k) - \mu(y_{k-1}) = -m(S_k)$$

where

$$S_k = \{P \epsilon S : y_{k-1} \leq f(P) < y_k\}$$

and

$$y_{k-1} m(S_k) \leq (L) \int_{S_k} f(P) dP < y_k m(S_k)$$

so that

$$-\sum_{k=1}^{N} y_{k-1}[\mu(y_k) - \mu(y_{k-1})] \leq (L) \int_S f(P) dP < -\sum_{k=1}^{N} y_k[\mu(y_k) - \mu(y_{k-1})].$$

Hence

$$(L) \int_S f(P) dP = -\int_0^s y d\mu(y).$$

By (1),

$$(L) \int_S f(P) dP = \int_0^s \mu(y) dy = (IR) \int_{-\infty}^{\infty} \mu(y) dy.$$

b) Remove now from a) the assumption of boundedness of f.
For every $P \epsilon S$, $\alpha \geq 0$, let

$$f_\alpha(P) = \begin{cases} f(P), & \text{if } f(P) \leq \alpha; \\ \alpha, & \text{if } f(P) > \alpha. \end{cases}$$

One easily checks that if $0 \leq y \leq \alpha$, then

$$\{P \epsilon S : f(P) \geq y\} = \{P \epsilon S : f_\alpha(P) \geq y\}$$

and hence, by a), for every $\alpha \geq 0$,

$$(L) \int_S f_\alpha(P) dP = \int_0^\alpha \mu(y) dy.$$

Thus [**2**, p. 568, §386] f is summable on S iff $(IR) \int_0^\infty \mu(y) dy$ converges, in which case,

$$(L) \int_S f(P) dP = (IR) \int_0^\infty \mu(y) dy = (IR) \int_{-\infty}^{\infty} \mu(y) dy.$$

c) Remove now from b) the assumptions that $f \geq 0$ on S and that $m(S) < \infty$.
Let

$$S^{(1)} = \{P \epsilon S : f(P) \geq 0\}, S^{(2)} = \{P \epsilon S : f(P) < 0\}$$

and set, for $r = 1, 2$,

$$S^{(r)} = \bigcup_{k=1}^{\infty} S_k^{(r)}, \text{ where } S_j^{(r)} \bigcap S_k^{(r)} = \emptyset \text{ whenever } j \neq k, \text{ and } m(S_k^{(r)}) < \infty, k = 1, 2, \ldots,$$

a representation which clearly exists [**3**, p.79, Theorem 109; p.6, Definition 8]. Also, for $r = 1, 2; k = 1, 2, \ldots$ and every real y, let

$$\mu_{rk}(y) = \begin{cases} m\{P \epsilon S_k^{(r)} : f(P) \geq y\} \text{ if } y \geq 0; \\ -m\{P \epsilon S_k^{(r)} : f(P) \leq y\} \text{ if } y < 0. \end{cases}$$

For $k = 1, 2, \ldots$, if f is summable over $S_k^{(r)}, r = 1, 2$, then

$$(L)\int_{S_k^{(1)}} f = (IR)\int_0^{\infty} \mu_{1k}(y)dy,$$

$$(L)\int_{S_k^{(2)}} f = -(L)\int_{S_k^{(2)}} -f = -(IR)\int_0^{\infty} m\{P\epsilon S_k^{(2)} : -f(P) \geq y\}dy = (IR)\int_{-\infty}^{0} \mu_{2k}(y)dy.$$

Thus, if f is summable over S,

$$(L)\int_S f = (L)\int_{S^{(1)}} f + (L)\int_{S^{(2)}} f = \sum_{k=1}^{\infty}(L)\int_{S_k^{(1)}} f + \sum_{k=1}^{\infty}(L)\int_{S_k^{(2)}} f = \sum_{k=1}^{\infty}(IR)\int_0^{\infty} \mu_{1k}(y)dy$$

$$+ \sum_{k=1}^{\infty}(IR)\int_{-\infty}^{0} \mu_{2k}(y)dy = \sum_{k=1}^{\infty}(L)\int_0^{\infty} \mu_{1k}(y)dy + \sum_{k=1}^{\infty}(L)\int_{-\infty}^{0} \mu_{2k}(y)dy$$

$$= (L)\int_0^{\infty} \sum_{k=1}^{\infty} \mu_{1k}(y)dy + (L)\int_{-\infty}^{0} \sum_{k=1}^{\infty} \mu_{2k}(y)dy = (L)\int_0^{\infty} \mu(y)dy$$

$$+ (L)\int_{-\infty}^{0} \mu(y)dy = (IR)\int_{-\infty}^{\infty} \mu(y)dy.$$

Conversely, suppose $\mu(y)$ is finite for every real $y \neq 0$ and $(IR)\int_{-\infty}^{\infty} \mu(y)dy$ converges. Then the last chain of equalities holds, as seen by reading it backwards.

Our proof of part a) follows that of a statement in [**3**, p.262, §4.1] which our Theorem greatly strengthens.

REFERENCES

1. G. Cross and O. Shisha, A new approach to integration, J. Math. Anal. Appl. **114** (1986), 289-294.

2. E. W. Hobson, "The Theory of Functions of a Real Variable and the Theory of Fourier's Series", Vol. I, Dover, New York, 1957.

3. H. Kestelman, "Modern Theories of Integration", 2nd Revised Edition, Dover, New York, 1960.

4. E. Landau, "Differential and Integral Calculus", Chelsea, New York, 1951.

TESTS OF EXISTENCE OF GENERALIZED RIEMANN INTEGRALS*†

Oved Shisha

Department of Mathematics
University of Rhode Island
Kingston, RI 02881, USA

1. The Perron integral I(f) and, equivalently, the restricted Denjoy integral [**13**,201,241,247] were given an equivalent definition, many years later, by J. Kurzweil and R. Henstock. This definition is merely a quite simple variation of a familiar definition of the Riemann integral. The terms generalized Riemann integral (integrable), abbreviated here GRI, refer to I(f) as defined by this variation. Thus GRI is at the same time very elementary but more powerful than Lebesgue and includes as special cases the Riemann, improper Riemann, Lebesgue and other integrals. It seems very sensible to make GRI the standard integral of the working analyst: [**3**] is a textbook essentially doing this. (It uses, however, instead of GRI, the term gauge integral). [**10**] is a Carus Monograph on the subject, while [**5**, **6**, **8**, **12**] are more technical monographs on GRI using for it also other names. The articles [**2**] and [**9**] introduce GRI and relate it to other integrals (e.g., improper Riemann). [**2**] contains also the simple proof of (1) below.

2. A major motivation for studying and using the Lebesgue integral has been the fact that it allows the important monotone and dominated convergence theorems, which the Riemann integral does not. Similarly, GRI allows unconditionally to retrieve a differentiable function from its derivative:

$$\int_a^x F'(t)dt = F(x) - F(a), \tag{1}$$

which the Lebesgue integral does not. The purpose of the present note is to give two other practical results which hold for GRI but not for Lebesgue. They are integral analogs of Dirichlet's and Dedekind's classical tests for convergence of infinite series [**7**, 315].

---

*AMS 1991 Mathematics Subject Classification: Primary 26A39.

†*Keywords and phrases:* Riemann, improper Riemann, generalized Riemann, Riemann-Stieltjes, Lebesgue, Perron and restricted Denjoy integrals; monotone, continuous, absolutely continuous and integrable functions; Dirichlet's and Dedekind's tests.

3. *Theorem* 1. *("Dirichlet's test"). Let* $-\infty < a < b \leq \infty$ *and let* f *be a real function, GRI on each* $[a,x]$, $a < x < b$. *Suppose*

A) $F(x) \equiv (GRI) \int_a^x f(t)dt$ *is bounded in* (a,b)*;*

B) g *is a real function, nonincreasing or nondecreasing in* $[a,b)$*, with* $lim_{x\uparrow b} g(x) = 0$.

Then (GRI) $\int_a^b f(x)g(x)dx$ *exists.*

Proof. By a known result [**11**, 332, Theorem 65.1], for every $x \in (a,b)$,

$$(GRI) \int_a^x f(t)g(t)dt = F(x)g(x) - \int_a^x F(t)dg(t), \tag{2}$$

the expression following the minus sign being a Riemann-Stieltjes integral of the continuous, in $[a,b)$, function F[**11**, 317]. By A) and B), the left-hand side of (2) has a finite limit as $x \uparrow b$, which is (GRI) $\int_a^b f(x)dx$ [**3**, 173, Theorem 42; 184, Theorem 6]. (The "singularity" in Theorem 42 is a, but the corresponding result for b clearly holds, too).

Theorem 2. *("Dedekind's test"). Replace, in Theorem 1, B) by*

B') g *is a real function, absolutely continuous in each* $[a,x], x \in (a,b); g'$ *is Lebesgue integrable on* (a,b) *and* $\lim_{x\uparrow b} g(x) = 0$.

Then again (GRI) $\int_a^b f(x)g(x)dx$ *exists.*

Proof. For every $x \in (a,b)$, (2) again holds and hence

$$(GRI) \int_a^x f(t)g(t)dt = F(x)g(x) - \int_a^x F(t)g'(t)dt,$$

which clearly converges as $x \uparrow b$, by A) and B'). (The last integral is Lebesgue).

Much simpler versions of Theorem 1 can be found in [**1**, 477; **4**, Theorem I; **6**, 21, Theorem 2.16].

Note that both Theorems 1 and 2 are false for Lebesgue integration, as seen by taking $f(x) \equiv sinx, g(x) \equiv 1/x, a = 1, b = \infty$; for the Lebesgue integral of $(sinx)/x$ on $(1,\infty)$ does not exist [**3**, 187].

REFERENCES

1. T. J. I'a Bromwich, *An Introduction to the Theory of Infinite Series*, Macmillan, 2nd edition reprinted, London, 1949.
2. G. Cross and O. Shisha, A new approach to integration, J. Math. Anal. Appl. **114**, 289-294 (1986).
3. J. DePree and C. Swartz, *Introduction to Real Analysis*, Wiley, New York, 1988.
4. G. H. Hardy, Notes on some points in the integral calculus, The Messenger of Mathematics **30**, 185-190 (1900-1901).
5. R. Henstock, *Theory of Integration*, Butterworths, London, 1963.
6. R. Henstock, *Lectures on the Theory of Integration*, World Scientific, Singapore, 1988.
7. K. Knopp, *Theory and Application of Infinite Series*, Blackie, London, 1949.

8. J. Kurzweil, *Nichtabsolut konvergente Integrale*, Teubner, Leipzig, 1980.

9. J. T. Lewis and O. Shisha, The generalized Riemann, simple, dominated and improper integrals, J. Approx. Theory **38**, 192-199 (1983).

10. R. M. McLeod, *The Generalized Riemann Integral*, Carus Mathematical Monograph No. 20, Mathematical Association of America, Washington, DC, 1980.

11. E. J. McShane, *Integration*, Princeton University Press, Princeton, 1957.

12. W. F. Pfeffer, *The Riemann Approach to Integration*, Cambridge University Press, Cambridge, 1993.

13. S. Saks, *Theory of the Integral*, 2nd revised edition, Hafner, New York, 1937.

REMARK ON A THEOREM ABOUT POLYNOMIALS

P. Vértesi[1]

Mathematical Institute
Hungarian Academy of Sciences
Budapest P.O.Box 127
H-1364

1. Let

(1) $$t_{mm} < t_{m-1,m} < \cdots < t_{0m}, \quad m = 1, 2, \ldots$$

be equidistant nodes on the real line. Consider the set $\mathcal{P}_n(m)$ of real polynomials $P(x)$ of degree at most n with $|P(t_{km})| \leq 1$, $0 \leq k \leq m$. Let

(2) $$\|P\| = \max_{t_{mm} \leq x \leq t_{0m}} |P(x)|,$$

(3) $$b(n,m) = \sup_{P \in \mathcal{P}_n(m)} \|P\|.$$

We shall investigate the behaviour of $b(n,m)$.

If $n > m$, then $b(n,m) = \infty$. So from now one can suppose that $n \leq m$ (when the "sup" in (3) can be replaced by "max"). Further, since $b(n,m)$ is invariant under linear transformations $u_{km} = at_{km} + b$, $0 \leq k \leq m$, $a \neq 0$, we shall consider convenient sets of equidistant nodes.

If $n = m$, then $b(n,n) = \Lambda_n$, the n-th Lebesgue constant of Lagrange interpolation based on equidistant nodes. Using relation

(4) $$\Lambda_n \sim \frac{2^{n+1}}{en(\log n + \gamma)} \quad \text{as} \quad n \to \infty,$$

($\gamma = 0.577\ldots$ is the Euler constant, cf. A. Schönhage [1], say) we get $\lim_{n\to\infty} b(n,n) = \infty$. However, if m is "large enough" compared to n, $\overline{\lim_{n\to\infty}}\, b(n,m)$ will be bounded. Indeed, again in [1] it is shown that, $\overline{\lim_{n\to\infty}}\, b(n,m) < \infty$, supposing $m > n^2$.

[1]Supported by the Hungarian National Science Foundation Grants Nos. 1910 and T7570.

This result was sharpened in a series of papers by H. Ehlich and K. Zeller [2], [3], [4]. Finally in 1992 D. Coppersmith and T. J. Rivlin [5] obtained the following result:
Theorem 1. *$b(n,m)$ is bounded as $n \to \infty$ iff*

$$\lim_{n\to\infty} \frac{m}{n^2} = \delta > 0. \tag{5}$$

(Actually, they considered even the *order* of $b(n,m)$, too.)

The aim of the present remark is to give another proof for the previous theorem using a fairly general statement obtained by P. Erdős more than 20 years ago.

2. To state Erdős' result let

$$x_{kp} = \cos\vartheta_{kp}, \quad 0 \le \vartheta_{kp} \le \pi, \quad k = 1,2,\dots,p, \quad p = 1,2,\dots, \tag{6}$$

be an arbitrary node-system in $[-1,1]$. Let $0 \le \alpha < \beta \le \pi$ and denote by $N_p(\alpha,\beta,X)$ the number of ϑ_{kp} satisfying $\alpha \le \vartheta_{kp} < \beta$. Let

$$(\vartheta_{i-1,p} <)\, \alpha \le \vartheta_{ip} < \vartheta_{i+1,p} < \dots < \vartheta_{jp} < \beta (\le \vartheta_{j+1,p}),$$

say. Now, for each $\eta > 0$ we define a subsequence $\vartheta_{i_1,p}, \vartheta_{i_2,p}, \dots, \vartheta_{i_u,p}$ of these ϑ's as follows. We choose $i_1 = i$ and if $i_1, i_2, \dots, i_{r-1}$ have already been defined, then let $\vartheta_{i_r,p}$ be the smallest $\vartheta_{\ell p}$, $i_{r-1} < \ell \le j$, with $\vartheta_{\ell_p} - \vartheta_{i_{r-1},p} > \eta/p$. Thus the distance between any two members of the subsequence is greater than η/p and any other ϑ_{kp} in $[\alpha,\beta)$ is at a distance $\le \eta/p$ from at least one $\vartheta_{i_r,p}$, $1 \le r \le u$. Put

$$N_p^{(\eta)}(\alpha,\beta;X) = u = u(\eta).$$

Let $c > 0$ be an arbitrary (small) number. For any fixed p we consider those polynomials $P_n(x)$ of degree n with $n(1+c) < p$, which satisfy

$$|P_n(x_{kp})| \le 1, \quad 1 \le k \le p, \quad p > n(1+c) \tag{7}$$

and

$$\max_{-1\le x\le 1} |P_n(x)| \le A(c) \tag{8}$$

where $A(c)$ depends only on c.

In [6, Theorem 1] P. Erdős proved
Theorem 2. *Let x_{kp} satisfy (6). Then the necessary and sufficient condition that, for any polynomial $P_n \in \mathcal{P}_n$, (7) should imply (8) is that there should be an $\eta > 0$ independent of p so that for every $\alpha_p < \beta_p$ with $p(\beta_p - \alpha_p) \to \infty$ $(0 \le \alpha_p < \beta_p \le \pi$, $p \to \infty)$*

$$N_p^{(\eta)}(\alpha_p,\beta_p,X) \ge (1+o(1))\frac{p}{\pi}(\beta_p - \alpha_p), \quad p \to \infty. \tag{9}$$

Remark. Condition (9) means that every interval large compared to $1/p$ contains asymptotically at least as many points ϑ_{kp}, no two of them are "too close", as $T_p(x) = \cos p\vartheta$, $x = \cos\vartheta$ (cf. [6, p. 136–137]).

3. In this part we show how to get Theorem 1 as a special case of Theorem 2.

(a) First we verify that (5) $\Rightarrow \overline{\lim_{n\to\infty}} b(n,m) < \infty$. Using the simple relation $b(n,M) \le b(n,m)$, $M \ge m$ (cf. [5, (iv), p. 971]), one can suppose

$$\delta n^2 \le m < \delta n^2 + 1, \quad 0 < \delta < 1/4, \quad n \ge n_0. \tag{10}$$

Let $r = 2n$ and consider the roots

$$y_{kr} = \cos \eta_{kr} = \cos \frac{2k-1}{2r}\pi, \quad 1 \le k \le r, \quad \text{of} \quad T_r(x).$$

Using (10), we have

$$y_k - y_{k+1} = 2 \sin \frac{\eta_{k+1} + \eta_k}{2} \sin \frac{\eta_{k+1} - \eta_k}{2} > 2\left(\frac{2}{\pi}\right)^2 \frac{2k-1}{2r}\pi \frac{\pi}{2r} =$$
$$= \frac{2}{r^2}(2k-1) \ge \frac{\delta}{2m}(2k-1) > \frac{4}{m} \quad \text{if} \quad k > d := \left[\frac{5}{\delta}\right], \quad y_{k+1} > 0.$$

So if we apply the normalization $t_{0m} = -1$ and $t_{mm} = 1$, every interval (y_{k+1}, y_k), $d < k < r/2$, contains at least three t_{im}'s, one of them, which will be denoted by $v_{km} = \cos \varrho_{km}$, is not farther from $u_{km} = (y_{k+1} + y_k)/2$ than $1/m$. The points $\{v_{km}\} \cup \{-v_{km}\}$, $d < k < r/2$, (using proper reordering) will form the system $\{x_{ip}\}$ with $p = r - 2d$ (cf. (6)). Condition (7) obviously holds true, so to get (8), we have to verify (9).

By construction

$$v_k - v_{k+1} = 2 \sin \frac{\varrho_{k+1} + \varrho_k}{2} \sin \frac{\varrho_{k+1} - \varrho_k}{2} \ge y_k - y_{k+1} - \frac{2}{m} > \frac{y_k - y_{k+1}}{2}$$

whence, as above,

$$\varrho_{k+1} - \varrho_k > \frac{2k-1}{r^2} \frac{1}{\varrho_{k+1} + \varrho_k} > \frac{2k-1}{2r^2 \eta_k} = \frac{1}{\pi r}, \quad d < k < \frac{r}{2}. \tag{11}$$

Formula (11) obviously yields the relation

$$\vartheta_{k+1,p} - \vartheta_{kp} > \frac{1}{\pi r} > \frac{1}{2\pi p} \quad \text{if} \quad n \ge n_0. \tag{12}$$

So choosing $\eta = \frac{1}{2\pi}$, we have for any proper interval-system $\{\alpha_p, \beta_p\}$ as follows

$$N_p^{(\eta)}(\alpha_p, \beta_p, X) = N_p(\alpha_p, \beta_p, X) \ge N_r(\alpha_p, \beta_p, Y) - 2d \ge$$
$$\ge (\beta_p - \alpha_p)\frac{r}{\pi} - 2\alpha - 1 = (\beta_p - \alpha_p)\frac{p}{\pi}(1 + o(1)),$$

which was to be proven. Relation $p > n(1+c)$ is obvious by $p = r - 2d = 2n - 2d$.

(b) Now we prove that (9) cannot hold if relation (5) is not valid. By $m = o(n^2)$ we can suppose that $m = \varepsilon_m n^2$ where $\varepsilon_m \searrow 0$ (we may take a proper subsequence). If there existed a proper matrix $X = \{x_{ip}\}_{i,p} \subset T = \{t_{km}\}_{k,m}$ (again, $t_{0m} \equiv 1$, $t_{mm} \equiv -1$) satisfying (9) with a proper η, we would have for the interval $(0, \beta_p)$, $\beta_p \le \pi/2$, say,

$$\frac{\beta_p p}{2\pi} < \frac{\beta_p p}{\pi}(1 + o(1)) \le N_p^{(\eta)}(0, \beta_p, X) \le N_p(0, \beta_p, T) \le \tag{13}$$

$$\le \frac{m}{2}(1-\cos\beta_p)+1 < 2m\sin^2\frac{\beta p}{2} < m\beta_p^2 = \varepsilon_m n^2\beta_p^2 < \varepsilon_m p^2\beta_p^2$$

where at the last step we used that p must be greater than n, otherwise (8) cannot be held (by Faber's theorem for Lagrange interpolation).

By (13) $\varepsilon_m > (2\pi p\beta p)^{-1}$, whence if $2\pi p\beta_p = 1/\sqrt{\varepsilon_m}$, say, we get $\sqrt{\varepsilon_m} > 1$, a contradiction. □

REFERENCES

1. A. Schönhage, Fehlerfortpflanzung bei Interpolation, *Numer. Math.* 3 (1961), 62–71.
2. H. Ehlich and K. Zeller, Schwankung von Polynomen zwischen Gitterpunkten, *Math. Z.* 86 (1964), 41–44.
3. H. Ehlich and K. Zeller, Numerische Abschätzung von Polynomen, *Z. Angew. Math. Mech.* 45 (1965), T20–T22.
4. H. Ehlich, Polynome zwischen Gitterpunkten, *Math. Z.* 93 (1966), 144–153.
5. D. Coppersmith and T. J. Rivlin, The growth of polynomials bounded at equally spaced points, *SIAM J. Math. Anal.* 23 (1992), 970–983.
6. P. Erdős, On the boundedness and unboundedness of polynomials, *J. d'Analyse de Math.* 19 (1969), 135–148.

CONFERENCE PARTICIPANTS

PARTICIPANTS

Jose Adell
Universidad de Zarogoza
Zaragoza 50009
SPAIN

George Anastassiou
Dept. of Mathematics
Memphis State University
Memphis, TN 38152

Bessy Athanasopoulos
Department of Statastics
University of California
Santa Barbara, CA 93106

Michel Balinski
1 rue Descartes
CNRS, Escole Polytechnique
Paris, FRANCE

Professor A. Basalykas
Institute of Math & Inform.
Akademijos 4
2600 Vilnius, LITHUANIA

Martin Bilodeau
Dept. of Math & Statistics
University of Montreal
CP6128, SUC.A
Montreal, H3C 3J7 CANADA

Paul L. Butzer
Lerhrstuhl A fur Mathematik
Westfalischen Tech Hochschule
Tempelgraben 55 Aachen
GERMANY

Stamatis Cambanis
Department of Statistics
University of North Carolina
Chapel Hill, NC 27599-3260

Bruce Chalmers
Mathematics Department
University of California
Riverside, CA 92521

Benny Cheng
Dept. of Statistics
University of California
Santa Barbara, CA 93106

Marco Dall'aglio
Dept of Statistics
University of Rome
Via Milazzo 42 Rome ITALY

Antonio da Silva
Mailbox 68530
Fed Univ of Rio de Janeiro
Rio de Janeiro, 21944

Jesus de la Cal
Universidad del Pais Vasco
Apdo 644
48080 Bilbao, SPAIN

Franz-Jurgen Delvos
Fachbereich Mathematik I
Universitat GH Siegen
HolderlinstraBe 3
D-5900 Siegen, GERMANY

Boyan Dimitrov
Concordia University
Sofia University
5080 Randall Ave
Montreal, CANADA H4V 2V1

Morteza Ebneshahrashoob
Department of Mathematics
California State Univ.
Long Beach, CA 90840-4502

Philip Feldman
Department of Mathematics
University of California
Santa Barbara, CA 93106

Raisa Feldman
Stat & Appl Probability
University of California
Santa Barbara, CA 93106

Richard Gayle
Dept. of Mathematics
University of Oregon
Eugene, OR 97403-1222

Matthias Gelbrich
FB Mathematik
Humboldt-Universitat zu Berlin
Unter den Linden 6, PSF 1297
0-1086 Berlin, GERMANY

Oleg Granichin
Department of Mathematics
St. Petersburg University
Bibliotechnaya pl. 2
St. Petersburg, Petrodvoretz
RUSSIA 198904

Gleb Haynatzki
Department of Statistics
University of California
Santa Barbara, CA 93106

Matthew He
Department of Math
Nova University
Ft Lauderdale, FL 33314

Tianxiao He
Department of Mathematics
Illinois Wesleyan Univ.
Blooming, IL 61702-2900

Christian Houdre
Dept. of Statistics
Stanford University
Stanford, CA 94305

Ying Kang Hu
Math & Computer Science
Georgia Southern University
8093 Landrum Box
Statesboro, GA 39460-8093

J.H.B. Kemperman
Dept. of Statistics
Rutgers University
New Brunswick, NJ 08903

Alexander Koldobsky
Dept. of Mathematics
University of Missouri
Columbia, MO 65211

J. Kozubowski
Dept. of Mathematics
University of Tennessee
Chattanooga, TN 37403

Alan C. Krinik
Cal Poly Pomona
3801 W. Temple
Pomona, CA 91768

Nagamani KrishnaKumar
Dept. of Statistics
University of California
Santa Barbara, CA 93196

Thomas J. Kunkle
Math Department
College of Charleston
Charleston, SC 29424

E. Bing Lin
Dept. M/C 249, Box 4348
Univ of Illinois-Chicago
322 Science & Engr Office
Chicago, IL 60680

Professor M. Maejima
Keio University
3-141, Hiyoshi, Kohoku-Ku
Yokohama 223, JAPAN

Dorothy Maharam-Stone
Math Department
567 Lake Hall
Northeastern University
Boston, MA 02115

Kurt Marti
Universitat der Bundeswehr
Munchen Fakultat LRT
Werner-Heisenberg-Weg 39
D-8014 Neubiberg, GERMANY

Pesi Masani
Dept. of Mathematics
University of Pittsburgh
Pittsburgh, PA 15260

Zuhair Nashed
Dept. of Mathematics
University of Delaware
Newark, DE 19716

Anna K. Panorska
Dept. of Math
University of Tennessee
615 McCallie Avenue
Chattanooga, TN 37403

D. Plachky
Inst. fur Mathematische
Statistick, WWU
Einsteinstr. 62
4400 Munster/Wesf., GERMANY

Wolfgang Preuss
Hochschule fur Technik &
Wirtschaft
Dresden, Fachbereich-List-Platz 1
Dresden 0-08010, GERMANY

Joao Prolla
IMECC-UNICAMP
Caixa Postal 6065
Campinas, SP 13081, BRAZIL

S. T. Rachev
Statistics & Appl. Probability
University of California
Santa Barbara, CA 93106-3110

Paolo Ricci
Rome University La Sapienza
Via Albertazzi 92
Rome 00137, ITALY

James Robertson
Dept. of Statistics
University of California
Santa Barbara, CA 93106

Ludger Ruschendorf
Inst. fur Mathematische
Statistik, WWU
Einsteinstr. 62, D-44
Munster, GERMANY

Yegnaseshan Sitaraman
Kentucky Wesleyan College
3000 Frederica St.
Owensboro, KY 42301

Boris Shekhtman
Dept. of Mathematics
Univ. of California
Riverside, CA 92521

Peter Singer
513 S. Catalina Ave.
Univ. of Southern CA
Redondo Beach, CA 90277

Milton Sobel
Dept. of Statistics
Univ. of California
Santa Barbara, CA 13106

Glen Swindle
Dept. of Statistics
Univ. of California
Santa Barbara, CA 13106

Oved Shisha
Dept. of Mathematics
Univ. of Rhode Island
Kingston, RI 02881-0806

Ziad Taib
Dept. of Mathematics
Univ. of Gothenburg
SWEDEN

Michael I. Taksar
Dept. of Appl Math & Stat
State Univ. of New York
Stony Brook, NY 11794-3600

Elzbieta Trybus
Dept. of Mathematics
CA State Univ., Northridge
Northridge, CA 91330

Patrick Van Fleet
Dept. of Mathematics
Sam Houston State Univ.
Huntsville, TX 77341

Peter Vertesi
Mathematical Institute
Hungarian Academy of Sci.
Budapest, Box 127
HUNGARY, 1364

Yannis Yatracos
Dept. of Mathematics
University of California
Santa Barbara, CA 93106

Xian Ming Yu
Dept. of Mathematics
Southwest Missouri St Univ
901 S. National Ave.
Springfield, MO 65304-0094

Julius Zelmanowitz
Dept. of Mathematics
University of California
Santa Barbara, CA 93106

INDEX